Advances in Intelligent Systems and Computing

Volume 788

Series editor

Janusz Kacprzyk, Polish Academy of Sciences, Warsaw, Poland
e-mail: kacprzyk@ibspan.waw.pl

The series "Advances in Intelligent Systems and Computing" contains publications on theory, applications, and design methods of Intelligent Systems and Intelligent Computing. Virtually all disciplines such as engineering, natural sciences, computer and information science, ICT, economics, business, e-commerce, environment, healthcare, life science are covered. The list of topics spans all the areas of modern intelligent systems and computing such as: computational intelligence, soft computing including neural networks, fuzzy systems, evolutionary computing and the fusion of these paradigms, social intelligence, ambient intelligence, computational neuroscience, artificial life, virtual worlds and society, cognitive science and systems, Perception and Vision, DNA and immune based systems, self-organizing and adaptive systems, e-Learning and teaching, human-centered and human-centric computing, recommender systems, intelligent control, robotics and mechatronics including human-machine teaming, knowledge-based paradigms, learning paradigms, machine ethics, intelligent data analysis, knowledge management, intelligent agents, intelligent decision making and support, intelligent network security, trust management, interactive entertainment, Web intelligence and multimedia.

The publications within "Advances in Intelligent Systems and Computing" are primarily proceedings of important conferences, symposia and congresses. They cover significant recent developments in the field, both of a foundational and applicable character. An important characteristic feature of the series is the short publication time and world-wide distribution. This permits a rapid and broad dissemination of research results.

More information about this series at http://www.springer.com/series/11156

Jerzy Charytonowicz · Christianne Falcão
Editors

Advances in Human Factors, Sustainable Urban Planning and Infrastructure

Proceedings of the AHFE 2018 International
Conference on Human Factors, Sustainable
Urban Planning and Infrastructure, July 21–25, 2018,
Loews Sapphire Falls Resort at Universal Studios,
Orlando, Florida, USA

 Springer

Editors
Jerzy Charytonowicz
The Angelus Silesius University of Applied
  Sciences in Walbrzych
Walbrzych, Poland

Christianne Falcão
Catholic University of Pernambuco
Recife, Brazil

ISSN 2194-5357   ISSN 2194-5365 (electronic)
Advances in Intelligent Systems and Computing
ISBN 978-3-319-94198-1   ISBN 978-3-319-94199-8 (eBook)
https://doi.org/10.1007/978-3-319-94199-8

Library of Congress Control Number: 2018947431

Printed on acid-free paper

This Springer imprint is published by the registered company Springer International Publishing AG
part of Springer Nature
The registered company address is: Gewerbestrasse 11, 6330 Cham, Switzerland

Advances in Human Factors and Ergonomics 2018

AHFE 2018 Series Editors

Tareq Z. Ahram, Florida, USA
Waldemar Karwowski, Florida, USA

9th International Conference on Applied Human Factors and Ergonomics and the Affiliated Conferences

Proceedings of the AHFE 2018 International Conference on Human Factors, Sustainable Urban Planning and Infrastructure, held on July 21–25, 2018, in Loews Sapphire Falls Resort at Universal Studios, Orlando, Florida, USA

Advances in Affective and Pleasurable Design	*Shuichi Fukuda*
Advances in Neuroergonomics and Cognitive Engineering	*Hasan Ayaz and Lukasz Mazur*
Advances in Design for Inclusion	*Giuseppe Di Bucchianico*
Advances in Ergonomics in Design	*Francisco Rebelo and Marcelo M. Soares*
Advances in Human Error, Reliability, Resilience, and Performance	*Ronald L. Boring*
Advances in Human Factors and Ergonomics in Healthcare and Medical Devices	*Nancy J. Lightner*
Advances in Human Factors in Simulation and Modeling	*Daniel N. Cassenti*
Advances in Human Factors and Systems Interaction	*Isabel L. Nunes*
Advances in Human Factors in Cybersecurity	*Tareq Z. Ahram and Denise Nicholson*
Advances in Human Factors, Business Management and Society	*Jussi Ilari Kantola, Salman Nazir and Tibor Barath*
Advances in Human Factors in Robots and Unmanned Systems	*Jessie Chen*
Advances in Human Factors in Training, Education, and Learning Sciences	*Salman Nazir, Anna-Maria Teperi and Aleksandra Polak-Sopińska*
Advances in Human Aspects of Transportation	*Neville Stanton*

(continued)

(continued)

Advances in Artificial Intelligence, Software and Systems Engineering	*Tareq Z. Ahram*
Advances in Human Factors, Sustainable Urban Planning and Infrastructure	*Jerzy Charytonowicz and Christianne Falcão*
Advances in Physical Ergonomics & Human Factors	*Ravindra S. Goonetilleke and Waldemar Karwowski*
Advances in Interdisciplinary Practice in Industrial Design	*WonJoon Chung and Cliff Sungsoo Shin*
Advances in Safety Management and Human Factors	*Pedro Miguel Ferreira Martins Arezes*
Advances in Social and Occupational Ergonomics	*Richard H. M. Goossens*
Advances in Manufacturing, Production Management and Process Control	*Waldemar Karwowski, Stefan Trzcielinski, Beata Mrugalska, Massimo Di Nicolantonio and Emilio Rossi*
Advances in Usability, User Experience and Assistive Technology	*Tareq Z. Ahram and Christianne Falcão*
Advances in Human Factors in Wearable Technologies and Game Design	*Tareq Z. Ahram*
Advances in Human Factors in Communication of Design	*Amic G. Ho*

Preface

The discipline of Human Factors and Sustainable Urban Planning and Infrastructure provides a platform for addressing challenges in human factors and engineering research with the focus on sustainability in the built environment, applications of sustainability assessment, demonstrations and applications that contribute to competitiveness and well-being, quantification and assessment of sustainable infrastructure projects, and the environmental, human, social, and economic dimensions of sustainable infrastructure. A thorough understanding of the characteristics of a wide range of people is essential in the development of sustainable infrastructure and systems and serves as valuable information to designers and helps ensure design will fit the targeted population of end users.

This book focuses on the advances in the Human Factors in Sustainable Urban Planning and Infrastructure, which are a critical aspect in the design of any human-centered technological system. The ideas and practical solutions described in the book are the outcome of dedicated research by academics and practitioners aiming to advance theory and practice in this dynamic and all-encompassing discipline.

A total of four main sections presented in this book:

I. Ergonomics in Material and Environment Design
II. Sustainable Urban Planning and Infrastructure
III. Ergonomics in Building and Architecture
IV. Construction Industry

Each section contains research papers that have been reviewed by members of the International Editorial Board. Our sincere thanks and appreciation to the board members as listed below:

Clinton Aigbavboa, South Africa
Agata Bonenberg, Poland
Wojciech Bonenberg, Poland
Bogdan Branowski, Poland
Alexander Burov, Ukraine

Alina Drapella-Hermansdorfer, Poland
Klaudiusz Fross, Poland
Anna Jaglarz, Poland
Bronislaw Kapitaniak, France
Ludmila Klimatskaya, Russia
Vladko Kolbanov, Russia
Robert Masztalski, Poland
Andrej Szpakov, Belarus
Romuald Tarczewski, Poland
Elżbieta Trocka-Leszczyńska, Poland
Joanna Tymkiewicz, Poland
Edwin Tytyk, Poland

We hope that this book, which is the international state of the art in Urban
Planning and Sustainable Infrastructure domain of human factors and ergonomics,
will be a valuable source of theoretical and applied knowledge enabling
human-centered design for global markets.

July 2018   Jerzy Charytonowicz
                                                                                      Christianne Falcão

Contents

Construction Industry

Ergonomics in Material and Environment Design

Urban Green Spaces: An Element of a City's Balance Between the Built and Natural Environments

Wojciech Bonenberg[1(✉)], Mo Zhou[1], and Shoufang Liu[2]

[1] Faculty of Architecture, Poznan University of Technology, Poznan, Poland
{wojciech.bonenberg,mo.zhou}@put.poznan.pl
[2] Liaoning Urban and Rural Construction and Planning Design Institute,
Shenyang Shi, China
liushoufang15940203503@hotmail.com

Abstract. The article presents a report on the possibility of using greenery as an element of natural balance in the city. The research was carried out in 2016 and 2017 with the aid of Poznan University of Technology Faculty of Architecture students and under the guidance of the authors. The reasons for the growing environmental problems in cities and proposals for remedial measures to prevent a breakdown of the ecological balance have been suggested. In this context, a model equilibrium in the natural spatial plan of the city was proposed, based on the quantification of the basic relationships between biotic and anthropogenic components. It emphasises the need to establish a dynamic balance in the urban environment. Particular significance to ecological engineering based on the principles for adapting biocoenoses to habitat conditions was assigned. Research within this scope focuses on three primary directions: adapting natural sites to the changed environmental conditions, ecosystem protection, formation of new natural ecosystems. The presented method for defining environmental balance was used in drafting detailed designs for rebuilding the natural environment in the Poznan Metropolitan Area.

Keywords: Urban greenery · Environmental balance · Anthropogenic activity

1 The Problem

Green spaces are an important component of urban structure. They provide a setting, where built up areas are uniquely interwoven with the fabric of a society. Natural conditions, social standard and investment activities all contribute to the state of green areas in cities.

The unique character of urban green areas takes root in the "urban planning process", which generates the structure of a city through complex environmental, social, organisational and technical interrelations combined with the actions of its residents.

The very diverse forms of urban green spaces are created with varying intensity. Sometimes they owe their existence to coordinated planning activities, whilst at other times to spontaneous growth. But regardless of whether they came about as a result of

© Springer International Publishing AG, part of Springer Nature 2019
J. Charytonowicz and C. Falcão (Eds.): AHFE 2018, AISC 788, pp. 3–13, 2019.
https://doi.org/10.1007/978-3-319-94199-8_1

natural phenomena or the efforts of urban planners, green spaces define the character of an urban environment.

1.1 Natural Constraints for Urban Growth

The interactions between humans and the natural environment they live in have assumed the characteristics of constraints, of a critical nature for further development of the settlement network. Nature's ability to spontaneously regenerate the intensively consumed resources has declined drastically. A deficit of natural resources is discernible in urbanised areas (which until quite recently were looked upon as an inexhaustible source of spatial development). Substances accumulate in the environment which exhibit harmful biological impact on life forms. Devastated areas, disused industrial areas and landfill sites are gradually taking up increasing swathes of land and in doing so they not only eradicate landscape qualities but also curtail settlement and recreation opportunities available to the ever growing urban populations [1].

Environmental problems attributable to the fact that more and more people are consuming increasing amounts of natural resources gave rise to serious fears for regional development perspectives. This is illustrated by Ehrlich's "ecological disaster" scenario, primarily driven by the disproportionate increase in the environmental burden resulting from population growth [7].

Carvalho considers incorrect use of technology to be behind the growing environmental problems in cities [6].

According to Maddox, the biggest conflicts with nature in developed, intensively urbanised areas do not stem from population growth, but rather incompetent use of technology [12].

Boulding points out that industrial production can be used as a measure for the loss of natural resources. The larger the economy, the more production is required to sustain it [5]. This subject is also associated with the rapid ageing of hi-tech products, energy and waste management. People are consuming increasing amounts of energy to obtain the latest models of advanced products. The planned lifetime period, which entails the destruction or disposal of an item which is still usable, has become the generally adopted means for growth, even though it is particularity harmful from the point of view of natural resource management (including spatial management) for the environment.

In this context, remedial measures to prevent a breakdown of the ecological balance have been suggested. The "spaceship earth" concept, put forward by Boulding is one of the most significant [4]. It emphasises the need to establish a dynamic balance in the environment. On a world scale this means an end to exponential growth, stabilisation of production and consumption and securing the quality of the primary natural resources. Maher emphasises that a shift from "quantity" to "quality" in spatial use will avert an ecological disaster [13]. Last assigns particular significance to ecological engineering based on the principles for adapting biocoenoses to habitat conditions [11].

Hough precisely defines the conditions which are required to ensure a decent quality of life [10].

"A Blueprint for Survival", co-authored by E. Goldsmith and D. Allen, is a text which not only spells out the general concepts for a sustainable (stabilised) ecology, but also refines plans for implementing given transition phases [8]. Based on the sustainable economics concept, the blueprint emphasises the need to appropriately orchestrate changes in spatial development management, aiming to reverse the current development directions which endanger the natural environment.

Bonenberg presents a compilation of measures and actions needed to restore the already damaged environmental balance [3].

Wenk presents an interesting approach to the environmental protection problem set [22]. The author makes no attempt at justifying the limited effectiveness of remedial measures used today and draws a conclusion that regeneration of natural environment is not possible without mankind's moral revival, especially when it comes to self-control.

Ecopsychology devotes a lot of attention to these concepts. This new discipline carries particular value, as it concerns the interpersonal dimension of health. Ecopsychology sets itself the task of defining the links between our health and well-being and the natural environment, where the primary focus is on green spaces. It is a type of a cultural personification of the bonds with nature. That new assessment of the bonds between mankind and the surroundings, has serious implications when it comes to spatial planning as it talks about individual preferences and behaviour of the residents with respect to green spaces. These preferences should be expressed through a subjective approach to the surrounding green areas, articulated by spatial policy with its interpersonal reflection in meticulous designs of green areas [2].

2 Environmental Balance in Cities

The "man - natural environment" relationship has become one of the primary factors limiting growth of urbanised areas. Human activity within these areas as well as the environment's ecological "quality" become entangled in an approximately inverse proportion. In a city centre for example, the former of the two factors is high, whilst the latter - low. On the other extreme, when we consider a location in a nature reserve, unspoilt by man, the aforementioned proportion will be inverted.

The idea to bond human activity with the environment's ecological "quality" was coined a long time ago. Architecture is one of the first disciplines, which considered the environment as an element inextricably linked with the activities of social groups as well as individuals. In the classic "De architectura libri decem", Vitruvius analysed the natural conditions for building new cities. In chapter four of Book One, he points out the impact of the climate and ways to choose a healthy site and in chapter six the effect of winds on the health of the residents. He devoted chapter six of Book Eight to testing for good water – he made a connection between the appearance of the inhabitants of a give region, the type of deposits in vessels and local vegetation with the properties of water. He presented ways for testing and avoiding harmful exhalations when digging wells [20].

References to the idea of unity between living organisms and the environment (and this includes unity between man and nature) may be found in subsequent geographical, biological and medical texts.

Whereas balance in natural ecosystems has been investigated quire thoroughly and unambiguously, the balance in artificial ecosystems still provokes discussions. Disputes primary revolve around theoretical research models and practical urban planning and architectural activities with reference to settlement areas.

In the context of nature, these are one of a number of spatially separate ecological systems found in the environment. It is emphasised that nature is a mosaic of ecosystems on different levels of succession, some already "mature" exhibiting high internal stability whilst the "younger" ones are less stable, with less species diversity and no internal balance. These are most often associated with human settlement activity.

Simmons identified four types of ecosystems which should make up a sustainable spatial system [17]:

(1) Artificial ecosystems, which include the "built-up environment" within settlements. In order to prolong their existence, these depend on various types of externally supplied energy and materials.
(2) Very productive, intensively used agricultural areas with the ability to generate high crop yields.
(3) Compromise areas, such as multiple-use forests, recreational areas, landscape parks or pastures.
(4) Mature non-agricultural ecosystems, or areas of vegetation unspoilt by man. These comprise important clusters of high biotic diversity and vital sites for gas exchange in the environment.

Odum presented a model of spatially separate types of environment needed by man. He used the given ecosystem's development stage and circulation of resources as the criteria [14].

The model encompasses the following ecological systems, grouped in accordance with the basic biological function:

(1) inanimate systems (the urban environment, urbanised areas),
(2) multi-use systems (intermediate type environment),
(3) developing systems (productive environment),
(4) mature systems (protective environment).

Maintaining an environmental balance requires an appropriative partial balance to be determined in each of the aforementioned areas.

This imposes particular requirements over the spatial planning process in order to ensure sustainable development of urbanised areas.

The theoretical description of the complex system of interactions between settlement activities and nature within those areas is still inadequate.

At first sight, such a statement may seem unjustified - is it not the case that numerous scientific disciplines have been probing the impact of mankind's economic activities on the environment? Nevertheless, the reasons for change, consequences of the negative impact of urban planning factors in nature require continuous research.

This is perhaps associated with the signification rate at which changes are taking place. Many failed, unsuccessful efforts in an urban environment, where the expected benefits where overwhelmed by the losses stand testament to the above. The reason behind such errors is most often found in the difficulty associated with drafting a reliable forecast as to the environment's reaction in urbanised areas and insufficient tried and tested methods for determining an acceptable level of anthropogenic interference. First and foremost, a lack of clearly defined criteria for an environmental balance in an urban environment should be emphasised.

3   Criteria for Environmental Balance in Cities

When talking about environmental balance in cities, one should define the balance level. It is obvious that reinstating optimal conditions for the development of mature natural biocoenoses within those areas is impossible as this would entail eradicating mankind's settlement and economic activities.

On the other hand, neither is it possible to continue the progressing degradation of natural vegetation complexes, which lead to them being entirely eliminated from an urban areas.

The desired balance level should be within the two extreme boundaries. And thus, spatial development plans should be drawn up for the green areas (vegetation) to retain their spontaneous regeneration capabilities with respect to the damage inflicted by settlement activities within territorially defined spatial units [3].

Environmental balance in a city can be described by the "man - natural environment" relationship. In order to obtain sufficiently accurate data to build a model environmental balance system within the scope of a city's spatial development plan, the basic relations between biotic and anthropogenic components were quantified. This quantification will make it possible to determine accurate proportions in terms of area and will provide guidelines as to the type of spatial development in a given areas. At the same time, it will facilitate a comparison of the implied state with the actual state, indicating spatial development directions which maintain or restore environmental balance.

The desired balance can be expressed using the following abstract, simplified notation:

$$NA = AC. \qquad (1)$$

where:

NA - stands for the quality of the natural environment within the boundaries of the given urban spatial unit,

AC - stands for anthropogenic activity within the boundaries of the given urban spatial unit,

$NA/AC = g$ - balance level index.

Equation (1) can also be expressed as:

$$NA/AC = 1. \tag{2}$$

When ecological balance within a given area is disrupted, then:

$$NA/AC < 1. \tag{3}$$

and $z = 1 - NA/AC$ indicates the scale of the danger.

There is extensive discussion in literature on the subject in question as to the selection of accurate indices describing the state of the environmental balance [3]. It depends on the specific nature of the area subject to the plan, scale and scope of available reports. A comprehensive suggestion within this scope has been put forward by the Mission to Save Earth Team as part of UN Agenda 21 on Sustainable Development. The indices have been selected on the basis of 7 criteria: measurability, innovation, adequacy, comparability, usability in design applications and effectives.

The two primary ecological system components - AC and NA are made up of sets of various types of values. For example, the NA component exhibits a system of (variable) characteristics, such as: energy flows at given trophic levels, plant species diversity, food chains, biomass quantity, etc. Similarly, the AC component may be described using the number of residents and the number of employed individuals in a given area, flow of materials, energy consumption, size of investment sites, etc.

3.1   Quality Characteristics of the Natural Environment

In spatial planning, when talking about green spaces we are dealing with a community of species organised in such manner, that it exhibits specific characteristic properties, not seen in individual specimen or populations which it is made up of. The community functions as a certain whole, through mutual metabolic connections. The term "biocoenosis" is used in most ecology texts to describe all biological components of such a community (fauna, flora and soil microbes). A "biotope" refers to the specific habitat conditions encountered in a biocoenosis.

Trojan cites the following characteristics of biocoenoses [19]: characteristic species composition, species composition richness, duration in time, area and boundaries. Biocoenoses exhibit a specific trophic structure, energy flow rate and the rate at which it accumulates, as well as properties such as: capacity, stability, diversity, succession stages. Many authors use the "biocoenosis" term for large, independent ecological units, such as forests. However, accordion to some opinions, biocoenoses may span areas as little as s few square meters. Odum makes a distinction between highly organised and relatively independent "large biocoenoses" and "small biocoenoses" which, to a large extent, are dependent on neighbouring ecological systems [14]. Wang associates the areas of biocoenoses which inhabit an ecosystem with the basic administrative units [21].

A similar diversity is seen when it comes to opinions of classifying biocoenoses. Such a classification is most often based on: major structural properties, physical environment of the biosciences or functional relations. As an example, we may cite

Perelman's classification attempts [15] based on basic biogeochemical properties and classification based on functional relations [18].

The presented diversity of opinions on the areas, boundaries and classification of biocoenoses, allows one to refer to a biocoenosis as a certain abstract value - rarely is there a clear distinction between biocoenoses, they often overlap, one blends into another. This provides some justification to link the areas and boundaries of biocoenoses with basic spatial units where sustainable growth is to be ensured.

In defining the variables associated with the quality of the natural environment NA, it was assumed (constituting a simplification of a kind) that they are linked to the biocenotic balance of green areas within settlement units. Partial environmental quality assessment criteria were selected to adequately reflect the basic requirements defining the balance level as set forth at the outset. Thus, variables were selected which are decisive to the largest extent possible in terms of:

– biocoenoses regeneration ability with respect to the damage done by production activities engaged in by the ecosystem's community,
– ability to compensate for the psychological and physical stress suffered by the residents, which, amongst others, is associated with biocoenoses landscape qualities.

One of the primary criteria for assessing the regenerative ability of vegetation is its stability.

The stability principle states that the energy flowing through every closed natural system changes it until self-adjustment mechanisms permanently adapt the system to the surrounding conditions. Upon reaching stability, energy changes occur within a system in a uniform, pre-determined manner and at a defined rate. A high degree of stability is a property of developed biocoenoses, which inhabit "mature", extensive forest, meadow or aquatic ecosystems.

However, it is known that a system is only stable within certain boundaries, outside of which controlled stresses result in destabilisation. Then the biocoenosis (flora) loses its ability to restore itself to the original state. The degree of floral stability depends on the changes taking place in the abiotic environment. These changes may stabilise or destabilise a system.

(a) A biocoenosis growing under natural ecological succession conditions will result in increased stability. An analysis of a succession trajectory, or the changes and order of biotic communities from the initial biotope inhabitation until the process of changes comes to an end and the final biocoenosis stabilisation are depicted in texts by Sahney and Benton [16].
(b) A reduction to biocoenosis stability is associated with the effects of various stresses, most often caused by urban planning and mankind's economic activities. Some effects of that activity, impact the flora and constitute a powerful stimulus disturbing the self-adjustment mechanisms. The most "sensitive" species are eliminated, and we have a progressively more barren, unstable biocoenoses. The phenomenon is widely discussed in many papers [9].

A general conclusion may be drawn from such research papers, that stability can be identified as a set of (variable) properties which describe biocenotic communities.

However, it is difficult to quantify most of these variables. For example, measuring an ecosystem's primary production requires the application of laboratory methods (isotope method or CO_2 and O_2 variation analysis). In the opinion of many authors, some properties listed in the table are correlated. Therefore, it is not necessary to measure all variables to assess the biocenotic stability of green areas, but only some, those which yield most easily to measurements. For example, Odum emphasises that species richness increases as the proportion of energy expenditures for breathing to the quantity of biomass decreases [14]. It has also been demonstrated that for a larger quantity of vertical habitat zones, the rate of exchange of nutrition components between the organisms and the surroundings increases. Similarly, resistance to external disturbances is stronger in biotic communities which exhibit higher species diversity and inhabit larger areas.

That relationship has been used as a basis for new industrial and communal waste disposal concepts. These methods take advantage of opportunities to utilise decomposable contaminants by appropriately controlling their supply to a biocoenosis. These substances, if supplied in moderate quantities, may increase the overall biocoenosis productivity and constitute a valuable source of mineral food components (e.g. phosphates, nitrates, carbonates, etc.).

These examples show, that sufficiently large and diverse (in terms of species) green complexes are able to maintain stability of an artificial ecosystem which includes settlement areas.

Quality characteristics of the natural environment should be easy to quantify. This is important for practical application in spatial planning.

The following quality characteristics of the natural environment in settlement regions have been selected on the basis of the depicted analysis [3].

s_1. Species richness index.

This is considered to be one of the simplest methods for detecting and assessing environmental pollution levels. The ability to identify species suffices to determine this index.

s_2. Species evenness index.

Species evenness refers to the distribution of individuals across all the present species. If the community is not very even, then most individuals represent a single species. The remaining individuals are distributed across the remaining species. This state prevails most often when some (most resistant) species are afforded significant growth opportunities. The growth of other (more sensitive) species is held back by the impact of limiting factors (e.g. pollution).

s_3. Spatial stratification - the number of vertical habitat zones.

Spatial stratification has been determined to be between 0 and 5 (trees, shrubs, herbs, undergrowth, litter and humus).

s_4. Number of indicator species.

Indicator species are rare species and their presence (or absence) is used to infer the quality of a natural environment.

These environmental risk indices are considered to be simple to acquire and reliable. For example, lichen is a good index for atmospheric pollution due to its sensitivity to acid rain.

s_5. This biocoenosis area (determined on the basis of planimetric readings and site observations).

3.2 Characteristics of Anthropogenic Activity

The following variables have been adopted to describe the AC component [3]:

a_1 - Number of residents within an urban unit.
a_2 - Number of people employed within an urban unit.
a_3 - Size of built-up area.
a_4 - Urban unit "catchment" area.

For sites located at various distances from the centre of the settlement activity, the aforementioned a4 decreases proportionally to that distance.

a_5 - Quantity of urban unit infrastructure components.
a_6 - Material balance - quantity of materials transported to and from the urban unit.

One should note that the variables describing AC are directly related to the given location. They depict mankind's current and planned urban and economic activity levels within each identified urban unit. They are easy to determine on the basis of analyses which encompass demographic forecasts, spatial development directions, the current and planned size of built-up areas, waste management effectiveness and energy consumption forecasts.

4 Measuring Environmental Balance

In accordance with the adopted assumption, ecological balance should include areas where the residents are active on a daily basis, and thus primarily urban functional and spatial units associated with work, dwellings and recreation.

Then, the desired balance for every urban functional and spatial unit, as defined by Eq. (1), assumes the following form [3]:

$$r(a_1, a_2, a_3, a_4, a_5, a_6) = w(s_1, s_2, s_3, s_4, s_5). \tag{4}$$

where:

r, w - location multipliers, associated with the specific nature of human activities and habitat conditions (climate, water and soil) within the urban functional and spatial unit.

The following procedure has been adopted to measure NA and AC:

– determination of partial indices for each area subject to the research on the basis of statistical data, planimetric readings and site visits,

- calculation of normalised values for the aforementioned indices by placing raw values of each index on an identical normalised scale (0/100),
- determination of a rank for each index; here given indices were compared using paired significance method
- calculation of NA and AC values as an arithmetic mean of the normalised and weighted partial indices values.

The presented method for defining environmental balance was used in drafting detailed designs for rebuilding the natural environment in the Poznan Metropolitan Area. The research area was divided into 344 functional and spatial units. AC and NA values were calculated for these areas which were then recorded on a city map. By linking areas where the aforementioned indices had similar values, "contour maps" were obtained depicting the level of anthropogenic activity (AC) and natural environment quality level (NA). Analogously, AC/NA were also marked on a city map, facilitating an identification of sites where environmental balance is most disrupted.

124 areas most at risk were identified on the basis thereof, and actions aiming to restore balance were suggested. The expected positive results stem from project suggestions entailing reinstating local environmental balance. And thus:

- in 15 functional and spatial units a reduction of local vehicle traffic was suggested in favour of safe pedestrian and bicycle access,
- in 3 functional and spatial units establishing water reservoirs (ponds) was recommended at sites which are currently home to abandoned substandard industrial and warehousing developments,
- in 45 functional and spatial units green corridors were recommended which comprise a cohesive ecological network interlaced with residential estates,
- in 16 functional and spatial units dense green belts were recommended on both sides of major trunk roads,
- in 30 functional and spatial units encompassing existing urban parks, enriching the greenery species mix was recommended,
- in 13 functional and spatial units, establishment of "city farms" was recommended to produce food for the local residents,
- in 2 functional and spatial units, establishment of extensive biocenotic communities was recommended, to filter municipal waste water and to compost organic wastes.

5   Conclusion

The "man - natural environment" relationship has become one of the primary factors when it comes to urban quality of life. Green areas are an important environmental balance component.

The suggested method for determining environmental balance in an urban setting provides significant support for the designs and planning procedure. The conclusions yielded by environmental balance analysis define the role and place of green areas in a city, the scope of necessary investments, how green spaces should be managed and the associated costs. In that context, green areas are an indispensable element of the "new urban planning culture" in urban design.

References

1. BenDor, T.K., Metcalf, S.S., Paich, M.: The dynamics of brownfield redevelopment. Sustainability **3**, 914–936 (2011)
2. Bonenberg, W.: Beauty and ergonomics of living environment. In: Vink, P., Kantola, J. (eds.) Advances in Occupational, Social, and Organizational Ergonomics, pp. 575–581. Taylor & Francis Group, Boca Raton (2010)
3. Bonenberg, W.: Przemysł w mieście. Ekologiczna metoda modernizacji zakładów przemysłowych zlokalizowanych na obszarach intensywnie zurbanizowanych, pp. 56–62, 97–106. Zeszyty Naukowe Politechnika Slaska 850, Gliwice (1985)
4. Boulding, K.E.: The economics of the coming spaceship earth. In: Jarrett, H. (ed.) Environmental Quality in a Growing Economy: Essays from the Sixth RFF Forum, pp. 3–14. John Hopkins University Press, Baltimore (1966)
5. Boulding, K.E.: What do we want to sustain? Environmentalism and human evaluations. In: Costanza, R.: Ecological Economics: The Science and Management of Sustainability, pp. 367–383. Columbia University Press, New York (1991)
6. Carvalho, A.C.V., Granja, A.D., Silva, A.G.: A systematic literature review on integrative lean and sustainability synergies over a building's lifecycle. Sustainability **9**(7), 1156 (2017)
7. Ehrlich, P.E., Holdren, J.P.: Impact of population growth. Science **171**(3977), 1212–1217 (1971)
8. Goldsmith, E., Allem, D., Allaby, M., Davoll, J., Lawrence, S.: Blueprint for survival. Ecologist **2**, 1–50 (1972)
9. Horn, R., Dahy, H., Gantner, J., Speck, O., Leistner, P.: Bio-inspired sustainability assessment for building product development - concept and case study. Sustainability **10**(1), 130, 2–25 (2018)
10. Hough, P.: Environmental security. Routledge, New York (2014)
11. Last, F.T., Hotz, M.C.B., Bell, B.G.: Land and its uses – actual and potential an environmental appraisal. Plenum Press, New York, London (1982)
12. Maddox, J.: The case against hysteria. Nature **235**, 63–65 (1972)
13. Maher, T.M., Baum, S.D.: Adaptation to and recovery from global catastrophe. Sustainability **5**, 1461–1479 (2013)
14. Odum, E.P.: Fundamentals of Ecology. W.B Sounders, Philadelphia, London, Toronto (1977)
15. Perelman, A.J.: Geochemia krajobrazu. PWN, Warszawa (1971)
16. Sahney, S., Benton, M.J.: Recovery from the most profound mass extinction of all time. Proc. R. Soc. B **275**, 759–765 (2008)
17. Simmons, I.G.: Ekologia zasobów naturalnych. PWN, Warszawa (1979)
18. Song, C., Kim, S.J., Moon, J.: Classification of global land development phases by forest and GDP changes for appropriate land management in the mid-latitude. Sustainability **9**(8), 1342 (2017)
19. Trojan, P.: Ekologia ogólna. PWN, Warszawa (1981)
20. Vitruvius: The Ten Books on Architecture (De architectura libri decem). Trans. by Hicky, M. Harvard University Press (1914)
21. Wang, N., Kang, N., Yu, Y.: Valuing urban landscape using subjective well-being data: empirical evidence from Dalian, China. Sustainability **10**(1), 36 (2018)
22. Wenk, E.: Margins for Survival. Pergamon Press, Oxford, New York (1979)

Open-Air Work Zones for Students at the Faculty of Architecture Depicted on the Basis of Pilot Student Projects

Dorota Winnicka-Jasłowska[✉], Joanna Tymkiewicz,
and Klaudiusz Fross

Faculty of Architecture, Silesian University of Technology, Gliwice, Poland
{Dorota.winnicka-jasloslowska,joanna.tymkiewicz,
klaudiusz.fross}@polsl.pl

Abstract. The following article pertains to a student project called: Open-Air Work Zone for Students of the Faculty of Architecture. It elaborates on the scope of pre-design research which preceded projects carried out by students. The context in which such places came to be, within the space of contemporary campuses, was also analyzed. Next, two concepts put forth by students were described as the result of end-of-semester papers completed at the Department of Design and Qualitative Research of the Faculty of Architecture at the Silesian University of Technology (Gliwice, Poland). The publication also contains a brief description of the changes made to the structure of the Polish higher education facilities, as a result of significant factors which influenced the new methods of making use of the space and its new quality.

Keywords: Higher education · Universities · Students' zone
Learning and teaching space

1 Introduction

The idea for the Open-Air Work Zone at the Faculty of Architecture of the Silesian University of Technology came into being a few years ago. There had been, of course, requests made by the students for places where one could spend their leisure time during breaks between classes, outside the building. Before the start of the project and at the very beginning, pre-design research was done together with students, which would diagnose their needs as to the projected space and would indicate the proper location, in the closest vicinity of the building of the Faculty where the open-air work zone would be located. In recent years, there have been lots of analyses made by students of the Architectural Faculty of the Silesian University of Silesia in the scope of pre-design research and concept designs connected with locations intended for spending leisure time between classes. During these breaks, one could either relax and socialize, or work and learn. In the course of research carried out by students in their own environment, it turned out that they wished to see a space for both work and leisure that would be located in the vicinity of the faculty's building and would function better than the contemporary one.

© Springer International Publishing AG, part of Springer Nature 2019
J. Charytonowicz and C. Falcão (Eds.): AHFE 2018, AISC 788, pp. 14–23, 2019.
https://doi.org/10.1007/978-3-319-94199-8_2

2 New Functional and Spatial Needs at Contemporary Universities

Common access to sources of knowledge possible by means of the World Wide Web led to the situation where forms of work changed along with the ways in which knowledge was acquired. What is more, new phenomena in the process of socializing also arose. At present, human beings, and in this case we are referring to students, may be active wherever they wish set foot; they could decide to work, learn and collect data of off the Internet at each place. They have the possibility to work at any given place, on the condition that they could find there appropriate working conditions. The possibility of using the Internet and its resources, at any place of the university, caused that new forms of work came into being along with possibilities to socialize at places where it was impossible in the past. [9, 10]. What is truly significant today for the shaping of new university structures and for the sake of innovative and functional solutions is the knowledge of the forms of work used today, as well as the cooperation in the scope of learning, and also the way people behave at that particular moment. According to OECD[1], the most basic form of the academic faculty's development process, along with their buildings (in terms of the shaping of the space) is the knowledge of users' needs. In the 21st century, OECD claims that we are facing a new image of space and facilities, based on the following assumptions: information and communication technology (ICT); students' expectations, their lifestyle (features of an information-based society); social interaction in the scope of the learning process [10, 11]. The aforementioned assumptions have a significant influence on the design of the space intended for learning. At present, one should no longer perceive the space of the university in a traditional way. Forms of learning and activities used have changed for the following models:

- Collaborative model which consists in actions based on team work, with a larger number of people.
- Immersive model which consists in individual work, requiring more focus.
- Mixed model which serves as a combination of group and individual work, such as works done at laboratories, where a high-level network connection is provided just like a proper space for group work.
- "Anywhere" model where knowledge is being acquired in both formal and informal spaces[2]. Inside the existing buildings of the university, more open and public-access spaces are being planned for. Their arrangement will allow students to work wherever they wish to. That model, according to the students who were interviewed, goes beyond the buildings themselves. As a result, the external space of the campus becomes almost equivalent in relation to the interiors of the buildings.

---

[1] OECD – Organization for Economic Cooperation and Development; OECD constitutes an international institution which deals with qualitative research into higher education systems. It assesses, supports and promotes good practice for the sake of the quality that the scientific environment should maintain. Source: A. Blyth: OECD Programme on Education Building. www.oecd.org/edu/facilities.

[2] The division made by OECD [13] D. Winnicka-Jasłowska's description based on the publication [12].

3   Statutory and Organizational Changes at Polish Academic Facilities

Polish universities of today have undergone a series of transformations in the scope of education and higher education system. Of course, there have also been some organizational and social changes made to the academic facilities. In 1989, Poland as a socialist country (until 1989) changed its political system into capitalism, which caused that legislature connected with education and higher education also had to undergo a thorough transformation. But the most important changes were made in 2004 when Poland became a member the European Union.

At first, the main changes were brought about by way of a reform passed in 2005. After that, a highly significant document called: *The Strategy for the Development of Higher Education in Poland until 2020* was prepared by Ernst & Young B.A and the Institute for Market Economics. That very document became the key strategic act and since 2012 it has been implemented at Polish academic facilities accordingly. Actions taken on the basis of that document are to help increase the quality across all fields of the university's activity, especially in education, scientific research and the relationship with the social and academic environment. According to the authors, raising the quality of the higher education requires Strategy [12], and action in the scope of six strategic goals:

- Diversity in the academic facilities and courses of studies;
- Mobility on the part of the academic staff and students;
- Competition in the higher education system;
- University's effectiveness in being able to take advantage of their own resources;
- Transparency in the activity of the facility;
- Openness of the academic facility to the social and economic surrounding.

Three out of the six strategies presented above can undoubtedly have an influence on the material resources of higher education institutions. These include: the openness, mobility and competitiveness [12].

Openness of universities requires a new approach to the shaping of the space within buildings and the campuses. When referring to the openness, the authors of the Strategy [12] determine the principles connected with the flow and transfer of knowledge between universities. However, the notion of openness also pertains to greater freedom and needs in the scope of relationship between various internal and external environments of academic facilities. Openness stands for, among others, actions that are aimed at closing the gap between external environments and the scientific environment. Through different modes of action as well as greater accessibility - academic facilities are becoming the organizers of numerous events of scientific and promotional character. Such an approach is currently very broad and requires new spaces within the structures and within the areas that belong to campuses. As many an event take place outside, the external space should be thought of differently than in the past.

Mobility of students and of the academic staff has become popular in Poland. As a result of that change, the quality of work and studies have significantly improved over the last 10–15, which means that Polish academic facilities have now found themselves in the same league as foreign higher education institutions. This pertains not only to the shaping of the space, the process of equipping the laboratories but also to the esthetics of the structures and buildings together with all academic areas. The fact that academic facilities can now compete against one another, which constitutes one of the conditions of the Strategy [12], has also had an influence on the above.

These three main strategic conditions - openness, mobility and competitiveness have significantly changed the methods of the functioning of academic facilities, along with their organization. They also influenced the ways the space is being used and improved its quality.

4   Silesian University of Technology - Revitalization of the Campus Area

Over the last couple of years, the Silesian University of Technology, has made numerous beneficial changes which caused that the quality of the usable space has improved. New structures equipped with common functions appeared, such as the Education and Congress Center (built in 2004), or the laboratory structure called the Center of New Technologies (2014). The latest investment made by the Silesian University of Technology was the reconstruction and functional and spatial modification of the whole area of the university's campus (2014). That undertaking was radical in its scope. It encompassed, above all, a project which involved the exclusion of the Akademicka Street from the vehicle traffic and the construction of a shared zone which now runs through the campus, from the eastern part to the western part. The pedestrian zone was significantly broadened and thus creates recreational spaces and offers a new image of squares in front of the entrances to the faculty's buildings. New landscaping elements were introduced such as: benches, fountains and pedestals with the names of the faculties. What is more, also the underground infrastructure for the installations was replaced.

The process of the campus reconstruction lasted for about two years. Changes which were made generated numerous subjective comments and assessments within the student and worker environment. These changes also influenced the decision as to whether the research should be done to make an assessment of the new solutions[3] (Figs. 1 and 2).

---

[3] Qualitative research at the campus of the Silesian University of Technology was done several times by different research teams. For the first time, the campus was examined in 2014/2015 by a team of students supervised by Doroata Winnicka-Jasłowska. That research was described at length in the authors scientific monograph, written in Polish [10].

Fig. 1. Campus of the Silesian University of Technology after the revitalization. View of the building of the Civil Engineering Faculty and in the background visible is one of the laboratory buildings of the Chemistry Faculty. Photo taken by D. Winnicka-Jasłowska

Fig. 2. Campus of the Silesian University of Technology after revitalization. View in the direction of the Faculty of Mining and Geology. Photo taken by D. Winnicka-Jasłowska

5 Pre-design Research and Student Projects Concerning the Open-Air Work Zone

Students have recently been speaking out about their own needs as to the space within and without the building of the Architectural Faculty, as well as the whole area of the campus. As part of classes[4] conducted at the Faculty, they made an assessment of the current condition and then they proposed a modification to the existing solution. They discussed the following problems: way-finding, security of people and property, ergonomic solutions in the scope of universal design. They also proposed open-air zones within the nearest vicinity of the building of the Architectural Faculty. During the research process, they precisely specified their needs concerning work and time spent during, after and in between classes. Moreover, they also specified their needs connected with the process of socializing. That aspect is currently more important than ever before. It is connected with the development of knowledge and science in the scope of sociological sciences and environmental psychology. On the basis of the conclusions drawn from the research process, the students carried out projects connected with changes made to the arrangements so that more places could be designed for learning and socializing out in the open space, in the vicinity of the building where the Faculty of Architecture is located.

The main method implemented in the pre-design research was the Post-Occupancy Evaluation[5]. At the beginning, site inspection was performed together with stock taking of the places deemed potentially suitable for designing an open-air work zone. After that, interviews were conducted with the members of the Student Government of the Architectural Faculty, on the basis of which the survey was composed. The survey was conducted among the students of the whole Faculty. The research questions were steered so that needs connected with the designed space could be recognized. The questions were formulated in such a way as to assess the current condition of the buildings in terms of places intended for learning. The most desirable elements of the equipment intended for the open-air work zone, in the eyes of the respondents, were pieces of furniture that included: relaxation poufs, sofas, arm chairs, couches, wide and large table worktops; comfortable chairs.

It should be added that students showed a lot of creativity in that respect. In teams of two, they presented some really interesting and diversified solutions - in terms of functionality, technicality and esthetics. The projects included: spatial development of the chosen area close to the entrance to the building of the Faculty, projections and views of the furniture along with their visualizations which best represented the whole concept. It should be highlighted that the most interesting concept belonged to Oliwia Kwaśniewska and Magdalena Wójtowicz (Fig. 3). It depicts a large table in the shape of piece of a jigsaw puzzle with stools placed around it. The whole thing stands under a roof in the

---

[4] The project was carried out in the scope of classes run by Joanna Tymkiewicz PhD. Eng. of Architecture, Dorota Winnicka-Jasłowska, PhD. Eng. of Architecture, under the patronage of the Dean of the Architectural Faculty of the Silesian University of Technology, Klaudiusz Fross, PhD. Eng. of Architecture in the academic year 2016/2017.

[5] Author's own methods for qualitative research based on POE, as well as examples of opinions concerning the structures and pre-design research carried out by the authors of the following publication were discussed in the following references [1–10, 13–18].

Fig. 3. Concept design for the Open-Air Work Zone for students of the Architectural Faculty. Visualization. Design and graphics made by Oliwia Kwaśniewska and Magdalena Wojtowicz.

Fig. 4. Concept design for the Open-Air Work Zone for students of the Architectural Faculty. Visualization. Design and graphics made by Natalia Płoskonka and Anna Krawczyk.

form of a canopy. A structure which has an intriguing shape makes for an attractive city sculpture. Thanks to the fact that it is made from concrete, one can be sure of its strength. That piece of furniture is located in a visible place, by the main square of the campus and the fountain, in order to decorate the spot and integrate the student environment.

The second project which also deserves recognition is the furniture complex for both sitting down and for work, which makes up the spatial composition based on kind of boxes. These crates play the role of a table, bench and flowerbeds. The whole structure has been composed into a square which is near the entrance to the building, in a quiet and secluded place. That project, unlike to previous one, is more suitable for work and relaxation in peace owing to its location, tables and separate spots for sitting down. The authors of the project are Natalia Płoskonka and Anna Krawczyk (Fig. 4)[6].

---

[6] Both projects have been completed under the supervision of Joanna Tymkiewicz, PhD. Eng. of Architecture and Dorota Winnicka-Jasłowska, PhD. Eng. of Architecture.

6   Summary

Contemporary users of the academic spaces expect to see new spatial solutions. This phenomenon is especially visible in the spatial organization of the academic space. Both buildings as well as entire campus areas, which belong to higher education institutions, are currently used in a different manner than in the past. New needs have given rise to new solutions within buildings and also in the scope of the development of campus areas. The Silesian University of Technology, as one of the most prominent technical and educational institutions in Poland, has also undergone a metamorphosis which caused that the quality of work and education stand at a very high level.

The following case study, concerning the design development for the competition for the Open-Air Work Zone for students of the Architectural Faculty, constitutes a good example of research by design as well as a didactic analysis. Besides the measurable design effect, students also appreciated the role of the pre-design research in the creation of proper architectural solutions.

All works carried out by the students - made up of posters and also models that go with them - were presented during an exhibition held inside the buildings of the Architectural Faculty. As expected, they drew a lot of attention on the part of the whole academic community (Fig. 5). During the said exhibition presented were also concept works completed as part of the didactic project which took place at the same time and was connected with the proposal to take certain actions (the so-called "soft interventions") which were aimed at breathing a new life into the Academic Zone.

Fig. 5. A fragment of the exhibition showing students' ideas for the open-air work zone, the Faculty of Architecture of Silesian University of Technology, June - September 2017, Photo taken by J. Tymkiewicz.

All examples of the concept solutions indicate that there is a high level of efficiency when it comes to pre-design qualitative research, useful in the manner of assessing the true needs of the users. Finally, it should be added that the design for the Open-Air Work Zone, described in this article, is one of the many projects currently being carried out in the scope of classes conducted at the Faculty of Architecture of the Silesian University of Technology. The scientific staff and all teachers from the Department of Design and Qualitative Research in Architecture are specialists in research on structures and urban space, with more than 20 years of experience and they have worked out both new didactic methods and individual techniques as well as research tools. Earlier experiments which took place with students were described, among others, in the following publications: [1–10, 13–18].

References

1. Fross, K.: Ergonomics in the practice of project architect on selected examples. In: Human-Computer Interaction. Theories, Methods, and Tools. 16th International Conference, HCI International 2014, 22–27 June 2014, Heraklion, Crete, Greece, Proceedings, Part I. LNCS, vol. 8510, pp. 77–85. Springer, Cham (2014). ISBN 978-3-319-07232-6, Print 978-3-319-07233-3
2. Fross, K.: Architect-researcher as a model combination of research and design practice on examples. In: Charytonowicz, J. (ed.) Advances in Human Factors and Sustainable Infrastructure. Proceedings of the 5th International Conference on Applied Human Factors and Ergonomics AHFE 2014, July 19–23 2014, Kraków, Poland, Las Vegas (2014). ISBN 978-1-4951-2092-3
3. Fross, K., Sempruch, A.: The qualitative research for the architectural design and evaluation of completed buildings – part 1 – basic principles and methodology. Archit. Civil Eng. Environ. (ACEE) 8(3), 13–19 (2015). Silesian University of Technology
4. Fross, K., Sempruch, A.: The qualitative research for the architectural design and evaluation of completed buildings – part 2 – examples of accomplished research. Archit. Civil Eng. Environ. (ACEE) 8(3), 21–28 (2015). Silesian University of Technology
5. Fross, K., Winnicka-Jasłowska, D., Gumińska, A., Masły, D., Sitek, M.: Use of qualitative research in architectural design and evaluation of the built environment. In: AHFE – HFSI 2015, Session: Ergonomical Evaluation in Architecture, Las Vegas (2015)
6. Fross, K., Winnicka-Jasłowska, D., Sempruch, A.: "Student zone" as a new dimension of learning space. Case study in Polish conditions. In: Charytonowicz, J. (ed.) Advances in Human Factors, Sustainable Urban Planning and Infrastructure. Proceedings of the AHFE 2017 International Conference on Human Factors, Sustainable Urban Planning and Infrastructure, 17–21 July 2017, Los Angeles, California, USA, 11 poz, pp. 77–83. Springer, Cham (2018)
7. Winnicka-Jasłowska, D.: Creating a functional and space program for new building of the Faculty of the Biomedical Engineering building, Silesian University of Technology. Archit. Civil Eng. Environ. (ACEE) 5(3), 41–50 (2012). Silesian University of Technology
8. Winnicka-Jasłowska, D.: Ergonomic solutions of facilities and laboratory work-stands at universities. Programming of ergonomic technological lines – case study. In: Stephanidis, C., Antona, M. (eds.) Universal Access in Human-Computer Interaction. 8th International Conference UAHCI 2014, held as Part of HCI International 2014, 22–27 June 2014, Heraklion, Crete, Greece. Proceedings. Pt. 4, Design for All and Accessibility Practice. LNCS, vol. 8516, Springer, Cham (2014)

9. Winnicka-Jasłowska, D.: Quality analysis of Polish universities based on POE method. Description of research experiences. In: Antona, M., Stephanidis, C. (eds.) 9th International Conference, UAHCI 2015 held as Part of HCI International 2015, 2–7 August 2015, Los Angeles, CA, USA. Proceedings. Pt. 3, Access to Learning, Health and Well-Being, pp. 236–242. Springer, Cham (2015)
10. Winnicka-Jasłowska, D.: Przestrzeń nauki współczesnego uniwersytetu. Rola badań przedprojektowych w programowaniu nowych funkcji wyższych uczelni. (Title in English: The Learning Space of a Contemporary University. The Role of Pre-design Research in The Programming of New Functions of Universities.) Monografia, Wydawnictwo Politechniki Śląskiej, Gliwice (2016)
11. www.oecd.org/edu/facilities
12. The Strategy for the Development of Higher Education in Poland until 2020–Ernst & Young Business Advisory, Instytut Badań nad Gospodarką Rynkową (2010). http://cpp.amu.edu.pl/pdf/SSW2020_strategia.pdf
13. Tymkiewicz, J., Bielak-Zasadzka, M.: Senior homes of the future in the eyes of students of architecture. Didactic experience from the application of the design thinking method. Archit. Civil Eng. Environ. (ACEE). 9(1), 49–56 (2016). Silesian University of Technology
14. Tymkiewicz, J., Bielak-Zasadzka, M.: The design thinking method in architectural design, particularly for designing senior homes. Archit. Civil Eng. Environ. (ACEE) 9(1), 43–48 (2016). Silesian University of Technology
15. Winnicka-Jasłowska, D., Jastrzębska, M., Tymkiewicz, J.: Ergonomics of laboratory rooms – case studies based on the geotechnical laboratories at the Silesian University of Technology. Archit. Civil Eng. Environ. (ACEE) 10(2), 35–41 (2017). Silesian University of Technology
16. Tymkiewicz, J., Winnicka-Jasłowska, D., Jastrzębska, M.: Pre-design studies on the example of modernization project of geotechnical laboratories. Archit. Civil Eng. Environ. (ACEE) 10(2), 43–52 (2017). Silesian University of Technology
17. Tymkiewicz, J.: Quality analyses of facades based on post occupancy evaluation. Research experience with students of architecture participation. In: Chova, L.G., Martinez, A.L., Torres, I.C. (eds.) 9th Annual International Conference of Education, Research and Innovation (ICERI), 14–16 November 2016, Seville, Spain. ICERI2016: 9th International Conference of Education, Research and Innovation. ICERI Proceedings, pp. 8831–8838 (2016)
18. Tymkiewicz, J.: Team work efficiency in finding innovative solutions. Experience with the design thinking method implemented into teaching at the faculty of architecture. In: Chova, L.G., Martinez, A.L., Torres, I.C. (eds.) 8th International Conference of Education, Research and Innovation (ICERI), 16–20 November 2015, Seville, Spain. ICERI2015: 8th International Conference of Education, Research and Innovation. ICERI Proceedings, pp. 5894–5902 (2015)

The Campus Space in Research and Student Projects

Joanna Tymkiewicz, Dorota Winnicka-Jasłowska,
and Klaudiusz Fross

Faculty of Architecture, Silesian University of Technology, Gliwice, Poland
{joanna.tymkiewicz,dorota.winnicka-jaslowska,
klaudiusz.fross}@polsl.pl

Abstract. As an inseparable part of the whole panel, entitled: "Teaching Methods in Architectural Ergonomics", the article shall shed some light on the methodology and the course of research, as well as on its detailed results connected with the issue of the academic zone, along with general remarks connected with scientific methods applied at the Faculty of Architecture of the Silesian University of Technology, which represents the "Silesian School of the Qualitative Research in Architecture", with its 20-year experience gained over the years as a result of series of research carried out within the buildings and urban spaces.

Keywords: Research in architecture and urban planning · Campus space
Student projects

1 Introduction

A few years ago, the central space of the campus of the Silesian University of Technology in Gliwice (Poland) did not come across as an impressive place. It was separated by the forever busy Akademicka Street along which there were parking lots, damaged sidewalks and squares, areas with benches surrounded by lush bushes and trees. The whole thing was in a dire need of a modernization (See Fig. 1). And that is what happened next. In the years 2012–2014, a thorough reconstruction of the Akademicka Street was done along with the areas adjoining the street, on the basis of a successful competition design. Moreover, a granite cobblestone alleyway, reserved for pedestrians only, came into being along with an oval area where an illuminated fountain could be found, surrounded by a gentle artificial hill, while around the place a surface of evenly trimmed lawns was designed with geometrically arranged composition of plants (see Fig. 2). The space was completed with new benches that bore the university's crest, while in front of the buildings one could find tall concrete "pylons" that inform people of the name of each faculty. An additional attraction is the so called "the path of light", which is a LED tract that connects the illuminated center of the town with the sports hall that was put up on the other side of the academic zone, the body of which is surmounted by a ring of light. It should be added that an integral element of the aforementioned tract is an interactive light sculpture.

The complete modernization has significantly improved the composition of the campus in terms of its order. It has also raised the esthetic value as well as its technical

© Springer International Publishing AG, part of Springer Nature 2019
J. Charytonowicz and C. Falcão (Eds.): AHFE 2018, AISC 788, pp. 24–35, 2019.
https://doi.org/10.1007/978-3-319-94199-8_3

quality. The space comes across as attractive, pretty and well taken care of. In addition, in 2017, it was equipped with a bicycle-sharing station. The whole modernization process was met with enthusiasm and received a distinction in the form of a special prize awarded by the Minister of Infrastructure and Development of the Polish government.

Fig. 1. The view in the direction of the main space of the campus of the Silesian University of Technology - as of November 2009; in the foreground one can see the Akademicka Street, in the background the edifice of the Faculty of Civil Engineering is visible (photo taken by Joanna Tymkiewicz)

One would have thought that a perfectly designed public pedestrian zone would be teeming with life. People were of the opinion that it would attract the citizens of Gliwice, and that parties for the academic community would be held at the central square, in front of the Faculty of Civil Engineering and the Faculty of Architecture. That, however, did not happen. On the basis of all the observations and conversations held, it was gathered that all those benches, which had been arranged in rows, did not encourage students to gather there to socialize. The concrete pylons definitely did not improve the sense of direction within the academic zone, while the "light trail" just like the "interactive light sculpture" have gone unnoticed. To cap it all, in spring and summer the fountain became occupied by noisy children who bathed in it. These were merely commonplace opinions which were subject to scientific methods of verification. Therefore, thanks to an initiative undertaken by the Dean, a project entitled: "The Experimental Project on 'Soft' Intervention Aimed at Enlivening the Academic Zone" was carried out at the Silesian University of Technology. It was conducted by students under the guidance of scientific and didactic workers, in the scope of classes in "Methodology of Scientific Work" conducted in the 2nd year of the second cycle studies.

The said project was aimed at students who would propose solutions that would enliven the architecture of the recently modernized academic zones, which in turn would entice students, senior citizens, school age children and youth as well as families with children to take advantage of those zones more actively. The project included scientific research as well as concept designs presented in the form of a visualization in which the broadly understood notion of solution ergonomics constituted a highly important aspect, which stood for - a human being – his/her comfort and convenience, physical and mental health which should be found within the architectonic and urban space.

Fig. 2. Roof view from the Faculty of Architecture in the direction of the main campus zone of the Silesian University of Technology, recently modernized - its condition as of March 2017. In the foreground one could notice an oval central plaza with a fountain; at the heart of it there is Akademicka street as well as edifices of the Faculty of the Electrical Engineering as well as Mining and Geology (photograph taken by Joanna Tymkiewicz)

2   State of the Art

The subject matter, which has been discussed in this article and signaled in the title, consists of three components to which one could, at the current state of research, refer three groups of publications: the first pertains to the problem of urban space, with special emphasis put on university space, the second - to research related issues, which include methods, techniques and tools used in research on structures and urban spaces carried out at the university in the scope of didactics.

In the first group of publications, which constituted the knowledge base during the preparation and execution of research, one could list the following papers [1–8].

In the second group of publications, this time on the topic of methodological issues - in the preparation of research, the reference titles which included the conditions and specificity present in Poland were as follows: [9–16].

The third group of publications, on the other hand, included analyses which depict the results of research done in cooperation with students as part of didactic classes, especially by scientists representing the "Silesian School of Qualitative Research in Architecture" [17–20].

3   Research Methodology

As regards opinion polls on the topic of academic zones of the Silesian University of Technology the following methods, techniques and research tools were used:

- the method of qualitative research, including independent variations of the POE (Post Occupancy Evaluation) as well as elements of the action research method, and also quantitative methods (at the indicative level, conducted on the basis of an unrepresentative, occasional sample);
- techniques: surveys, participatory observations, site inspections, targeted questionnaires, focus meetings;
- tools: Internet surveys, lists of criteria, mood boards, filming, collection of photographic documentation.

The students had an influence on the choice of the method, techniques and tools connected with research. Scientific and didactic workers of the university were there to see that all the research was done in a correct way.

4   The Course of the Research Process

Each of the groups of students analyzed one topic, chosen from the group of problems put forth by Dorota Winnicka-Jasłowska - an expert in the scope of the quality of buildings and academic spaces, an author of an important scientific monograph [21]. These topics included:

- Way finding, visual identification,
- Accessibility, the needs of the disabled, universal design, ergonomics,

- Service networks (buildings and outside), influence of the city on the Campus,
- Networks designed for self-study, conducive to building social contacts, networks of work stations,
- Design out Crime,
- The Image of the University,
- The Academic Zone seen as city space - external users.

Due to the breadth of the aforementioned subject matter, only research and projects connected with the notion of "The Image of the University", carried out under the supervision of Joanna Tymkiewicz, will be presented in this article. The author has already raised a similar subject matter in her research [22, 23].

The first step which would prepare students to initiate scientific research, which would give them some food for thought and would commit them to the proposed subject matter was the survey, prepared by Joanna Tymkiewicz, taken by students during didactic classes (in the amount of 20 people), where initially they assumed the role of respondents, not researchers. The questions on the survey were both closed and open questions, connected with the perception of the external image of academic buildings. One of the questions read: "How would you describe buildings of each of the Faculties to a person who never saw them before and would ask you how to get to one of them?" To formulate the answers, students were advised to use a technique where they would have to finish a questions, for example: "The Building of the Faculty of Architecture is the one....". The task that the students were given was to write the first thing that came to their mind. The later analyses of the answers received made it possible to assess whether the provided definitions had a negative or rather a positive overtone. Do these definitions point to any characteristic features of the buildings, or no? To make the subject matter complete, a lecture about methods, techniques and tools applied in research on architecture and urban planning was given. Apart from the aforementioned survey, it was also possible to gather information on the way the surveyed students take advantage of the space within the campus.

Another stage of the research process belonged to students and consisted of research planning, selection of methods, techniques and tools. Students decided to carry out an online survey, by means of a tool which they were/are using on a daily basis with dexterity. Of course, these tools were smartphones. In their opinion that was supposed to be the best way to reach out to the largest part of the peer academic community – students of different faculties of the Silesian University of Technology. They were not wrong. The survey was taken by 234 respondents. The questions on the questionnaire were repeated and were taken from the aforementioned survey proposed by Joanna Tymkiewicz. The results of the survey were approved, summarized in the form of a multimedia presentation and elaborated on during classes. In addition, students were able to seek out inspiring examples coming from different campuses around the world where problems, analogous to those which were raised during research, were solved. Field inspections which took place at the campus of the Silesian University of Technology as well as all the observations that accompanied them were both significant in the whole process. All that provided a basis for the next stage of research, the predesign research which meant brainstorming sessions.

The aforementioned research resulted in a list of propositions of urban and architectonic "interventions" within the space of the campus that would make it more attractive and would also enlighten it and serve as a guideline for the concept designs carried out by students. In the next stage, on the basis of the guideline(s), the said concept designs came into being. That was the stage of "research through design", a method described by Klaudiusz Fross, in, among others, [24]. Examples of methods for qualitative research and an assessment of the structures was discussed in the references [25, 26].

5   Results and Conclusions

As it was mentioned at the beginning - the following article pertains to only one of the topics, which is "The Image of the University". The research conducted showed that the appearance of the academic buildings oftentimes generates negative connotations ("gray, run-down"), with a facade showing no characteristic features, or is dominated by the facade of the neighboring building ("Oh, that is the one next to the building of the Civil Engineering Faculty, isn't it"), or it denotes its function incorrectly ("it looks like a cloister").

Due to the fact that changes to the facades are not planned for, the students in the scope of the raised subject matter, created a list of "soft" interventions, which, in their opinion, would enliven the academic zone and would positively influence "The Image of the University". These interventions constituted elements located in front of the entrances to the buildings of the faculties, whose aim was to advertise the advantages that each faculty is famous for. These could include exhibitions of pieces of work prepared by students in front of the buildings or urban furniture designed and made by students. Below are three exemplary propositions put forth by the students (Figs. 3, 4 and 5).

Fig. 3. Visualization of the fragment of the central space of the campus with new extra elements designed by students: multi-function boxes and multimedia system on the existing information pylon, with a renewable energy source; authors - students: Joanna Basek, Karolina Chodura, Piotr Górny; consultations: Joanna Tymkiewicz.

Fig. 4. A multimedia "photoplasticon" - a box in which all the achievements of one the Faculties have been presented; authors - students: Anna Pietkiewicz, Monika Rakowiecka; consultations: Joanna Tymkiewicz

Fig. 5. Visualization with a designed box intended for presenting works done by students in front of the Faculty of Civil Engineering; authors - students: Anna Pietkiewicz, Monika Rakowiecka; consultations: Joanna Tymkiewicz

On the basis of conclusions drawn from the research done, concept projects were made to show what more could be done and how to make the central space of the campus, which is located between the main buildings of the faculties, even more attractive. Among the suggestions made, one could find the following:

- creation of a social space for students and workers of specific faculties;
- transformation of the main alleys and squares, which belong to the university, into important parts of the city's space that would also serve the residents of the city of Gliwice – "the second market square";
- completion of the campus with missing functions - food courts, shops and cafes, etc.

All projects were consulted by scientific and didactic workers in terms of the ergonomic correctness, also in its broader sense.

The whole didactic project was capitalized by a presentation of all the results of the students' research, there were concept projects presented in front of the authorities of the university as well as an exhibition of those works (Fig. 6).

Fig. 6. Exhibition of works prepared by students in one of the buildings of the Architectural Faculty (photograph taken by Joanna Tymkiewicz)

A review of all the design solutions allowed to choose one for a follow-up analysis. Authors of the bench covers, which broaden the functionality of benches, have been invited to take part in design workshops with a company (a partner of the Faculty of Architecture) that produces, among others, city seats (Fig. 7).

Fig. 7. Colorful extra elements for benches that increase their functionality; authors: Hanna Haczek, Aleksandra Targiel; consultations: Joanna Tymkiewicz

Thanks to the project conducted in the scope of didactics, our students could test and "familiarize" themselves with the Post Occupancy valuation method as well as all research techniques that were applied as part of that project. They had a chance to design research scenarios, they could choose the tools and also tackle all difficult problem connected with selecting the respondents.

It should be noted that the proposed solutions constitute a very important concept stage which summarizes the pre-design research stage. Moreover, the goal was to interest and inspire the Authorities of the University (the investor) so they could take the decision about investing in the revitalization of the public space of the academic zone. Potential follow-up to the works will consist in indicating and selecting solutions which would be noteworthy and in assessing its feasibility.

The didactic project at hand is one of the many projects carried out at the Faculty of Architecture of the Silesian University of Technology, at the Department of Design and Qualitative Research in Architecture. For example - parallel with the described project, scientific studies connected with the open-air work zones for students were realized. The scientific and didactic staff of the aforementioned department are all experts on research done in architecture with more than 20 years of experience gained at the "Silesian School of Qualitative Research". Engaging students in this type of research broadens their knowledge as well as the competence of the future architects, necessary in the professional practice. We are hopeful that it will generate more sensitivity in relation to the needs of the users, especially in the scope of ergonomics, in its broad sense.

References

1. Fross, K., Winnicka-Jasłowska, D., Sempruch, A.: "Student Zone" as a new dimension of learning space. case study in polish conditions. In: The 8th International Conference on Applied Human Factors and Ergonomics AHFE 2017, 17–21 July 2017, Los Angeles, USA, pp. 77–83. Elsevier (2017)
2. Neary, M., Harrison, A., Crellin, G., Parekh, N., Saunders, G., Duggan, F., Williams, S., Austin, S.: Learning Landscapes in Higher Education. Centre for Educational Research and Development. University of Lincoln (2010)
3. Piaścik, F., Stangel, M.: Campus development and downtown regeneration perspectives for katowice. Archit. Civil Eng. Environ. (ACEE) **9**(1), 35–42 (2016). Silesian University of Technology
4. Project for Public Spaces (PPS): How to Turn a Place Around, Michigan University (2000)
5. Stangel, M., Cielińska, A., Harat, Ł.: Silesian conurbation as a polycentric structure of urban districts. Archit. Civil Eng. Environ. (ACEE) **10**(2), 21–33 (2017). Silesian University of Technology
6. Winnicka-Jasłowska, D.: Quality analysis of Polish universities based on POE method - description of research experiences. In: Margherita, A., Stephanidis, C. (eds.) Universal Access in Human-Computer Interaction. 9th International Conference, UAHCI 2015 held as Part of HCI International 2015, 2–7 August 2015, Los Angeles, CA, USA. Proceedings. Pt. 3, Access to Learning, Health and Well-Being, pp. 236–242 Springer, Cham (2015)
7. Palus, K., Zabawa-Krzypkowska, J.: Contemporary public spaces as meeting places. Archit. Civil Eng. Environ. (ACEE) **9**(2), 21–28 (2016)

8. Gumińska, A.: Modern revitalizing interferences in the historical urban development. Archit. Civil Eng. Environ. (ACEE) **8**(2), 5–10 (2015)
9. Fross, K., Winnicka-Jasłowska, D., Gumińska, A., Masły, D., Sitek, M.: Use of qualitative research in architectural design and evaluation of the built environment. In: Ahram, T., Karwowski, W., Schmorrow, D. (eds.) Proceedings of the 6th International Conference on Applied Human Factors and Ergonomics 2015 and the Affiliated Conferences. AHFE 2015, 26–30 July 2015, Las Vegas, USA, pp. 1625–1632. Elsevier (2015)
10. Nawrocki, T.: The usefulness of mental maps for sociological research of the city. Archit. Civil Eng. Environ. (ACEE) **10**(3), 19–31 (2017). Silesian University of Technology
11. Niezabitowska, E., Masły, D.: Research projects in the field of architecture - experiences of the faculty of architecture at the Silesian University of Technology in Gliwice, Poland. In: Chova, L.G., Belenguer, D.M., Torres, I.C. (eds.) 3rd International Conference of Education, Research and Innovation, Proceedings. ICERI2010, 15th–17th November 2010, Madrid, Spain, pp. 1983–1990. IATED (2010)
12. Stangel, M., Witeczek, A.: Design thinking and role-playing in education on brownfields regeneration. Experiences from Polish-Czech cooperation. Archit. Civil Eng. Environ. (ACEE) **8**(4), 19–28 (2015). Silesian University of Technology
13. Szewczenko, A., Benek, I.: Quality research of healthcare facilities for the elderly. In: Chova, L.G., Martinez, A.L., Torres, I.C. (eds.) 8th International Technology, Education and Development Conference (INTED), 10–12 March 2014, Valencia, Spain. INTED 2014: 8th International Technology, Education and Development Conference. INTED Proceedings, pp. 2458–2467 (2014)
14. Tymkiewicz, J., Bielak-Zasadzka, M.: The design thinking method in architectural design, particularly for designing senior homes. Archit. Civil Eng. Environ. (ACEE) **9**(1), 43–48 (2016). Silesian University of Technology
15. Tymkiewicz, J., Bielak-Zasadzka, M.: Senior homes of the future in the eyes of students of architecture. Didactic experience from the application of the design thinking method. Archit. Civil Eng. Environ. (ACEE) **9**(1), 49–56 (2016). Silesian University of Technology
16. Tymkiewicz, J.: Team work efficiency in finding innovative solutions. Experience with the design thinking method implemented into teaching at the faculty of architecture. In: Chova, L.G., Martinez, A.L., Torres, I.C. (eds.) 8th International Conference of Education, Research and Innovation (ICERI), 16–20 November 2015, Seville, Spain. ICERI2015: 8th International Conference of Education, Research and Innovation. ICERI Proceedings, pp. 5894–5902 (2015)
17. Masły, D., Sitek, M.: New ideas and tools in the educational process of students of architecture. The introduction of a new subject to the curriculum - new technologies and methods in architecture design. In: Chova, L.G., Martinez, A.L., Torres, I.C. (eds.) 9th Annual International Conference of Education, Research and Innovation. ICERI2016. Proceedings, 14th–16th November 2016, Seville, Spain, pp. 7374–7379. IATED (2016)
18. Tymkiewicz, J., Winnicka-Jasłowska, D., Jastrzębska, M.: Pre-design studies on the example of modernization project of geotechnical laboratories. Archit. Civil Eng. Environ. (ACEE) **10**(2), 43–52 (2017). Silesian University of Technology
19. Tymkiewicz, J.: Quality analyses of facades based on post occupancy evaluation. Research experience with students of architecture participation. In: Chova, L.G., Martinez, A.L., Torres, I.C. (eds.) 9th Annual International Conference of Education, Research and Innovation (ICERI), 14–16 November 2016, Seville, Spain. ICERI2016: 9th International Conference of Education, Research and Innovation. ICERI Proceedings, pp. 8831–8838 (2016)

20. Winnicka-Jasłowska, D., Jastrzębska, M., Tymkiewicz, J.: Ergonomics of laboratory rooms – case studies based on the geotechnical laboratories at the Silesian University of Technology. Archit. Civil Eng. Environ. (ACEE) **10**(2), 35–41 (2017). Silesian University of Technology

21. Winnicka-Jasłowska, D.: Przestrzeń nauki współczesnego uniwersytetu. Rola badań przedprojektowych w programowaniu nowych funkcji wyższych uczelni. (English title: The Learning Space of a Contemporary University the Role of Pre-Design Research in the Programming Of New Functions Of Universities), Monografia, Wydawnictwo Politechniki Śląskiej, Gliwice (2016)

22. Tymkiewicz, J.: Facades and problems in correct recognition of the functions that buildings perform. Archit. Civil Eng. Environ. (ACEE) **5**(1), 15–22 (2012). Silesian University of Technology

23. Tymkiewicz, J.: Funkcje ścian zewnętrznych w aspektach badań jakościowych. Wpływ rozwiązań architektonicznych elewacji na kształtowanie jakości budynku, (English title: Functions of the Exterior Walls of Buildings in View of Quality Analyses. The Impact of Architectural Design Solutions of Facades on the Quality of Building), Monografia, Wydawnictwo Politechniki Śląskiej, Gliwice (2012)

24. Fross, K.: Architect-researcher as a model combination of research and design practice on examples. In: Charytonowicz, J. (ed.) Advances in Human Factors and Sustainable Infrastructure, Proceedings of the 5th International Conference on Applied Human Factors and Ergonomics AHFE 2014, 19–23 July 2014, Kraków, Poland, pp. 31–39, Las Vegas (2014)

25. Fross, K., Sempruch, A.: The qualitative research for the architectural design and evaluation of completed buildings – part 1 – basic principles and methodology. Archit. Civil Eng. Environ. (ACEE) **8**(3), 13–19 (2015). Silesian University of Technology

26. Fross, K., Sempruch, A.: The qualitative research for the architectural design and evaluation of completed buildings – part 2 – examples of accomplished research. Archit. Civil Eng. Environ. (ACEE) **8**(3), 21–28 (2015). Silesian University of Technology

Public Spaces - For People or Not for People?

Klaudiusz Fross[(✉)], Joanna Tymkiewicz,
and Dorota Winnicka-Jasłowska

Faculty of Architecture, Silesian University of Technology, Gliwice, Poland
{klaudiusz.fross,joanna.tymkiewicz,
dorota.winnicka-jaslowska}@polsl.pl

Abstract. Public spaces are, as a matter of principle, meant for people. It does not matter whether it be local residents, incoming guests or tourists. Just as our lives and the manner in which people spend their leisure time change, so do the requirement as to the public spaces. One could even say that these requirements are constantly growing. At present, a mere bench, tree or a fountain will just not serve. One could even raise a question as to how the contemporary public spaces should look? What functions and attractions should they include? How should the architects meet those new requirements? Answers to all those specific questions shall be answered. The authors shall subject some exemplary public spaces to the process of evaluation. Conclusions drawn shall be prepared on the basis of completed qualitative research. At the close, a recipe for a public space, both model and attractive, shall be created.

Keywords: Architecture · Public space · Ergonomics · Qualitative research

1 Introduction – Spaces and Their Users

The built environment is a frequently applied term used to describe an artificial surrounding, designed by human beings, as well as all the relations which take place within it. This environment is made up of buildings, complexes of building as well as their surroundings. An important element which constitutes part of the built environment are the public spaces. They serve to meet various human needs such as: entertainment, relaxation, sports, leisure etc. Designers give public spaces forms while the form shapes the users. For users in general, the quality of the built environment has an enormous significance. Moreover, relations between the built environment and the natural one are also of key importance. All elements of the built environment are designed, built and then used. Among them, one can find admirable and remarkable pieces of architecture, "ordinary" and properly planned pieces of architecture, pieces that meet the needs and pieces that bring satisfaction. However, there also buildings and other spaces that function incorrectly and cause numerous problems. Such structures and spaces are often referred to as "sick". An inefficient structure does not always have to be old, dilapidated and unable to meet today's expectations. It mightas well be modern and newly built. One could say that structures which have already been completed constitute a long-lasting record of the success and failure of design. Structures are being built by humans so they serve them and their needs. The main user

J. Charytonowicz and C. Falcão (Eds.): AHFE 2018, AISC 788, pp. 36–46, 2019.
https://doi.org/10.1007/978-3-319-94199-8_4

together with the intended use have a great influence of the parameters of the structure. People not only stay within public spaces and take advantage of them, they also assess them. Therefore, both the built environment and the users constitute a research field and a potential source of knowledge, information of the built environment, their advantages, strengths as well as disadvantages and drawbacks (Fross 2012), or other publications: [1–3, 5].

All users have the right to make an assessment. They decide whether they will use the given space or not. Of course, space created by an architect constitutes a specific offer as well as a proposition. It also possesses specific solutions. At the end of the day, however, it is the user who makes the final choice. His or her approval constitutes a confirmation of a well-prepared functional and utility program and a correct recognition of user groups together with their needs. It is common knowledge that there are both well designed spaces, which often become cultural meeting places, as well as badly designed spaces which constitute asocial, uninteresting and unacceptable structures. The most important conclusion which may be drawn from the above - the users assess the structures which they make use of.

Public spaces are, as a matter of principle, intended for people. It does not matter if these are local residents, some visitors or tourists. Just as our life or the way we spend our leisure time change, so do the requirements concerning public spaces. One might even claim that these requirements are constantly on the rise. Today, a bench, a tree or a fountain will not suffice. One might ask what the contemporary public spaces should look like. What functions and attractions must they contain? How can the architects meet these new expectations?

One should also specify what creates quality in public spaces, and what specifies the norm. One might claim that the quality of space is created by a set of functional features which are looked at in technical, functional, behavioral (including esthetic), organizational and economic terms. The standard of the space, however, which is seen as an average model of that quality and constitutes a point of reference for all spaces, is specified by means of normative requirements (e.g. of the Building law) as well as by specified requirements of the market and needs of the users. Spaces which do not meet the requirements of the standard are deemed substandard and will require either significant modernization, a redevelopment or demolition. Above-standard spaces, on the other hand, will always offer an above-average quality in terms of qualitative parameters (Fross 2012, 2015). The scientific staff from Department of Design and Quality Assessment in Architecture are specialist in "design by research", for examples publications: [4, 6–8].

2 State of the Art

In scope of the subject matter, the state of the art is significant. Research on public spaces has been the subject of numerous important conferences. Professional literature on the subject is incredibly extensive. Titles of central importance for the professional literature are connected with the methodology of qualitative assessments of the built environment and they include the following:

In the first group of publications that constituted the basis of knowledge during the preparation and execution of research, one may find:

- Fross and Sempruch [2].
- Fross et al. [3].
- Fross et al. [4].

Studies and analyses which describe the effects of research conducted together with students as part of their classes make for an interesting complement to the literature and include:

- Tymkiewicz et al. [6].
- Tymkiewicz et al. [7].
- Winnicka-Jasłowska [8].

It is worth mentioning that the Silesian Metropolis has undertaken an important initiative. On November 16th 2017, during an inaugural conference – "Forum Przestrzeni" (The Space Forum), held at the Marshall's Office in Katowice, a letter of intent was signed between the Silesian Voivodeship and 8 Stakeholders: The Silesian University of Technology, Silesian University in Katowice, University of Economics in Katowice, Academy of Art in Katowice, The Association of Polish Architects, The Association of Polish Town Planners, The Silesian Regional Chamber of Polish Architects, The Silesian Association of Communes and Poviats. Both the Silesian University of Technology and the Faculty of Architecture were represented by the dean of the faculty - Klaudiusz Fross. One of the forms of cooperation will include participation in the execution of "The Space Forum".

Preamble of the letter: "*Being aware of the role and significance of the exceptional value that the resources of space and landscape offer in the process of a harmonious social and economic development of the region and striving for their potential to be used as best as possible, the Leader and Signatories of the following letter declare that they will undertake to cooperate for the sake of a rational use of the resources of the space and the landscape of the Silesian voivodeship.*"

During the press conference, Mr. Klaudiusz Fross, the Dean, stated that: "*...One of the most important elements is the public space. The life and well-being of the residents depends on its quality. As shown by scientific research, the best way to promote the city or the commune is to create compelling and attractive structures and public spaces. These places are often seen as the hallmarks of a given place. But what does a public space mean to a resident? A public space is a place which he or she wishes to visit, a place where he or she will feel good, which meets his or her needs and expectations, where there are numerous attractions, it is safe and there are also numerous facilities (toilets, food courts, parking lots, etc.). However, today's expectations are constantly on the rise and by planning ahead we must meet them. At present, one would even expect a bench to be something more than an ordinary bench. One would like to see in it some additional functions like, for example, a smartphone charger. Such a bench was designed and built by the students of the Silesian University of Technology. It was also equipped with a solar panel. One should keep in mind that we design for all age groups. Proper planning should be started with an identification of user groups and their needs. Next, a program and a concept should be created and only then can we* *proceed to civil engineering designs. It is also worth mentioning that the Silesian School of Qualitative Research, created at the Faculty of Architecture, at the Silesian University of Technology, is already 20 years old - we should take advantage of that."* (Fross 2017).

3   Research Methodology

Research methodology as well as examples of already executed studies have been described time and time again in author's publications:

- Fross and Sempruch [1].
- Fross [5].

In his assessment of the public space, the author applied reliable and long-used research methods for an "8-step" pre-design assessment (Fross 2012) devised on the basis of Post Occupancy Evaluation. Prepared on the basis of multiple cases (over 15 years), a research scheme allows people the possibility of a quick assessment based on specific criteria [1, 5].

Various techniques and research tools, in various configurations, have been applied. They included: participant observations, surveying, site inspections, targeted questionnaires, spontaneous conversations, way finding, photographing of the manner of use and user behavior, graphic analyses, attractiveness rankings of a given place, elements of small architecture.

The main criteria of the evaluation included: meeting the needs of various user groups, user satisfaction, universality, safety, functionality, comfort, multi-functionality, esthetics, technical quality, costs of investment, solutions economics, adequacy of solutions etc.

4   The Course of the Research Process

Public spaces which have recently been completed within towns and communes were chosen for research. The author performed an on-the-spot assessment by making an observation of its use and the behavior of the users. He went on to compile photographic records and performed a simplified qualitative assessment for the following categories: technical, functional, behavioral and economic quality. Next, he attempted at finding an answer to the following question: is the space attractive or multi-functional, does it meet the expectations and needs, is it adequate for the place where he/she currently is? In selected cases, more detailed assessments were performed, for example, by applying the technique of an unstructured interview in the form of a casual, spontaneous conversation. At each time there was an attempt to specify the correctness of the investments, program assumptions and the final effect that was produced through the prism of meeting the user needs. As the open-access space within a city, district, a settlement or a commune serves, as matter of principle, all users (both the residents and "visitors") then a specially designed and developed public zone, built in the form of square, park etc. and offering a new quality, should also serve all groups of users. Due to that fact, in the research process the focus was directed towards that very aspect, towards the needs of various user groups: little children, mothers with children, youth, adults, the elderly, the disabled, couples, multi-person groups etc (Fig. 1).

Fig. 1. Examples of the public spaces. What like or what not like people? Examples from Poland (EU): Marklowice "Tropical Island", Katowice "European Congress Center", Tychy-Paprocany, Rydułtowy "Fikołkownia" (Foto K. Fross 2012, 2017).

Fig. 2. Examples of the public spaces. What like or what not like people? Examples from Japan: Osaka (Foto K. Fross 2013).

Fig. 3. Examples of the public spaces. What like or what not like people? Examples from EU: Budapest (Hungary), Milano (Italy) (Foto K. Fross 2013, 2014).

At each time, during the inspection of a site/place, in order to find some answers, the author asked the following questions:

– *why would I come here?*
– *what would I do here?*
– *would I like to visit this place again?*
– *would I recommend this place as a noteworthy place?*

This question was asked in the context of trying to understand different users, not only as a personal assessment of the place/site. However, an expert assessment made by the author was of great importance in the matter (Fig. 2).

In the next part, in the form of a photograph, the collection of author's individual assessments of selected public spaces was presented. As part of a qualitative and observation-based assessment, it was supplied with descriptions and remarks (Figs. 3 and 4).

Fig. 4. Examples of the public spaces. What like or what not like people? Examples from Poland: Marklowice, Gliwice, Tychy, Marklowice (Foto K. Fross 2017).

5 Recapitulation and Conclusions

Questions provided in the previous point: *Why would I come here? What would I do here? Would I like to come here again? Would I recommend this place a noteworthy place?* constitute typical questions, often asked by users. Whenever children or teenagers are faced with the question: *Would you like to go there?* We often receive an answer which is a question: *What will we do there?* These questions contain the whole essence of design, adequate and well-suited to the expectations of the public space. Public space is perceived as a place which meets the needs, a place which we often revisit and where we enjoy staying. Were there positive answers to all questions, then one could conclude that the public space had been designed correctly. We design for people so when the end users of a piece of architecture are happy, that means a job well-done. It was the author's intention not to touch upon the issues of esthetics, image, composition, urban context, reference to the existing building development etc. nor the elements of urban and architectonic design. It is because these elements should always be "there". They constitute the elements of the design workshop. Nevertheless, an interesting and intriguing composition or a fashionable design should never blot out the essence. "Graphic" design (as one of the users of a pretty but impractical space called it) cannot supplant the design based on fulfilling needs. At present, a park or a square which is merely "pretty" is definitely not enough.

At the time when there are multiple attractive forms of spending one's free time, at the time of information science development which means that everybody possesses a smartphone, which enables staying up-to-date, staying in touch with the whole family, friends, which enables access to information, films, games and music, etc. and at the time when the expectations and requirements as to the forms of spending one's free time are constantly growing, it is not easy to meet expectations. Active and attractive forms of spending one's free time are becoming more and more popular as people now value their free time.

And What About Public Spaces?

Sometimes, spaces which are incredible in their composition and design, attractive for architects, look well on the cover of magazines, are in fact empty, impractical and nobody wishes to use them. They often constitute an artistic vision of the designer, which is totally out of connection with the needs – one could call them a wasted opportunity for designing a genuine public space, a space for all people, a space which would satisfy all their needs and expectations.

Everything comes down to the planning and programming phase. In the world of today, without professional knowledge of any discipline, it is really difficult to reach success, especially in investments where one must take into consideration not only the risk, competition, a galloping pace of technological development, ever-changing expectations, but also the end user, above all. The end user is the one who will assess the product (structure, space), the work of architects, as well as the assumptions of the investor. Investment decisions as well as decisions connected with the design, which are based on surface premises or just intuition, might not be appropriate and could be encumbered with the risk of multiple errors, as well as complete failure. An ill-prepared, erroneous in its intent and targets, functional and spatial plan of the building or any other structure may not bring about the anticipated effects and, as a result, can put paid to the resources and means used (Fross 2012, 2015).

Since all users are more or less familiar with the public spaces which they use, then why not take advantage of that knowledge at the planning, programming and deigning stage to prepare that new investment correctly and rid it of any potential threats and risks of failure (lack of approval on the part of the users/occupants).

What are the elements of a "well-prepared" public space? For a certainty, these elements will include: play and recreational attractions adjusted to various age groups, (something suitable for all groups including interactive, innovative, intriguing and educational facilities), various sitting places, elements which ensure the feeling of safety (CCTV, security, fencing, division, sanitary equipment for example WC, influence on the comfort in use and its service-time), well groomed and organized green space, functional and decorative night lighting (which prolongs service-time and ensures the feeling of safety). To improve the quality of the recreational space, elements which were never considered in the past, such as fencing, security, a WC, should now be provided. They have an influence on the growth of comfort, safety and also on property protection which helps maintain high quality. What is also necessary is the cost analysis, the cost of maintenance and the cost of maintaining cleanliness. Moreover, one should also ensure protection and preventive measures against all acts of vandalism and devastation (Fross 2015) [2].

Investment plans, business targets, functional programming as well as the process of design all require professional knowledge which may only be acquired through qualitative research. At present, *Design by Research* is not an option, it is a duty and necessity. Qualitative research constitutes universal methods of assessing all types of structures, and at the same time it helps recognize the needs of users. Thanks to the results of the assessment, we now are able to take correct decisions which minimize the risk of making an error. In the design practice (at the pre-design stage, in programming). This is the perfect way to take advantage of the norms and standards of the best practice and the best way to avoid incorrect solutions.

References

1. Fross, K., Sempruch, A.: The qualitative research for the architectural design and evaluation of completed buildings – part 1 – Basic principles and methodology. Archit. Civ. Eng. Environ. ACEE **8**(3), 13–19 (2015)
2. Fross, K., Sempruch, A.: The qualitative research for the architectural design and evaluation of completed buildings – part 2 – examples of accomplished research. Archit. Civ. Eng. Environ. ACEE **8**(3), 21–28 (2015)
3. Fross, K., Winnicka-Jasłowska, D., Gumińska, A., Masły, D., Sitek, M.: Use of qualitative research in architectural design and evaluation of the built environment. In: Ahram, T., Karwowski, W., Schmorrow, D. (eds.) Proceedings of the 6th International Conference on Applied Human Factors and Ergonomics 2015 and the Affiliated Conferences, AHFE 2015, Las Vegas, USA, 26–30 July, 2015, pp. 1625–1632. Elsevier (2015). bibliogr. 19 poz
4. Fross, K., Ujma-Wąsowicz, K., Wala, E., Winnicka-Jasłowska, D., Gumińska, A., Sitek, M., Sempruch, A.: Architecture of absurd. In: 18th International Conference, Human-Computer Interaction, HCI International 2016, Toronto, Canada, 17–22 July 2016, Springer (2016)
5. Fross, K.: Architect-researcher as a model combination of research and design practice on examples. In: Charytonowicz, J. (ed.) Advances in Human Factors and Sustainable Infrastructure, Proceedings of the 5th International Conference on Applied Human Factors and Ergonomics AHFE 2014, Kraków, Poland, 19–23 July 2014, Las Vegas (2014). ISBN 978-1-4951-2092-3

46   K. Fross et al.

6. Tymkiewicz, J., Winnicka-Jasłowska, D., Jastrzębska, M.: Pre-design studies on the example of modernization project of geotechnical laboratories. ACEE Archit. Civ. Eng. Environ. **10** (2), 43–52 (2017). bibliogr. 14 poz
7. Tymkiewicz, J., Winnicka-Jasłowska, D., Jastrzębska, M.: Ergonomics of laboratory rooms - case studies based on the geotechnical laboratories at the Silesian University of Technology. ACEE Archit. Civ. Eng. Environ. **10**(2), 35–41 (2017). bibliogr. 17 poz
8. Winnicka-Jasłowska, D.: Quality analysis of Polish universities based on POE method - description of research experiences. In: Antona, M., Stephanidis, C. (eds.) 9th International conference on Universal Access in Human-Computer Interaction, UAHCI 2015 held as part of HCI International 2015, Los Angeles, CA, USA, August 2–7, 2015. Proceedings. Pt. 3, Access to learning, health and well-being, pp. 236–242. Springer, Cham (2015). bibliogr. 11 poz

Success Analysis in Architectural Design Competitions in Terms of Design Quality

Wojciech Bonenberg[(⌧)]

Faculty of Architecture, Poznan University of Technology, Poznań, Poland
wojciech.bonenberg@put.poznan.pl

Abstract. The article presents a report on research pertaining to the possibility of success in architectural design competitions. The research was carried out in 2016 and 1017 with the aid of Poznań University of Technology Faculty of Architecture students and under the guidance of the author. The significance of architectural design competitions in building spatial systems and improving the built environment was highlighted. The significance of architectural design competitions for the growth of architectural creative activities and in particular educating architecture students was also pointed out. As a result of the research, architectural design quality attributes were identified, which are decisive to the greatest extent when it comes to success in professional architectural design competitions.

Keywords: Architectural design · Competitions · Success

1   The Research Problem

The objective of architectural design competitions is to obtain the best design solutions possible. They aim to ensure a high quality of the built environment with the public interest in mind and to establish conditions conducive for the development of architectural creative works. Designated judges - authority figures in the field of architecture, commissioned by the ordering entity, select the best architectural design within the scope of an architectural design competition. A strong belief prevails amongst architects, that architectural design competitions are an effective way for acquiring architectural creative ideas for further investment execution. The competitions objective is to select the best solutions, solutions which not only exhibit high aesthetic virtues, but ones which are also functional and thus effective. The opportunity to publicly compare awarded works with those which were not successful is very important. This constitutes a basis for further discussions amongst the interested parties and stimulates architectural critique which furthers the improvement of the built environment.

Interest in architectural design competitions stems from the fact that people evaluate their surroundings and this process contributes to satisfaction or discontentment. Thus, it is so important to be able to perform variant simulations of development visions, leading to a consensus as to the chosen design solutions, prior to interfering with the surroundings. This objective is associated with the general belief, that there are clear relations between the qualities of the built environment, professionalism, skill and creativity of architectural creators.

© Springer International Publishing AG, part of Springer Nature 2019
J. Charytonowicz and C. Falcão (Eds.): AHFE 2018, AISC 788, pp. 47–55, 2019.
https://doi.org/10.1007/978-3-319-94199-8_5

Many myths and half-truths have arisen around architectural design competitions. Some competition participants believe that success in a competition depends on luck and is random. Others think that it is who you know that matters or that the jury has to be favourably inclined to you. There are also opinions which say that well known architectural studios, the so called "architecture stars" are most likely to win the competitions. Despite these doubts, the lion's share of architects taking part in the competitions consider that design quality virtues are decisive when it comes to competition success. However, a question does arise, as to which architectural elements (attributes) are most decisive when it comes to competition success. Is it functional correctness, stylistic innovation or perhaps an attractive graphical presentation of the design? In other words, what makes the awarded designs stand out from the crowd?

The conducted research aims to determine a relationship between architectural design competition success and attributes (properties) of designs entered for the competitions. Design suggestions presenting solutions for the same competition problem differ from one another. Partial quality attributes of given designs are the element which makes them stand out. Thus, it is so important to identify those attributes and to estimate the significance of each attribute for competition success.

2   Research Plan

A group of 20 Poznan University of Technology Faculty of Architecture students took part in the research. Stage one of the research entailed selecting a set of representative designs awarded in domestic and international competitions.

Then the competition designs were analysed. Within the scope of the analysis, partial attributes (features) characteristic for the designs subject to the research.

The characteristics of competition designs are to provide an answer to the question of why some architectural competition designs are awarded, and others are not? Which competition design attributes determine whether the given design is awarded? Which attributes are most common amongst awarded designs?

Thus, we are looking to identify those design elements which the jury awards the highest mark to.

15 characteristic partial attributes of an architectural design pertaining to its quality have been identified on the basis of subject literature [1, 3, 6, 7].

Then the "intensity" of the occurrence of the identified attributes in each competition design part of the research was measured. As a result, a ranking of the attributes characteristic for the given competition design was obtained. Each competition design was characterised using the "intensity" of the occurrence of the given attributes [8].

Then the designs subject to the analysis were ordered according to competition jury assessment. And thus, projects which won 1st, 2nd or 3rd place in architectural design competitions were grouped together.

Then an analysis of the attribute "intensity" distribution in each of the three competition design groups was performed. This made it possible to identify those attributes, which are decisive to the largest extent when it comes to top spots in a competition. As a result, a standard for the intensity of quality attributes which are decisive in terms of success in architectural design competitions was obtained.

3   Research Methodology

A research sample spanned 108 domestic and international architectural design competitions held in between 2011 and 2017. Competition designs were analysed, with particular emphasis on identifying attributes decisive in terms of competition success. The research methodology comprised the following steps [4]:

A. Selection of a representative sample.
B. Identification of partial quality attributes which characterise each design in terms of functional, structural and composition virtues (properties) and the design graphical presentation method.
C. Categorisation of competition designs according to the awarded prize (1st, 2nd or 3rd).
D. Analysis of partial attributes associated with design specific virtues in awarded designs (functionality, composition, context, etc.) and the design graphical presentation method. In other words, this is a determination of the impact of partial quality attributes on competition success. This analysis resulted in a tanking of partial attributes which contribute to architectural design competition success.

A representative sample contains designs which won 1st, 2nd or 3rd place in domestic and international competitions. The following criteria were taken into account in selecting designs: competition prestige, competition jury composition (share of judges which are considered to be authority figures amongst professionals), scope of competition design, design presentation type (architectural creative concept).

15 attributes characterising an architectural design were identified:

$x1$ - relevance - whether designers followed competition guidelines, emphasis on the performance of objectives and a functional programme,

$x2$ - context, urban planning conformation - location character reflected by references to the surrounding structures, colours, elements; attention focused on the land development aspect,

$x3$ - symbolism and illusoriness - dominance of abstract forms with numerous semantic references, a large number of architectural metaphors, use of obscure forms,

$x4$ - functional arrangement - meticulous solution of functional connections inside a building, interior design invention, creation of an appropriate entourage for the planned function, invention in terms of creating architectural interiors,

$x5$ - structure, material solutions - use of modern structural materials and systems; exposing structural elements as architectural composition elements, original building façade interpretation [5], x6 - uniqueness - exhibiting values such as traditions of the site, mood of urban interiors, unique expression, cosy scale, interesting adaptation into the local context; individualised spatial design character,

x7 - legibility - combining composition, functional and structural virtues in an understandable form; easily understood meaning of architectural forms in an urban setting; legibility creates a clearer environment, one which provides more emotional stimuli,

x8 - cultural identification - a system of symbolic and emotional values, established as a result of identifying with the location's culture and traditions,

x9 – "en vogue" - conforming to fashion, solutions refer to the latest trends,

x10 - prestige - use of uncommon materials, refined design, elegance, dominance of cohesive formations in a conservative form,

x11 - innovation - esprit, brilliance, imagination, original thinking; use of rare technologies, materials and means of expression; pioneering idea for the execution of the given subject,

x12 - form/spatial form - highly irregular building spatial form, use of atypical shapes (including parametric forms) [2],

x13 - styling - emphasis on creating a visually attractive and stylistically different building,

x14 - ecology/energy efficiency - use of ecological motifs and associations; use of natural materials and greenery on walls and roofs, use of solar panels, heat pumps, mini wind turbines; building orientation with reference to cardinal directions meticulously thought-out,

x15 - design presentation method - competition boards visual message expression, meticulousness or execution, graphic design attractiveness.

For the research, design attribute was an independent variable X, whereas the places in a contest were dependent variables. Then the relationships between design attributes (independent variables X) and architectural design competition success were analysed. In other words, attributes (or a combination of attributes) which contributed to success in the competition were identified.

All variables are of a qualitative character. One should note that the relationships between design attributes and the place in a competition are of a probabilistic character.

Assessments by 20 competent judges (expert assessments) were used in the research, which improved the reliability of assessments. An assessment entailed determining the "intensity" with which each of the identified attributes occurred in the awarded designs.

The procedure was applied to each awarded design in the representative sample. The research was conducted through the use of a questionnaire interview. To measure the relationships between design attributes (independent variable X) and competition success (independent variable) a five-level Likert scale was used [9]. Expert assessments (based on their intuitive belief) were quantified in the 1–5 range, indicating the

Table 1. Questionnaire specimen used for design expert assessments in terms of partial attributes occurrence intensity (example).

Architectural design partial quality attribute	Attribute intensity				
	None (1)	Low (2)	Medium (3)	High (4)	V. high (5)
x1 – relevance				X	
x2 – context, urban conformation			X		
x3 – symbolism and illusoriness				X	
x4 – functional arrangement					X
x5 – structure, material solutions		X			
x6 – uniqueness	X				
x7 – legibility		X			
x8 – cultural identification			X		
x9 – "en vogue"				X	
x10 – prestige	X				
x11 – innovation		X			
x12 – form/spatial form					X
x13 – styling				X	
x14 – ecology/energy efficiency			X		
x15 – design presentation method				X	

intensity of a given attribute in awarded designs. The experts were asked to complete a pre-prepared questionnaire. A specimen questionnaire used to assess a product is shown in Table 1. The advantages of this technique include rate of data acquisition and the possibility to process it further.

As a result, the designs part of the representative sample were characterised. Each design was analysed in terms of partial attributes characterising its quality.

The expert assessment answered the question as to the degree (with what "intensity") with which partial attributes characterise the awarded designs. In other words, the significance (rank) of a given partial attribute in the overall design idea message was determined.

Synthetic research results have been shown on graphs. Graph 1 (Fig. 1) depicts the attribute ranking for designs which won prizes in domestic competitions.

Graph 2 (Fig. 2) depicts the attribute ranking for designs which won prizes in international competitions.

Graph 3 (Fig. 3) depicts the attribute ranking for all designs which won prizes.

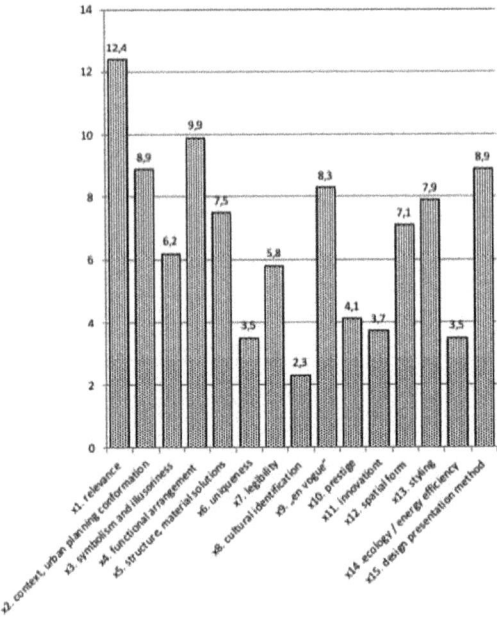

Fig. 1. The attribute ranking for designs which won prizes in domestic competitions.

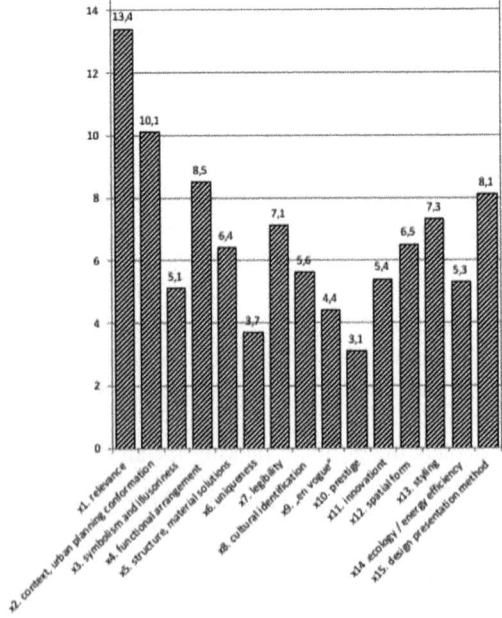

Fig. 2. The attribute ranking for designs which won prizes in international competitions.

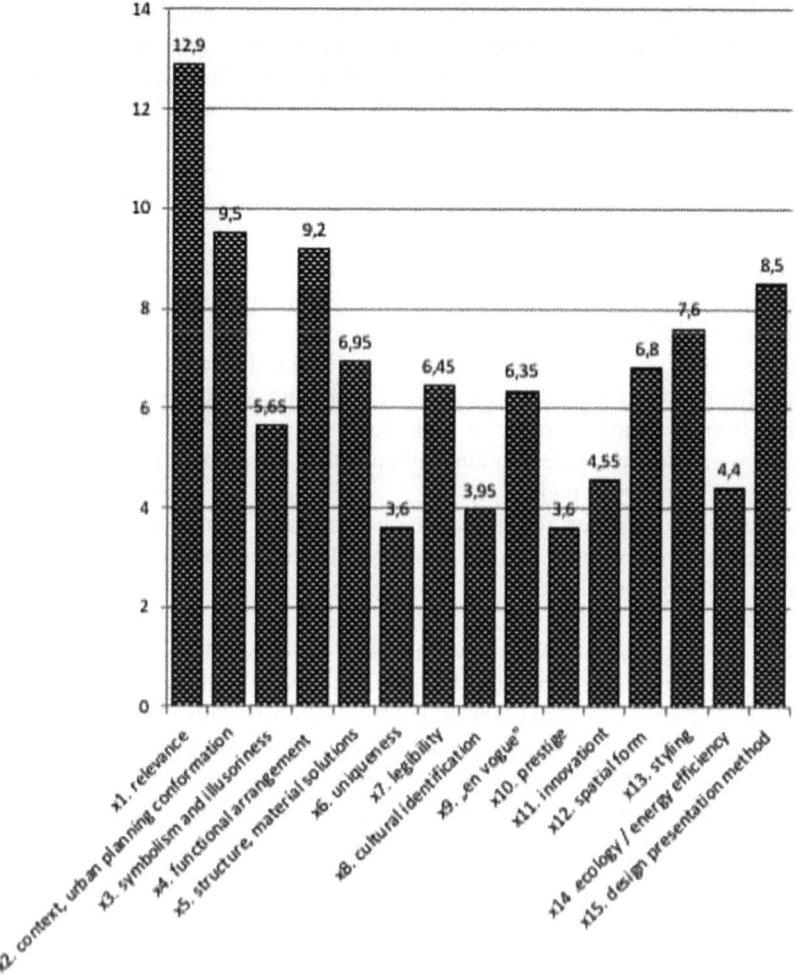

Fig. 3. The attribute ranking for all designs which won prizes.

4 Conclusions

The conducted research shows, that the attributes weights are similar for designs on domestic as well as international levels. This indicates a certain standardisation of competition assessments. And thus, designers from Poland are able to successfully take part in foreign competitions, just as foreign firms stand a chance in Polish competitions.

Out of the identified quality attributes, relevance is most important - whether designers followed competition guidelines, emphasis on the performance of objectives and a functional programme (attribute x1). Context and urban planning conformation (attribute x2) was among the top spots in the ranking.

The research demonstrated significantly higher importance than expected of design presentation method - competition boards visual message expression, meticulousness or execution, graphic design attractiveness (attribute x15). This is interesting inasmuch that this attribute is not linked with actual qualities of the space shaped by the design. Attributes which were lower down in the ranking are also noteworthy, and in particular the surprisingly low position of ecology and energy efficiency (attribute x14) as compared with styling (attribute x13) and the manner in which the building form/spatial form is shaped (attribute x12).

The design presentation method can be summarised in greater detail - competition boards visual message expression, meticulousness or execution, graphic design attractiveness (attribute x15). In most cases, the winning works contained sunshine and a cloudless, blue sky. Approximately 1/3 of the works showed the designs on cloudy days or after rain. In most cases the use of such measures emphasised the character of a building (urban complex), made it look more rugged, further accentuated by the materials used for the façade. Building illumination was shown in the evenings, just after sunset and rarely at night.

Works which depicted designs as a photorealistic visualisation or an ordinary render were staunch winners. Works which employed the collage technique made up just over 15%, with just 2 works out of more than 100 subject to assessment featured hand drawings.

Most winning works were shown in pastel colours. Most often bright colours appeared on boards in the form of a strong accent or lead motif, reflected in the overall board colour scheme and visualisations. Dark colours were prevalent in designs, where the lead motif entailed an interior, monument or shelter. Less than 10% of the works were presented in monochrome.

Architects may find this synthetic characterisation of presentation methods for winning designs very useful. The fact that in 2017 five out of the 20 students taking part in the research were awarded prizes in international student competitions stands testament to its usefulness.

References

1. Bonenberg, A.: Ergonomic experimentation in the architectural design process – types of test environments. In: Charytonowicz, J. (ed) Advances in Human Factors, Sustainable Urban Planning and Infrastructure, Advances in Intelligent Systems and Computing, vol. 600, pp. 199–206. Springer International (2017)
2. Bonenberg, A.: Facades and multimedia screens in contemporary architecture–ergonomics of use. In: Advances in Social and Organizational Factors, pp. 122–127. CRC Press Taylor & Francis Group (2012)
3. Bonenberg, A.: Place brand-building. Urban empathy as an evaluation method. In: Antona, M. (ed.) Universal Access in Human-Computer Interaction Methods, Techniques, and Best Practices, pp. 150–160. Springer International Publishing, Switzerland (2016)
4. Brown, T.: Change by Design. Harper Collins Publishers, New York (2009)

5. Ceconello, M., Bisson, M., Boeri, C., Vignati, G.: Colour plan for urban design. In: Ontologies for Urban Development: Conceptual Models for Practitioners. 2nd Workshop COST Action C21 – Towntology, pp. 150–161. Politecnico di Milano, Dipartimento INDACO, Turin (2007)
6. Koos, U., Richter, K.: The Book of Design. Keiser Verlag, Neusass (2001)
7. Melss, P.: Elements of Architecture, From Form to Place. Chapman and Hall, London (1992)
8. Miller, D.: Wpływ wiedzy projektanta na formułowanie problemu projektowego. Polska Akademia Nauk. Instytut Filozofii i Socjologii. Ossolineum, Wrocław (1990)
9. Norman, G.R.: Likert scales, levels of measurement and the "laws" of statistics. Adv. Health Sci. Educ. **15**(5), 625–632 (2010)

Computer Lab - Space Organization and Environmental Conditions: Case Study Assessment Compared to the Theoretical Model

Michał Sitek[✉]

Faculty of Architecture, Silesian University of Technology, Gliwice, Poland
michal.sitek@polsl.pl

Abstract. There are many different patterns of spatial organization and job settings within the computer classrooms. The organization of these spaces should be adapted to the working methods of the group. This article constitutes a description of a case study and a critical assessment of the working conditions in the existing classrooms. The conclusions presented in the summary should be used to make some changes in order to improve the status quo. Description of the suggested changes reflect the author's many years of experience and interest in comparative literature studies.

Keywords: Computer lab · Ergonomic · Environment parameters

1 Introduction

The dream of each educational facility manager is a modern, well equipped computer laboratory. Since the moment that computers became the part of the teaching process, educational units have been facing and dealing with a difficult task of organizing the space prepared and equipped for specific functions. Computer labs constitute rooms of significant technical requirements. They must guarantee proper microclimatic conditions, lighting and an access the utilities. A properly designed and equipped computer lab should support the works of teachers and instructors and should ensure a convenient, safe and ergonomic use of workplace equipped with a computer. Unfortunately it is not always the case. The article constitutes a description of the author's experiences, his long use of various rooms which had served as computer workshops. These experiences have not always been positive. Failures, technical problems, design faults and mistakes connected with making use of spaces saturated with technology were just common everyday problems the administrator of the structure had to face. Improvement of the existing conditions rests with practitioners but sometimes it is also included in the scope of the researcher's activity. By means of a critical assessment of the existing state, there is a possibility of recording and analyzing of both positive and negative experiences. The aim is to eliminate design and functional errors.

© Springer International Publishing AG, part of Springer Nature 2019
J. Charytonowicz and C. Falcão (Eds.): AHFE 2018, AISC 788, pp. 56–67, 2019.
https://doi.org/10.1007/978-3-319-94199-8_6

2 The History of the Computer Laboratory at the Faculty of Architecture of the Silesian University of Technology

The first premises for the computer laboratory were launched at the Faculty in 1993. It was nothing more than just one room on the second floor, adjusted to a new function. The equipment mentioned above and the new furniture made it possible to prepare computer workstations. The laboratory had its own supervisor while maintenance was done by an IT specialist, hired full-time. Equipment of the laboratory included 3 workstations, with desktop PCs with Intel 486 processors, operated by DOS system. Autodesk CAD software was installed on those computers. These were two copies of AutoCAD 12 (software released in 1992). Peripherals which were purchased together with the whole computer equipment included a digitizer and a plotter. The hardware and software served a small group of academic workers of the unit. The lab was also used in teaching, to carry out optional classes thematically connected with computer aided design. First groups of students acquired knowledge and skills on their own, working on their private computers, from home.

Successful receipt of a subsidy for the execution of the TEMPUS project (1996–99) by the Faculty of Architecture, constituted another qualitative and quantitative change. Financial means obtained from the EU made it possible to purchase the first, fully equipped computer lab, solely designed for educational purposes. The laboratory had been prepared from scratch and had been equipped with a structural network - Ethernet with a transfer of 10 Mbps and a modernized electrical system. Apart from a dozen computer stations, dedicated to work with students, the Faculty's laboratory was equipped with the first ever file server and a multimedia projector. The Siemens PC was equipped with Windows NT 4.0, AutoCAD 14 and a Microsoft Office 95 software package.

Rooms for the computer laboratory were equipped with power sockets with anti-surge protection, overload protection as well as structural sockets for RJ 45 mounted on the plaster. Access to network was possible thanks to one fiberglass connection with a transfer of 100 Mbps. All windows of the laboratory were exposed to the northern side and were equipped with internal, horizontally mounted, blinds which reduced the amount of natural light in order to improve the work conditions when using the multimedia projector. Additional piece of equipment, found in the laboratory, was a projection screen placed on a stand and an HP DesignJet 350 C ink plotter with a paper feeder. After several years of usage, in 2004 the authorities of the Faculty decided that the didactic computer laboratory would be transferred to newly renovated rooms on the top (the sixth) floor of the building. The authorities managed to organize two separate rooms which enabled simultaneous classes in two groups, a dedicated room for the control room, technical room and a workstation for the IT specialist who was responsible for the local network and all software. Right after the relocation, the laboratory was extra equipped with new computers and software in order to carry out classes. Each of the classrooms was equipped with ten computer workstations. Classes were conducted in small groups with up to eight people. One of the stations was dedicated to conducting classes and was equipped with a multimedia projector. The tenth station was equipped with a scanner which was used as a support or an auxiliary. Unfortunately, the size of the classes was extremely disadvantageous.

Being long and narrow (11.35 m × 2.87 m and 14.65 m × 2.87 m), these rooms could only accommodate two computer workstations in a row (there were five rows) Il.1. Such arrangement of the stations generated a lot of inconvenience due to too large a distance between the last row and the surface of the screen on which didactic content was to be presented. Students who worked at far-off stations complained about difficulties with reading information presented on the screen. Functional zones designed for students were lighted with natural light from the southern and northern side. Both parts were equipped with roller blinds. These roller blinds were mounted internally for the northern side and externally, with manual regulation, for the southern side. Unfortunately, in summer time, the laboratory was exposed to high temperatures while in winter one would always complain about the cold. Lack of sufficient insulation as well as thin wall divisions on the shell did not guarantee adequate thermal comfort. The only possible way to improve thermal comfort in the summer was by airing out the classrooms. It was done by tilting the windows in order to create a draft that aired out the rooms in an intensive fashion. The system of natural ventilation did not guarantee an efficient and controllable exchange of air in the classrooms (Fig. 1).

Fig. 1. Third laboratory of the Architectural Faculty, sixth floor - its condition as of 2009. Source: author's archive.

There had been attempts to acquire investment means to modernize the system of heating and ventilation. Unfortunately, the means granted only allowed the authorities to modernize the heating system. The planned installation of the mechanical ventilation supported by air conditioning was never realized. In 2008, equipment was purchased and installation works were done to enable the assembly of Wi-Fi based on the **"eduroam[1]"** system. With its range, it covered all student zones and it also allowed authorized users to access the local academic network as well as the global network. All one had to do was log in.

---

[1] **Eduroam** - educational roaming - an initiative which derives from European academic computer networks. The aim of the eduroam service is to provide a secure connection within the academic environment. Access to eduroam is co-created by all institutions that make use of it. Academic staff as well as students who take advantage of eduroam, with the alma mater's consent, are able to gain access to the Internet on the premises of all the facilities associated with eduroam (both in Poland and abroad).

3 New Quality - Why Should We Design Together with the User

In 2009, on the strength of the decision made by the Rector of the Silesian University of Technology, the Dean of the Architectural Faculty received a newly renovated building, adjusted to educational purposes. Following a reconstruction and modernization, it was handed over to the people of a new specialization - Internal Design, launched in 2008. Within the limits of the first floor, a classroom which served as a computer laboratory was designed. The design concept envisioned that the laboratory be placed in the central part of the building and that 16 workstations be organized in four rows – red color, Il. 2. During the works connected with the preparation of the detailed design, the idea of building a corridor was dropped for the sake of enlarging the laboratory - green color, Il. 2. Special space was also designed inside the floor where the power supply system together with connectors for the structural network could be lodged. The plan envisaged that the connection boxes should be placed parallel to the walls of the room, eight boxes in each of the two rows (Fig. 2).

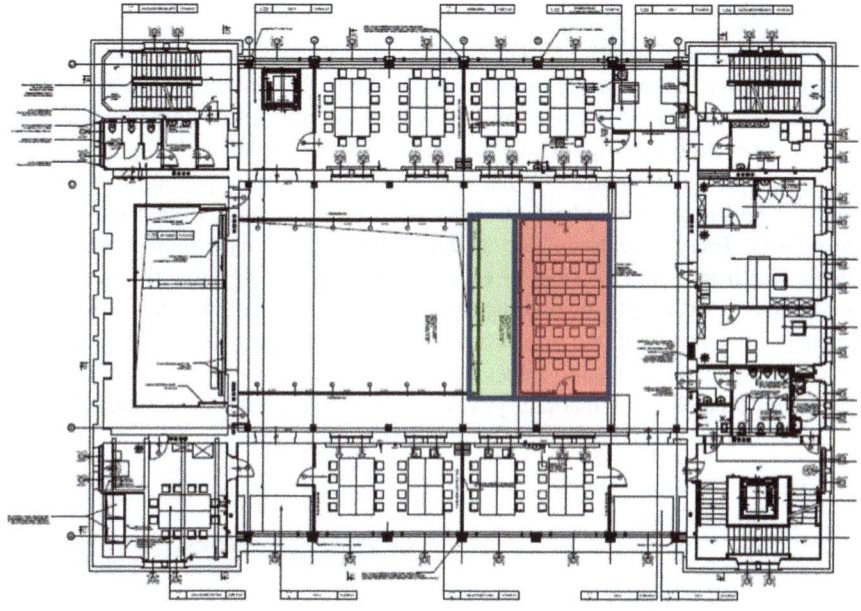

Fig. 2. Location of the fourth laboratory of the Architectural Faculty (color - red), plan of the first floor. Source: author's archive.

The project also envisaged that the workstations for students should be organized in two lines, arranged back-to-back. In accordance with the design, the connecting point located on the floor ensures the connection of one computer workstation (two DATA sockets and two RJ 45 sockets). The number of workstations that the room was designed for counted 16. That number also applied to the users in the rooms. In order to handle such a load, the zone was equipped with HVAC system. Balanced load within the room, thanks to the energy recovered from users and computer hardware, had been calculated and included when selecting the amount of air supplied from the air conditioning unit in order to ensure adequate temperature conditions, humidity and quality of air within a room that has no natural ventilation. During the preparation of the detailed design and implementation, the project team and contractors did not consult the future users. Designers did not have any notion of the expectations expressed by future users as to the manner in which work was supposed to be done with the students or as to the arrangement of the space [3, 4]. The room was handed over to the users without the Audio-Video system. Having completed the construction works and after the handing over, a decision was made to purchase 24 computer workstations and place them within the aforementioned space. That also meant purchasing furniture which enabled the possibility of installing as many workstations as possible within the space of the laboratory. The furniture was custom-made. High concentration of workstations in the room, which had initially been planned for sixteen users, led to some negative situations, which have been described below.

3.1   The Space

Too large a number of workstations within the room of the computer laboratory led to a series of mistakes in the manner in which it was used. The first problem was connected with the access to specific stations from the side of the main entrance. Moving between the rows is difficult and often causes discomfort due to students bumping into each other, or people already seated. What is more, it forces other people to move or to change the position of their chair. Students who arrive for classes bring into the room large objects. These objects include drawing tubes or e.g. hard folders which could range in size from the A1 format to even A0 format. Moving around poses an additional problem for people who bring such large objects into the lab Il.3. Another problem is connected with the infrastructure connecting points. They are located in the floor of the room. Their location does not allow for a direct connection of the equipment already located in the laboratory. The manner in which the workstations had been arranged led to an alternative, unergonomic fixing of the utilities (Fig. 3).

Fig. 3. Density of the workstations as well as large personal objects hinder the passage within the laboratory. Source: author's archive.

The problem was solved by purchasing ACAR extension cords and by providing a network wiring system, distributed in bunches across the Room, Il. 4. That solution was implement only temporary. It has served for eight years and it generated a series of complications (Fig. 4).

 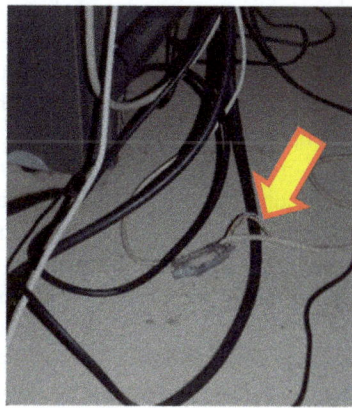

Fig. 4. Distribution of utilities - electricity and network. Dangerous and susceptible to damage. Source: author's archive.

That solution hinders the passage within the room and leads to frequent damage to cables. Carelessly placed chair of the workstation, where the user is seated, leads to cuts and can crush the cables. Users and cleaners can trip over the cables as well. Despite placing the RJ45 sockets in the floor and securing them with covers, they were pulled out on several occasions. Covers which close the connecting points on the floor do not secure the socket in a proper way. Their construction and load capacity were not selected in a way that would allow chairs to be put on top of them. Such situations lead

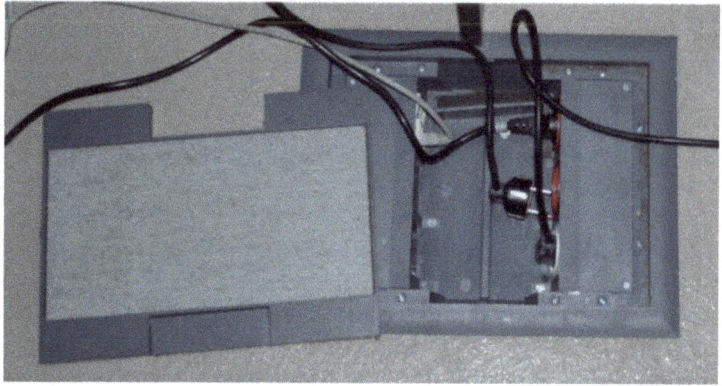

Fig. 5. Damaged floor cover for the connecting points. Source: author's archive.

to damage dealt to the covers. They can cave in or be pulled out of the floor Il. 5. There were also cases of short circuits in the electric sockets caused by flooding or by using an excessive amount of water when mopping the floor (Fig. 5).

3.2   Workstations Ergonomics - Furniture, Peripherals

The laboratory is equipped with furniture which was custom-made. The necessity of placing such a number of workstations forced the authorities to purchase non-standard furniture. The furniture supplier suggested that specific stations be divided by installing an elevated lateral board. Its height corresponds to the height of the upper edges of the monitors operating at the workstations. Unfortunately, the working surface of specific stations is minimal and it only allows students to work with the keyboard and the computer mouse. There is a lack of space for personal belongings, notes or drawings, specific to the academic major, there is no room for drawing or sketches which are later converted into a digital form during the classes. During its operation, the room was also equipped with two sets of multimedia projectors and screens. Work performed from the position of the lecturer makes it possible to project different image combinations on the screen. There is a possibility of submitting video signal from two desktop computers or a laptop. In addition, the systems functionality enables a concurrent viewing of the same image on both screens. It is also possible to view different content from one or even two sources. Multimedia projectors were placed under the ceiling in an inverted position. Unfortunately the screen with which the image was projected had been suspended low due to the local drop in height caused by the building's load bearing beam. Lowering of the screens in relation to the line of eye sight from the computer workstations causes that the lower part of the projected vision is blocked by monitors and walls/boards that separate the stations. On frequent occasions, the users are heard to complain about poor visibility of the content being projected at the bottom of the screens. That situation forces people to change their position in relation to the image being projected. The location of the power switch for the multimedia projectors - next to power-off switches for the lights in the laboratory, turned out to be an ergonomic and operational problem. During work with the students, there were oftentimes mistakes connected with students switching the projectors off which also led to their shorter service life (decalibration of the vision resulting from high level thermal stress inside the casing of the projector caused a permanent loss of sharpness), disturbances during classes (long period of time before the projector could be switched back on as a result of a lack of adequate cooling of the components). The room does not possess an integrated audio system. A portable system of computer loudspeakers, which makes it possible to amplify the laboratory, is used on an ad hoc basis.

3.3   Lighting

The room of the computer laboratory does not require to be lit by means of natural light. Nevertheless, a well scattered/dispersed??? natural light can reduce the demand for energy which is necessary to light up the space by means of artificial light. Lighting based on artificial light can be provided from light frames suspended under the ceiling in three lines, with four points per each line. The system had been zoned out and allows of an independent light control based on a 2 + 1 system (two lines closer to the wall with screens and the line behind the local ceiling lowering). In case of the laboratory at hand, there has been no success in trying to protect the space against the negative impact of direct insolation Il. 6. In spring and summer, the low sun hanging over the horizon can penetrate into the space of the first floor. If it is not blocked, it may hinder the work performed by the users in front of the monitors and may also limit the clarity of the image projected on the screens. There are not any blinds or shutter to reduce solar impact. Solutions of this type are widely described in the literature [5, 6] (Fig. 6).

Fig. 6. Direct sunlight limits the legibility of information presented on the screens by means of multimedia projectors. Source: author's archive.

3.4   Noise

Computer laboratories require numerous electronic devices which are made of noise generating components. As a result, it is necessary to implement solutions that will limit the influence of noise. Computer sets which were purchased by way of a tender procedure met all the conditions set out in the order. However, during their exploitation, a serious fault was found. The base stations were equipped with really loud graphic card cooling systems. Simultaneous activation of several workstations generated a lot of irritating noise. That inconvenience was reduced at stage one, through a replacement of cooling systems in some of the cards. At stage two, 12 graphic cards were exchanged for new models, more efficient and quieter. At present, as many as 18 workstations are in use. Noise generated by multimedia projectors constitutes part of the background hum which has a negative influence on the people who are present inside as it is continuous and hazardous. In its first year, the laboratory was also plagued by one more unexpected noise. It appeared during the activation of the ventilation system. During its operation, one could hear the air whistling near the duct which received the used air. The regulation blind which is located on the collecting duct should be regulated during the system's calibration process. After a year, the author of the article inspected the blinds and revealed that following the initial regulation works it remained closed and did not perform its task correctly. The author reported the situation to the facility manager which resulted in a correction of the system setups, improved the freshness of the air inside the room and eliminated the cumbersome noise. Users who are exposed to such an inconvenience for a longer period of time should be protected against this hazardous factor. These users are people who conduct classes and stay inside the laboratory longer than three hours at a time. Another problem resulting from the elevated level of noise consist in impeded communication between the tutor running the class and his/her students. In order to be well heard and understood, the tutor must raise his or her voice. This often leads to problems with vocal organs and chronic inflammations which result in occupational diseases in teachers, also academic teachers.

3.5   Microclimate

Modern buildings are buildings that react to changeable needs of the user. In response to that demand, one can now find various systems of building management, systems of intern environment management and systems of user behavior observation. There is only one goal - to improve working, relaxation and learning conditions. Properly designed systems, regardless of the level of complexity and integration at the BMS[2] level are supposed to provide us the possibility of adjusting the conditions of the microclimate to the expectations of the users. These elements include:

- temperature regulation option,
- air flow velocity,
- smell,
- humidity,
- concentration of hazardous substances,
- access to natural light,
- regulation options – to control the proportions and the amount of excessive light, both natural and artificial.

The building and the computer laboratory located within, are equipped with a controllable and programmable ventilation, heating and cooling system. However, designers advised the installation of a mechanical ventilation in selected zones only, as a result of cost reduction policy connected with the modernization and exploitation of the structure. These zones include lecture rooms (huge auditorium designed for 160 people and a small auditorium for 80 listeners) didactic rooms which students/users often call "the aquarium" - computer laboratory as well as a didactic room on the second floor. The system made it possible to control the aforementioned zones online. I was possible for the zones to be adjusted to the timetable and to be synchronized with programmed operation parameters of specific ventilation units. A handy remote panel also enabled a manual steering of the basic parameters connected with the devices' operation. Despite numerous attempts at setting up the parameters of the system's operation, it still caused multiple problems. Achieving stable microclimatic conditions proved problematic due to a high level of the system's inertia. Proper preparation of the

---

[2] BMS – Building Management System, known as a building automation system (BAS), is a computer-based control system installed in buildings that controls and monitors the building's mechanical and electrical equipment such as ventilation (HVAC), lighting, power systems, fire systems, and security systems.

laboratory room for work with a large number of users requires an earlier activation of the system. Users and facility managers, despite training, did not follow the rules and guidelines provided by the designer. That resulted in various problems. The most frequent mistake, connected with using the space of the laboratory, was leaving the door open. Despite installing on the doors a self-closing mechanisms, the users often blocked them and they were stuck in an open position. In such conditions, the ventilation system could not operate in accordance with the assumptions set out by the designer. The ventilation unit which functioned within the laboratory (ZNW3), made by FRAPOL, was installed and equipped with ventilators located in the cellars of the building. The unit is made up of two ventilators, a water heater, an auxiliary electric heater, radiator, ball recuperator and sets of filters equipped with sensors.

4 Emergency, What then?

In June 2016, there was a serious failure of the mechanical ventilation system which operated within the whole building. Due to economic, legal and organizational reasons, works connected with fixing the failure did not commence until December 2016. The company responsible for repairing the failure revealed a series of irregularities and design flaws in the existing system. Works continued until January 2017. From October 2016 to January 2017, classes were conducted in a room with no ventilation (natural or mechanical). In such a situation, the only possibility to deliver air into the room was through the open door. Airing of the room was supported by means of an independent fan which blew "fresh" air from the corridor into the room. During classes conducted as part of the "winter semester" of 2016/17, some disturbing situations were observed. Students complained of difficult work conditions while the instructor/teacher was losing control over the group. Students complained of drowsiness and too high temperature in the room. These problems inspired the author to conduct a series of measurements and observations in order to present the situation to the authorities of the Architectural Faculty. In January 2016, southern Poland saw some really fine weather conditions. High temperatures during the day and night, high level of insolation only intensified the feeling of discomfort in the laboratory. Being in possession of a certified device for registering microclimate conditions, it was possible to make multiple measurements. Data collected made it possible to document the difficult conditions that were found in the non-ventilated room used by as many as 16 people. Measurements were made by means of special measurement set which was composed of:

- Testo 435-4, multimeter and recorder,
- IAQ probe used for measuring the quality of air in the rooms, CO_2 concentration, humidity, temperature and absolute pressure.

The measurements included three basic microclimate parameters:

- temperature,
- humidity,
- CO_2 concentration.

Averaged parameters per the day on which the measurement was made, at 8.00 AM, before the classes started. They showed the following values:

- temperature – 21, 3 °C
- CO_2 – 400 ppm.

Exemplary results obtained during the measurements are shown in Table 1. Level of humidity measured in the room showed an average value of 32%. Air temperature blown out of the computer chassis oscillated between the value of 28 °C. Obtained results of measurements, which were made more than once, confirmed a very high concentration of CO_2. Each measurement, made within an hour of the commencement of the classes, exceed the value of 1100 ppm. The measurement cycle was repeated three times during the day, for four days and it was done in the same location in the room. A key element in the observations was the time in which the change to the conditions found within the room occurred. With only 15 min into the class, the people who were in the room were reported to experience highly adverse conditions of the environment in which they had to remain for some time. High temperature inside the room as well as the high CO_2 concentration negatively influenced the microclimate inside the computer laboratory as well as the users' comfort. Having analyzed the values obtained during the measurements and having compared them with data coming from references, one could comprehend users' feeling of dissatisfaction and the general negative mood. Comparative values coming from reference-based data [1, 2] and their influence people's frame of mind have been presented in Tables 2 and 3.

Table 1. Data obtained during one of the measurements - conditions found in the laboratory during the class of the second group (the first group had classes from 9.00 to 11.00, then the room was aired out for 30 min, using a fan).

Measurement time calculated from the start of the classes (duration of one class - 90 min)	Temperature	Level of CO_2
11.30 start of classes	23.4 °C	961 ppm
11.15 – after 15 min	24.2 °C	1141 ppm
12.00 – after 60 min	24.3 °C	1157 ppm
12.15 – after 75 min	24.1 °C	1361 ppm

Table 2. Limit values and the extent to which they were exceeded in relation to the CO_2 level [1].

What are safe levels of CO_2 in rooms	
CO_2	
250–350 ppm	Normal background concentration in outdoor ambient air
350–1,000 ppm	Concentrations typical of occupied indoor spaces with good air exchange
1,000–2,000 ppm	**Complaints of drowsiness and poor air**
2,000–5,000 ppm	Headaches, sleepiness and stagnant, stale, stuffy air. Poor concentration, loss of attention, increased heart rate and slight nausea may also be present
5000 ppm	Workplace exposure limit (as 8-h TWA) in most jurisdictions
>40,000 ppm	Exposure may lead to serious oxygen deprivation resulting in permanent brain damage, coma, even death

Table 3. Minimum ventilation rates in breathing zone – ASHRAE standards [2].

Occupancy category	People outdoor air rate Rp		Area outdoor air rate Ra	Default values			Air class
				Occupant density	Combined outdoor air rate		
	Cfm/person	L/s × person	Cfm/m^2	#/1000 ft^2 or #/100 m^2	Cfm/person	L/s × person	
Computer lab	10	5	0,6	25	15	7,4	1

5 Summary and Conclusions

Computer laboratories are often associated with new technology, innovation, advanced systems as well as a conscious and responsible user. One should always bear in mind one important principle connected with the design of all architectonic structures and rooms within. One should always design together with the user. One should not design spaces which are full of technology, not being fully aware of the necessity to make changes to the initial assumptions. Technological progress as well as the expectations of the users are changing with increasing speed. Therefore, modern building should be prepared for changes. Moreover, they should be not only economical but also flexible. The room of the computer lab described in this article had been designed for a smaller number of users but it was adjusted to an increased number of computer workstations. The consequences of that change were not subjected to an analysis so the users must pay the price. They were forced to agree to a lower functional standard and resign from ergonomics. Such situations are of course undesirable and should be eliminated.

References

1. The Engineering ToolBox. http://www.engineeringtoolbox.com/co2-comfort-level-d_1024.html
2. ANSI/ASHRAE Standard 62.1-2016: Ventilation for Acceptable Indoor Air Quality. https://www.ashrae.org/technical-resources/bookstore/standards-62-1-62-2
3. Tymkiewicz, J., Winnicka-Jasłowska, D., Jastrzębska, M.: Pre-design studies on the example of modernization project of geotechnical laboratories. ACEE Archit. Civ. Eng. Environ. **10** (2), 43–52 (2017)
4. Winnicka-Jasłowska, D., Jastrzębska, M., Tymkiewicz, J.: Ergonomics of laboratory rooms - case studies based on the geotechnical laboratories at the Silesian University of Technology. ACEE Archit. Civ. Eng. Environ. **10**(2), 35–41 (2017)
5. Masły, D.: Daylighting simulation studies - new ways of sustainable, high performance buildings design. In: Proceedings of the 8th International Conference of Education, Research and Innovation, IATED, pp. 2400–2404 (2015)
6. Masły, D.: New design philosophies in architecture as a way of achieving substantial improvements of office buildings' quality in consideration of sustainable development. In: Human Factors of A Global Society. A System of Systems Perspective, pp. 61–68. CRC Press/Taylor & Francis Group, New York (2014)

Quality of the Built Environment from the Point of View of People with Autism Spectrum Disorder

Agnieszka Bugno-Janik[✉] and Maria Bielak-Zasadzka

Faculty of Architecture, Silesian University of Technology, Gliwice, Poland
{agnieszka.bugno-janik,maria.bielak-zasadzka}@polsl.pl

Abstract. The features of the built environment that pose obstacles to people with Autism Spectrum Disorder or Asperger Syndrome have been evaluated in recent times to a certain degree, however, the awareness of these problems are still not common among architects. An effective way that could change the social awareness seems to be the dissemination of direct personal experience of contact with people that have a different perception and response to the built environment. Such contact can evoke the emotional reaction of sympathy and desire to understand their specific problems, which should entail a permanent change in the awareness of those involved. In view of the above a participatory action research experiment has been launched to enable students of architecture to investigate selected problems of the design of the environment with teenagers with ASD/AS, and, at the same time, assist them in their direct experience of space.

Keywords: Participatory action research · Asperger syndrome
Autism spectrum disorder · Built environment features · Quality of space
Users' needs

1 Context – People with ASD in Architectural Space

The Autism Spectrum Disorder (ASD) is a developmental disorder which is being diagnosed with an increasing frequency in recent years. In tandem with a growing number of cases recognized early on, one can observe an increase in social actions taken in order to raise and change the awareness of the problems that people with this developmental disorders must face in everyday life.

As shown by scientific research[1] and the experience of families and therapists involved, some features of the built environment can influence the life and work of people with ASD differently than in the case of neurotypical people. It is very important to incorporate the knowledge of these features into the architectural design practice, especially in the process of designing educational facilities, where students with ASD can encounter numerous problems resulting from various deficiencies. These difficulties pertain not only to the influence of the architectural environment on the

---

[1] For current state of research which constitutes basis for this text please refer to our other text "Shaping the space for persons with Autisms Spectrum Disorder" in this publication.

© Springer International Publishing AG, part of Springer Nature 2019
J. Charytonowicz and C. Falcão (Eds.): AHFE 2018, AISC 788, pp. 68–78, 2019.
https://doi.org/10.1007/978-3-319-94199-8_7

students' ability to concentrate during classes but also, due to the specific developmental differences in people with ASD, to the broadly defined notion of well-being. These difficulties, e.g. due to excess sensory stimulation, may cause or worsen the anxiety, feelings of disorientation or feeling of being overwhelmed, which can also influence the quality of communication skills as well as interpersonal relationships. Knowledge of the influence of the built environment on people with ASD is improving all the time. However, its penetration into various professional groups which, just like architects, are not in touch with ASD problems on daily basis, is still not enough to bring about realistic and institutional changes that would improve the space for everyday life and social activity of people suffering from ASD.

In Poland there is no organized support system for people with ASD that would be based on legal regulations. Few educational facilities are truly ready to accept and take in children and teenagers with ASD. Support in all spheres of life (in the process of diagnosing, in therapy, in finding ways to solve everyday life problems) is only provided by non-governmental organizations. At public educational facilities, children and teenagers with ASD cannot hope that environmental conditions will be adjusted to their needs. In the scope of architectural activity, one can observe few pioneer cases of actions taken for the sake of "autism friendly" design which would include needs expressed by ASD people, for example as part of the approach called "Universal Design"[2].

At the same time, the current state of research into the features of the built environment related to the needs of people with ASD made it possible for many developed countries to introduce various organized actions, which incorporate the knowledge of the needs of people with dysfunctions such as the ASD into architectonic design. Nevertheless, despite all informational actions and efforts made by social organizations in Poland, stereotypical perception of what autism spectrum disorder stands for still persists. Moreover, there is still a lot of animosity and lack of understanding when it comes to the nature of that phenomenon.

2 Research Background

Research presented in this article was very limited but its significance can be based on the pioneering character (as per our local setting) and participatory experimental nature. The situation opened new field of research for our team and broadened our search for design solutions which in later years effected in the form of new courses and master thesis projects, publications and presentations. Above all, the results of research action introduced the needs and understanding of the significance and specificity of situations of the people with ASD into our awareness and experience. The main aim of the research was to sensitize young architects-researchers to the possibility of a different space perception by means of a direct experience which consisted in assisting teenagers with ASP in their exploration and evaluation of the architectural space[3].

---

[2] Recently, the students from Gdańsk, Poland, made an interesting research on playground places for children with ASD, Herkt et al. [1].

[3] Similar experienced students have gained during participatory research workshop with elders, which were organised by our team, see Bielak-Zasadzka and Tymkiewicz [2].

The construction of the research project stemmed from the scientific interests of both the authors of the text. On the one hand, this is the pre-design research based on ideas of Universal Design (Maria Bielak-Zasadzka, PhD) and, on the other hand, this is doing and disseminating research in compliance with the idea of Participatory Action Research (Agnieszka Bugno-Janik, PhD). The common denominator for these two approaches is the social sensitivity which guides the research with people discriminated because of their disability or social inequality. It consists in investigating and trying to change their social situation. Therefore, a discriminatory spatial environment constitutes the subject matter of the actions and investigations carried out by the both authors of this article.

Participatory Action Research (PAR) main assumption is to make research which can influence positively the researched situation by empowering discriminated people in a way that help them actively influence the change of their own situation. The process of scientific investigation of the social problem is interwoven with the process of acquiring new competences by the people from the investigated community. The participants have the position of co-researchers and have their own active role in planning and conducting the research. In our case two cooperating groups – architects and people with ASD – created the participatory action research situation focused on evaluation of educational space quality in one of the University building.

3 Assumptions, Goals, Methods

In the presented research situation the intention was not to discover new significant features of a built environment affecting users with ASD. The main intention of the experiment was to create a possibility of establishing a relationship between students of architecture and a group of teenagers with autism spectrum disorder, who for the first time came into contact with an entirely new educational space within a newly modernized building which had been dubbed as a distinctive example of modern architecture. The teenagers with ASD who had been invited to take part in the project were treated – in compliance with the PAR approach - as experts in their field.

The research was carried out at the beginning of 2014 by the group of 4th year students of Architecture. Planned as participatory action, directed at the problems of people with ASD, were initiated by an alumnus of the Architectural Faculty - young architect Marta Stachurska and her mentor, dr Joanna Ławicka, the president of the PRODESTE Foundation, which acts for the improvement of the situation of people with ASD.

The goals of the research were specified in three areas:

1. scientific:

   - to compare the evaluations of basic features of the built environment of educational facility, carried out by invited teenagers with ASD with the knowledge from experiments and research carried out around the world,
   - to test usability of created research tools for architectural space evaluation, working within the PAR framework with people with ASD

2. social:

- to strengthen competency for future influence of the changes in the built environment by people with ASD,
- to sensitize young architects-researchers to the problems of people with ASD through direct contact with teenagers and their problems, in a specific situation in a real building.

3. educational:

- training of the participatory methods of research on the built environment.

Organization of the research was, on the one hand, supposed to provide students/researchers with the possibility to collect information about how people with ASD assess the indicated features of a building, on the other hand, it was supposed to allow respondents/teenagers with ASD – to familiarize themselves with the architectural terminology, architects' way of thinking about the space and the methods of space evaluations. Such interaction and cooperation were aimed at changing both parties of the research process - the teenagers with ASD could acquire new skills allowing them to better communicate their observations, needs and difficulties related to the space of a building, while the students of architecture could experience, by assisting teenagers wit ASD, what it is like to have untypical needs in neurotypical space[4].

The research was conducted during carefully prepared workshop which lasted six hours. Before the workshop preparation the students were introduced into the problems of autism spectrum by Joanna Ławicka, PhD, by several-hour long lecture followed by discussion. The lecture was focused on the most important features of ASD type of mental development which may cause problems in normal life within a typical built environment and typical social situations and also treated the issues related with the specificity of the communication and social relations of people with ASD, necessary for an adequate preparation of the research tools for the workshops to come.

The role of co-researchers was given to a group of 10 teenagers aged 13–19, dr Joanna Ławicka's protégés, together with volunteers and therapists from the PRO-DESTE foundation who work with that group of people on a daily basis.

The group of researchers counted 20 4th-year students from the architectural faculty and authors of the text.

Typical research activities, within the PAR method, require long period of time, necessary to build a deepened relationship between the co-researchers, which enable better understanding and broadening of experience. Due to the fact that the workshops lasted only a day – (around 6 h) and due to specificity of the group of co-researchers (ASD), a solution was chosen that made it possible to establish a relationship between the participants in small teams, using tools, specially designed to facilitate communication during the execution of research tasks.

---

[4] Similar approach was demonstrated by Ian Scott in his experiment with design the ideal classroom with children with ASD [3].

Each team consisted of 2–3 students of architecture, 1–2 teenagers, who played the role of experts in terms of how people with ASD function within the building, and helpers from the PORDESTE Foundation.

The first part of the workshops was an introductory lecture for the teenagers co-researchers, prepared by the students. The lecture demonstrated selected features of the built environment in an easy to understand way, with definitions and explanations as to the set of characteristic features of a building subjected to research.

The set of features, which could significantly influence the wellbeing of an ASD person and, at the same time, proves difficult to change, was agreed on the basis of an analysis of the state-of-art and the situation of the building chosen for research[5]. The features chosen for research, are these which need to be taken into consideration at the building's design stage, which make it difficult to change after the building is finished. At the same time, identification of these features does not require special qualifications and is possible following a short theoretical introduction.

Research stage of the workshops was planned in form of a field game. Subsequent teams had to locate a given room, assess it (people with ASD played the role of both co-researchers and experts, responsible for the assessment, while the students of Architecture played the role of assisting researchers as they had knowledge of the building) on the basis of strict instructions (for ASD people precision as well as good task specification are of great importance) in a given time and according to a plan which differed depending on the team. The gaming aspect of the activity was introduced as entertainment and relation building mean.

The building in which the research was done constitutes an interesting example for that type of research due to the following reasons:

- it was recently modernized in an untypical way for a building of an educational facility in Poland - a potentially interesting experience for young people who had not had any contact with such designed buildings. The building was rewarded for its interesting design. Because of his specific design it constitutes a good educational tool to present numerous functional, technical and esthetic problems of the modern architecture;
- diversity of architectonic features of didactic rooms - rooms of different sizes and proportions, lighting quality, transparent walls, with finishing materials creating different interior climates (moods);
- unclear circulation with wayfinding problems (historical structure of a building which served previously as a cinema and students' club, later redeveloped into an educational facility make real maze for new users), rendering the site challenging for the teenagers invited to research organized as a field game (Fig. 1);

For the purposes of evaluation eight didactic spaces were chosen, significantly different in the scope of:

lighting, access to external view, transparency of inner and outer walls, the finishing materials (colors, texture and patterns), equipment, and the general climat/mood:

---

[5] The selection of features to investigate was based on dr Joanna Ławicka personal experience and knowledge and state of the research review, especially of Simon Humphreys experience [4].

Evaluated spaces:

1. large lecture hall (014) - for about 200 people, with proportions similar to a cube, without natural light, with glazing at the upper floor level along the circumference of the room, uniform artificial warm lighting, not too bright, finished in warm tones, lateral walls finished with perforated brick (for better acoustics);
2. small lecture room (015) - semi-open (border between the room and corridor difficult to define, dark with no access to natural light, with cold artificial lighting, longitudinal, high;
3. large classroom (107) – a narrow, high room with a large glazing on the outer wall, all internal walls made of glass or with internal windows, a lot of natural light (which may sometimes be blinding);
4. computer lab (110) - a small room equipped with 25 computer workstations, 3 m high, with three glazed internal walls, brick-finished walls, lack of natural light, dark, with quite dim (but warm) artificial light, tables densely arranged for individual work with screens, dark furniture and finishing;

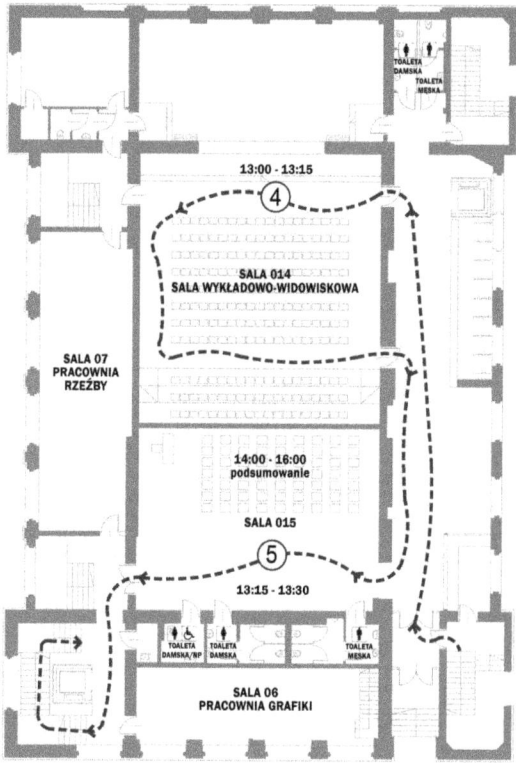

Fig. 1. First floor of evaluated University building with field game route.

5. two small classrooms (112 and 113) with identical features - small, with 4 fully glazed walls, including the internal wall, with lots of daylight, gray (neutral) finishing and furniture;
6. art classroom (202) - a longitudinal, medium-sized room, with natural lighting of high windows with limited external view;
7. small lecture room (207) - medium-sized room, with 3 glazed rooms, indirect natural light, gray equipped, lack of access to external view of the building;

Features which were subject to evaluation:

- quality of the natural and artificial lighting (intensity, regularity, glare);
- size (size, height) and proportions of the rooms (length to width, floor area to height);
- method of divisions of the room (transparency, zoning, shape of the walls)
- interior finishes
    divisions – number, regularity, continuity
    colors – saturation, brightness, combination
    texture – roughness, gloss, pattern
- details – open question "which detail is mostly annoying or causing deconcentration?"
- room equipment – number, distribution
- equipment finishes – material, color, texture,
- climate – outer sounds, inner sounds, smell, temperature.

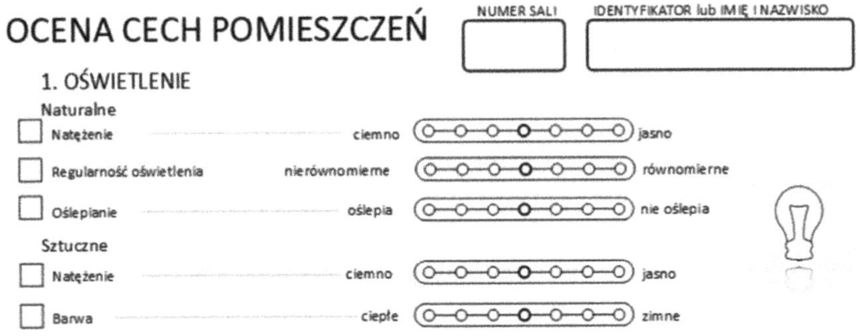

Fig. 2. Part of evaluation form with semantic differential.

The evaluations consisted of two elements:

- the basic impression, evaluated as positively, neutral or negatively (+,0,−) which influences the way in which a given feature is expressed (left column on evaluation form), and
- evaluation of the symptoms of a specific feature, expressed by means of the Semantic Differential (Fig. 2),

Research was supposed to find answers to the following questions:

1. will the texture of the exposed brick walls, either full or perforated, found at significant places of some of the rooms, be recognized as an element exceedingly stimulating (sharp, visible pattern of bricks r) which may hinder concentration or distract participants?
2. what intensity and type of lighting are preferred by the persons with ASD - bright or rather dim, natural or artificial?
3. will the large or disproportionate rooms (e.g. very long or very tall) be evaluated negatively?
4. will the internal and external glazed walls be viewed as significant sources of discomfort or distraction?
5. are bigger rooms preferred over smaller ones, or vice versa?
6. what type of color arrangements will be evaluated as better: the neutral grays of the brightly, naturally lit rooms, or warm, natural materials colors of artificially lit rooms?
7. which combinations of the features listed above will be perceived as positive?

4 Results

The research delivered several sets of information:

1. results of the assessments, written in the assessment forms,
2. remarks concerning the rooms from the conversations held during workshops,
3. general assessment, approved during the final discussion held among the partici-pating teenagers with ASD and students of architecture,

The analysis of the results showed a high degree of convergence resemblance conformity in the assessments, in several matters:

Size and Proportion. The large lecture hall (014) was evaluated unanimously as the best. Critical remarks were directed at elements which had been damaged; at the large lighting fixtures, narrow aisles and uniform lighting. Cubic proportions of the room were perceived as good and the climate of interior was also evaluated highly. Its size did not make a negative impression on the majority of the evaluators (we supposed that its height might seem overwhelming). Perhaps, this could be connected with larger proxemic needs of people suffering from ASD, as noted by Humphreys [4].

Patterns as Source of Distraction. In positively assessed large lecture hall, the most characteristic features include walls with a visible pattern of perforated. When planning research we focused on the fact that that very pattern could cause irritation as well as distraction. We were curious to see in which rooms it would be negatively perceived. In that room the finishing was jointly assessed as rather uniform (although few answer differed).

Only two people out of eight [6] perceived the brick pattern of that room as a negative phenomenon. Perhaps, all the positive opinions were influenced by the effect of scale, the texture of the perforated brick lining might not seem dominant in relation to the size of the room. It is also possible that the lateral walls which are located at a significant distance in relation to the center of the room are not capable of distracting one's attention away from the large screen located in the center.

In the much smaller room of the computer laboratory (110), only one person out of nine described the brick bond pattern negatively, two people expressed neutral opinions while as many as six delivered positive reviews. In case of room 202, a room where the respondents spent most time during the lecture, the assessment of the brick pattern on the wall where the projection screen is located were neutral (4) and positive (5). There was only one remark saying that the "bricks cause a distraction, as you are forced to look at them during the presentations". To recap, for most of our young experts the brick bond pattern on the wall was not as annoying as we had expected.

And yet, the co-researchers were able to find more patterns at places which we did not take into consideration - the glazed wall partitions. We observed correlation between the responses at places where one could find some sort of lack of organization (e.g. there was disorder in the arrangement of the furniture or the furniture itself was damaged) or mismatched patters. In places perceived as not ordered there were fewer positive opinions about patterns. Possible explanation could be related to over-whelming aspect of disorder, which affected the perception of patterns visible for evaluators.

Invisible Border. What was really interesting, was the discovery made during the assessment of a small lecture room (015), whose characteristic feature was that it lacked one wall, which had been criticized during everyday use by neurotypical students and by teachers. Our co-researchers with ASD did not pay any attention to undefined border (lack of a dividing wall on the side of the corridor, with just a row of columns), perhaps due to the fact that the corridor had not been, at that time, used by any people from outside of the workshop group.

That very room, which from the perspective of neurotypical people (architects) seems "boring", dark, gloomy, long and slightly too high ('bad' proportions), was assessed as quiet, well-ordered and non-distracting (facilitating concentration). It has neutral medium-gray walls, dim cold artificial lighting and average, evenly distributed chairs. It has no windows and only meager access to day light through the mat glazing of the door. Dark ceiling does not attract too much attention. In the summary, at the end of the workshops, one of evaluators said "that is irony: we do prefer such boring rooms". Also the proportions of that room were assessed as rather positive, which probably could have been caused by the unclear ("invisible") border of that room.

---

[6] Not all rooms were evaluated by all teams. It resulted from the specificity of the game in which the assumption was that inside a small room there might only be 1 team to avoid disturbances during evaluation. The attractiveness of the game for teenagers, and the fact there was no rush to do the evaluations were more important than the number of results.

Further conclusions

- Those features which was assessed as clearly visible by the neurotypical architects, were definitely more often assessed as positive. Ambiguous situations, expressed irresolutely, received mixed opinions (positive or negative).
- Disorder, trash, signs of petty vandalism were viewed negatively and as irritating in additional remarks
- Translucent internal and external walls were viewed by some researchers as irritating only if what happens behind these walls attracts too much attention and caused distractions.
- Features which were listed as irritating in the comments section of the questionnaire or during conversations (due to the fact that they are exceedingly absorbing and distracting):
  - glass panes if it is possible to see people who pass behind them;
  - light if it gets too dark or to bright
  - colors if they are to intense;
  - distant sound in small glazed rooms;
  - dearth of light in some rooms;
  - too bright daylight;
  - narrow aisles between chairs, rows and furniture;
  - large, prominent, untypical elements - enormous lighting fixtures in abundance, large mechanical ventilation pipes, damaged elements (chipped and scratched tables), trash, lack of order in the elements of the equipment.

5   Assessment of the Results of the Research

Research was conducted on the basis of PAR, however, the undertaking itself was too short to provide a deep insight into the situation of the ASD people (which constitutes the essence of PAR). Nevertheless, the assumed social targets were met. After almost 4 years, a student-researcher (currently professionally active architects) still claim that that experiment was extremely impressive and say that it was a significant experience for the way in which their professional stance came into shape.

For the members of the co-researchers group, that experiment was one of many organized in the scope of the activity performed by PRODESTE foundation called "Autism Friendly Space". Directly after the workshop, they also stated that the experiments were interesting and educative.

Space evaluation results, which were obtained as part of that research, do not constitute an unequivocal indication for designing. They do, however, indicate a direction of further research. It will not be easy as the organization of the workshops required a lot of effort, especially on the part of the PRODESTE foundation. We would like to continue research as they revealed numerous weak spots in the understanding of the reality which people with ASD space must face at educational institutions. Thanks to them we seem to have realized that going deeper into that subject may yield significant results, also for the practice of the architectural design.

References

1. Herkt, K., Bucała, M., Kielak, E., Jarosz, N., Jagiełka, W.: Przestrzeń placów zabaw dostosowana do potrzeb dzieci z autyzmem, Badania Interdyscyplinarne w Architekturze 2, tom 4, Gliwice (2017)
2. Bielak-Zasadzka, M., Tymkiewicz, J.: Senior homes of the future in the eyes of students of architecture. Didactic experience from the application of the design thinking method. Archit. Civ. Eng. Environ. ACEE 9(1), 49–56 (2016). Silesian University of Technology, Gliwice
3. Scott, I.: Analysis of a project to design the ideal classroom undertaken by a group of children on the autism spectrum and students of architecture. In: GAP, vol. 12 (2011)
4. Humphreys, S.: Autism and Architecture. http://www.autismlondon.org.uk/pdf-files/bulletin_feb-mar_2005.pdf. Accessed 28 Feb 2018

Sustainable Housing Environment: Form and Territory

Wojciech Januszewski[✉]

Faculty of Architecture, Wroclaw University of Science and Technology,
Boleslawa Prusa s 53/55, 50-317 Wrocław, Poland
wojciech.januszewski@pwr.edu.pl

Abstract. The problem of sustainability in housing cannot be limited only to issues like energy saving and environmental impact but should address the primary purpose of the organization of habitable environment. The goal is to harmonize the opposing needs of people: privacy and collective life, isolation and integration, locality and openness to the world, the quality of the home and the quality of the city. The expression of these aspirations is a territorial structure shaped by an architectural form. The ecological form of housing architecture balances the individual and community needs. The living space should be legibly defined by buildings. The configuration of space in a residential environment should provide a hierarchy of privacy. The formation of transition zones should ensure establishing social contacts at various levels.

Keywords: Housing · Residential architecture · Sustainability
Urban form · Territory · Habitat

1 Introduction

Housing architecture is the basic matter of the city and the city is the basic form of the ecosystem shaped by the man. In the debate on sustainability, the housing problem is one of the key issues. The effectiveness of that environment directly determines the stability of the social structure and the dynamics of spatial transformations. Housing architecture arises in diverse cultural, economic, social, and political contexts. Is it possible to create a recipe for the sustainable housing architecture? This question cannot be limited only to issues like energy saving and environmental impact but should address the primary purpose of the organization of habitable environment. The goal is to harmonize the opposing needs of people: privacy and collective life, isolation and integration, locality and openness to the world, the quality of the home and the quality of the city. The expression of these aspirations is a territorial structure shaped by an architectural form. One can distinguish three basic aspects of this issue: defining the space – how the living space is shaped by physical elements; configuration – how the living space is related to other spaces in a greater spatial structure; *transition zones* – how the physical relation among the spaces is shaped. The right solution to these aspects is the basis for the economic management of the city space.

© Springer International Publishing AG, part of Springer Nature 2019
J. Charytonowicz and C. Falcão (Eds.): AHFE 2018, AISC 788, pp. 79–89, 2019.
https://doi.org/10.1007/978-3-319-94199-8_8

2 Defining the Space

The visual perception of spatial systems in urban complexes shapes the territorial relations of the housing environment. Let us consider how diverse formations and spatial configurations affect the creation of territories.

The concept of space can be understood as an abstract and unlimited mathematical space or as a place – concrete and empirical space. The latter category includes the housing space. A habitat is a place on earth that is a reference point to the man's existence. The organization of the living space is, in the most general sense, the creation of here categories, in opposition to the shapeless somewhere. The creation of existential space is, in the first place, its separation from the continuum by defining the boundaries, which is the function of the architectural form [1].

Two models of perceiving the relation between space and architectural form can be distinguished:

– a free-standing form in a negative space
– a form that limits a positive space (Fig. 1).

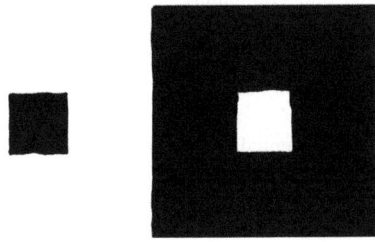

Fig. 1. Two types of space-form relation: a free-standing form in a negative space (left), a form that limits a positive space (right). Author's drawing.

In the first model, the architectural form creates a distinctive object, while outer space (void) is the background (negative). In the second system, it is the opposite – space is the proper object, while the form remains the background [2, pp. 176–179].

The positive space model builds an urban interior and clearly defines the boundaries of a territorial zone. It seems to be the most appropriate one for the housing environment as a response to the needs of locality and security. However, one must remember that this is only one side of human needs in relation to space. The abuse of the space closing strategy results in a claustrophobic environment, an example of which was the nineteenth-century city with corridor-like streets and extremely narrow backyards [3, pp. 11–15].

The reaction to that type of environment was the extreme scattering of building development in modernist urbanism and the dominance of the negative space model.

Let us consider the system of free-standing buildings located at relatively large distances from one another (Fig. 2). The reception of such space is dominated by architectural forms. The outer space is negative: it is shapeless and homogeneous. Within that space, it is difficult to visually identify smaller spatial units that may have

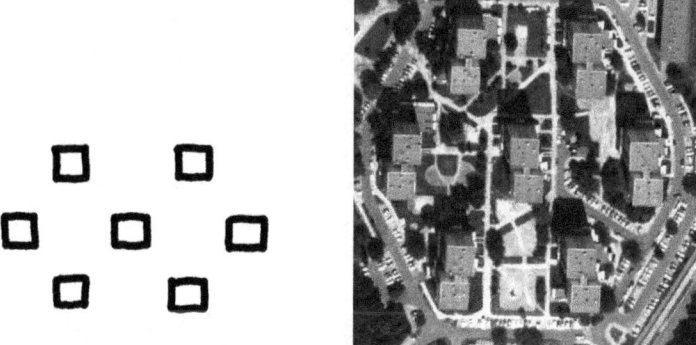

Fig. 2. Free-standing urban form: diagram (left); example: residential development Kozanow in Wroclaw, Poland (right). Image Courtesy of Google Maps.

territorial significance. The only spatially defined territory is the interior of the building which is an island of privacy in the surrounding public space [4, pp. 90–122].

The traditional urban tissue is based primarily on positive spaces. Two basic types of such space can be distinguished: a courtyard (static space) and a street (directional space). Both models shape diverse territorial relations (Fig. 3).

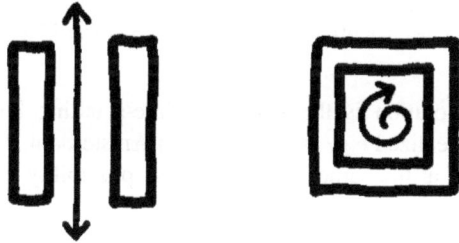

Fig. 3. Two types enclosed space: "a street" (left), "a courtyard" (right). Author's drawing.

The interior of the courtyard is the space separated from the background of the surrounding area. It can naturally create an enclave, controlled by inhabitants, with a limited access to outsiders in a physical or psychological sense. Territorial relations in the case of the street are different. The street space combines static and dynamic aspects. The street is open to traffic and flow – it is, therefore, a public space, but at the same time it is a partially defined space and enables control by its users, including the inhabitants. The degree of street privacy depends on many details in its shape, such as the dimensions of space, the formation of threshold spaces, the type and intensity of traffic, etc. The free-standing form, the street, and the courtyard are the basic types of spatial organization occurring in various variants and connections.

A common form of a middle-way solution is the so-called open block composed of the perimeter arrangement of free-standing structures (Fig. 4). Despite the occurrence of openings, this system is perceived as closed, in accordance with the general tendency of human perception to complete the shapes. The model was widely used in garden city type of development such as neighborhoods designed by Raymond Unwin and Clarence Stein. It was a way to improve the conditions of ventilation and insulation of building blocks [5, pp. 319–359]. In the post-modern urbanism, the model was promoted by Christian de Portzamparc as a contemporary compromise between the model of the nineteenth century and the functionalism [6, pp. 222–223].

Fig. 4. An open block: diagram (left); example: the urban block designed by Christian de Portzamparc in Nantes, France (right). Image Courtesy of Google Maps.

Another possible solution is the system of free-standing buildings in a greater enclosed block space, realized e.g. in Funenpark in Amsterdam. This system separates the territorial zone and at the same time creates the possibility of a free architectural composition (Fig. 5).

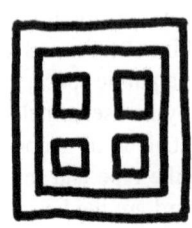

Fig. 5. Free-standing form enclosed in a perimeter block: diagram (left); example: Funenpark, Amsterdam (right). Image Courtesy of Google Maps.

The housing environment requires the balance of enclosure and openings. The postmodern criticism of the modern idea of open space seems to be exaggerated when one takes into account that modernism turned against the claustrophobia of the nineteenth-century city. The mandatory return to closed quarters is by no means the postulate of common sense. It seems rather that one should postulate for a maximum opening while maintaining the legibility of the territorial structure and the composition of the urban space.

3 Configuration

The character of architectural space is defined both by its individual features and its location in a larger spatial configuration. Among others, Bill Hillier drew attention to that problem in his research on the "syntax" of space. [7, pp. 26–51] Housing environment is a holistic problem – it will be difficult to reduce it only to the issue of individual buildings or housing estates. The difficulty in defining the limits of the habitual zone is specific for housing issues. It is expressed in the ambiguity of colloquial statements we use when talking about these matters. The sentence "Here is my home" or "I live here" may refer to the building but also to the street, district, city, country, etc.

The authors of the Charter of Athens, studying the structure of an urban organism, adopted the method of functional analysis by making a basic division of the city into the zones of living, working, and recreation. In the newly designed urban plans, a principle was established for the division of these purposes and the separation of mono-functional housing estates [8, pp. 73–91]. The housing function itself was treated rather utilitarian – as a creation of an adequate number of apartments providing proper hygienic conditions (insulation, ventilation, etc.). The doctrine of functionalism was criticized by the younger generation of architects from the later Team 10. With regard to housing problems, the term habitat – derived from biological sciences – was understood as the holistic environment of human life. The basic structure, described by habitat, is the "hierarchy of human associations" [9]. This idea was further developed by subsequent concepts based on the criticism of functionalism – the concept of

Fig. 6. Perimeter urban block: diagram (left); example: residential development for mineworkers Nikiszowiec, Katowice, Poland 1908–1918. Image Courtesy of Google Maps.

defensible space by Oscar Newman [10, pp. 1–21], degrees of publicness by Christopher Alexander [11, pp. 192–196], and the territorial structure by John Habraken [12, pp. 143–160].

These models describe the housing environment as a hierarchical arrangement of spaces sorted according to the degree of privacy and accessibility (territorial depth) – e.g. personal, private, semi-private, semi-public (or social) spaces, and finally – public space. It is also a postulate for the designed settlements to have hierarchy organized in a clear way, ensuring the implementation of diverse needs: for privacy and participation in social life.

The most widespread example illustrating the territorial hierarchy is the perimeter urban development (Fig. 6). It is a combination of street and courtyard rules. In that type of an urban system, there is a clear hierarchy of privacy and the division of different ways of utilization. The courtyard is a more private space, dedicated to the residents of a building block. The street is a public space. The legible street space is the element that brings individual quarters together into a larger urban entirety [13, pp. 81–100].

In modernist urbanism, implemented in accordance with the CIAM doctrine, the model of a street along a continuous frontage of buildings was rejected. The structure was shaped independently of the driveways, as free-standing elements – usually repetitive and standardized residential buildings. The method used to overcome the monotony was often a creative approach to urban composition, which resulted in a large variety of housing forms.

Fig. 7. Ambivalent interiors in modern urban planning: diagram (left); example: residential superblock in Brasilia. Image Courtesy of Google Maps.

It would be a mistake to state that modernist urbanism operated only in negative space. The formal repertoire contained various types of urban interiors ensuring a relative visual intimacy of the residential domain. At the same time, these interiors did not create a legible territorial hierarchy. The so-called adjacent interiors, where the buildings functioned as walls separating identical spaces; or ambivalent closing-opening systems were used as frequent solutions [14, pp. 74–81]. It is difficult to identify public, semi-public, and semi-private spaces in such arrangements. It is also difficult to assign specific exclusive spaces to specific groups of buildings, and therefore to specific user groups. Despite the clear spatial articulation, the environment remains undivided in the terms of territory (Fig. 7).

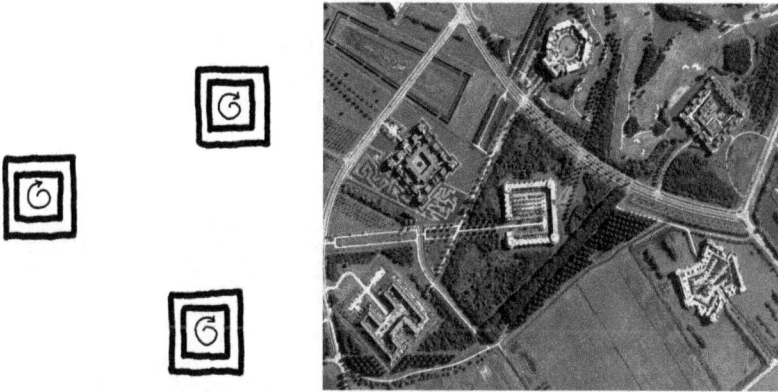

Fig. 8. Isolated residential courtyards: diagram (left); example: housing development Haverleij, Den Bosch, The Netherlands. Image Courtesy of Google Maps.

The lack of hierarchy may occur in spatially closed systems, for example in the system of courtyards that as such are free-standing objects in the landscape. The surroundings do not have the character of a built area. It is not strictly the public space, it is rather the natural or semi-natural background for the housing development. Neighbourhood spaces are not connected to the public space but are isolated enclaves (Fig. 8). Such an arrangement is rarely the result of conscious design (the example on the picture is an exception). Island systems are usually the manifestation of chaotic, unplanned urbanization. The isolation of housing enclaves has a number of negative consequences for the quality of a living place, the cohesion of urban space, and ecology.

Fig. 9. Connected courtyards: diagram (left); example: Rabenhof housing development in Vienna, 1928. Image Courtesy of Google Maps.

Another example of a non-hierarchical system is a layout of connected residential courtyards. It is in a sense the opposite to the modernist open composition (Fig. 9). These solutions are characteristic for some of the interwar "workers' yards" (Arbeiterhof) and for theoretical projects from the circle of neo-rationalism as a reaction to the modernist open form [15, pp. 71–81]. The external space in this system is completely closed. Similarly to the open system, space is devoid of hierarchical diversity. Subsequent urban interiors do not differ in territorial depth. This certainly applies to an ideal situation in which the structure has no beginning or end. In reality, with a limited development size, the spaces located closer to the periphery are less private than those located inside.

Fig. 10. Residential street: diagram (left); example: Het Dorp, housing development for disabled people designed by Jaap Bakema, 1963, Arnhem, The Netherlands. Image Courtesy of Google Maps.

An interesting example is the so-called residential alley composed of intimate static and directional spaces (Fig. 10). This system was often used in late-modernist residential architecture [16, pp. 266–268]. The feature of these developments was an intimate character of the interior and an informal shape of the building line. It was an expression of a return to the idea of the street as a social space. Theoretically, the space of the street is accessible to the public but its scale and shape give it a semi-public character.

4 Transition Zones

As it has been stated, the architectural space is defined by the relation to the neighboring spaces. This applies to the spatial configuration discussed above, as well as the physical shape of the perimeter of a given space and its contact with the surrounding. The proper design of intermediate zones is of particular importance for creating harmonious connections among different zones in the habitual environment. It is important to ensure a proper balance between the separation and the connectivity of space through the view and the physical access.

The boundaries, territorial gates, and threshold spaces are important for this matter. Territorial boundaries are spatial elements that articulate the change of control zones. These can be various types of fences, surface changes, or building walls. The territorial gate is a break in the continuity of the territorial border, allowing the transition to a different territorial level. The concept of the threshold space (or doorstep) connects to the idea of the boundary and the gate. It has been researched in architecture since the

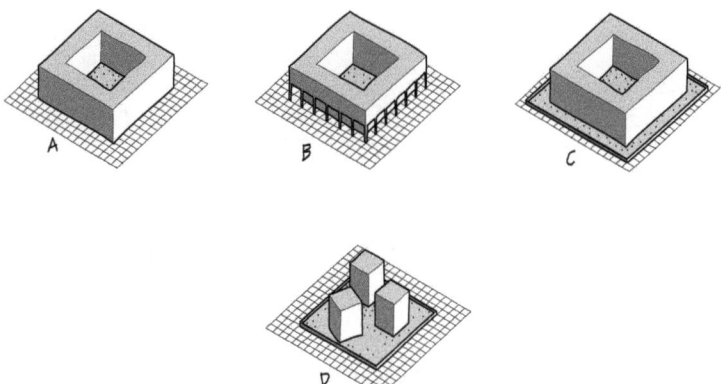

Fig. 11. A territorial border in relation and the wall of a residential building. Author's drawing from: Januszewski, W.: Wybrane aspekty organizacji przestrzeni habitatu: forma i terytorium. In: Bac, Z. (ed.) Moj piekny Habitat, pp. 74–89. Oficyna Wydawnicza Politechniki Wrocławskiej, Wrocław (2016)

post-war times and Team 10 group, mainly by Aldo van Eyck [17]. The interest in the problem was born with the undermining of the doctrine of functionalism, characterized by strict separation of functions. The threshold space occurs in the joint of spaces with different levels of privacy. Its task is to mitigate the rapid transition between different territorial zones, e.g. public space of the street and private interiors of the house (e.g., entrance zone). The created threshold space signals the change of the territorial zone – it warns an unauthorized person and welcomes the guest. The threshold space – as a neutral zone – is also a convenient space for establishing social relations.

The layout of gates, boundaries, and threshold spaces defines to a large extent the character of adjacent spaces. Let us consider the relation between the territorial border and the wall of a residential building:

- the border in the building line (Fig. 11A) - the outer line of a building is at the same time the boundary of the private territory or the territory assigned to a neighbourhood community;
- withdrawn border (Fig. 11B) – public space is introduced into a block, e.g. in the form of arcades or generally accessible service premises; In this variant, the street becomes more public.
- extended border (Fig. 11C) - foreground with the greater degree of privacy than the street is created in front of the building line; The street gets more of a neighborly character.
- secondary border (Fig. 11D) - the boundary is unconnected with the buildings and clearly moved away from them; The street loses its relation to the buildings and loses its importance as a social space.

Different forms of boundaries significantly influence the relations between private and public spaces. This problem affects the threshold spaces, i.e. the transition zones between various territorial levels. Developing this issue, let us look at the examples below, illustrating different solutions.

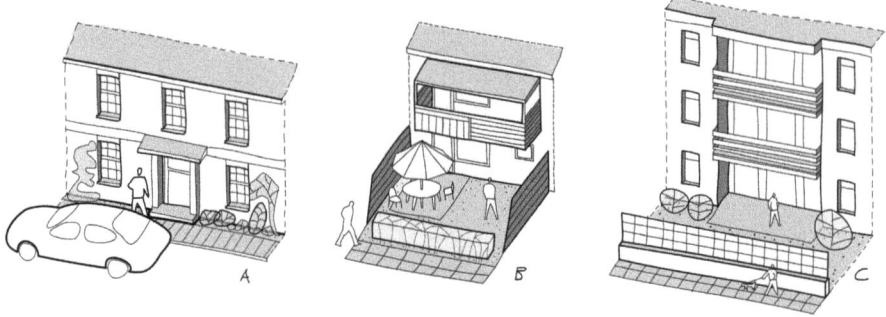

Fig. 12. Transition spaces in a residential environment. Author's drawing from: Januszewski, W.: Wybrane aspekty organizacji przestrzeni habitatu: forma i terytorium. In: Bac, Z. (ed.) Moj piekny Habitat, pp. 74–89. Oficyna Wydawnicza Politechniki Wrocławskiej, Wrocław (2016)

In the first example, a residential house is shifted to the street, practically without any foreground. Only a narrow strip of green covering the width of the entrance threshold is under the control of residents who, according to their own preferences, decorate it with flowering shrubs and vines. The facade of the house is representative, the entrance zone is clearly accentuated. This solution emphasizes the public character of the street (Fig. 12A).

In the next example, the street meets the garden, which is separated by a high fence from the neighbors and a low hedge from the pavement. The building is "opened" with large windows, a terrace, and a balcony, which is its informal backside. There is the view from the public space to the private space, which makes the character of the latter more intimate, focused on neighborly relations, and therefore semi-public (Fig. 12B).

The third example shows a building separated from the street by a high, non-transparent enclosure. In this case, it is difficult to refer to any relationship between the street and the apartment building. The street functions here neither as a real public space nor as a neighbourhood. It is a "no man's land", where nobody exercises natural control, and its meaning is limited to the function of communication (Fig. 12C).

The proper shape of the transition zone should, therefore, take into account the nature of the external space, towards which the residential building is oriented. It should be possible to easily view and to access directly from the side of the private space, which makes the public space naturally controlled [18, pp. 97–99].

5   Conclusions

The analysis of different aspects of form and territory allows formulating the general characteristics of the sustainable housing architecture.

The living space should be legibly defined by the architectural form. This enables to develop the territorial structure. At the same time, there should be a balance between closures and openings to avoid the impression of being trapped or lost in the space. Defining the space should not take place at the expense of ventilation, insulation, and contact with nature. The configuration of space in a residential environment should provide a hierarchy of privacy: public space, semi-public space for a limited neighborhood group, and private space. The formation of transition zones should ensure the possibility of maintaining privacy and establishing social contacts at various levels.

The ecological form of housing architecture balances the individual and community needs. It provides the right balance of the private domain, creates home comfort, and the quality of the public zone, ensuring the cohesion of urban organism. It is characterized by the economic use of space, through which the idea of a compact city is realized. In times when urban planning often loses to urbanization, this seems to be the right interpretation of the problem of sustainability in housing architecture.

References

1. Heidegger, M.: Bauen, Wohnen, Denken, in Vorträge und Aufsätze, Gesamtausgabe, Bd. 7. Frankfurt a.M. (1951)
2. Carmona, M., Tiesdell, S., Heath, T., Oc, T.: Public Spaces: Urban Places. Routledge, London (2010)
3. Chwalibog, K.: Ewolucja struktury zespołów mieszkaniowych. Panstwowe Wydawnictwo Naukowe, Warszawa (1976)
4. Panerai, P., Castex, J., Depaule, J.: Urban Forms: The Death and Life of Urban Block. Architectural Press, Oxford (2004)
5. Unwin, R.: Town Planning in Practice: An Introduction to the Art of Designing Cities and Suburbs. Fisher Unwin, London (1909)
6. Firley, E., Stahl, C.: The Urban Housing Handbook. Wiley, Chichester (2009)
7. Hillier, B., Hanson, J.: The Social Logic of Space. Cambridge University Press, Cambridge (1984)
8. Mumford, E.: The CIAM Discourse on Urbanism, 1928–1960. The MIT Press, Cambridge (2002)
9. Smithson, A., Smithson P.: The Doorn Manifesto. In: Smithson, A. (ed.) Team 10 Primer, p. 96. Studio Vista, London (1968)
10. Newman, O.: Defensible Space: Crime Preventiom Through Urban Design. Collier Books, New York (1973)
11. Alexander, C., Ishikawa, S., Silverstein, M.: A Pattern Language. Oxford University Press, New York (1977)
12. Habraken, N.J.: The Structure of the Ordinary: Form and Control in the Built Environment. The MIT Press, Cambridge (2000)
13. Gehl, J.: Life Between Buildings: Using Public Space. Island Press, Washington (2011)
14. Ciechanowski, K.: Podstawy kompozycji architektonicznej. Oficyna Wydawnicza Politechniki Wroclawskiej, Wroclaw (1974)
15. Schenk, L.: Designing Cities: Basics-Principles-Projects. Birkhauser, Basel (2013)
16. Wejchert, K.: Elementy kompozycji urbanistycznej. Wydawnictwo Arkady, Warszawa (1984)
17. van Eyck, A.: Doorstep. In: Smithson, A. (ed.) Team 10 Primer, p. 96. Studio Vista, London (1968)
18. Gehl, J.: Miasta dla ludzi. RAM, Kraków (2014)

Modelling of Runway Infrastructure Operations in an Effort to Increase Economic and Environmental Sustainable Development

Julio Roa$^{(\boxtimes)}$ and Junqi Hu

Civil and Environmental Engineering Department, Virginia Tech, Blacksburg,
VA 24061, USA
{julioroa, junqi93}@vt.edu

Abstract. In the effort to increase economic and environmental sustainability of the airport system; longitudinal spacing of aircraft approaching a runway has been identified as a major obstacle for runway efficiency. As demand approaches capacity at main hub airports, research efforts have focused on safely reducing current aircraft separations, which are considered a bottleneck in the effort to increase the utilization of existing airport facilities. This research uses modelling and computational algorithms to demonstrate the potential for aircraft separation reductions based on discrete scenarios comprising environmental conditions, aircraft-dependent parameters, and aircraft operational capabilities. Results from this research show that separation reductions can be obtained for aircraft such as the A320 and E170, corresponding to groups D and E in RECAT I. This results in single runway airport efficiency increase and gains are further translated into reductions of emission and fossil fuel.

Keywords: Sustainable development · Air transportation
Airport sustainability · Wake separations

1 Introduction

Airports play a vital role in the development of local and national economies; but this development results in a significant environmental impact in terms of noise pollution, emissions and waste water; as well as social impacts such as community development and traffic disruptions and congestions [1].

Airlines continue to merge, and air traffic operations increasingly rely on main hub airports to serve the flight demand. Runway capacity is becoming critical to serving the increasing operational demands and helping airports recover from long queues created by non-favorable weather conditions that result in underutilized economic and environmental resources.

Wake vortex separations have become a major influence for runway throughput because they determine the time that the runway will be unutilized between each pair of sequential arrivals. When the separation between arrivals is large enough, air traffic controllers may allow departures in between, but these departures also rely on wake separation standards for safety [2].

© Springer International Publishing AG, part of Springer Nature 2019
J. Charytonowicz and C. Falcão (Eds.): AHFE 2018, AISC 788, pp. 90–99, 2019.
https://doi.org/10.1007/978-3-319-94199-8_9

In an effort to increase sustainable development of airport operations wake vortex separations become a major parameter to study in order to understand the potential for further safe reductions in aircraft separations. Dynamic wake separations is a concept that proposes the use of dynamic conditions and their impact on wake behavior in order to dynamically adapt such separations according to actual conditions. This would result in reductions from current wake separations distance between aircraft [3].

1.1 Objective

The focus of this research is to develop a computer model of runway operations in order to simulate aircraft wake separation reductions, develop a methodology and present simulation conditions and available operational data that could be used in order to further reduce wake separation and increase airport economic and environmental development.

Parameters for sustainability development in infrastructure utilization such as aircraft emissions and noise are used as variables into the design of the simulation.

The research simulates aircraft operations during final approach utilizing RECAT I.5, the dynamic separations of RECAT III, and real data recorded for fleet mix, runway occupancy time (ROT), operational buffers, and aircraft approach speed. The research calculates and compares the RECAT I.5 and RECAT III gains in runway capacity, according to their proposed reductions in aircraft wake separations.

This research utilizes a methodology to calculate dynamic wake separations dependent on aircraft configuration, such as weight, and environmental conditions, such as turbulence, wind speeds, and temperature. These factors could be considered in airport-specific wake separations under RECAT III [4] (Fig. 1).

ENVIRONMENT   **OPERATIONS**

SUSTAINABLE DEVELOPMENT

COMMUNITY   **ECONOMY**

Fig. 1. Sustainability development for infrastructure operations.

2 Methodology

The proposed methodology for calculating runway throughput for the airport runways under study is based on Monte Carlo constructive simulation of runway operation procedures. A computer program has been created to simulate real-world conditions of aircraft and airport operations during approach, arrival, and departure conditions. This method accounts for static and dynamic wake vortex separations, ROTs, aircraft approach speeds, aircraft wake circulation capacity, environmental conditions, and operational error buffers.

2.1   RECAT I, RECAT I.5, RECAT III

Wake vortex categorizations are groupings of aircraft with similar wake vortex generation and endurance capabilities. The Federal Aviation Administration created such categorizations in the 1960s, and they remain practical and safe today.

To increase operational efficiency of airports, especially for airports under capacity constraints, the FAA led an effort to revise legacy wake separations. These efforts increased legacy wake separations from a five-by-five matrix to a six-by-six separation matrix called RECAT I. RECAT I was implemented at Memphis International Airport (MEM) at the end of 2012 and has expanded to Louisville International Airport (SDF), Cincinnati/Northern Kentucky International Airport (CVG), Hartsfield-Jackson Atlanta International Airport (ATL), George Bush Intercontinental Airport (IAH), and Charlotte Douglas International Airport (CLT).

RECAT 1.5 was developed as a modification of RECAT I; its results are still being assessed. RECAT 1.5 was implemented at the beginning of 2016 at George Bush Intercontinental Airport (IAH) and continues to expand to airports such as Charlotte Douglas International Airport (CLT), John F. Kennedy International Airport (JFK), Newark Liberty International Airport (EWR), La Guardia Airport (LGA), O'Hare International Airport (ORD), and Denver International Airport (DEN) (Table 1).

Table 1. RECAT I FAA N7100.659A to N7110.659. Wake turbulence separation table for "on approach".

Leader	Follower					
	A	B	C	D	E	F
Old – RECAT I						
A		5NM	6NM	7NM	7NM	8NM
B		3NM	4NM	5NM	5NM	7NM
C				3.5NM	3.5NM	6NM
D						5NM
E						4NM
F						
New – RECAT "1.5" as of april 2015						
A		5NM	6NM	7NM	7NM	8NM
B		3NM	4NM	5NM	5NM	7NM
C				3.5NM	3.5NM	6NM
D						4NM
E						
F						

This research focuses on a proposed methodology development for future wake separations called RECAT III dynamic separations.

ASPM. In order to identify airports that operate under congested conditions, Aviation System performance metrics (ASPM) records from the Bureau of Transportation Statistics (BTS) are used. Late arrival delays, operational delays and National Airspace (NAS) delays and weather delays were studied and an understating of the current delay conditions and causes was obtained.

Late arrival delays were used to understand the delays between the origin-destination for each aircraft pair.

Operational delays include delays due to weather, high volume of traffic and reduced runway capacity; experienced by individual flights. These delays result in the air traffic controller (ATC) detaining an aircraft at the gate before the final approach fixed, short of the runway, on the runway, on a taxiway, and/or in a holding configuration.

National airspace delays are delays that occur after Actual Gate Out. This kind of delay is within the control of the National Airspace System (NAS) may include non-extreme weather conditions, airport operations, heavy traffic volume, air traffic control.

Weather delays are caused by extreme weather conditions that are present or forecasted on the point of departure, en-route, or on point of arrival.

ASDE-X. To provide validity and increase the reality of the simulation results, this research uses detailed data of movement on the final approach, runways, and taxiways, based on Airport Surface Detection Equipment Model X (ASDE-X). This data allows for the calculation of aircraft approach speed, runway occupancy times, and traffic control buffers. ASDE-X data for all runways at the airport under study, for the months January to November of 2016, has been analyzed (Fig. 2).

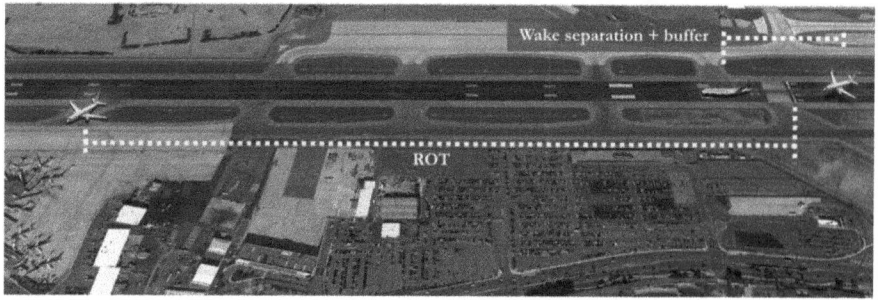

Fig. 2. Wake separation and operational buffer on approach.

Runway Fleet Mix. One of the most important factors influencing runway capacity calculations is fleet mix. The methodology for the aircraft mix selected for simulation is based on ASDE-X recordings of operating aircraft for each runway. Basically, those aircraft that represent between 85% and 90% of operations at the runways under study

Table 2. Approach speed for aircraft ID according to ASDE-X records.

Aircraft ID	A319	A320	B738	CRJ2	CRJ9	E145	E170	MD82	MD83
*Aircraft Mix	4	9	26	8	16	23	8	2	4
**Approach Speed	149	155	158	151	145	154	147	148	150

* Aircraft Mix in Percent.
**Approach Speed Knots.

are selected to represent the runway fleet mix. The simulation consists of launching 1,000 aircraft per runway, which is equivalent to 24 h of peak-hour operation at a typical runway, with 10 operations every 15 min (Table 2).

Runway Fleet Mix. Estimating dynamic wake separation between leader and following aircraft requires consideration of a wake vortex threshold that is known not to adversely affect an aircraft under wake influence flight operation. This threshold considers the maximum wake vortex effect that a follower aircraft can endure at minimum wake separation standards from the wake generator aircraft, referred to as the maximum wake circulation capacity (MCC). The methodology in this research is considering FAA RECAT II wake separations and a 95% confidence interval in the wake behavior CDF curves [5].

Common Approach Path. Final approach fix is a specified point of approach that identifies the commencement of the common approach path or final segment. The common approach path is a specified distance in the FAA terminal procedures charts; this distance is between the final approach point and the runway threshold. The data recordings in ASDE-X show that the length of the common approach path is calculated for each operation; the methodology selected for this simulation applies a fixed common approach path of 8.5 nm (Fig. 3).

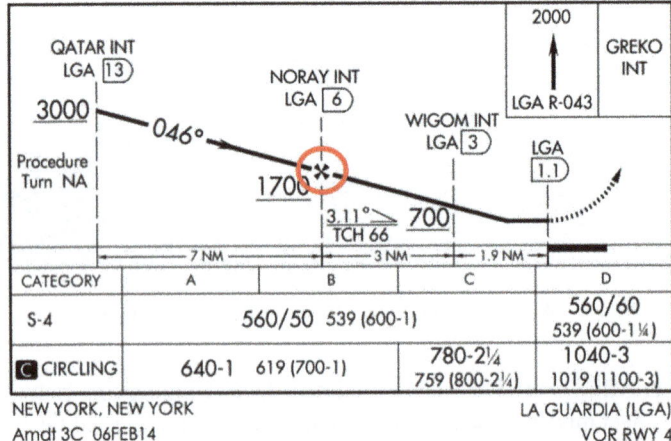

Fig. 3. Common approach path length in VOR chart at La Guardia (LGA).

Operational Buffer Times. Air traffic controllers use operational buffers to modulate aircraft separations at every moment. The purpose of the buffer is to guarantee that controllers do not violate minimum wake separations standards. Operational buffers account for pilot operational errors, delays in communication, and reaction times during pilot-controller interactions.

The proposed methodology calculates operational buffers based on more than ten months of operational data from ORD. The buffers are calculated by measuring inter-arrival time separations and subtracting them from RECAT 1.5 time separation standards. The buffer is calculated in the final, common approach path utilizing a fixed distance of 8.5 nm from the runway threshold. The resulting buffer is the closest point of approach (CPA) between each pair of aircraft during the entire approach path length.

During this analysis, negative buffer values are the result of runway approach operations under Visual Flight Regulations (VFR), where aircraft-to-aircraft separations are below those of Instrument Flight Regulations (IFR) due to pilot self-separation procedures made possible in favorable environmental conditions.

To calculate operational buffer, peak-hour operations were parsed from the ASDE-X data utilizing a baseline of 10 or more operations in 15-min periods. Only peak-hour periods were considered because inter-arrival separation times, and the resulting operational buffers, during non-peak-hours buffers are larger by nature, even though the current wake vortex minimum separation rule is dictated by the RECAT 1.5 separations.

Runway Occupancy Times. When decreasing wake vortex separation during the final approach path, ROT becomes a critical next step in the runway capacity analysis. If a runway must service more aircraft, measured ROTs have to be studied to understand how much of the theoretically calculated capacity increment of RECAT III is operationally feasible on currently available airport infrastructures (Table 3).

Table 3. ROT calculated values for specified aircraft.

Aircraft ID	A319	A320	B738	CRJ2	CRJ9	E145	E170	MD82	MD83
ROT seconds	58	57	57	55	60	53	54	56	57

Environmental and Aircraft Parameters. Environmental and aircraft parameters are key factors in RECAT III dynamic separation calculations. These calculations vary environmental factors such as temperature, crosswind speeds, and environmental turbulence every 15 min. Values for temperature range between 283 and 299 K, values for crosswind speeds range between 0 and 12.86 m/s, and EDR values follow the distribution shown in Fig. 4.

In this research, aircraft operational performance is simulated utilizing the Base of Aircraft Data (BADA), an aircraft performance model developed and maintained by EUROCONTROL in cooperation with aircraft manufacturers and airlines. This database has been designed for simulation and aircraft trajectory prediction The BADA performance model is based on the Total Energy Model, which equates the forces acting on the aircraft with the rate of increase in potential and kinetic energy [7].

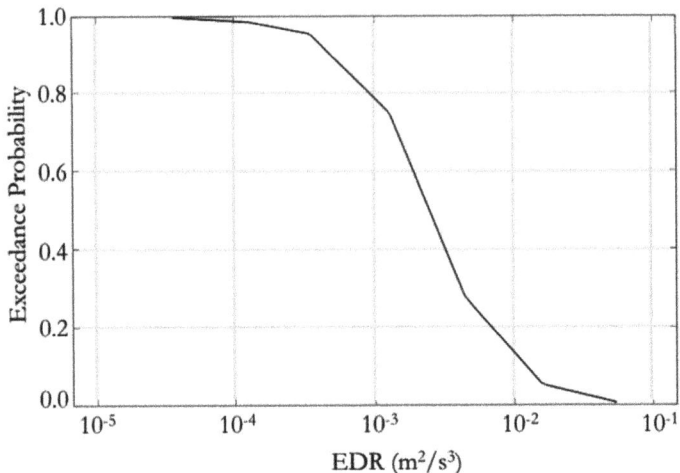

Fig. 4. Distribution of exceedance probabilities at 40 m altitude from DFW 8/97 – 12/98 [6].

Aircraft weight, a critical parameter of wake initial circulation strength, is simulated based on empirical operational data for typical airline operations.

3   Results and Conclusion

Runway capacity improvement is one of the main benefits that make wake separation attractive for sustainable airports.

Results from this research show that separation reductions can be obtained between all aircraft under study but greater gains are obtained for aircraft on different RECAT 1 groups such as the A320 and E170, corresponding to groups D and E in RECAT I. This results in single runway airport efficiency increase. Gains are further translated into reductions of emission and fossil fuel consumption for which the calculation methodology and preliminary results are presented.

Results show that a higher runway utilization; produce fewer aircraft in queue in the airspace surrounding the airport infrastructure and queuing times for aircraft taxing are reduced due to higher runway throughput.

The complexity of implementing pairwise dynamic wake separations, such as in RECAT III, requires advanced decision support tools that can help controllers coordinate and direct arriving and departing flights. There is a need to develop reliable tools that can help the controllers in this task [8].

Dynamic wake separations require more advanced LIDAR capabilities and technologies that allow for wake measurements during flight.

In this simulation, the dynamic wake separation required for a follower aircraft is calculated by using the National Aeronautics and Space Administration (NASA) Aircraft Vortex Spacing System (AVOSS) Prediction Algorithm (APA) model, a semi-empirical wake behavior model developed by NASA that predicts wake decay as a function of atmospheric turbulence and stratification. Wake behavior under each set of dynamic conditions is show in Fig. 5 [9].

The lifetime of the vortex depends on its initial strength, which depends on aircraft weight and wingspan as well as the ambient weather conditions. It is for this reason that the simulation changes environmental conditions every fifteen minutes and aircraft weight in every operation [10, 11].

The number of arrival operations simulated is one thousand. Departures are launched in between successive arrivals if the wake gap for each pair of aircraft is larger than wake minimum separation standards. The simulation assumes 2 nm plus an additional buffer to be the minimum separation between arriving and departing aircraft.

Simulated conditions for ROT and approach speeds are in accordance with the runway under study and follow the distribution according to the ASDE-X data studied. The common approach path length is considered to be 8.5 nm.

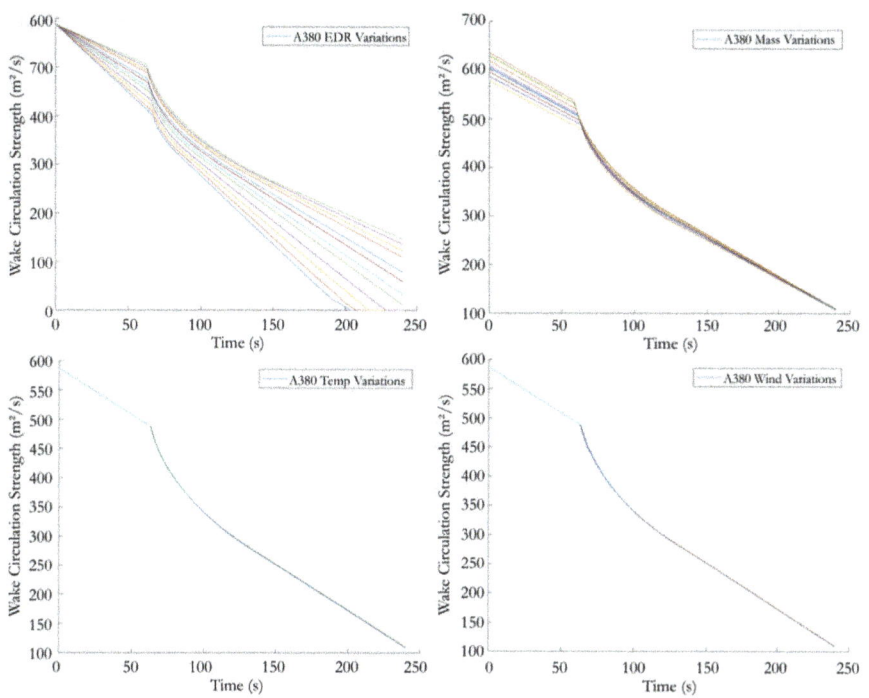

Fig. 5. Wake vortex behavior for dynamic separations.

The approach to calculating the percentage of operations restricted by ROT taking into account the buffer does the following:

If the operational wake separation between leading and following plus operational buffer is less than the ROT for the leading aircraft; then separation is adjusted to ROT plus buffer otherwise use operational wake separation (Fig. 6).

Fig. 6. Aircraft measuring and reporting aircraft parameters and environmental conditions.

The focus of this research is to develop runway operation computer simulation in order to implemented aircraft wake separation reductions and asses the emissions and noise due to less aircraft circling the airspace because of the lack of runway capacity. Dynamic wake separations require airport management tools in order to guarantee proper planning and coordination; as runway throughput increases runway occupancy times and gate utilization become critical factors in the system; it is for this reason that they are the proposed focus of a future study [5].

Acknowledgments. This research has been made possible thanks to the research contributions of the Federal Aviation Administration Wake Program, National Aeronautics and Space Administration (NASA) wake program, EUROCONTROL, Japan Aerospace Exploration Agency, WakeNet meeting members, MIT Lincoln Laboratory, and private industry research labs such as the MITRE Corporation.

Disclaimer. The contents of this material reflect the views of the author only. Neither the Federal Aviation Administration nor the United States Department of Transportation nor National Aeronautics and Space Administration nor Virginia Tech. Virginia Tech does not make any warranty or guarantee, or promise, expressed or implied, concerning the content or accuracy of the views expressed herein.

References

1. FAA and NextGen: NextGen Implementation Plan 2016. F.A. Administration (2016)
2. De Visscher, I., Bricteux, L., Winckelmans, G.: Aircraft vortices in stably stratified and weakly turbulent atmospheres: simulation and modeling. AIAA J. **51**(3), 551–566 (2013)
3. Johnson, S.C.: Simplified aircraft-based paired approach: concept definition and initial analysis, vol. 217994, 2013-217994. National Aeronautics and Space Administration, Langley Research Center, Hampton, Virginia (2013)
4. Feuerle, T., Steen, M., Hecker, P.: A new concept for wake vortex hazard mitigation using on-board measurement equipment. In: 2013 Aviation Technology, Integration, and Operations Conference. American Institute of Aeronautics and Astronautics (2013)

5. Tittsworth, J., et al.: The development of wake turbulence recategorization in the United States. In: AIAA Aviation. The American Institute of Aeronautics and Astronautics (2016)
6. Perras, G.H., Dasey, T.J.: Turbulence Climatology at Dallas/Ft. Worth (DFW) Airport - Implications fr a Departure Wake Vortex Spacing System. NASA L-4, Lincoln Laboratory, Langley Research Center, Hampton, VA (2000)
7. EUROCONTROL: Base of Aircraft Data Aircraft Performance Model (BADA) (2016)
8. Soares, M., et al.: Wake turbulence research: an esoteric field of study that pays big dividends. Volpe, The National Transportation Systems Center (2015). http://www.volpe.dot.gov/news/wake-turbulence-research-esoteric-field-study-pays-big-dividends
9. Ahmad, N.A.N., Van Valkenburg, R.L., Pruis, M.: NASA AVOSS fast-time wake prediction models: user's guide. Langley Research Center, Hampton, Virginia (2014)
10. Proctor, F.H., et al.: Meteorology and wake vortex influence on American airlines FL-587 accident. NASA (2004)
11. Pruis, M., et al.: Summary of NASA wake and weather data collection at Memphis international airport: 2013–2015. American Institute of Aeronautics and Astronautics, AIAA 2016-3274 (2016)

BIM in Prefabrication and Modular Building

Wojciech Bonenberg[✉], Xia Wei, and Mo Zhou

Faculty of Architecture, Poznan University of Technology, Nieszawska 13C,
60-021 Poznan, Poland
wojciech@bonenberg.pl, wei.x.1028@gmail.com,
zhoumo6141@hotmail.com

Abstract. "Prefabrication and Modular Building" and "BIM" were two major buzzwords in the industry of construction in 2017. If the Prefabrication and Modular Building refers to a revolution in building production, **BIM (Building Information Modeling)** [2] application will act as an important technical means to promote the revolution. BIM technology can be applied in the design, construction, operation and maintenance, demolition and other full life cycles of construction field. As a newly applied technology, BIM is able to digitally virtualize and informatively describe the elements of various buildings as well as achieve the collaborative design of information, the visual assembly, the interaction of engineering information and the simulation and verification of node connection. Prefabricated buildings can use BIM technology to integrate various industrial chains and realize an all-round information integration. In the era when the connection between industrial elements and informational elements becomes closer and closer, BIM technology will be integrated with the prefabricated buildings in a perfect way so as to promote the innovation and development of the construction industry and even overturn the traditional construction industry.

Keywords: BIM · Prefabrication · Modular building

1 Introduction

1.1 The Definition of "Prefabrication and Modular Building" and "BIM"

Prefabrication [7] refers to the process of firstly assembling components of a structure in a factory or other manufacturing sites, and then transporting the complete assemblies or sub-assemblies to the construction site where the structure is to be located. The term is used to distinguish itself from the more conventional construction practice of transporting the basic materials to the construction site where all assemblies are carried out. Modular buildings [6] and modular homes are partially prefabricated buildings or houses composed of multiple sections called modules. "Modular" is a method of construction different from other methods of building. The module sections are constructed at an off-site (sometimes remote) facility, and then delivered to the intended site for application. Complete construction of the prefabricated sections is conducted on site. The prefabricated sections are sometimes lifted and placed on basement walls using a crane. The module sections are set onto the foundation of building followed by

© Springer International Publishing AG, part of Springer Nature 2019
J. Charytonowicz and C. Falcão (Eds.): AHFE 2018, AISC 788, pp. 100–110, 2019.
https://doi.org/10.1007/978-3-319-94199-8_10

being joined together to make a single building. The modules can be placed side-by-side, end-to-end or in stack, allowing a wide variety of configurations, building styles and layouts.

Building information modeling (BIM) [5] is the process to generate and manage the digital representations of physical and functional characteristics of places. Building information models (BIM) are files (often but not always in proprietary formats and containing proprietary data) which can be extracted, exchanged or networked to support decision-making with regard to a building or other built asset. Current BIM software is used by individuals, businesses and government agencies who plan, design, construct, operate and maintain diverse physical infrastructures, such as water, refuse, electricity, gas, communication utilities, roads, bridges, ports, tunnels, etc.

1.2   The Development of Prefabrication and Modular Building in Various Countries in the World

Industrialized Prefabrication and Modular Building originated in Europe after the World War II and prevailed in the 1970s.

For example, Germany suffers the largest reduction in building energy consumption in the world. After the World War II, the industrialization level of multi-storeyed residential buildings in East Germany reached 90% in the 1970s. In recent years, it has been proposed to develop passive energy-consuming buildings. New villas and other buildings pursue for the basic assembly of steel (wood) structure. Powerful prefabricated building industry chain, universities, research institutions and enterprises provide technical support for research and development. Construction companies and mechanical equipment suppliers work closely. Mechanical equipment, materials and logistics get advanced, exceeding the fixed modulus size limit.

In China, however, the prefabricated buildings were suddenly stagnated in the late 1980s and then disappeared very quickly. Yet these buildings reemerged experiencing over 30 years of silence. The Chinese government has put forward requirements in the Outline for the Development of Modernization of Construction Industry, indicating that the proportion of fabricated buildings in new buildings will be over 20% by 2020 and the number will be over 50% by 2025.

2   Research Problem

This article explores the application of BIM in prefabrication and modular buildings, with the aim to answer the question that "What are the advantages of analyzing the application of BIM in Prefabrication and Modular building?" and "What problems can be solved by such solutions?" In this paper, BIM shows its wide range of benefits at the production phase of prefabricated modular elements of building structure.

3   Research Methodology

3.1   BIM in Prefabrication and Modular Building

BIM Method for prefabrication and modular building design.

The new and prefabricated building is a systematic and integrated building adopting the "five-in-one", namely design, production, construction, renovation and management rather than the "traditional production + assembly" strategy. Assembled buildings should present the five major features of the new building industrialization which are "standardized design, factory production, assembly-type construction, integrated decoration and information management." The core of an assembly building lies in "integration," and the BIM approach exactly meets the main line of integration. Different from the traditional construction methods, designs for fabricated buildings seems more prominent. Assembly building is a highly integrated product, taking design as the core. The application of BIM information technology will take into account the three-dimensional refinement of the design, the components of production, construction technology, maintenance and other aspects of the entire industrial chain. At present, design remains a relatively weak link in the assembly-type industrial chain. And to complete the entire process of assembly-type construction in an independent and high-quality way can seldom achieved without a wealth of relevant design experience [3, 4].

3.2   Methodology

Key Technical Methods

(a) Use BIM to refine modeling and splitting model in the design stage;
(b) Use BIM modular design to design reusable components, coupled with the establishment of a free combination of the module library;
(c) BIM technology is used to refine the factory production by means of the automatic statistical functions and processing functions map;
(d) BIM technology is used to achieve on-site lifting and construction simulation, so as to optimize the assembly-type construction.

Standardization, Modularity

(a) A key feature of prefabricated buildings lies in producing standard prefabricated components at the factory which are then transported to the construction site for assembly and disassembly.
(b) While being applied to the design of prefabricated buildings, BIM can be used to simulate the prefabrication process to build a BIM component library for each assembly component on the one hand and digitally assign and integrate the number, type and specification of each component on the other hand.

(c) Following the building standards, modular system can achieve cost-effective components of prefabricated production and facilitate the assembly as well as be equipped with the traditional convergence precision. In addition, it can standardize the specifications of related building materials category, and finally achieve the integration.

Industrialization

(a) BIM model of building components can directly complete production via the information and component processing drawings in the BIM model. In this process, both the two-dimensional relationship of the traditional drawings and the relationship between the complex spatial sections can be clearly expressed. At the same time, the discrete two-dimensional drawing information can be collected into a model which can be closely implemented and prefabricated through Factory collaboration and docking.

(b) In the process of production and processing, BIM information technology can intuitively illustrate the spatial relationship and various parameters of the component, and automatically generate component unloading orders, dispatching orders and mold specification parameters. Also, in virtue of BIM technology, workers are able to better understand the design intent through visual expressions. Besides, auxiliary materials such as simulation animation, flow chart, and explanatory diagram of BIM can be formed so as to help improve the accuracy and quality of workers.

(c) The BIM information data input device allows you to achieve the mechanically automated production. This way of digital construction can greatly improve work efficiency and production quality.

Information Technology

(a) The construction schedule into the BIM information model and the spatial information and time information integrated in a visual 4D model ensure you to intuitively and accurately reflect the construction of the entire building process.

(b) Through collision detection and analysis, we are able to collect and correct imperceptible mistakes in the traditional two-dimensional mode.

(c) Through the construction simulation, the complex parts and key construction nodes are rehearsed to increase the workers' familiarity with the construction environment and construction measures so as to further enhance the construction efficiency.

(d) The integration of civil engineering and decoration as an industrialized mode of production is beneficial to improving the production efficiency of the whole process. The standardization of decoration stage integrating into the program stage can effectively allocate production resources.

(e) Indoor rendering with the help of visualization. The purpose is to ensure the quality of indoor space and help designers refine and optimize the design.

4 Case Study

How to use BIM to make prefabrication and modular building? As a well-known Chinese construction company, Vanke said in the project of BIM and prefabrication building, "Dalian Vanke City Industrialization Phase II Project" [1] that the perfect combination of BIM and prefabrication can greatly enhance the project cycle.

(a) Efficiency: 30% reduction in design costs and 40% reduction in design cycles.
(b) Cost: Standardized design controls costs, industrialization reduces construction costs, reduces modulus and increases template utilization.
(c) Drawings: Compared with the traditional CAD drawing mode, BIM improves the depth of drawing and makes the expression clearer.
(d) Quality: Through the standard to enhance the quality of products, customers are more satisfied.

We will examine the use of BIM and prefabrication modular buildings through a practical project (Fig. 1).

Fig. 1. Shandong Binhe new home public rental housing project. Bird view (author: Xiaolei Wang).

Overview Project

(a) Project name: Shandong Binhe new home public rental housing project;
(b) Design company: Shandong Tong Yuan Design Group Co.Ltd;
(c) Project location: Jinan City, Shandong Province, China;
(d) Project time: 2015
(e) Planning area: 98,000 m^2;
(f) Type: High-rise buildings;

g) Industrial design range: Unitized modular design, prefabricated facade, prefabricated staircases, prefabricated slabs, lightweight interior partitions, integral bathroom, bathroom;

(h) Monomer assembly rate: 70%.

The project is located in Licheng District of Jinan City, west of Damwang Road and south of Chenjiadu Road. The total planned land area is 3.51 hectares. The planning and construction land area is 28,598.76 m^2 and the landform is flat and the traffic is very convenient. Surrounding the project planning schools, kindergartens, food markets and other ancillary facilities. A total of 1,303 public rental units are planned for the project, along with underground garages and related public buildings such as property management, commercial services and community neighborhood committees, with a total construction area of 98073.58 m^2 and ground construction area of 71944.21 m^2 and floor area ratio of 2.52 (Fig. 2).

Fig. 2. Shandong Binhe new home public rental housing project. Street view caption.

Project Analysis

(a) The shape of the building square, easy to freely combine;

(b) Simple structure, through the modular design award, easy to prefabricated production;

(c) The number of households can be flexibly added and subtracted according to the needs, any combination, with greater flexibility and adaptability;

(d) The internal elevator, staircase, tube wells, kitchen, bathroom, etc. are standard modules to improve the modeling efficiency;

(e) There are 20 kinds of prefabricated family libraries;

(f) 13 kinds of embedded parts, one kind of integral sanitary ware, sleeve-type connection up and down of prefabricated components.

The Technical Scheme of the Project

BIM Prefabricated Component Library

Create different standard BIM prefab libraries in REVIT. Relying on the standard library BIM components, according to different needs of the prefabricated design. Standard BIM prefabricated components, both to meet the scale of the factory, automated processing, but also to meet the high-efficiency prefabricated site requirements (Fig. 3).

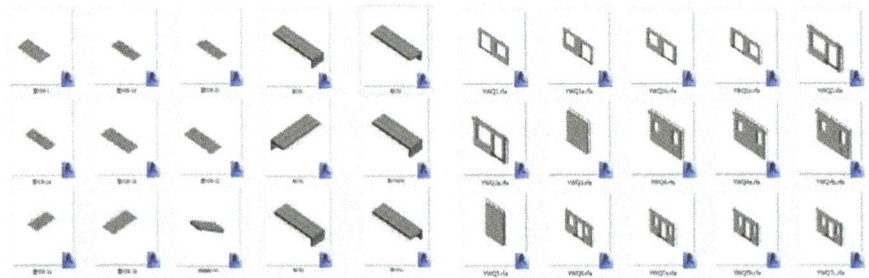

Fig. 3. Shandong Binhe new home public rental housing project. Prefabricated floor and wall panels.

Modular Design

The kitchen module, bathroom module, balcony module and apartment layout of the module to do different layout, in order to get the most reasonable layout plan (Fig. 4).

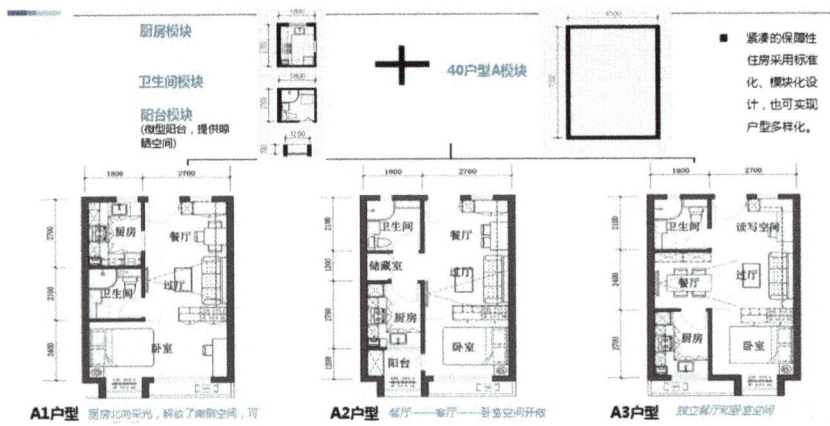

Fig. 4. Shandong Binhe new home public rental housing project. Module combination form.

Full Professional Collaborative Design Based on BIM
Through a unified BIM model, this model contains information such as architecture, structure, electromechanical, etc., and can achieve full professional collaborative design and optimization of fabricated modular buildings (Fig. 5).

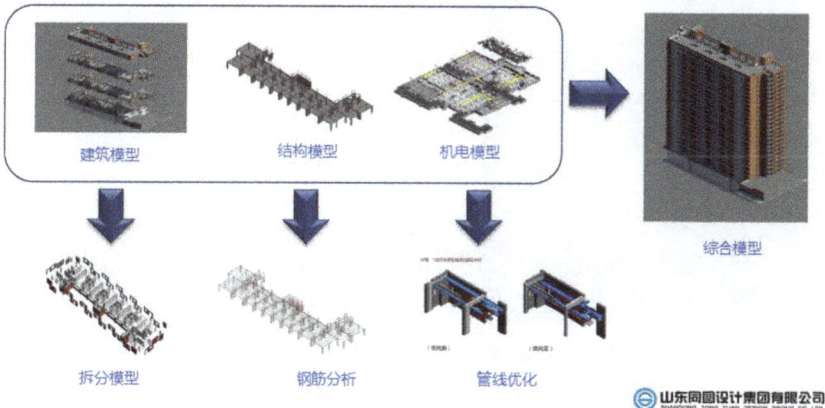

Fig. 5. Shandong Binhe new home public rental housing project. BIM model of the full professional information.

BIM Fine Design
The refinement of the component design minimizes the waste of reinforcing steel and enables the prefabricated components to be installed correctly at the construction site (Fig. 6).

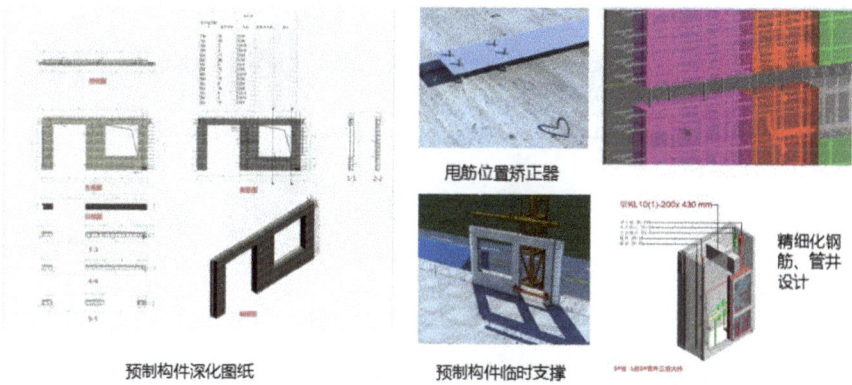

Fig. 6. Shandong Binhe new home public rental housing project. BIM's fine design.

BIM Project Statistics

Through the classification of statistics for the amount of engineering analysis to achieve the initial cost control (Fig. 7).

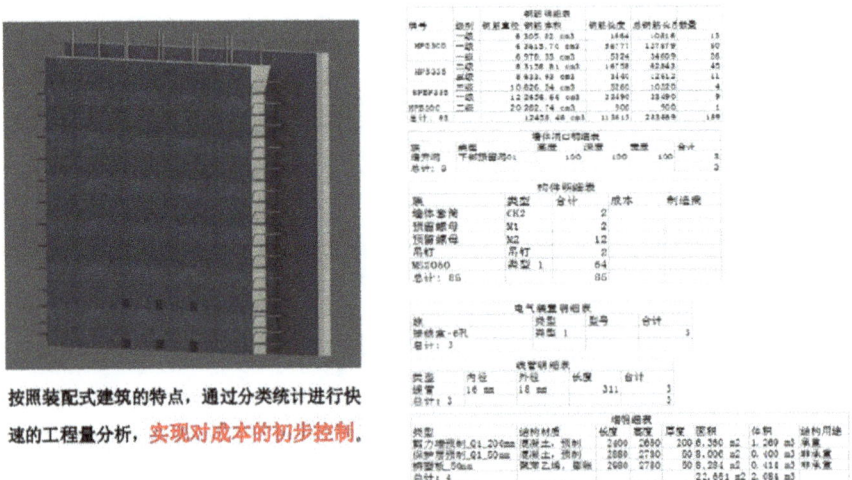

Fig. 7. Shandong Binhe new home public rental housing project. BIM project statistics.

BIM Technology Construction Simulation

Using **BIM** software to simulate the construction, eliminating hidden dangers in the field, optimizing the construction process and realizing efficient construction management (Fig. 8).

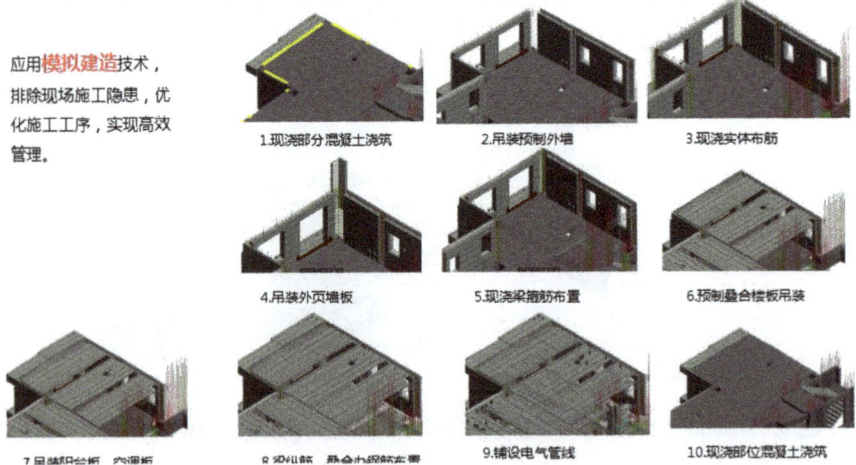

Fig. 8. Shandong Binhe new home public rental housing project. BIM technology construction simulation.

BIM Standardization of Technology and Conventional Design Comparison
By BIM standardized design, greatly reducing the types of prefabricated components. There are 10 types of BIM prefabricated façade components, while more than 30 prefabricated façade components are conventionally designed (Fig. 9).

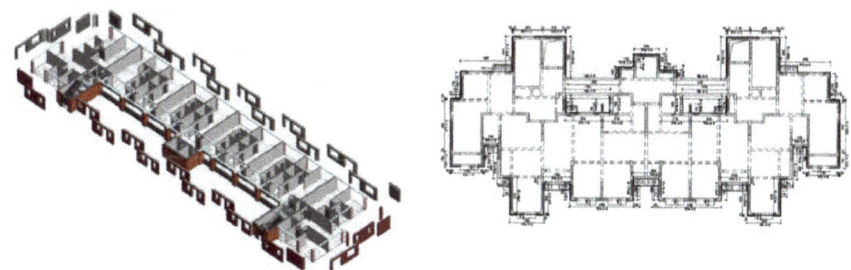

Fig. 9. Shandong Binhe new home public rental housing project. BIM standardization of technology and conventional design comparison

5   Conclusion

The presented example shows that the use of BIM (Building Information Modeling) is an effective tool for the prefabrication and modular buildings.

Through the actual project analysis, we can draw conclusions. Through the combination of BIM technology and modular building, BIM takes the standardization, industrialization and informationization as prerequisites. BIM technology optimizes all the details in advance through the entire life cycle of the industrialization of the building to ensure the post-manufacturing and construction of the assembled modular building of the accuracy, and can use BIM technology to simulate the construction of the site to guide the late management of all the information traceability.

Through BIM technology component visualization, collision inspection, design and construction of construction progress simulation, optimization of site layout, improve design efficiency, quality of drawings, test the progress of the construction of scientific and instructive construction workers have the important significance of the correct construction.

Looking to the future with the development of digital technology, green building technology, 3D printing and the Internet, 3D laser scanning and 3S technology lay the technical foundation for smart cities, they must also be equipped with wings for industrialization.

References

1. BIM 188com. http://bim.co188.com/info/d1091.html
2. BIM Definition: Frequently Asked Questions About the National BIM Standard-United States - National BIM Standard - United States. Nationalbimstandard.org. Archived from the original on 16 Oct 2014. Accessed 17 Oct 2014

3. Bonenberg, A.: Ergonomic experimentation in the architectural design process – types of test environments. In: Advances in Human Factors, Sustainable Urban Planning and Infrastructure, AHFE 2017. Advances in Intelligent Systems and Computing, vol 600, pp. 199–206. Springer International (2017)
4. Bonenberg, A., Zabłocki, M.: Residential architecture for health and longevity. Universal kitchen design. Space Form | Przestrzeń i FORMa 31 (2017)
5. Building information modelling. https://en.wikipedia.org/wiki/Building_information_modeling
6. Modular building. https://en.wikipedia.org/wiki/Modular_building
7. Prefabrication. https://en.wikipedia.org/wiki/Prefabrication

Reviewing the Negative Impacts of Building Construction Activities on the Environment: The Case of Congo

Mbuyamba Mbala[✉], Clinton Aigbavboa, and John Aliu

Sustainable Human Settlement and Construction Research Centre,
University of Johannesburg, Johannesburg, South Africa
jpmbala007@gmail.com

Abstract. The construction industry plays a significant role in the provision of physical infrastructure to meet the yearning needs of the society. It is well known that mankind has continually manipulated the natural environment in a bid to improve standard of living by the creation of amenities including buildings, highways, bridges amongst others. This paper investigates the major impacts of building construction activities on the environment in the Democratic of Congo, Kinshasa. Several possible impacts of construction activities on the environment were identified from extant literatures. Through a structured questionnaire survey, the views of respondents were elicited. The sample for this study consisted of one hundred and thirty-four (134) respondents drawn from professionals in the construction industry. The study revealed that dust generation from machinery, air pollution, land pollution, waste generation and noise pollution are among the major impacts of construction activities on the environment in Congo. This paper therefore recommends that the government alongside stakeholders in the industry are to come up with certain standards, codes or legislations relating to sustainable construction practices specific to Congo's construction environment to ensure its proper and effective implementation.

Keywords: Construction activities · Construction industry · Environment
Congo · Impact

1 Introduction

Environmental degradation as a result of construction activities has been one of the most discussed subjects locally, nationally and globally [1].

Infact, Langston and Ding [7] asserts that the world is currently in a crucial environmental catastrophe. This is a result of an increase in population and the need to develop the built environment, which has led to global warming, ecosystem destruction, ozone layer depletion, and resource depletion amongst others. This hazards and consequences have placed the construction industry and the built environment under a microscope, since its activities significantly influence the environment and its constituents.

© Springer International Publishing AG, part of Springer Nature 2019
J. Charytonowicz and C. Falcão (Eds.): AHFE 2018, AISC 788, pp. 111–117, 2019.
https://doi.org/10.1007/978-3-319-94199-8_11

The main objective of this study is to identify the major impacts of various construction activities on the environment in the Democratic of Congo. The study sought to identify the perceptions of professionals (architects, quantity surveyors, civil engineers, construction managers, project managers, and other professionals) regarding the impacts of construction activities on the environment in Congo and to suggest possible ways of minimizing the impacts.

2 Literature Review

The state of affairs of the construction industry in Congo is not necessarily different from other developing countries. The focus of the Congolese construction industry is mainly based on economic growth and improving the quality of life of the people whilst environmental protection is utterly downgraded. According to Uher [15], the construction industry has a significant impact on the environment as a result of its on-site and off-site activities. The consumption of raw materials by the industry increases on a daily basis which results in the depletion of natural resources as well as increased environmental impacts. Levin [8] state that structures (buildings) are considerably large contributors to environmental degradation. It is clear that certain actions are needed to ensure that construction activities and the built environment are more sustainable [5, 10, 13]. Therefore, this subject of the impact of construction activities on the built environment may need to be viewed from an extremely critical viewpoint [11].

As one of the largest exploiters of both renewable and non-renewable natural resources, the construction industry relies heavily on the natural environment for the supply of raw materials like timber, water, sand and aggregates for its activities [4, 15]. The extraction of these natural resources result in an irreversible change to the natural environment, both from an ecological and a scenic point of view [6]. The subsequent dissemination of these into various construction sites leads to consumption of energy as well as an increased amount of particulate matter into the atmosphere. Likewise, the extraction of raw materials also contributes to the accumulation of pollutants in the atmosphere [12]. Other harmful gases, such as chlorofluorocarbons (CFCs), which are used in insulation, air conditioning, refrigeration plants and fire-fighting systems have seriously depleted the ozone layer [6]. Similarly, these pollutants are also responsible for causing serious land and water contamination due to on-site negligence. This leads to degradation of land and interference with the environment's capacity to provide a naturally balanced ecosystem.

Furthermore, the industry produces a whole variety of different wastes. The types of waste depend on factors such as the stage of construction, type of construction work and construction practices on site. Building activities also transforms arable lands into physical structures including roads, buildings, dams and other projects [15]. Arable lands are also lost through mining and quarrying of raw materials to be used in construction. The construction industry also leads to the loss of forests through the timber used for building activities. Both deforestation and burning of fossil fuels contribute directly to air pollution and global warming. These processes contribute to the depletion of natural resources and produce atmospheric emissions. Chen et al. [3] and Cardoso [2], divides sources of pollution and hazards from construction activities divided into seven different types: harmful gases, solid and liquid wastes, dust, noises, fallen objects, ground movements and others. Furthermore, Shen and Tam [14] and Chen et al. [3] classified construction impacts under eight types: soil and ground contamination, production of substantial volume of waste, underground water contamination, noise and vibration, dust, hazardous emissions and odors, wildlife and natural features impacts and archaeology impacts.

3   Research Methodology

This study utilized a descriptive survey design because it was effective in providing both numeric and quantitative description of the respondents for the study. The target population were construction professionals in the Congo construction industry namely; architects, quantity surveyors, civil engineers, constructor managers, project managers, and other professionals. The selection of these construction professionals was made because they were more likely to have sufficient understanding to contribute to the aims of this study. The study adopted the random sampling technique because it gives all participants an equal opportunity of being selected for the study with the same criteria. Hence, a total of 134 respondents took part in this study. The instrument of data collection was a well-structured questionnaire, which was designed by the researcher following a review of extant literatures. The questionnaire was further validated by giving some to experts in the construction industry before adopting it for the study. Closed-ended questions was used the questionnaire design because they are less-time consuming and they provide participants with an option of multiple choices. The collected data was analysed using descriptive statistics.

3.1   Mean Item Score

This study adopted the mean item score (MIS) to analyse collected data. The mean ranking of each item was presented to provide a clearer picture of the agreement reached by the respondents. Following the mathematical computations, the criteria were then ranked in descending order of their relative importance.

4   Findings and Discussions

4.1   Background Information About Participants

The participants were requested to indicate the extent of the degree of impact of construction activities on the environment based on a four point Likert scale (No extent = 1, small = 2, moderate, = 3, Large = 4). One hundred and thirty-four complete questionnaires were received signifying a 74.4% response rate. Findings from the 134 usable questionnaires revealed that 70. 7% of the respondents were male and 29. 3% were female.

Also, 7.6% of the respondents were in the age group of 21–25 years old, 15.2% of the respondents were in the group age of 26–30 years old, 27.3% of the respondents were in the group age of 31–35 years old, 17.4% of the respondents were in the group age of 36–40 years old, 6.8% of the respondents were in the group age of 41–45 years old, 6.1% of the respondents were in the age group of 46–50 years old,12.9% of the respondents were in the age group of 51–55 years old and 6.8% of the respondents were in the age group of 56 and above.

It was also revealed that 1.6% of the total respondents had no work experience, 5.5% had a year experience while 14.2% had two years of construction experience. It also revealed that 33.9% had experience that ranged from three to five years, 17.4% had experience that ranged from six to eight years, 9.4% had experience that ranged from nine to eleven years, and 18% of the respondents had experience of eleven years and above. The years of experience of respondents were deemed sufficient to provide useful responses to achieve the purpose of the study.

4.2   Mean Item Score for Ranking of the Various Negative Impacts of Construction Activities on the Environment

A summary of the test result is shown in the table below. The mean for each variable included the standard deviation.

Table 1. Result of mean item score

Negative impact	Mean	Standard deviation	Rank
Dust generation	2.96	0.559	1
Air pollution	2.96	0.559	1
Land pollution	2.95	0.584	2
Effect on biodiversity	2.94	0.557	3
Waste generation	2.93	0.661	4
Noise pollution	2.93	0.646	4
Soil erosion	2.89	0.640	5
Spread of undesirable disease	2.84	0.697	6
Water pollution	2.84	0.668	6
Raw material consumption	2.82	0.653	7
Landscape alteration	2.78	0.606	8
Natural disaster	2.76	0.707	9
Radiation exposure	2.73	0.633	10
Chemical pollution	2.71	0.620	11

Source: Author's work

Table 1 reveals the respondents' rankings of the various negative impacts of construction activities on the environment. The table shows that, with a mean score (M) of 2.96 and standard deviation of (SD) = 0.559, 'Dust generation from machinery' and 'Air pollution' were jointly ranked as the most detrimental impact of construction activities on the built environment. 'Land pollution' was ranked second with (M = 2.95;

SD = 0.584); 'Effects on biodiversity' was ranked third with (M = 2.94; SD = 0.557); 'Waste generation' and 'Noise pollution' was jointly ranked fourth with (M = 2.93; SD = 0.661) and (M = 2.93; SD = 0.646) respectively; 'Soil erosion' was ranked fifth with (M = 2.89; SD = 0.640); while 'Spread of undesirable diseases' and 'Water pollution' were jointly ranked sixth with (M = 2.84; SD = 0.697) and (M = 2.84; SD = 0.668) respectively. Also, 'Disturbance of the temperature' was ranked seventh with (M = 2.82; SD = 0.653), 'Landscape alteration' was ranked eighth with (M = 2.78; SD = 0.609), 'Natural disaster' was ranked ninth with (M = 2.76; SD = 0.707), 'Radiation exposure' was ranked tenth with (M = 2.73; SD = 0.633) and 'Chemical pollution' was ranked the least with (M = 2.71; SD = 0.620).

5 Summary and Implications of Finding

From the empirical study, it was revealed that the respondents admitted that raw materials (timber, water, sand and aggregates) consumption is one of the most significant environmental impact. Developing the built environment also result in vegetation removal, loss of edaphic soil as well as interference with the ecosystem which all falls under the 'effects on biodiversity'. This was corroborated by the various construction professionals. Also ranked highly were noise and vibration generation; this may be due to personal experience of respondents who face this situation in their day to day activities. There is also sufficient evidence to back the claims that various construction activities are responsible for the generation of dust, vibration and noise. Furthermore, most of the respondents ranked highly the effects of atmospheric emissions, including volatile compounds (VOCs), chlorofluorocarbons (CFOs), nitrogen and Sulphur oxides. They are predominantly released during the production and transportation of construction materials and have caused serious threat to the natural environment. Other harmful emissions like chlorofluorocarbons are used in air conditioning, insulation, refrigeration and fire-fighting systems and have led to the increased depletion of the ozone layer.

6 Lessons Learnt and Conclusion

The construction industry plays a significant role in the provision of physical infrastructure to meet the increasing needs of the society. On the other hand, our built environment and its interactions with natural surroundings are complex and have a significant impact on the world around us. This study was aimed at investigating the negative impacts of construction activities on the environment in Congo. The study sought the views of architects, quantity surveyors, civil engineers, constructor managers, construction project managers and project managers. The study showed that out of the various environmental impacts identified, the top ten most important environmental impacts factors agreed by all the respondents are as follows: Dust generation from machinery, Air pollution, Land pollution, Effects on biodiversity, Waste generation, Noise pollution, Soil erosion, Spread of undesirable diseases, Water pollution and Raw material consumption.

In conclusion, the Congolese government need to intervene such that construction strategies and sustainable construction designs becomes the norm in the country. This paper therefore recommends that the government alongside the stakeholders in the construction industry come up with certain standards, codes or legislations relating to sustainable construction practices specific to Congo's construction environment to ensure its proper and effective implementation. Also, there is the need for a periodic review of the national building regulations to take account of environmental regulations. Furthermore, the government can enforce environmental laws such as mandatory Environmental Impact Assessment (EIA) and stress environmental monitoring and compliance. The communities should also be briefed on the impending dangers of environmental dilapidation which affects, not only human beings but also the ecosystem. The rules and regulations that govern urban construction should also be adhered to strictly by communities residing within the urban settings. This can be emphasized by the relevant urban authorities including the environmental officers, physical planners and estate agents. These steps are necessary because construction activities are increasingly threatening the built environment, hence the need for actions to be taken by the government, industry professionals and legislators.

References

1. Bentivegna, V., Curwell, S., Deakin, M., Lombardi, P., Mitchell, G., Nijkamp, P.: A vision and methodology for integrated sustainable urban development: BEQUEST. Build. Res. Inf. **30**(2), 83–94 (2002)
2. Cardoso, J.M.: Construction site environmental impact in civil engineering education. Eur. J. Eng. Educ. **30**(1), 51–58 (2005)
3. Chen, Z., Li, H., Wong, C.T.C.: Environmental planning: analytic network process model for environmentally conscious construction planning. J. Constr. Eng. Manag. **131**(1), 92–101 (2005)
4. Curwell, S., Cooper, I.: The implications of urban sustainability. Build. Res. Inf. **26**(1), 17–28 (1998)
5. Holmes, J., Hudson, G.: An evaluation of the objectives of the BREEAM scheme for offices: a local case study. In: Proceedings of Cutting Edge 2000. RICS Research Foundation, RICS, London (2000)
6. Langford, D.A., Zhang, X.Q., Maver, T., MacLeod, I., Dimitrijeic, B.: Design and managing for sustainable buildings in the UK. In: Ogunlana, S.O. (ed.) Profitable Partnering in Construction Procurement, CIB W92 (Procurement Systems) and CIB23 (Culture in Construction), Joint Symposium, pp. 373–382. E & FN Spon, London (1999)
7. Langston, C., Ding, G.K.C.: Sustainable Practices in the Built Environment, 2nd edn. Butterworth Heinemann, Oxford (2001)
8. Levin, H.: Systematic evaluation and assessment of building environmental performance (SEABEP). In: Proceedings of Second International Conference, Building and the Environment, June, Paris, pp. 3–10 (1997)
9. Lenzen, M., Treloar, G.J.: Embodied energy in buildings: wood versus concrete-reply to Börjesson and Gustavsson. Energy Policy **30**, 249–255 (2002)
10. Morel, J.C., Mesbah, A., Oggero, M., Walker, P.: Building houses with local materials: means to drastically reduce the environmental impact of construction. Build. Environ. **36**(10), 1119–1126 (2001)

11. Ofori, G., Briffett, C., Gang, G., Ranasinghe, M.: Impact of ISO 14000 on construction enterprises in Singapore. Constr. Manag. Econ. **18**, 935–947 (2000)
12. Rohracher, H.: Managing the technological transition to sustainable construction of buildings: a sociotechnical perspective. Technol. Anal. Strateg. Manag. **13**(1), 137–150 (2001)
13. Scheuer, C., Keoleian, G.A., Reppe, P.: Life cycle energy and environmental performance of a new university building: modelling challenges and design implications. Energy Build. **35**, 1049–1064 (2003)
14. Shen, L.Y., Tam, V.W.Y.: Implementation of environmental management in the Hong Kong construction industry. Int. J. Project Manag. **20**(7), 535–543 (2002)
15. Uher, T.E.: Absolute indicator of sustainable construction. In: Proceedings of COBRA 1999, pp. 243–253. RICS Research Foundation, RICS, London (1999)

Architectural Design in the Context of Sustainable Development

Beata Majerska-Palubicka[✉]

Department of Housing and Public Architecture Design, Faculty of Architecture,
Silesian University of Technology, ul. Akademicka 7, 44-100 Gliwice, Poland
beata.majerska-palubicka@polsl.pl

Abstract. Currently, the sustainable development paradigm is the component of architecture, which in the future will certainly affect urban planning and architecture to a much greater extent. On the one hand, an issue of great significance is the need to integrate sustainable system elements with the spatial structure of environmentally friendly architectural facilities and to determine their influence on design solutions. Another problem to be solved is how to design buildings, housing estates and towns so that their impact on the environment will be acceptable, i.e. will not exceed the possibilities of natural environment regeneration, which is the basis of sustainable development. In this broad interdisciplinary context an increasing importance is being attached to design strategies. The above topics are the subjects of the research presented in this paper.

Keywords: Sustainable development · Sustainable design in architecture
Integrated Design Process (IDP)

1 Introduction

Architectural design is a creative process leading to the creation of an original solution. A result of the above-named process depends on the designer's intellect, their knowledge, experience, skills and views.

In the design embedded in the context of sustainable development, where architectural, building, technical, social, cultural, ecological and economic aspects are equally important, it is necessary to approach design-related tasks in a multi-dimensional, nearly interdisciplinary manner. Architects are primary decision-makers when it comes to: the setting of a newly-designed object in the context, land development, spatial-functional, technological and structural-material solutions, aesthetic aspects, environmental and ecological effects, effects related to the preservation or destruction of natural resources and the emission of greenhouse gases, social effects in the cultural, historic, comfort and safety-related context as well as economic aspects. In view of the foregoing, having such a large number of diversified yet interdependent process components, it is difficult to define explicitly the final solution. The process itself should be aided by developed accelerating systems and tools that could facilitate analytical, decision-making and concept-creating processes, as well as, first and foremost, communication among members of design teams.

© Springer International Publishing AG, part of Springer Nature 2019
J. Charytonowicz and C. Falcão (Eds.): AHFE 2018, AISC 788, pp. 118–127, 2019.
https://doi.org/10.1007/978-3-319-94199-8_12

2   In Search of Sustainability in Design Strategies

The sustainable development paradigm is present in all spheres of life, including architectural design, where it involves the search for architectural solutions and concepts enabling the development and reinforcement of the present and future potential allowing for needs of present and future generations. Interviews made in 2001 by Brian Roberts (the editor of the Architectural Design) with some contemporary architects concerning the definition of principles governing sustainable architectural design revealed that many well-known creators approached the issue very seriously. According to Norman Foster, sustainable design signifies achieving 'the most' at the lowest expense. In turn, Richard Rogers, when searching for sustainability in design processes, reaches out for the definition of sustainable development and emphasizes the close correlation between these two notions. He claims that sustainable design aims to satisfy today's needs in a manner enabling future generation to make use of natural resources [1]. A holistic approach to social, environmental, spatial and technical problems is of great significance also in terms of sustainable design strategies. Equally important is the aesthetic overtone, reference to the form of architectural expression and beauty, i.e. elements indispensable for architecture. Thomas Herzog believes that successful sustainable design depends on functional values, which can be summarised and defined as sustainable, however, beauty is of equal significance as utility. Beautiful buildings enrich our surroundings. The introduction of technologies using renewable energy provides opportunities of creating new forms of architectural expression, closely tied to local conditions including microclimate, topography, natural resources and the cultural heritage of a given region.

As regards sustainable development, the optimum solution is the one where the building sector remains neutral in relation to the natural environment. However, the aforementioned situation is purely hypothetical as it is impossible to eliminate the impact of construction works, operation, adaptation, recycling of buildings and waste disposal on the natural environment. The life cycle of a residential building in Europe is estimated at approximately one hundred years. The foregoing depicts the scale of environmental hazard and showcases the need for undertaking all necessary activities to identify the permissible impact level. This, in turn, requires the specification of the manner and range of effects as well as the determination of their limits or boundary values [2].

For many years, environmental impact surveys has been conducted and developed in order to facilitate the design and operation of technological and industrial buildings which pose an evident hazard to the environment. Because of the fact that technological processes taking place in the aforementioned facilities are stable, predictable and well-defined, the development of environmental impact surveys has not posed any major difficulty. In turn, public utility buildings, office buildings and, in particular, residential buildings, with their users being an important factor, are technologically less complex, yet their environmental impact depends on a considerable number of unknown variables, which makes the related assessment far more complicated.

Developed systems of multi-criteria assessment and certification constitute tools supporting processes leading to the obtainment of increasingly good and sustainable architectural, technical and building engineering solutions. The above-named systems are used to find solutions and create buildings which could be environmentally efficient, more affordable in use, more durable and characterised by high quality, in other words, to obtain a superior end product in the form of a sustainable building. Such systems help their users plan, construct and manage sustainable buildings. Assessment systems and tools enable a multi-criteria building-related evaluation in the context of environmental (effect on the natural environment), economic (reduction of costs) and social (improvement of living standards and possibility of participation) issues, which are the basis of sustainability in the building sector. This makes it possible to develop short and long-term policies requiring thorough analysis as regards the number of criteria taken into account when performing analysis (quantitative aspect) and in relation to the objective, i.e. sustainable development (qualitative aspect).

The significance of systems enabling the complex assessment and certification of buildings is also essential in the world where the term of 'environmental friendliness' is abused in many aspects of life and economy as a marketing trick. It is necessary to convincingly indicate and promote a product (building) truly satisfying sustainability-related requirements.

It should also be noted that sustainable design should not be limited to the scale of a building but, extending beyond that, refer to a larger and more important scale involving urban development. Preferable features of sustainable designs include energy saving, renewable energy sources, adaptability, multi-functionality, 'alliance with nature' as well as the use of local traditions and technologies.

Not all architects accept such architecture. In many cases the architectural profession is more interested in architecture dominating 'time and space' rather than architecture acting as the background 'in the context of a space (a place)'. Sustainable architecture, drawing primarily on tradition and a widely defined local context, remains indifferent to abstract trends and fashions.

3   Design Process Taking into Consideration the Sustainable Development Paradigm - Integrated Design Process (IDP)

In accordance with the sustainable development paradigm, which gives hope for the reduction of the negative impact on the surroundings/environment and the improvement of living standards, architectural and building activities cannot be considered in isolation from the natural environment. The obtainment of harmony between the built environment, elements of nature, social needs and economy requires the revaluation and optimisation of needs, views, solutions, dissemination of knowledge, imperatives and achieved results, which inevitably leads to the determination of new procedures in urban and architectural design.

As can be seen from performed analyses [3], the cost efficiency related to interventions, solutions and design-related decisions is the highest if these issues are approached at an early stage of design. This means that within the strategy of sustainable development, the earliest design and concept stages should include decisions resulting from the analysis of numerous criteria concerned with the effect of a designed object on the built environment, the natural environment and human beings. The decisions should be characterised by a holistic approach expressed by the integration of environmental, social, cultural, ethical, ideological, normative, functional, spatial, technical and economic issues. As a result, entirely new elements of the design process appear and, at the same time, this leads to the transformation of classical elements which previously constituted the design process and now are confronted with new requirements.

The ever-increasing diversity of criteria and multi-directionality of activities necessary within the sustainable design require a system approach based on an acceptable hierarchy of values [4]. Indeed, it is favourable where undertaken activities are based on the methodology of the general systems theory [5], consisting in the implementation of the holistic approach to the design strategy and involving the treatment of a design as multidimensional and defined by a significant number of determinants. The aforesaid approach enables the obtainment of the synergistic effect through the preservation of the cohesion of all components, assumed objectives as well as the structure and function of a designed object.

In the context of a significant number of elements requiring analysis there is a concept concerned with the graded selection of design solutions based on ecological, economic, social and spatial criteria [6]. According to this concept, ecological criteria, consisting in the elimination of the negative impact on the natural environment, should be based on the use of optimum economic and cultural solutions (in given circumstances); economic criteria should take into account an increase in capital expenditure in comparison with conventional solutions and the return of additionally incurred costs over a specific period of time; social criteria, related to the dissemination of knowledge and education, should aim to encourage the use of new technologies, whereas spatial criteria should prefer renovation, modernisation and regeneration of existing spatial structures and technological systems. All solutions should be subjected to optimisation.

The design process, becoming composed of increasingly many elements and tasks as well as being a wide spectrum of activities and correlations, requires expertise in many areas, including architectural, building engineering, technological, social, environmental, cultural, natural, ecological, power engineering, economic issues, etc. Inter-professional collaboration previously used in the traditional design is becoming insufficient and fails to provide the appropriate integration of teams and the optimum result in the form of an effective and sustainable object. The concept of Integrated Design Process (IDP), also referred to as Programme C-200, was developed in the 1990s in Canada as an antidote addressing all of the above-presented issues. It was also developed in the USA in the form of studies by the Rocky Mountain Institute and in Europe as a research programme Annex 23 developed by the International Energy Agency [7].

Initially, innovative principles of collaboration, viewed as undermining well-established relations between design team members, were not welcomed by designers. In addition, the aforesaid principles were complicated, hence considered as unclear. Presently, the IDP is becoming a platform for close collaboration within a wide range of participants in a process aimed to obtain the built environment constituting the minimum possible burden to the natural environment.

The appropriate optimisation of design processes performed using, among other things, multi-criteria methods of assessment and certification including the Life Cycle Assessment (LCA) analysis, cannot be conducted without the overall understanding of advantages, losses, reasons and results [8] and, primarily, without understanding the reason for their generation. The understanding of correlations between various conditions of the internal and external environments at various stages of the design process facilitates the perception of how the architectural form is determined by the structure and function and, vice versa, how material characteristics determine the detail and affect perception, how the cycle of building life affects durability and sustainability etc. Samuel Mockbee from Auburn University [9] was one of the first designers to develop the principle of sustainable design based on the following:

- understanding a place as the basis of sustainable design, interpreted in the context of insolation, topography, the preservation of the natural and build environment, the flow of energy and resources preceding and following the construction of the object, thus the creation of patterns integrating the place with its surroundings;
- understanding nature by becoming aware of a position occupied in the natural environment;
- understanding the environmental impact in the context of seeking balance between the destructive effects of the building sector on the environment as well as activities neutralising the aforementioned negative effects;
- understanding people in the context of widely defined cultural heritage.

Social participation is another important criterion of the IDP. The build environment is created by people and their participation in the sustainable design, particularly in everyday life and changes of artefacts. Consequently, the condition of the natural environment depends on society's behaviour.

The integrated design in relation to methods of the assessment and the certification of buildings adopted as a tool supporting the design process should consist in the selection of criteria. In terms of the IDP, it is assumed that a set of criteria should be adjusted so that objectives could be achieved on the 'top down' basis (the term used in reference publications), i.e. involving the selection of systems and criteria of evaluation adjusted to the primary goal of the assessment. It is important to clearly and unequivocally specify and select design targets at the initial stage of the process.

Architectural and town-planning solutions (location, land development, transport-related, functional and spatial, structural, material, technological and aesthetic requirements) should be subordinated to the requirements of the sustainable environment, analysis of life cycle assessment (LCA), life cycle cost (LCC) economic analysis and requirements of other certification/assessment systems. All of these elements, following the principle of 'strong sustainability', must remain in the state of sustainability or be compensated in accordance with the principle of 'weak sustainability', assuming that some cannot be solved at the cost of others [10]. The objectives include the obtainment of high efficiency, or even energetic self-sufficiency of the designed built environment and a zero ecological footprint.

The quality of the end product, i.e. the built environment, depends on the quality of each stage of the design process. The end product quality is the resultant of experience, professionalism, knowledge, awareness, determination as well as communication and cooperation skills of design team members. Knowing the current state of the art and experience, it seems impossible to obtain the sustainable built environment without the Integrated Design Process.

4 Differences Between the Traditional Process and Integrated Design Process (IDP)

Each design decision results from existing circumstances and context, yet it also affects the context and other decisions. Each change triggers subsequent changes. In fact, a building is a system of combined solutions creating variable conditions sufficient for needs, situations and a context. It should also be noted that the context conditions, environmental conditions, with which a designed object enters into a relationship and correlation, also constitute a set of variable parameters. Because of this, it is extremely difficult, if not impossible, to create unified design principles and standards suitable for each situation and location on a global scale. Nevertheless, it cannot be excluded on a local scale, which is characterised by less diversified criteria and more predictable behaviours and reactions. This was already demonstrated by attempts to standardise methods of the assessment and certification of buildings.

In Poland, in spite of the increasingly high complexity of issues, in most cases the design process continues to follow a traditional path and is based on the fragmentary solving of consecutive design problems at analytical, conceptual, design and building stages. From the beginning to the end, the process is of a rectilinear nature and is perceived as irreversible. In turn, in the context of sustainable development, greater complexity of the process and a larger number of criteria require repetitions and returning to previous stages, therefore the linear process often fails to produce satisfactory results and becomes insufficient. The development of IT and the evaluation of the manner of communication provide innovative design tools enabling the replacement of the traditional linear process with a more integrated one. According to Richard Foquè, representing a new approach to the design process in the context of Life Cycle Assessment (LCA), successive design stages should be defined as various levels of the set of information in relation to consecutive integrated levels of the design process, i.e. programming, concept, design, building and use/operation [11].

The concept of Integrated Design Process (IDP) is based on assumptions, the characteristic of which is not to close decisions made at consecutive stages of the process until the moment, when an optimum result, in relation to a given design solution, has been obtained as a result of interdisciplinary simulations, analyses, comparisons and iterations. According to M. Hegger, the integration of interdisciplinary approaches within the IDP should be developed multi-directionally (horizontally and vertically), unlike the unidirectional linear traditional process, where decisions are taken and closed subsequently one after another. The horizontal direction should constitute a level of interdisciplinary programming and designing, whereas the vertical direction should apply to the duration of successive stages of the life cycle of a building, i.e. use, modernisation and recycling, respectively [12].

By contrast with the conventional process, the integrated design process, in addition to primary criteria, including location, functional and spatial, structural and material as well as installation-related solutions, also takes into consideration pro-ecological aspects, the socio-cultural context as well as energetic and economic improvements and involves the optimisation of solutions through selecting a solution being the most favourable and feasible in relation to a given space, situation and needs.

Such an approach requires a different sequence of design tasks, different activities of designers at successive stages of the process, changes in the duration and complexity of consecutive stages, and, consequently, the significant expansion and changes within the composition of a design team.

Methods of collaboration involving a large number of specialists facing a greater number of more complicated elements to be analysed and solved require a change in designers' habits and an escape from routine. R. Pelli compares the existing situation to the need for the 'redesign of design processes' [13].

According to the author of this paper, the proper IDP and the obtainment of an intended sustainable objective requires that the design team recognise the following facts:

- initial stage of the project (a programme concept) is the most effective stage as regards interventions, changes and optimisation,
- from the very beginning a building needs to be designed in accordance with the assumptions of sustainable development in the context of Life Cycle Assessment (LCA) and has to satisfy ecological, social, cultural, spatial, environmental and economic (Life Cycle Costs, LCC) criteria;
- each stage of design work should have clearly and unequivocally specified visions, targets and tasks, leading to the obtainment of the primary goal assumed at the beginning of the design process.

Typically, in Poland the design process is composed of four stages including the performance of preliminary analyses, the conceptual design, the building design (being the basis for the obtainment of a planning permission) and the detailed design, usually followed by, but often accompanied by, the process of construction, finishing with the acceptance and the approval of an object for use. Usually, the collaboration of the design team finishes with the acceptance and the approval of a building for use, but sometimes as early as after receiving a planning permission. The traditional design process includes the stage of preliminary analyses, yet the latter bears little significance and is often performed superficially. Information gathered is not subjected to in-depth analyses, hence the lack of conclusions which should constitute the basis for further actions.

Related studies revealed that particularly the first stage, i.e. the stage of programming, involving the identification of the objective to be obtained and principal assumptions related to the potential of a site (location in the environmental, social, cultural, ecological and economic context) is essential. The above-named objective concerns, among other things, functional properties, durability, energetic parameters as well as costs, and is of great, nearly principal, significance as regards the optimisation of solutions. For this reason, in the IDP the stage of programming is treated as the initial point and the basis enabling the determination of the site potential in many aspects. Preliminary studies and analyses lead to conclusions serving as a material facilitating the decision-making process at later stages. Therefore, this early stage requires strict collaboration of experts representing many various areas and cooperating within the so-called 'charrette' workshops, during which work takes the form of interdisciplinary design workshops and is supported by innovative design tools facilitating the process of communication and decision-making.

The IDP seeks to create a situation where applied solutions are sufficiently flexible so that they can be adapted to specific conditions related to the context of the natural and built environment, community and economy, as is the case with methods of assessment and certification. Solutions should be easy to use and effective as regards the final outcome. In view of the foregoing, the IDP has been provided with an entirely new element, i.e. Post Occupancy Evaluation (POE), providing the possibility of monitoring (during use) the aptness and efficiency of solutions applied at the design stage. The principal of the POE is to provide the design team with feedback enabling the verification of experience, the acquisition of new knowledge and the improvement of competence [14].

In Poland, the satisfaction of legislative requirements, the achievement of conformity with the local development plan, the providing of utilities, the obtainment of conformity with Construction Law regulations and technical conditions to be satisfied by buildings and their location as well as with other sectoral orders remain tasks which must be performed both in the traditional and integrated processes. However, in each of the aforesaid processes the above-named activities take place at different stages.

5 Summary

Interdisciplinary analyses including all of the stages related to the creation and the life of building structures (planning, design, erection, fitting, use and operation, allowing for functional flexibility, adaptability, demolition and recycling) constitute the basis of the Integrated Design Process (IDP). Interactions between the natural and the built environment, resulting from architectural interventions, noticed sufficiently early (at the programming or conceptual design stage), should stimulate innovative and unconventional solutions as regards the shaping of the built environment in the social, economic and ecological context. Because of the foregoing, the IDP should be characterised by continuity and dynamism, involve multi-criteria comparisons and simulations aimed to optimise solutions and not finish when the building is put into operation. A novelty in principles governing the collaboration and the strategy of design process is the common work of all design team members from the stage of programming, an early concept, through the multi-stage verification of assumed parameters to the obtainment of the final version of a design, construction and the possibility of verifying adopted solutions when monitoring the building during its operation. Related analyses revealed that the traditional elements of the design process do not cease to exist in the IDP but are treated in a different manner, are not closed at successive stages of the process and are subjected to multiple and multi-criteria simulation, analysis optimisation and improvement.

The integrated approach to the design process expands the area of studies and decisions, entails the analysis of a significant number of criteria concerning the interaction of the built environment, the natural environment and the human being and, consequently, requires:

- change in mentality from the 'transformation of nature' to the 'transformation of the society', leading to better correlations between the built environment and the natural environment, aimed to achieve higher living standards;
- revaluation and the readjustment of needs and assumptions to the economic and social context;
- ordering of ecological, economic, social and spatial criteria by, e.g. a system approach based on an acceptable hierarchy of values;
- finding means of expression – an architectural concept taking into consideration the effective use of resources and land as well as directions of the technological and institutional development strengthening the present and future potential, allowing for needs of present and future generations and related to spatial order, aesthetic qualities, form, comfortable use and function, i.e. elements indispensable for architecture;
- development of strategies, methods and legislation enabling the obtainment of assumed design solutions.

As a result, elements constituting the design process must satisfy new requirements. In addition, new process-supporting elements, e.g. Life Cycle Assessment, LCA), systems of the multi-criteria assessment and certification of buildings (BREEAM, LEED, DGNB, HQE, SBTool etc.), methods of material certification etc. are put in place. Unlike the conventional design process., the IDP is characterised by dynamism, in the context of multi-criteria simulation and iteration aimed to optimise solutions and by continuity, in the context of monitoring and maintenance following the commissioning of an object.

Obviously, the primary criterion used when assessing the built environment can no longer be solely based on aesthetics, an attractive design, functionality or structural stability or even energy-related aspects. It is necessary to introduce the criterion of sustainability. On the one hand, changes in the design process are driven by notable ecological, social and economic needs, whereas, on the other, they are necessitated by the improvement in communication and collaboration efficiency of design teams confronted with a significant number of criteria to be analysed and a large number of process participants to be reckoned with.

Presently, in order to change existing design strategies, Poland needs to develop and adapt methods, strategies and multi-criteria simulation programmes used in other countries. Decisions, which significantly affect the design, should be analysed and made by a wide range of specialists as early as at the programming and conceptual stages. Sadly, such decisions continue to be made excessively late, thus extending the process and generating additional costs.

References

1. Edwards, B. (ed.): Green Architecture. Architectural Design, vol. 71, no. 4. Wiley, London (2001)
2. Schneider-Skalska, G.: SUSPURPOL Projektowanie zrównoważone. Środowisko mieszkaniowe – Housing Environment, nr. 4/2006. KKŚM, Kraków (2006)

3. Majerska-Pałubicka, B.: Zintegrowane projektowanie architektoniczne w kontekście zrównoważonego rozwoju. Doskonalenie procesu. Wyd. Pol. Śl., Gliwice (2014)
4. Baranowski, A.: Projektowanie zrównoważone w architekturze, s. 151. Wyd. Pol. Gdańskiej, Gdańsk (1998)
5. Stabryła, A.: Generalne formuły postępowania badawczego w procesie projektowania. In: Zeszyty Naukowe Małopolskiej Wyższej Szkoły Ekonomicznej w Tarnowie, Tarnów, vol. 20, no. 1, s. 171 (2012)
6. Baranowski, A.: Projektowanie zrównoważone w architekturze, s. 110–112. Wyd. Pol. Gdańskiej, Gdańsk (1998)
7. Kujawski, W.: Zintegrowany Proces Projektowy, czyli jak możemy projektować lepiej. W: Zawód: Architekt, #19, 1_11, s. 65–66 (2011)
8. Zimmerman, A.: Integrated Design Process Guide. CMHC, Ottawa (2008)
9. Mikoś-Rytel W.: O zrównoważonej architekturze ekologicznej i zarysie jej teorii. Zeszyty Naukowe Politechniki Śląskiej. seria Architektura z. 41, Gliwice, no. 1602, p. 75 (2004)
10. Kronenberg, J., Bergier, T. (red.): Wyzwania zrównoważonego rozwoju w Polsce, p. 71. Fundacja Sendzimira, Kraków (2001)
11. Foqué, R.: Building Knowledge in Architecture, pp. 126–127. University Press Antwerp UPA, Brussels (2010)
12. Hegger, M., Fuchs, M., Stark, T., Zeumer, M.: Energy Manual, Sustainable Architecture, p. 187. Birkhauser, Basel (2008)
13. Snoonian, D., Gould, K.L.: Architecture rediscovers being green. Archit. Rec. (06), 94 (2001)
14. Niezabitowska, E. (red): Potrzeby użytkownika a standard budynku inteligentnego. Budynek inteligentny, vol. I. Wyd. Politechniki Śląskiej, Gliwice (2005)

Sustainable Urban Planning
and Infrastructure

Shaping the Space for Persons with Autism Spectrum Disorder

Maria Bielak-Zasadzka[✉] and Agnieszka Bugno-Janik

Faculty of Architecture, Silesian University of Technology, Gliwice, Poland
{maria.bielak-zasadzka,agnieszka.bugno-janik}@polsl.pl

Abstract. In the last twenty years the significant growth in the number of people suffering from disorders which belong to the Autism Spectrum can be observed. In this article we present basic theoretical assumptions in tandem with the current state of research into the issues related to the shaping of the space intended for persons suffering from ASD. The recapitulation of the following analysis will be rendered in the form of innovative examples showing how space for people with ASD can be created. These examples are depicted in master theses by students of the Architectural Faculty, at the Silesian University of Technology, Department of Design and Qualitative Research in Architecture.

Keywords: Architectural design · ASD autistic spectrum disorder
Sensory sensitive approach · The neurotypical approach
Evidence-based design

1 Essential Features of Autism

In recent years more and more actions has been taken in order to improve the state of knowledge concerning design of space for people with the Autism Spectrum Disorder. Creation of an environment where people suffering from ASD would feel safe and comfortable constitutes a complex issue. When creating such environments, one should take into consideration the wide scope of knowledge coming from various fields, such as environmental psychology, psychotherapy, medicine, architecture as well as physical phenomena.

Autism is a highly complex and multi-lateral medical and psychological term. It refers to a disorder of our body's functioning at neurodevelopmental level and is often referred to as *ASD or autism spectrum disorder*. The term "spectrum" refers to a wide range of autistic disorders that could appear.

The creator of that term was Leo Kanner, who many see as the creator of children's psychiatry. In 1943 he described the disorders found in a group of 11 children. The term "autism" refers to a Greek word *autos,* which stands for "by oneself". The children who were examined by Kanner showed a tendency toward social isolation.

Today, autism is diagnosed on the basis of the concurrent presence of three main groups of disorders [1]:

- impaired social interactions;
- problems with language and communication;
- repetitive behaviors with a lack of imagination.

© Springer International Publishing AG, part of Springer Nature 2019
J. Charytonowicz and C. Falcão (Eds.): AHFE 2018, AISC 788, pp. 131–139, 2019.
https://doi.org/10.1007/978-3-319-94199-8_13

This constitutes the so-called autistic triad. It encompasses a highly diverse scale of difficulty as well as levels of functioning. Moreover, in autistic people also other symptoms with a varying level of intensity can be found. As a result of that and after observing that fact in 1988, the notion of ASD - *autistic spectrum disorder* was introduced. That notion includes [1]:

- childhood autism,
- Asperger syndrome,
- Rett syndrome,
- atypical autism,
- childhood disintegrative disorder,
- and other overall developmental disorders.

Causes of autism have not been found yet. It is presumed that the key role in autism's development is the genetics as well as various environmental factors. Its genetics is highly complex and it is not clear yet whether it is the result of rare mutations or a combination of genetic variants. Autism cannot be cured but there are therapies which could alleviate the effects of the disorders and facilitate the adaptation of autistic persons to the best possible level of functioning.

2 People with ASD as the Users of Built Environment

Due to various environmental factors as well as incorrect information, there have been some erroneous generalizations and stereotypes concerning people's suffering from autism. The most popular generalizations claim that autistic persons are not independent and show a low level of intelligence, other stereotype is related to the people with Asperger Syndrome who often demonstrate high operational intelligence stereotypically associated with savant syndrome (as presented in the movie "Rain Man"). Moreover, they are likely to suffer from mental diseases, could be aggressive or even dangerous due to their undaunted temperament and obsessive shyness. Most of these stereotypes might relate to individual symptoms of autism, without their general point of reference and without deep understanding of the causes.

The level of the needs and way of functioning of people with ASD can vary just like the profile of each person. Many people with ASD are able to live and work independently within the society. Many will require guidance (often by therapist) in trying to comprehend the surrounding world and the prevailing principles. Some will require constant professional help and specially prepared spaces. There are many features which are characteristic for ASD which occur to a greater or a lesser degree. Nevertheless, a general list of the features could still be compiled [1]:

- language and communication difficulties,
- difficulties with building interpersonal relationships which lead to sense of loneliness,
- inability to recognize unclear or complicated social rules, which lead to difficulties to behave "properly" among a specific group of people,
- easy sensual overstimulation,

- strongly expressed passions and interests,
- inability to concentrate on one thing for a longer period of time,
- motor disorders,
- obsessive behaviors, strict routines and repeatability.

In order to shortly summarize the profile of an ASD users' needs, one could note that people on this developmental path often encounter problems with an excessive stimulation coming from the surrounding world. They require a friendly and familiar individual space to calm down and relax or concentrate and develop their own interests. They also need a safe space which would be adjusted to their individual behaviors and motor disorders.

3 The Scale of Research and Architectural Design Answers

The magnitude of problems and dysfunctions connected with autism and its complicated medical and sociological nature causes that few architects have decided to delve into the issue of design adjusted to people with ASD.

One of the first was UK architect Simon Humphreys, who lived with his autistic brother for many years, designed buildings for people with ASD, and in 1989 completed research thesis on architecture for autism [2].

Humphreys, basing on his professional experience, [2–4] suggests important features of space to consider while designing for people with ASD, which is a sense of calm and order, clarity and simplicity, classical proportion and natural harmony, restraint of too many perceptible details; carefully considered sensual stimulation (acoustic, light, texture, ventilation, cool colors); sense of safety; larger proxemics needs, passages allowing easy, comfortable movement; highly adaptable spaces, with built–in places to escape; clear distinctions between spaces for different activities, use of limited number of good quality materials.[1]

Another, well known approach to design based on research, is Magda Mostafa ASPECTSS™ Index. She conducted the series of research during her doctoral studies at the Cairo University in order to specify the principles of design of space for autistic children. She listed 7 principles which had been presented as basic guidelines for design for autism. These were factors which are supposed to have an influence on the duration of the focus given to a specific task/subject matter/person. In addition, they may also reduce the response time and have a positive influence on the temperament of people stricken with ASD. These principles include: [5]

- Acoustics
- SPatial Sequencing
- Escape Space
- Compartmentalization

---

[1] Humphrey's suggestion were used as a theoretical base for our own Participatory Action Research experiment in 2014 with teenagers on ASD evaluating educational space of university building. The outcomes of this experiment are described in a text "Evaluating Quality of the built environment for the people with Autism Spectrum Disorder" further in this publication.

- Transition Zones
- Sensory Zoning
- Safety.

These principles were applied by Mostafa in a redevelopment project of the existing multi-family building into a school for children with special educational needs: *Advance School for Developing Skills of Special Needs Children* [6] based in Cairo. Experience gained in this project as well as acquired knowledge are described in several articles. In *An Architecture for An Autism: Concepts of Design Intervention for the Autistic User* the "Sensory Matrix" tool is presented. Sensory Matrix helps to evaluate and organize the interrelations between the features of the built environment and specific sensory needs of people with ASD [7].

Mostafa's design principles were also applied in Morris-Union Jointure Commission's (MUJC) Developmental Learning Center (DLC) in Somerville, New Jersey, designed by USA Architects [8], *New Struan Centre for Autism* in Alloa, Scotland design by Aitken Turnbull architect Andrew Lester [9], have been based also on similar qualities.

The research trend related with architectural design for ASD people is evolving in recent years. Many new buildings (schools, pre-schools or rehabilitation and education centers for children) supports the needs of people with ASD in many different ways. A group of architect-specialist in the field of design for users with ASD is growing gradually.

The issue of housing adjusted to the needs of adults with ASD is still not very well studied. Research carried out by Sherry Ahrentzen and Kimberly Steele (from Arizona State University) was an important achievement in that scope. In 2009, during the research, a report was prepared with preliminary general conclusions and design guidelines [10]. As a final result of their work, a book entitled: "At Home with Autism. Designing Housing for the Spectrum" was published in 2016 describing the problem of housing for ASD people in a form of guidelines for its design.

4 Architectural Design and Building Programming Approaches

As shown by Henry [11] in both the theory and practice of designing space for people with ASD, there are now present two somewhat contradictory approaches: *the neurotypical approach* and *the sensory sensitive approach*, both directly based on different therapeutic approaches and different views on possible therapeutic situation created in building.

Neurotypical approach. is based on assumption that people with ASD, who often suffers with sensory overstimulation in "real life" situations, can better learn generalization skills in the environment close to neurotypical. This approach in therapy defines a set of actions which confront the problems that autistic people must face in everyday life. It provides a generalization of different types of human behaviors and kind of functional spaces in the world that surrounds autistic people [11].

Sensory sensitive approach. Therapeutic actions which belong to that approach are aimed at ordering and understanding the impulses, behaviors and environments which surround autistic people, using the help of specific activities and controlled spaces. This type of therapy allow autistic people to acquire or sharpen their skills of recognizing types of situations, verbal and non-verbal forms of communication as well as the surrounding. This approach tends to rather indirect form of explaining and understanding one's own body and the space. It also serves to improve mood and to get rid of fears and feelings of anxiety. *Sensory sensitive approach* is often aimed at establishing a relationship with people with severe symptoms of autistic isolation. Space in which autistic people acquire new skills is really significant for sensory sensitive therapy. It is necessary to limit sensory stimuli and facilitate the ability to focus on activities which take place in space [11].

The concept of *the sensory room* is also part of the sensory sensitive paradigm. This *notion* refers to a wide range of specialist therapeutic rooms. They had been prepared as a therapeutic response to the problems with deciphering and understanding sensory impulses. Sensory rooms are aimed at creating an isolated atmosphere that promotes relaxation and the feeling of safety, necessary for emotional health of the users.

They offer a series of activities which provide better chances of understanding ones feelings and better chances of relating them to the actions and changes taking place within the environment. As a result of that, users of sensory rooms see an improvement in their eye-hand coordination, they better understand and interpret the new space and are able to deal with stress better. That tranquilizing surrounding is also a place where bonds could be built between people who take advantage of that space. Sensory rooms come in a lot of different forms. The most simple of them are acoustically isolated rooms equipped with systems and devices which allow one to change the color or even the light's intensity, which in turn gives the possibility of lighting up simple formulas and symbols on the surfaces of the walls. It also allows one to control the sounds and music. They are usually equipped with soft floors, mattresses, arm-chairs and balls made of different material structure. More extended and complex rooms include, among other things, swimming pools and hydro-therapeutic systems.

Both approaches presented above are close to the concept of *healing environment*, as shown by Roger Ulrich and Craig Zimring within the evidence-based design framework (EBD) [12, 13][2]

Regardless of the terminology, people who have been researching the subject of space have also been delivering new scientific evidence which support the idea of designing therapeutic spaces. This evidence gave rise to specific architectonic modifications to the organization, the functioning and the shaping of the space, aimed at improving not only the mood and the feeling among the users (patients, their families and employees) but also at improving the efficiency of the therapeutic process. Therapeutic environments do not constitute a manifestation of a new design philosophy, they constitute a manifestation of a new approach to the users, their experiences,

---

[2] Concept of healing environment was used by Magdalena Jamrozik-Szatanek in her research on quality of children hospitals, Jamrozik-Szatanek [16].

well-being and to the structure, health and therapeutic facilities within the community and the urban space.

Another concept, which is related to the problem discussed above, is the learning space (or learning environment) concept [14], which can be used in design of educational buildings. We can assume that all described concepts can have significant value for designing space for people with ASD as not exclusive but complementary. From architect perspective it would be difficult to design educational or healthcare complexes based only on one of the above concepts, it is rather obvious that in bigger facilities different zones can have different "load of stimuli" and different role to play in learning or therapy processes.

5 Examples of Creating Space for People with ASD

The issue of architectural response to the needs of people with ASD has been introduced to the curriculum of Faculty of Architecture of Silesian University of Technology in 2014[3]. Below we present examples of master degree designs based on research prepared by students at the Department of the Design and Qualitative Research in Architecture [15].[4]

Presented master thesis were created on the basis of universal design principles and current research-based evidence.

Concept design for the Center for Children Suffering from the Autistic Spectrum Disorders. by Wolniak 2016 [18] Care center for ASD children includes in its structure educational, residential, recreational and integrative functions.

The Center Project basic assumption was the design of a "hybrid" -on the one hand, the complex of buildings was supposed to create an enclave which constitutes both a house and a shelter for children with ASD, on the other hand, it is supposed to integrate children with the community (within it's public, open-access space). A functional core of the complex was made up of structures which feature residential functions. They are accompanied by green recreational spaces, which are gradually being turned into a sensory garden (Figs. 1 and 2).

Straight out of the residential part, one can pass into the multi-function facility. The ground/first floor of the building is extension of a public space. Here the main concept of design is revealed: an integration of users, stimulation of creativity, a space in which each individual would have a possibility to develop their talents (laboratories of arts, music, technic etc.), available for both residents and external users.

---

[3] The general issue of evidenced-based design and "design for users' needs" in architectural design is one of the main didactic concept introduced to architecture curriculum at our Faculty starting in 1996. The basic principle of "first research than design" resulted in many interesting evidenced-based projects, e.g.of redevelopments or new buildings, examples can be found at ACEE Journal, Bugno-Janik [17] .

[4] Another example of the research and didactic experiment can be found in an article "Quality of the built environment from the point of view of people with Asperger Syndrome and High Functional Autism Spectrum Disorder" further in this publication.

Fig. 1. Visualization of the center for children suffering from autistic spectrum disorder. [18]

Fig. 2. Visualization of the multi-functional structure - entrance zone. [18]

Another important function is served by the "hiding places", distributed at the entire center, where one can hide from the stimuli, which are impossible to be fully eliminated from the surrounding in public or social places. Their design is diverse, for different needs of different users.

The form and function of the complex has been designed on the basis of the concept of a gradient, which is visible in numerous aspects of the project, ranging from the growing vertical green wall and the changeable size of the building up to the availability gradation of the structure, the land form, loudness of specific spaces, the number of incoming external stimuli and the greenery.

6 Summary and Conclusions

Support tools for people with ASD are commonly understood as behavioral therapy, specialist care, medical treatment and even medicaments. Understanding architecture as an important method of improving the quality of life and the condition of people with ASD is not yet widespread. Meanwhile, the specificity of autistic disorders causes that the manner in which the built environment is shaped has a direct influence on the well-being, the quality of life and educational/therapeutic progress made by people with ASD. A space which is inadequate for specific needs of people with ASD, may lead to deepening the sense of fear, disorientation, detachment or simply physical or mental discomfort to a much larger extent than in the case of the majority of the society. This applies to spaces of therapy and education, specialist diagnostic and therapeutic centers and special educational facilities as well as places of residence.

Architectural design for people suffering from ASD requires an especially sensitive approach. ASD people should be offered space which is friendly and helpful, evoking rather the feeling of safety, peace and convenience, with great emphasis on their users' dignity.

References

1. Pisula, E.: Autism Przyczyny – Symptomy – Terapia: Causes - Symptoms – Therapy. Harmonia Publishers, Gdansk (2010)
2. Humphreys, S.: Autism and Architecture. http://www.autismlondon.org.uk/pdf-files/bulletin_feb-mar_2005.pdf. Accessed 28 Feb 2018
3. Humphreys, S.: Autism and Architecture. http://researchautism.net/publicfiles/pdf/Simon%20Humphreys%20Autism%20Show%202016.pdf. Accessed 28 Feb 2018
4. Humphreys, S.: Creating autism-friendly spaces. http://www.autism.org.uk/professionals/others/architects/top-tips.aspx
5. Mostafa, M.: Architecture for Autism: Autism ASPECTSS™ in School Design, Cairo (2014)
6. Mostafa, M.: An Architecture for Autism – A New Dimension in School Design, Cairo (2006)
7. Mostafa, M.: An architecture for autism: concepts of design intervention for the autistic user. Archnet-IJAR Int. J. Archit. Res. **2**, 189–211 (2008). https://archnet.org/system/publications/contents/5107/original/DPC1837.pdf?1384788342
8. Design for Autism. https://www.scottishautism.org/about-autism/research-and-training/design-autism. Accessed 28 Feb 2018
9. Quirk, V.: An Interview with Magda Mostafa: Pioneer in Autism Design. http://www.archdaily.com/435982/an-interview-with-magda-mostafa-pioneer-in-autism-design. Accessed 28 Feb 2018
10. Ahretnzen, S., Steele, K.: Advancing Full Spectrum Housing: Design for Adults with Autism Spectrum Disorders. https://d3dqsm2futmewz.cloudfront.net/docs/stardust/advancing-full-spectrum-housing/full-report.pdf. Accessed 28 Feb 2018
11. Henry, C.N.: Designing for Autism: The 'Neuro-Typical' Approach, ArchDaily, November 2011. Accessed 28 Feb 2018

12. Ulrich, R., Zimring, C.: The Role of the Physical Environment in the Hospital of the 21st Century: A Once-in-a-Lifetime Opportunity (2004). https://www.healthdesign.org/system/files/Ulrich_Role%20of%20Physical_2004.pdf. Accessed 28 Feb 2018
13. Ulrich, R.: View through a window may influence recovery from surgery. Science **224** (4647), 420–421 (1984)
14. Henriksen, K., Kaup, M.L.: Supportive Learning Environments for Children with Autism Spectrum Disorders. http://www.kon.org/urc/v9/henriksen.html. Accessed 28 Feb 2018
15. Bielak-Zasadzka, M.: Metodologia pracy badawczej. Zastosowanie metod badawczych w pracach magisterskich. (Methodology of Research Works. Application of research methods in MA theses). Silesian University of Technology, Gliwice (2015)
16. Jamrozik-Szatanek, M.: Research on social space in rehabilitation hospital for children in Radziszow, Poland. ACEE Civ. Eng. Environ. **6**(2), 5–11. Silesian University of Technology, Gliwice (2013). http://www.acee-journal.pl/cmd.php?cmd=download&id=dbitem:article:id=274&field=fullpdf
17. Bugno-Janik, A.: Building programming as an element of changes of the culture of an organization. ACEE Archit. Civ. Eng. Environ, **1**(1), 17–24. Silesian University of Technology, Gliwice (2008). http://www.acee-journal.pl/cmd.php?cmd=download&id=dbitem:article:id=6&field=fullpdf
18. Wolniak, M.: Projekt koncepcyjny ośrodka dla dzieci ze spectrum autyzmu. Projektowanie przestrzeni terapeutycznej, mieszkalnej i integracyjnej dla osób ze spektrum autyzmu. (Concept design for a center for children suffering from autism spectrum disorder. Design of therapeutic, residential and integrative space for autism spectrum people). Master thesis (2016). Supervised by - Bielak-Zasadzka, M.: Silesian University of Technology, Gliwice

A Stakeholders Perspective on the Causes of Poor Service Delivery of Road Infrastructure

Nokulunga Mashwama[(⊠)], Clinton Aigbavboa[(⊠)],
and Wellignton Thwala[(⊠)]

Department of Construction Management and Quantity Surveying,
University of Johannesburg, Johannesburg 2028, South Africa
{nokulungam, caigbavboa, didibhukut}@uj.ac.za

Abstract. This study adopted a quantitative approach in order to investigate on a stakeholder's perspective on the causes of poor service delivery of road infrastructure. Structured questionnaires were circulated to 75 stakeholders in construction industry in Gauteng Province, which were registered with various approved councils, construction professionals and contractors. 50 came back completed and eligible to use. Random sampling method was used to select the respondents in various organizations. Research findings revealed that community unrest and land proclamation were the highest ranked causes for poor service delivery. Followed by, time, financial constraints, cash flow, lack of proper planning, resources, delivery of material, plant and equipment, shortage of skilled labourers, lack of equipment, lack of materials, performance guarantees, project duration/period, cost overruns were the major causes of poor service delivery of roads infrastructure according to stakeholders perspective. Therefore, mitigating the causes would be by having proper management planning, skills transfer, education and training.

Keywords: Challenges · Construction industry · Road infrastructure
Stakeholder

1 Introduction

The roads make our life easier in many ways as they link province to province even into other neighboring countries of South Africa [1]. In generally roads boost the country economy and simplify people's life. Furthermore, it creates employment, income for people and construction entails a complex interplay of client, consultants, contractors, tool, equipment's and materials [2, 3]. Moreover, roads boost the economy in terms of transporting goods, mineral resources in mining, farmers, and improve the access of different facilities such as schools, hospitals, shopping centers, work places and recreation centers [3]. If roads are in good conditions, they also reduce travel times and people save fuels on their vehicles, reduce production costs for the ever-growing number of goods shipments [1, 4]. The construction industry is the main distinct part that provides vital components as business, high recruitment of people and developing entrepreneurs for economy development [5].

© Springer International Publishing AG, part of Springer Nature 2019
J. Charytonowicz and C. Falcão (Eds.): AHFE 2018, AISC 788, pp. 140–150, 2019.
https://doi.org/10.1007/978-3-319-94199-8_14

However, the road construction has been facing some challenges that cause poor service delivery. Road construction challenges causing poor service delivery are examined to be one of most repeatedly problem in the construction industry in South Africa. Most of the road construction challenges are continuous and there are recently new added challenges occurring in the industry due to new technology and other influences [1]. There is a need to unlock or resolves these challenges within the industry [6]. The challenges on roads construction are experienced on infrastructure which already incorporate, existing roads and the new infrastructure that is mushrooming which needs to take traffic into consideration [5, 7].

Therefore, it is vital to carefully take into consideration the road challenges that cause poor service delivery for better future modernizes quality roads in South Africa [4]. Many projects experience comprehensive challenges as results of exceeding initial scheduled time and cost estimates due to change of scope, procurement system. Stakeholders have positive and negative impact on road construction projects and hence, they contributes to the success or failure of the projects [8, 9]. Department of Roads and Transport of Gauteng Provincial Government (GDRT) is obligated to road infrastructure network that interconnects the Gauteng provincial roads within the province and connects the Gauteng province with other provinces and countrywide. Hence, the study will focus on the causes of poor service delivery in the infrastructure projects in the Gauteng Province of South Africa where most of the roads construction projects are taking place.

2 Literature Review

2.1 Overview of Gauteng Provisional Road Network

Table 1 reflects the roads in Gauteng, 5,846 km provincial roads under the jurisdiction of the Gauteng Province Department of Roads and Transport (GPDRT), of which 4456 km is paved and 1390 km is unpaved [10]. The Road Infrastructure Strategic Framework for South Africa (RISFSA) classification was used, for the development of the South African Road Classification and Access Management Manual (RCAM) classification. RCAM deals with both rural and urban roads and also including the spect of access management. Both RISFSA and RCAM were used in all Gauteng provincial roads [11].

Table 1. Network length by pavement type, 2015

Road type	Length (km)	Length (%)
Paved roads	4,456	76%
Unpaved roads	1,390	24%
Total	5,846	

2.2   Condition of the Surfaced Roads in Gauteng

2.2.1   Visual Condition Index (VCI)

Figure 1 above represents the VCI (Visual Condition Index) distribution (by length) for the road network. The condition of the paved GPDRT roads is acceptable but borderline with 10% in the poor and very poor condition categories. According to a [11] recommendation, a maximum 10% of the road network shall be allowed to be in a 'poor to very poor' condition. The associated costs of roads in very poor' condition is excessive [10]. Road users driving on these deteriorated roads pay extra for increased maintenance, time and fuel costs. Preventive maintenance, such as reseals, cannot improve the functionality and performance of these roads and large capital expenditure is required to rehabilitate "very poor" roads to a "good" and functional condition [10].

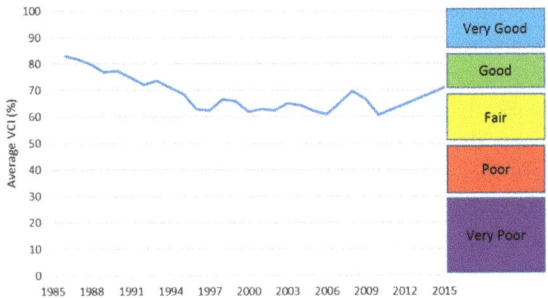

Fig. 1. Average VCI trend paved roads, 1985 to 2015

Only 1% (32 km) of the paved GPDRT roads is in "very poor" condition. These roads will require heavy rehabilitation [10]. 386 km (9%) of paved GPDRT roads are rated in a "poor" condition in 2015. Roads in a "poor" category will typically require rehabilitation within the next 5 years [10]. In some cases preventive, reseal treatments with extensive pre-treatments such as pothole and structural failure repairs are still possible provided the intervention be done soon and the roads are not allowed to deteriorate much further. The window of opportunity to maintain these roads with cost-effective measures is unfortunately very brief [10, 11].

1,399 km (32%) of paved GPDRT roads were rated to be in a "fair" condition in 2015. The large percentage of roads in a "fair" condition remains a concern because they have the short-term potential to deteriorate into the "poor" category if appropriate preventive treatments are not applied timeously [10]. A substantial portion of the road network thus currently operates within the fair condition category, placing a huge demand on reseal needs to prevent further deterioration and delay subsequent expensive future rehabilitation costs [10, 11].

2.3   Construction Industry

The civil engineering construction industry focuses on the development of infrastructural works. [4] Identified four primary areas distinctive of the contracting sector. The first area being the construction of roads, railway, bridges, and tunnels, the second area being the erection of harbors, docks, waterways, dams, reservoirs and sea defense and land reclamation works, just to name a few [12]. The construction industry creates and maintains the built environment that underpin all modern human endeavor [4].

The built environment is the reflection of the developmental progress as well as the physical foundation for economic and social advance into the future. The construction industry creates and maintains this foundation in a process that must deliver value to clients and society. The construction industry, as part of the business sector, is one of the main instruments in the task of working for a better life; It is an engine of reconstruction and development. Without the participation of, and the expertise and capacity of the construction industry and its related professions, backlogs obstacles will be difficult to overcome [1, 2, 10].

2.4   Stakeholders in Road Construction Projects

In South Africa, community leaders and municipalities are the very respective stakeholders in the project for political leaders to gain majority of votes in times of election [9, 13]. However, those leaders 80% of them they are not technically qualified as they have occupied political posts; civil engineers at the national political level are rare which poses a challenge in the decision making [8]. Stakeholders are Individuals and organizations that are actively involved in the project or whose interests may be affected as a result of project execution or project completion [8]. In construction, Stakeholders can be a client, consultant, contractor, suppliers, community leaders, service providers [8, 9]. Some researchers say stakeholders are internal and external, some say stakeholders are inside and outside and others say stakeholders are primary and secondary [9]. Figure 2: Is identifying the potential stakeholders in construction industry adopted from [8]:

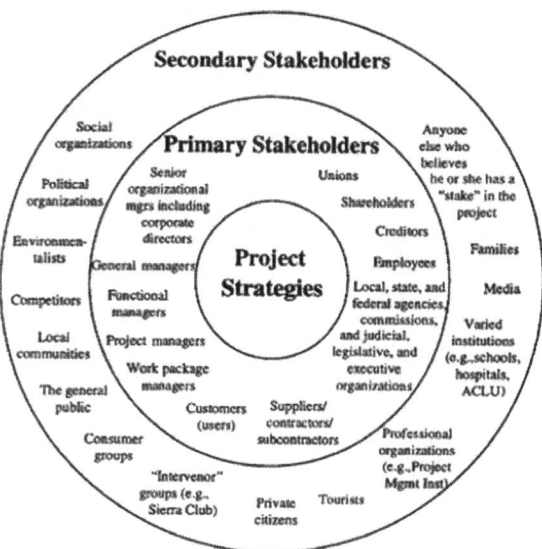

Fig. 2. Potential Stakeholders for construction projects

2.5   Challenges Behind the Causes of Poor Road Construction Projects

[4] states that the South African construction industry, is supposed to go through a high growth phase this year, owing to an increased number of construction projects, and a greater focus on housing projects and large-scale infrastructure projects, is facing severe problems regarding construction project delivery. These problems stem from a lack of capacity, skills shortage and quality standard [14, 15]. Furthermore, there are indirect challenges that affect construction and cause poor service delivery such as proclamation or landscaping issues that involves legal issues. Properties along the way where the road needs to be constructed and the proclamation were not done properly therefore the legality process comes in if the owner of the property is not satisfied [15]. Other challenges are government regulations and policies which differ from one department to another. For instance, the roads construction can be done by department of roads and transport, and other government departments that need their protocol to be followed [7, 10]. Below are the causes of poor service delivery or constraints from contractor, consultant, client/general that may adversely affect the progress or failure to any road construction project in the Gauteng Province, South Africa:

Poor Causes of Poor Service Delivery- Design Related Problem as per [1, 7, 15–17]

- Lack of detail specifications;
- Poor design reflection;
- Unrealistic specification;
- Cost overruns;
- Barriers to the uptake of new knowledge;
- Lack of essential skills necessary for the management of projects;
- Undermining of engineering skills in municipalities;
- Poor project scoping and specification;
- Lack of knowledge relative to skills and resources required to implement projects;
- Contract awards based mostly on poorly defined tender processes,
- Lack of experienced employees capable of managing projects;
- Unnecessary time lags between tender submission and award.

Challenges Faced by Contractors as per [16–22]
Following are some of the challenges faced by the contractor in the roads projects

- Lack of proper planning;
- Inadequate Construction programme of works;
- Resources;
- Improper equipment selection & faulty equipment;
- Poor performance by the contractors;
- Performance guarantee;
- Cash flow;
- Delivery of materials, plant and equipment;
- Time and Financial constraints;

- Absenteeism;
- Tiredness;
- Legal provisions;
- Procurement method chosen;
- Lack of Risk management;
- Existing services and land/properties proclamation;
- Traffic volume.

2.6   Mitigation the Causes of Poor Service Delivery of Road Infrastructure

Communication. Good communication among teamwork should be monitored, clear, honest, open and frequent but not excessive [1, 23, 26]. Communication in project management is one of the ten-knowledge area that employs the processes required to ensure timely and appropriate generation, collection, distribution, storage, retrieval and ultimate disposition of project information [1].

Education and Training in Construction Industry. All three parties (client, consultant and contractor) should perform continuous training courses to improve capability of personnel and professionals carrying out designs, supervision and construction of all levels [19]. Civil Engineering Body of Knowledge advise that civil engineers must always seek broad knowledge base on empowering themselves in thinking out of the box on how civil engineers do business, including how to manage projects, risk management routine incorporation, handling of contracts, procurement and legal issues [1, 7]. Most of the contractors have related qualifications towards their field of work however, they did training at short period of time of which more internal training are needed [23].

Skills Transfer. It is necessary for skills transfer to the youth as number of youth receive qualification without skills on the ground to use those qualifications on site. According to [21, 24], nearly half a million skilled workers in construction will go for pension in the next five years therefore, to overcome skills gap, it is good that companies and government sectors employ talented workforce to assist in skills transfer to benefit the organization [18, 23].

Proper Planning Management. Community affected by the construction as their properties are along road reserve needs to be proclaimed and noticed in time during planning stage of the project [19, 24, 25]. Stakeholders organizations such as Telkom, Eskom, rand water, city power should be involved during the initial or planning stage of the project to coordinate and corporate with relocation and relocate of services [19, 24]. By involving those stakeholders' organizations, it will help the recent contractor or organization to find the existing underground existing services by producing accurate and clear as-built drawings other than mapping underground services which were unknown [24]. A well-planned project, which is cautious monitored have a direct positive impact on good performance, profitability of the contract and the company [19, 25].

Timeframe to Respond. The timeframe to respond in different aspect of the project is required. The sensitive aspects such as payment response, delivery of materials, site instruction by resident engineer to the contractor, decision making by the project stakeholders and client response in every enquiry [26]. The Public Finance Management Act of 1999 (PFMA) determines that all contractual obligations (and accounts) must be settled within 30 days from its receipt [Sect. 3 8(1) (f) read with Part 4, Regulation 8.2.3 of the Regulations in South Africa therefore the client needs to be take into consideration the payment response within 30 days [26].

3   Methodology

3.1   Research Approach and Design

This study adopted a quantitative approach as the purpose was to investigate the causes of poor service delivery in the road infrastructure projects. A well-structured questionnaire was distributed to different construction companies in Gauteng Province, amongst construction professionals such as civil engineers, project managers, directors, quantity surveyors, construction managers and resident engineers. The questionnaire were sent via e-mails, some were delivered to the known construction companies by the researcher and some were distributed during site clarification meetings of contractors and consultants bidders on Gauteng Department Roads and Transport roads tenders.75 Questionnaires were distributed and 50 came completed and eligible to use and reflects 67% response rate. It was difficult to gather questionnaires as the professionals are always busy, some of them returned questionnaire after scheduled time, and others apologized of not sending the completed questionnaire back. The study was conducted from reliable scholarly sources such as articles, journals, books, publications, websites and site experience on the field.

3.2   5- Point Linkert Scale

5- point linkert scale was adopted for the study which gave a wider range of possible scores and increase statistical analyses that are available to the researcher [21]. The 5 point scales were transformed to mean item score abbreviated as (MIS) for each of the challenges faced in the road construction projects, the mitigations taken and the impact of stakeholders in the roads construction projects evaluated by the different respondents within the roads construction industries [21].

3.3   Computation of the Mean Item Score (MIS)

The computation of the mean item score (MIS) was calculated from the total of all weighted responses and then relating it to the total responses on a particular aspect. The formula is used to rank the causes of poor service delivery of road construction projects based on frequency of occurrences as identified by participants [21].

$$\mathbf{MIS} = \frac{1n1 + 2n2 + 3n3 + 4n4 + 5n5}{\sum N}$$

Where;

n1   = number of respondents for strongly disagree
n2   = number of respondents for disagree
n3   = number of respondents for neutral
n4   = number of respondents for agree
n5   = number of respondents for strongly agree
N   = Total number of respondents

4 Findings and Discussions

4.1 Causes of Poor Service Delivery

Table 2 reveals that community unrest/disruption was ranked the highest with (MIS = 4.10; STD = 1.093); Existing services and land/property proclamation was ranked second with (MIS = 3.80; STD = 1.050); time and financial constraints was ranked number third with (MIS = 3.72; STD = 1.05) followed by cash flow ranked fourth with the (MIS = 3.70, STD = 0.91); Lack of proper planning ranked fifth with the (MIS = 3.64; 0.851); Cost overruns was ranked sixth with (MIS = 3.52; STD = 1.035); Resources in terms of staff/personnel was ranked seventh with (MIS = 3.44; STD = 1.072); Delivery of materials, plant were ranked eight with (MIS = 3.44; STD = 0.972);Lack of detail specification ranked nineth with (MIS = 3.34; STD = 1.272); Shortage of skilled workers was ranked tenth with (MIS = 3.32; STD = 1.291); Lack of equipment and materials was ranked eleventh with (MIS = 3.28; STD = 1.246); Performance guarantees was ranked twelve with (MIS = 3.26; STD = 1.291); Project duration corrupt officials was ranked thirteen and second last unrealistic construction programme was the second last factor with (MIS = 3.02; STD = 1.152) and poor performance by the contractor was ranked the last factor contributing to challenges faced by stakeholders with (MIS = 2.90; STD = 0.544).

Table 2. Challenges faced by the stakeholder in roads projects

Challenges faced by the contractor in roads construction	MIS	STD. DEV	Rank
Community unrest/disruption	4.10	1.093	1
Existing services and land/properties proclamation	3.80	1.050	2
Time and financial constraints	3.72	1.051	3
Cash flow	3.70	0.909	4
Lack of proper planning	3.64	0.851	5
Cost overruns	3.52	1.035	6
Resources	3.44	1.072	7
Delivery of materials, plant & equipment	3.44	0.972	8
Lack of details specifications	3.34	1.272	9

(*continued*)

Table 2. (*continued*)

Challenges faced by the contractor in roads construction	MIS	STD. DEV	Rank
Shortage of skills labourers	3.32	1.203	10
Lack of equipment and materials	3.28	1.246	11
Performance guarantees	3.26	1.291	12
Project duration/period	3.24	1.061	13
Corruption by officials	3.24	1.287	13
Unrealistic construction programme	3.02	1.152	14
Poor performance by the contractor	2.90	0.544	15

4.2 Mitigation to the Challenges Facing Roads Construction Industry

Table 3 below reveals that proper planning management was ranked first with (MIS:4.46; STD = 0.676), second was Skills transfer with (MIS: 4.36; STD = 0.851), Timeframe to respond by consultant and contractor ranked third on the table with (MIS :4.32; STD = 0.683). Fourth was education and training with (MIS = 4.30;STD = 1.020). Second last was good communication with (MIS: 4.24; STD = 1.061). Legality of contract ranked last on the table below with MIS (4.00; STD = 0.948).

Table 3. Respondents response on mitigations to the challenges

Mitigations	MIS	STD. DEV	Rank
Proper planning management	4.46	0.676	1
Skills transfer	4.36	0.851	2
Timeframe to respond by client, consultant and contractor	4.32	0.683	3
Education and training	4.30	1.020	4
Good communication	4.24	1.061	5
Legality of contract	4.00	0.948	6

5 Conclusion

Community unrest was the highest ranked factor which is a major concern to all stakeholder (client, consultants and contractors). Community unrest such as strikes, stoppage of project by business forums, councilors interruptions disrupt the project spending more time negotiations of the community grievances delay project and sometimes total abandonment of project sites is the biggest threat to road construction projects and causes poor service delivery. Moreover, the existing services and land proclamation also pose a huge challenge to delivering a project on time, it needs to be taken into consideration during early stage of the project. Therefore, proper planning, skills transfer and timeframe to respond would aid in the poor service delivery. Good communication with the various communities and different department must take place especial during the preliminary stage, and transparence, honesty must be key to every individual who is involved in the project.

6 Recommendations

It is recommend that all three parties (client, consultant and contractor) should perform continuous training courses to improve capability of personnel and professionalism its encouraged to think out of the box. It is necessary for skills transfer to the youth as number of youth receive qualification without skills on the ground to use those qualifications on site. It's recommended that companies and government sectors employ talented workforce to assist in skills transfer to benefit the organization.

References

1. Mahamid, I., Bruland, A., Dmaidi, N.: Causes of delay in road construction projects. J. Manag. Eng. ASCE **28**, 300–310 (2012)
2. Levinson, H.: Highways, people and places: past, present and future. ASCE Int. J. Civ. Environ. Eng. **130**, 406–411 (2004)
3. Mofokeng, T.G.: Assessment of the causes of failure among small and medium sized construction companies in the Free State Province, pp. 1–268 (2012)
4. De Beer, M., Sallie, I.M., Van Rensburg, Y.: Revision of TRH 11 (1999–2000). Recovery of road damage – discussion document on a provisional basis for possible new estimation of mass fees – under review for TRH 11 (2000) – final summary report V1.0. Contract Report, p. 93 (2009)
5. Government Gazzete Regulation Gazette (2014). GDRT Road Network; 2015/2016 REF 111363, vol.584, no. 10113 (2014)
6. Bowen, P., Edwards, P., Cattell, K.: Corruption in the south african construction industry: a mixed methods study. In: 28th Annual ARCOM Conference, pp. 521–531 (2012)
7. Oyewobi, L., Windapo, A., Cattell, K.: Competitiveness of construction organizations in south africa. In: Construction Research Congress, pp. 2063–2073 (2014)
8. Assefa, F., Worke, Z.T., Mohammed, M.: Stakeholders impact analysis on road construction project management in Ethiopia: a case of western region. Int. J. Eng. Technical Research (2015)
9. Chinyio, E., Olomolaiye, P.: Construction Stakeholder Management. Wiley-Blackwell, Oxford (2010)
10. Gauteng Department of Roads and Transport, Road Asset Management System (RAMS), pp. 1–70 (2015)
11. TRH, 26: South African Road Classification and Access Management Manual (2012)
12. CIDB: Construction Industry Indicators Summary Results, pp. 1–13 (2008)
13. Edwards, P., Bowen, P., Hardcastle, C., Stwewart, P.: Identifying and communicating project stakeholders risks. Building a sustainable future. In: ASCE, pp. 776–785 (2009)
14. Inuwa, I., Wanyona, G., Diang, A.: Identifying building contractors' project planning success indicators: The Case of Nigerian Indigenous Contractors. Applied Research Conference in Africa (ARCA), pp. 468–479 (2014)
15. Malongane, D.D.: Challenges facing emerging contractors in Gauteng, pp. 1–92 (2014)
16. Chan, W.M., Kumaraswamy, M.M.: Contributors to construction delays. J. Constr. Manag. Econo. **16**, 17–29 (2007)
17. Assaf, S.A., Al-Heijjie, S.A.: Causes of Delays in large Construction Projects. UKUR BAHAM Quantity Surveying (2005)

18. Bob Muir, P.E.: Challenges facing today's Construction Manager. Suppl. Reading for CIEG, pp. 486–010. Construction Methods & Management, pp. 1–9 (2005)
19. Kamanga, M.J, Jyd, W., Steyr, M.: Causes of delays in road construction projects. J. South African Inst. Civ. Eng. pp. 2309–8775 (2013)
20. Sunjka, B.P., Jacob, U.: Significant causes and effects of project delays in the Niger Delta Region, Nigeria, pp. 641-1–641-13 (2013)
21. Eke, C., Aigbavboa, C., Thwala, W.: An exploratory study of the causes of failure in construction industry, South Africa, pp. 1055–1062 (2015)
22. Ngosong, F., Tounga, T.: Investigation of problems or challenges faced by the procurement and delivery of quality construction projects in Africa and Cameroon, pp. 1–16 (2015)
23. Azhar, S.: Building Information Modelling (BIM): Trends, benefits, risks, and challenges for the AEC industry. Leadership and Management in Engineering, pp. 241–252 (2008)
24. Inuwa, I.I., Saiva, D., Alkizim, A.: Investigating Nigerian indigenous planning in construction procurement: an explanatory approach. Int. J. Civ. Environ. Eng. IJCEE-IJENS **14**(04), 16–25 (2014)
25. Ogunde, A., Fagbenle, O.: Assessment of effectiveness of planning techniques and tools on construction projects in Lagos state, Nigeria. In: AEI, pp. 397–408 (2013)
26. Muszynaska, K.: Communication management in project teams –practices and patterns. In: Joint International Conference, pp. 1359–1566 (2015)

Affordable Housing as a Spatial Planning Tool for Shrinking Cities. Case of Poland

Agata Twardoch[⊠]

Faculty of Architecture, Chair of Urban Design and Spatial Planning,
Silesian University of Technology, Ul. Akademicka 7, 44-100 Gliwice, Poland
Agata.Twardoch@polsl.pl

Abstract. According to recent studies, almost all Polish medium and big cities (apart from Bielsko-Biała, Kraków, Olsztyn, Rzeszów, Warsaw and Zielona Góra) are shrinking. Paradoxically, population outflow does not cause an increase in availability of homes, and due to withdrawal of investors and lack of funds for modernisation, it leads to degradation of resources and actually results in decreased availability of homes in the affordable housing sector. This article concerns the housing policy in shrinking cities. It presents the results of evaluation of the housing policy in selected Polish cities (Bytom, Sosnowiec, Sopot) affected to various degrees by depopulation. The evaluation was performed on the basis of the results of studies on shrinking cities in Europe and in the USA.

Keywords: Shrinking cities · Affordable housing · Depopulation
Housing policy

1 Introduction

According to demographic forecasts, by 2050 Poland will have lost 4.5 million inhabitants, i.e. slightly more than 12% of the population [5]. The projected demographic losses will not be evenly distributed: in some of the voivodeships the decrease will exceed 20%, while in Mazowieckie [Mazovian] and Pomorskie [Pomeranian] Voivodeships there will be no decrease, or it will be less than 3%. A change in the demographic structure of the society is also forecast: by 2050, people aged 65+ will constitute 1/3 of the population, and their number as compared to 2013 will increase by 5.4 million. On the other hand, the population of children, the youth and working people will shrink significantly. Even though over 54% of the global population live in urban areas and according to projections by 2050 this number is expected to reach 70%, in Poland the process of shrinking is mostly observed in cities. At present only 6 of Polish cities with the population exceeding 100,000 people do not lose their inhabitants, but by 2050 even the capital: Warsaw will be losing people (2.2% as compared to 2015). Around the year 2020, the total population decrease in cities is to be 100,000 per year, and after 2030-140,000 - as if we deleted one and a half of the 23 Polish cities of 100,000 inhabitants from the map every year. The fastest shrinking cities are those which - mainly due to deindustrialisation - lose their social and economic functions; in the next 33 year Zabrze, Bytom and Tarnów will lose almost 50% of population [6].

© Springer International Publishing AG, part of Springer Nature 2019
J. Charytonowicz and C. Falcão (Eds.): AHFE 2018, AISC 788, pp. 151–163, 2019.
https://doi.org/10.1007/978-3-319-94199-8_15

Globally, the shrinking of cities has been observed for decades (in Europe since the 1960s, in North America and Canada - since the 1950s), however, for Poland and other post-socialist countries this is a relatively new phenomenon. Our population has been growing since WWII. If we compare the census data from 2002 and 2011, the total increase was still about 0.7%, however, a decrease in urban population was also observed (approx. 1%). It is assumed that the processes of de-urbanisation started in Poland around the year 1999, when advancing transformation finally put an end to the social and economic process of the socialist system.

The problem of shrinking cities is permanently present in the discourse concerning urban development (e.g. [7–9]). There are studies on American cities [1], European cities [2, 3], or global studies on the subject [10]. There have been more and more studies on shrinking cities in Poland [6, 11–13], and some of them also concern the strategies used to counteract depopulation [14–16]. A separate category of publications are those concerning housing in the context of depopulation [1, 4, 17].

2   Polish Context and Reasons Behind the Shrinking of Cities

There are several reasons for depopulation in Poland, which can be similar to those in other parts of Europe [18]:

(1) economic reasons: collapse of the labour market (deindustrialisation, privatisation of large work establishments), stagnation and economic recession, land rent and speculations,
(2) changes in distribution of functionalities: the basic functions in city centres are displaced by banks, public administration and business, the other functions are allocated to industrial parks, shopping malls or functional zones (e.g. the Culture Zone in Katowice);
(3) suburbanisation and spatial changes in population distribution;
(4) migration to bigger cities and abroad;
(5) natural population decrease (more deaths than births), aging society.

Apart from those aspects, we must remember that there are also some specific features which do not always allow for implementation of good practices used else-where. Socialist history of Poland has influenced the specificity of the process of shrinking cities. The cities that lose their inhabitants the quickest are those that were strongly industrialised after WWII, where a lot of people came from the countryside [19] and large housing estates were built for them in industrialised technology, to the detriment of the historical urban tissue. After the factories were closed, some of those people simply went back home. The structure of development in historical city centres, influenced by the socialist system, is also a specific problem. After the war, city centres were nationalised and handed over to city administration. Flats were let at cost prices, and so no ongoing repairs were possible, therefore today, in the historical cities centres, in the most precious architectural tissue, we can find underinvested council homes. After shifting to the free market system, some of the flats were privatised - they were sold to the tenants at only a token rate, some remained with the city. As a result of that process we have a mosaic ownership structure, hard to manage. In shrinking cities where - due to lack of investor interest - the resources have not been commercialised, there is also a problem of keeping the resources at an appropriate level (In contrary, in successful cities, the flats are being redeem, and affordable flats in city centres vanishes).

Specific situation of Polish depopulating cities is also affected by:

- large housing estates built in industrialised technology, whose social status is much more diversified than e.g. in Paris. Some of them were built in very good locations;
- in 1999, as a result of an administrative reform, the number of voivodeships - and consequently of voivodeship capital cities - was reduced from 49 to 16. For 33 cities this meant a huge decrease in significance and loss of administrative functions;
- after Poland joined the EU in 2004, many attractive labour markets were opened. As soon as in 2006, one million Poles migrated abroad. According to the Polish Central Statistical Office in 2016 there were 2.5 million Polish emigrants;
- as a result of competition of new homes, local plans and land use plans are prepared in excess by the municipalities. According to the "Report on Economic Losses and Social Costs of Uncontrolled Urbanisation in Poland" [12] demographic capacity of land according to local regulations in 2013 was 229,000,000 people, i.e. over 6.5 times more that the demographic projections for Poland for 203;
- lack of cadastral tax and no chances for its introduction;
- Warsaw-centric approach to housing policy. Central regulations still mainly concern an increase in the number of new flats;
- relatively small saturation of the housing market (according to Eurostat, in 2015 over 40% of Poles lived in over-crowded flats, which is more than double the EU average [33]), poor condition of a lot of housing resources.

3   Susceptibility of Affordable Housing to Depopulation

The housing sector is sensitive to demographic changes [1, 4, 17]. Paradoxically, population outflow does not cause an increase in availability of homes, and due to withdrawal of investors and lack of funds for modernisation, it leads to degradation of resources and actually results in decreased availability of homes in the affordable housing sector [2]. Moreover, a certain group of activities used to counteract depopulation may deepen the housing problems of impecunious people. Studies on American shrinking cities: Detroit, New Orleans, Cleveland, and Pittsburgh [1] show that such a negative impact is caused e.g. by unskilful revitalisation: if the revitalisation processes are not combined with social activities and with special care for the existing neighbouring structures, they may cut the inhabitants off the benefits of the modernisation, push them to the peripheries and lead to loss of the social capital linked to the established social structure. Richard Florida himself, 15 years after he published his famous book *The Rise of the Creative Class* [20], where he pointed to the key role in urban development planning addressing the creative class - i.e. people of creative professions and the bohemia - admits in his latest book *The New Urban Crisis: How Our Cities Are Increasing Inequality, Deepening Segregation, and Failing the Middle*

Class—and What We Can Do About It [21] that focusing only on the creative class may ultimately lead to social stratification and actually aggravate the problems of cities. Prioritising activities aimed at spectacular infrastructural, cultural or commercial investments, may result in limiting the funds that should be dedicated to affordable housing. In Poland the problem is, paradoxically, intensified by EU subsidies, with housing investments being excluded. As a result, some of the investments, particularly showy entertainment and sports facilities (white elephants) and infrastructural facilities, are unnecessary and only generate high costs of maintenance for the future [22]. These costs are then a burden to shrinking cities. Moreover, if revitalisation is not combined with reduction of urban sprawl and a regional housing policy as a factor preventing outflow of people, it can be inefficient. Uncontrolled fight for the inhabitants which cannot be reduced without redistribution of a certain part of local taxes, leads to the sprawl of residential development to the suburbs: open and potentially environmentally valuable.

In the case of co-existence of growing and shrinking cities, the uniform central housing policy becomes futile; the cities with decreasing population require different tools than growing cities [23, 24]. An example here can be the central housing Policy in Poland, with the construction of new homes as a priority. The tool that can be useful in Warsaw is detrimental to regions undergoing depopulation. It is easier for the developers to construct new buildings on suburban plots instead of modernising the existing ones in the cities.

4 Housing Policy Strategies Against Depopulation

As there is no single model nor cause for the shrinking of cities, there is also no single, reliable cure. Activities undertaken by cities against depopulation can be basically divided into three groups: (1) adaptive, (2) growth-oriented [15] and (3) ignoring the phenomenon. The first two groups of activities can be performed simultaneously. The first group includes all such strategies as: planning for shrinkage [25], shrink smart or right sizing, which are based on a belief that the shrinking of cities is a natural phase of development, and with appropriate management it can improve the quality of life in cities. An exceptionally good example of a consistent strategy against depopulation is Leipzig - an industrial city in eastern Germany.

4.1 The Leipzig Strategy

The outflow of inhabitants from Leipzig started in 1966 and it accelerated significantly in 1989 after German reunification, when the population decreased by 12% in only 10 years. As a result of undertaken activities, in 1999 depopulation stopped, and then, in 2001, the number of inhabitants started growing. In years 2011–2013 the increase was 2.8% per year [30]. Even though after the fall of the Berlin Wall the process of city shrinking actually affected all the eastern German cities, Leipzig was the first to accept it and it immediately started acting to change the apparently negative trend into an opportunity [26]. The other cities, e.g. Dresden or Halle, ignored the phenomenon (cf. strategy 3) until the year 2000 when the general debate on shrinking cities (*schrumpfende Städte*) started in Germany. Activities related to depopulation in Leipzig focused around three main axes: (1) preservation of architectural heritage, (2) creation of public spaces and green areas in abandoned places and demolishing vacancies (dilapidated housing estates) and (3) creation and support for neighbourhood local centres [14]. From the housing policy's perspective, the most interesting axis of actions, the so called 'patchwork urbanism', is the most interesting [16]. In years 1997–2007, as a result of strategy implementation, 11,390 homes were demolished in Leipzig [26]. It was financed under federal (e.g. *Stadtumbau Ost*) and central programmes, and according to their assumptions, no permanent building structures could be erected in those sites for the next 10 years. Thus, the city was implementing the 'greener through fewer houses' strategy. Therefore, post-demolishing sites were turned into grasslands, small parks, artistic installations (e.g. sponsored by the Urban II EU programme) and gained temporary social functions, such as pop up [18]. A significant part of demolished buildings were large prefabricated housing estates built in the post-war period, therefore the process was accompanied by efforts to increase the standard of the remaining homes. In parallel, a programme entitled "Home Guardians" was carried out in Leipzig, under which the city facilitated re-occupation and use (sometimes temporary) of abandoned homes. The city would bring together the owners and potential tenants and provided legal assistance for temporary tenancy agreement signing. Even though Leipzig strategy is criticised for a certain dose of chaos, lack of a coordinated demolishing plan and to high costs [14], it should be appreciated for efficiency, and we must also remember its pioneering role and learn from its mistakes.

4.2 Collaborative Housing as an Efficient Tool for Depopulation – The Example of Rotterdam

An solution which seems to perfectly address the problem of depopulation is collaborative housing, which is a form of obtaining homes where the basic rules include: (1) the non-for-profit idea, (2) participatory and (3) co-operative-based nature of the undertaking from the beginning of the designing process at least until the moment of occupying the premises and (4) group initiation by future tenants[1]. Collaborative housing can be executed through participation in a construction group, in a small housing co-operative or in co-housing.

The use of collaborative housing in the process of regeneration and revitalisation of building that are degraded not only in technical terms but mainly morally degraded, can be presented on the example of a regeneration case study of the Wallisblok block in the emptying neighbourhood of Spangen in Rotterdam (the Netherlands). It is important that due to the investment the quality and security in the neighbourhood were improved significantly, the outflow of the inhabitants was stopped, and new ones were attracted, without causing the negative processes of gentrification: a significant number of local inhabitants remained.

---

[1] It concerns the very group initiation, it does not mean that the conditions for initiative development cannot be established by third parties, e.g. the city or a non-for-profit institution.

Wallisblok, De Dichterijke Vrijheid 07 Rotterdam. In 2003 the city of Rotterdam decided to solve the problem of the degraded, dangerous and depopulating neighbourhood of Spangen. It started from analysing the Wallisblok block from the 1930s, located near the Schie canal. This dilapidated and partly abandoned complex, originally with very interesting architecture, comprised 75 small, mostly abandoned labourers' homes. It was in a catastrophic condition: the windows were smashed, the roof was leaking, the foundations were disturbed and soaked with water and so it was impossible to attract private investors. Preliminary financial analyses showed that the overhaul outlays - apart from the purchase price and foundation repair cost - would be equal to the value of the building after renovation. Apart from the costs, potential investors were deterred by the neighbourhood's bad reputation. In this difficult situation, the architects that were engaged to evaluate the condition of the complex (Hulshof Architecten) proposed to give the flats to future inhabitants and to use the potential of collaborative housing to regenerate the development. The Collectief Particulier Opdrachtgeverscha (CPO) scheme - i.e. the Dutch model of a building group - was adopted for the undertaking.

In October 2004 the authorities of Rotterdam announced that they were distributing flats for free, however, future inhabitants had to fulfil certain conditions:

- they had to invest at least 1,000 euro in every square metre of a flat;
- the flat could not be sold or rented for at least a year after investment finalisation;
- the owners of each flat had to get involved in the planning and designing process;
- the designing process had to be supervised by a manager and by architects appointed by the city;
- the building permit design and implementation had to meet the specified quality standards;
- the construction had to start no later than within a year, and the overhauls had to be finished within 6 months from their beginning.

Even after taking all the conditions into account, the offer was attractive: after termination, the process promised a relatively inexpensive high-standard flat in acentral location and a strong group of neighbours. On its part, the city repaired the foundations, and it allowed to totally reconstruct the internal façade of the block even though the building was under building preservation protection (which allowed to increase the interior and improve energy efficiency). Ultimately, 35 out of 200 interested families were selected for the process. The supervision over the undertaking was entrusted to Frans van Hulten from the Steunpunt Wohnen studio (today's Urbannerdam) and architect Ineke Hulshof from Hulshof Architects. The architect and the manager took the assisting position in the inhabitants' process of self-organisation and of specifying their needs and capabilities. At first, they defined a set of rules as well as quality and spatial standards.

As it turned out, it was difficult to match expectations of particular families with the existing structure, however, in the end, a compromise was reached. The proposed model of block reorganisation assumed maximum diversity of spatial solutions, best suiting the needs of future inhabitants: flats of different sizes, studios, office spaces and segments with gardens. Attics were also used as residential area. It was determined that every flat would have an independent exit to the outside, new stairs, thermal insulation, central heating and installations. The inhabitants agreed to share a garden and parts of the terrace on the roof. Future inhabitants were divided into groups which were made responsible for particular construction-process elements (garden, construction, finance). The inhabitants could perform the construction works by themselves, provided that they maintain the established standards, or they could commission the works to the team that was working on the whole complex. The costs of renovation of a single home were calculated at the level of EUR 70 K for a small apartment up to EUR 200 K for a four-storey house. Finally, 41 flats were prepared with the area ranging from 55 to 300 m^2 which - at the end of the process - became privately owned homes belonging to the inhabitants (privatisation was justified in that case, as city homes were dominant in that neighbourhood, so some ownership diversity was needed).

At present the inhabitants have been living there for over 10 years, they are well-integrated and very satisfied with their place of residence. The city of Rotterdam continued such a scheme for consecutive years. Similar works have been carried out for more blocks all over Spangen and so the neighbourhood became a safe and appreciated living area. The key to success seem to be the preliminary conditions established by the city: the fact that the flats were handed over to people on condition that they invest in them and maintain the standards and that they work under the supervision of professionals. The support of the city was also important: patronage over the talks with the bank on loans and with infrastructural companies supervised by the city as well as procedural facilitation. For the sake of the future inhabitants' sense of security, it was important that all the group moves in at the same time. This action also had positive effects: the security evaluation of the block and then of the whole neighbourhood grew on a 1 to 10 scale from 4 to 7. The investment was awarded the Job Dura Prijs 2006 prize awarded every other year for activities changing Rotterdam to the better.

4.3   Other Elements

Some other interesting elements in the strategies related to housing, introduced in the depopulation periods, include:

- 'don't move improve' - a bottom-up initiative to revitalise Bronx, carried out in years 1970–2012, where e.g. 320 tenements burned in fires were reconstructed;
- the integrated 'GhettUp' programme, combining the social policy approach with revitalisation in emptying and aging Genova [18];
- British planning basis according to which 80% of new investments must be located in brownfields, which prevents urban sprawl and city centre depopulation.
- financial penalties in German and French cities for misusing or not using homes. Such activity prevents the negative phenomenon of keeping vacancies for speculation purposes;
- support in council home exchange, e.g. by use of simple applications, such as the Dutch Huisjehuisje [31] which is similar to Tinder dating app and which gets tenants with different needs together.
- adding lifts to residential homes (in China, as council funds for such activities are scarce, commercial companies are allowed to fit lifts in buildings and then to collect small fees for their usage).

5 Housing Strategies in Polish Cities Against Depopulation

The studies were carried out on three shrinking cities, two of them belongs to Silesian region, that is endangered with depopulation to the greatest degree (Bytom and Sosnowiec) and one from Pomerania, littoral region with observed population increase (Sopot).

Sosnowiec and Bytom are located in Upper Silesia - an industrial conurbation in the south of Poland. Due to industrial heritage and deindustrialisation which occurred in that area after 1989, the whole region has been experiencing a significant outflow of inhabitants. In years 1988–2015 about 340,000 people left the region, which ranks Silesia first among the depopulating voivodeships in Poland. It is projected that by 2020 another 128,000 inhabitants will depart. Population losses are so significant mainly because of loss of jobs in sectors related to industry.

Depopulation processes, even though they mainly occur internally, are particularly severe in Sosnowiec [11]. According to projections, by 2035 as compared with 1988, about 37.8% of people will leave the town, turning Sosnowiec into the most shrinking town in the Silesian Voivodeship. The outflow of people from Bytom is also internal, but also a large share of emigration abroad is involved. It is projected that by 2035, as compared with 1988, the outflow will reach 31.6% of people. In Bytom the problem was exacerbated by the closure of 4 out of 6 hard coal mines being the main employers in the town (in 2012 the unemployment rate in Bytom reached 19.9%) as well as by severe development degradation due to mining damages. As a result of mining operations, the surface of the whole town lowered by 40 m, several housing estates were destroyed and there are periodical rock bursts and subsidence (e.g. in 2011 Karb housing estate was demolished and so were about 50 different tenement houses in 2016). What is interesting though, is that the outflow rate of people from Sosnowiec is higher than from Bytom even despite lower unemployment and smaller degradation of the urban tissue.

Sosnowiec and Bytom Housing Policy Assessment. According to an analysis of academics from the University of Silesia [19] almost all the towns in the region, including Bytom and Sosnowiec, have got their strategic and urban planning documents assuming mainly their development; the possible reduction of spatial, demographic, economic structures and infrastructural elements that are included therein are immediately balances by provisions on modernisation, compensation, replacing "the old" with "the new"; no advancing process of growing share of derelict spaces or social and economic structures being under the influence of advanced regress is assumed. In the whole strategy for Bytom the word 'depopulation' is mentioned only once.

In their strategies, both Bytom and Sosnowiec [27, 28] underline the need to privatise housing resources, pointing at the same time to the problem of affordability of homes. What is quite striking is that both these towns show that even despite loss of inhabitants the number of people waiting for a council home has not decreased. Bytom also has a problem with housing market stagnation.

Both Towns Emphasise the Need to Retain Young People. Bytom does it through an action called: "Homes for the Young" [32] where they provide council homes for a promise of an overhaul, however the flats are in a very poor technical condition (very often with no bathroom nor heating), and second, there are only a few such flats available every year. Too few for the action to improve the situation. Sosnowiec runs an action called "A Home for a Graduate", where talented university graduates can rent redecorated homes at a moderate rent price. Due to the form of rental and not sale, the flats remain within the municipality's assets and they can be let to next generations. Sosnowiec also runs an action called "Self-overhaul" - where flats are handed over for an overhaul, however, it is open to all the inhabitants. Moreover, Sosnowiec offers permanent assistance in reducing the debt of indebted inhabitants of council homes. The town also looks for non-standard solutions that can help with using the potential of affordable housing. In February 2018 it started its cooperation with the Faculty of Architecture at the Silesian University of Technology in order to verify the possibility of implementing the above described collective housing model, as in Rotterdam, in unoccupied facilities owned by the town.

In the policies of Sosnowiec and of Bytom, concerning depopulation, there is also a difference in the form of town management. In Bytom the authorities work with a smaller involvement of the stakeholders from the social and economic zone. The problems are intensified by the fact that an acting mayor was dismissed as a result of a referendum twice. In Sosnowiec, on the other hand, there is stricter cooperation between the town and the economic zone. There are certain actions carried out to encourage investors to invest in the brownfields, which are quite numerous after deindustrialisation. The processes of reindustrialisation are also supported on an ongoing basis.

Sopot is a depopulating city whose situation is, however, completely different than that of the two towns described above. As opposed to Silesia, the region of Trójmiasto (Tri-city) that Sopot is a part of (it includes Grańsk, Sopot, Gdynia and adjacent municipalities), is characterised by a constant growth in the number of inhabitants. Problems with the collapse of the shipbuilding industry in the region is compensated by new, emerging employers, largely related to the tourism potential of the region. The loss of inhabitants in Sopot is therefore mainly related to negative natural growth and suburbanisation: it is the so called apparent depopulation, where the inhabitants leave the core city and move to the suburbs. However, they remain within the metropolitan area and so they do not reduce the potential of the region. The main problem with Sopot is aging population, because the outflow to the suburbs mainly concerns relatively young people. It is projected that by 2050 the natural growth in Sopot will be the smallest in the voivodeship, reaching the rate of −10 per 1,000 inhabitants. The average age in Sopot will be than 52.5 years.

Sopot Housing Policy Assessment. An analysis of Sopot's housing policy [29] shows that the city is trying to respond to emerging phenomena on a current basis. Its priority goals include an increase in the percentage of young people in the city's housing resources and undertaking activities that could materialise this goal. Some actions are also undertaken to improve the quality of life of seniors. Some of the most interesting activities under the housing policy in Sopot include:

- promoting the model of co-living (of students, young professionals and elderly people) in larger apartments;
- zoning of the city and keeping separate housing policies for particular neighbourhoods, facing different challenges;

- an internet database of homes for exchange - supporting exchange of flats between tenants of council resources (according to the above described mechanism used in the Netherlands). What is more, the system of exchange is also co-ordinated with the neighbouring municipalities;
- preferring families with children, at least for council homes for rent
- exchange of large apartments without a lift, in old buildings, to smaller ones, but better suited to the needs of elderly people (purchase and sales);
- active search for lonely people occupying too large, underinvested flats in order to exchange them for smaller and more comfortable ones;
- limiting privatisation of flats. Sopot, as a tourist destination, is exposed to the loss of residential function and dominance of short-term lease premises as well as homes purchased for speculation purposes.

Conclusions. A comparison of the housing policy in Sosnowiec and in Bytom in the aspect of depopulation leads to several conclusions. First of all, Bytom, even though it is in a much worse situation (higher unemployment rate and much more degraded housing resources), initially shows a lower tendency for depopulation than Sosnowiec. Robert Krzysztofik from the University of Silesia [11, 19] points to the fact that the strength of Bytom is its strong local identity related to a large number of inhabitants identifying themselves with the town. Sosnowiec, on the other hand, has a much less clear identity; a significant number of its inhabitants are people who came to the town after WWII.

Bytom starts losing its beneficial position as a result of a policy which ignores the processes of depopulation (type three of depopulation strategy). The latest data shows that unemployment in Sosnowiec, as opposed to Bytom, is decreasing and depopulation can be slowed down a little bit, as compared to the assumptions.

Sopot, despite a better situation than Bytom and Sosnowiec, has a much broader and much better-informed housing policy. The activities performed by the city seem to be aimed at changing the depopulation trend (type two of the depopulation strategy). Observation of depopulation policies of cities exposed to a different degree to depopulation confirms the above noted regularity: cities that are less affected by depopulation seem to react to it quicker. It is noted that local authorities chosen in periodical democratic elections try to belittle or ignore serious problems of depopulation. The reason for this is that after the end of the term of office it is impossible to show that the problem has been reduced. These fears are to some degree justified - the problem of depopulation, particularly in such regions as Upper Silesia, which is affected as a whole, cannot be solved within a 4-year term. If the solution is perceived as the revers of the trend - it will probably never be solved. The observation was confirmed by a situation experienced in 2013 by a scientific consortium from the Silesian University of Technology, the University of Silesia and the Katowice University of Economics, which was trying to raise external funds for a vast research and development programme concerning depopulation of large-panel prefabricated housing estates: none of the Silesian cities affected by the problem of loss of inhabitants was interested in cooperation, even though it did not involve any financial outlays on their part. The municipalities were simply afraid to admit that the depopulation is their problem.

6 Summary

After decades of recovery activities, about 40% of European cities with the population of more than twenty thousand people are still shrinking [9]. In view of global demographic changes, even in Poland the trend cannot be reversed completely in predictable future. Still, ignoring the situation seems to be the worst possible strategy. A potential of slowing down the outflow of people from cities undoubtedly seems to be provided by changes attributable to the second demographic transition. Non-traditional households: singles, partnerships, single-gender households or non-family households create a new fashion for urban lifestyle. Gradually, housing preferences are changing, dominated for a certain period of time by dreams of a suburban house with a garden.[2] In order to use the potential of that trend, it is necessary, however, to run a well-informed housing policy against depopulation. A policy which will focus not only on increasing the quality of homes, provide access to green areas and high quality public areas, access to good schools or high quality public transport - i.e. to elements that are traditionally taken into account in revitalisation processes - but also on caring for the balance and social diversity, broadening the possibilities and the number of ways to get a flat (e.g. the potential of collective housing and stopping privatisation of council resources), or inclusion of the inhabitants in activities for the benefit of their own neighbourhood (as in the 'don't move improve' action). Moreover, it seems to be necessary to run a housing policy at all levels: central, local, but also at the completely neglected in Poland regional level.

As city activities in the field of housing will always be a political issue, it is of key importance to inform the society about the problem, its consequences and potential strategies, and awareness of the authorities is also very important. Regardless of whether we consider the shrinking of cities as a natural process which can be turned into success or a negative phenomenon which should be fought down - we cannot remain indifferent to it.

References

1. Silverman, R.M.: Affordable Housing in US Shrinking Cities : From Neighborhoods of Despair to Neighborhoods of Opportunity?. Policy Press Shorts Research, Bristol (2016)
2. Haase, A., Bernt, M., Großmann, K., Mykhnenko, V., Rink, D.: Varieties of Shrinkage in European Cities. Eur. Urban Reg. Stud. **23**(1), 86–102 (2016)
3. Haase, A.: Emergent spaces of reurbanisation: exploring the demographic dimension of inner-city residential change in a european setting. In: Population Space and Place, vol. 16, no. 5 (2009)
4. Cameron, S.: From low demand to rising aspirations: housing market renewal within regional and neighbourhood regeneration policy. In: Housing Studies, vol. 21, no. 1 (2006)
5. Waligórska, M., Kostrzewa, Z., Potyra, M., Rutkowska, L.: Prognoza ludności na lata 2008–2035. Zakład Wydawnictw Statystycznych, Warszawa (2014)
6. Śleszyński, P.: Delimitacja miast średnich tracących funkcje społeczno - gospodarcze (2016)

---

[2] In Poland that period was in years 1990–2010.

7. Oswalt, P. (ed.): Shrinking Cities, vol. 1. Hatje Cantz, Berlin (2006)
8. Wu, T.: Martinez-Fernandez, C.: Shrinking cities: a global overview and concerns about Australian cases. In: Pallagast, K. (ed.) The Future of Shrinking Cities - Problems, Patterns and Strategies of Urban Transformation in a Global Context. IURD, Berkeley (2009)
9. Turok, I., Mykhnenko, V.: The trajectories of european cities, 1960–2005. Cities 24(3), 165–182 (2007)
10. Martinez-Fernandez, C.: Shrinking cities: urban challenges of globalization. Int. J. Urban Reg. Res. 36(2), 213–225 (2012)
11. Krzysztofik, R.: "Zagłada Miast" - projekt Shrink Smart - The Governance of Shrinkage within an European Context na Uniwersytecie Śląskim. In: Szajewska, N. (ed.) Zarządzanie rozwojem miast o zmniejszającej się liczbie mieszkańców, pp. 45–56. Kancelaria Senatu, Warszawa (2013)
12. Nowicki, et al.: Opinie i ekspertyzy na konferencję o ekonomicznych stratach i społecznych kosztach niekontrolowanej urbanizacji w Polsce, Warszawa (2014)
13. Szajewska, N., Lipińska, M. (eds.): Zarządzanie rozwojem miast o zmniejszającej się liczbie mieszkańców. Chancellery of the Senate, Warszawa (2013)
14. Florentin, D.: The "Perforated City:" leipzig's model of urban shrinkage management. Berkeley Plan. J. 23(1), 83–101 (2010)
15. Hollander, J.B., Pallagst, K., Schwarz, T., Popper, F.: Planning shrinking cities. Prog. Plan. 72(4), 223–232 (2009)
16. Ryan, B.: Rightsizing shrinking cities: the urban design dimension. In: Dewar, M., Manning, J., (eds.). University of Pennsylvania Press, The City After Abandonment (2012)
17. Oguz, S.C., Velibeyoglu, K., Velibeyoglu, H.: Vulnerability of housing in shrinking cities: case of Izmir, presented at ENHR 2010. In: 22nd International Housing Research Conference, Istanbul, 4-7 July 2010
18. Haase, A.: No one-size-fits-all. O różnorodności kurczących się miast. In: Szajewska, N., Lipińska, M. (eds.) Zarządzanie rozwojem miast o zmniejszającej się liczbie mieszkańców. Chancellery of the Senate, Warszawa (2013)
19. Krzysztofik, R., Runge, A., Runge, J.: Kantor - Pietraga I.: Miasta konurbacji katowickiej. In: Stryjakiewicz, T. (ed.) Kurczenie się miast w Europie Środkowo-Wschodniej. Bogucki Wydawnictwo Naukowe, Poznań (2014)
20. Florida, R.: The Rise of the Creative Class. Basic Boks, New York (2002)
21. Florida, R.: The New Urban Crisis: How Our Cities Are Increasing Inequality, Deepening Segregation, and Failing the Middle Class—and What We Can Do About It, 1st edn. Basic Books, New York (2017)
22. Olbrycht, J.: Shrinking cities - problem globalny, problem europejski. In: Szajewska, N., Lipińska, M. (eds.) Zarządzanie rozwojem miast o zmniejszającej się liczbie mieszkańców. Chancellery of the Senate, Warszawa (2013)
23. Twardoch, A.: Centralna, regionalna i lokalna polityka mieszkaniowa w kontekście prognozowanych zmian demograficznych. Wybrane skutki przestrzenne, społeczne i gospodarcze. In: Stud. Ekon. Zesz. Nauk. Uniw. Ekon. w Katowicach, no. 223, pp. 21–31 (2015)
24. Twardoch, A.: Regionalna polityka mieszkaniowa - wyzwania w obliczu zmian demograficznych. In: Stud. Ekon. Zesz. Nauk. Uniw. Ekon. w Katowicach, vol. 290, pp. 271–279 (2016)
25. Wiechmann, T., Volkmann, A.: Making places in increasingly empty spaces: causes and outcomes of demographic change in Germany. In: Martinez-Fernandez, C., Kubo, N., Noya, A., Weymann, T. (eds.) Demographic Change and Local Development: Shrinkage, Regeneration and Social Dynamics, Paris, pp. 57–64. OECD (2012)

26. Rink, D., Haase, A., Bernt, M., Arndt, T., Ludwig, J.: Urban Shrinkage in Leipzig and Halle, the Leipzig-Halle Urban Region. Helmholtz Centre for Environmental Research, Leipzig (2010)
27. Polityka Mieszkaniowa Gminy Sosnowiec na lata 2010–2020, City Hall, Sosnowiec (2010)
28. Strategia Rozwoju Bytomia Na Lata 2009–2020, City Hall, Bytom (2009)
29. Polityka mieszkaniowa Miasta Sopotu na lata 2015–2019, City Hall, Soppot (2015)
30. http://population.city/germany/leipzig/
31. https://www.huisjehuisje.nl
32. http://www.bm.bytom.pl/category/lokale-mieszkalne/wykazy/
33. http://ec.europa.eu

Kurdish Garden as an Example for Middle East Botanical Garden: New Approach and Aspect

Kardo N. Kareem[1,2(✉)], Wojciech Bonenberg[1],
and Bahram K. Maulood[3]

[1] Institute of Architecture and Planning, Faculty of Architecture,
Poznan University of Technology, Poznan, Poland
[2] Horticulture Department, College of Agriculture,
University of Salahaddin-Erbil, Erbil, Kurdistan Region, Iraq
kardon.kareem@gmail.com
[3] Hawler Botanical Garden Erbil Government, Erbil, Kurdistan Region, Iraq

Abstract. Botanical Garden (B.G) is becoming a multi-purpose garden in contrast to traditional gardens or parks. However, it's almost absent in whole Middle East countries so far. The landscape study for this project is a necessary need at the moment to create, improve and recreate botanical gardens in Kurdistan. Hawler Botanical Garden (H.B.G), with various landscape designs involves samples of almost all known gardens of the world. However, Kurdish garden will be forwarded for the first time and scientifically will be studied. Kurdish gardens will be the guide line for any progress in this aspect. Any way throughout the present project natural Kurdistan gardens will be considered in contrast to other gardens of other parts of the world. One must refer to that, the goal of Hawler Botanical garden will be to establish a Kurdistan flora, Kurdish Garden, Kurdish national herbarium and seed bank all of which are not exist so far.

Keywords: Kurdish garden · Botanical garden · Urban planning

1 Introduction

Gardening in Kurdistan unlike other parts of the world may go back to more than2000 years. The background precise knowledge on Kurdish Garden (K.G) historically is absent in comparison to different spot in the other parts of the world. Quite many reasons are behind this that includes social, political, educational and even scientific ones. For the best of our knowledge, no documents about Kurdish garden in literature exist so far [1].

Culture and history of Kurdistan in respect to garden may be compared and contrasted with that of Arab world in one side and Persian culture on the other, still the gardens hear may be effected by Ossmanian and Turkish culture. In spite of that there is evident that K.G, still kept its Owen specialty in respect to culture, design and landscape architecture. In fact the region belongs to Indo-European origin; geographical Kurdistan lies in the land between Iran, Iraq, Turkey and Syria [2].

© Springer International Publishing AG, part of Springer Nature 2019
J. Charytonowicz and C. Falcão (Eds.): AHFE 2018, AISC 788, pp. 164–173, 2019.
https://doi.org/10.1007/978-3-319-94199-8_16

The present assay will deal with more detail properties of gardens and orchards in Kurdistan. For the best of our knowledge, this is a first attempt to deal with such garden in this part of the world.

In fact, topography of Iraqi Kurdistan includes plane land, hills and mountains across altitudes from less than hundred to more than 3000 m A.S.L. plane land in the area is known to be one of the most fertile land of the area [3, 4].

Geologically it lies upon upper Bakhtiari plane, in addition to dolomite, sedimentary rocks and conglomerate stones. Volcanic and igneous rocks may also be found [5, 6]. Soil varies from fertile ones in Hawler and Shahrazor plane to organic rich brown soils of natural forests, Soil generally has PH above (7) and dominated with calcium and potassium [5, 7]. Hydrologic ally water is almost available all over the year in different form rain, snow, surface and underground water with alkaline properties [8].

The climate is semi-arid type [9]; subtropical [10] belongs to Irano-Turanian climate [11] which is characterized by the wide range of annual and diurnal range of temperature [12]. Climate wise occurrence of the three seasons: a cold rainy winter, a mild growing period of spring and a hot dry summer all these are distinctive properties of this region [13, 14].

Ecosystems in Kurdistan may be fall under three distinct categories: steppe (Plane land) vegetation, forest vegetation that includes hills, foothills and mountain in lower altitudes. Whereas high mountain vegetation (Alpine vegetation) include vegetation in high altitudes. However, the zonation of Iraqi Kurdistan may be found in [15].The vegetation in the proposed Kurdish garden are demonstrated within the present landscape investigation which includes trees, shrubs, flowers and grasses.

The present investigation is an attempt to reduce the existing gap of the knowledge on Kurdish garden. It focuses on some specialty of Kurdish garden in comprises to world other gardens in respect to landscape architecture, greenery, flora and even irrigation.

2   Materials and Methods

Hawler Botanical Garden (H.B.G.) were planned to be established on an area of 200 Donum (1 donum = 2500 m^2) which is equivalent to 500 000 m^2 (50 ha). Forestry department ministry of agriculture cared of the land before. It is situated along the road to Kasnazan-Koya facing the main Martyrs monument on an arid land around 14 km from Erbil (Hawler) city center [1].

The project started since 2014. Surveying more than 1000 spots so far (Fig. 1B) as part of flora of Kurdistan project team in order to collect, identity and establish Kurdistan flora. Identification of plants was carried out in H.B.G. and samples were preserved in its herbarium.

Kurdish garden that what the present paper is concerned with is chosen within H.B. G it has an approximately of 9 dounm (22,500 m^2) (2.25 ha) and located close to the west part of botanical garden.

The design and photographs of various natural gardens in all different parts of the country from the previous mentioned surveys were taken in consideration, in the present project. The fences, outdoor landscape and architecture were all conceded in

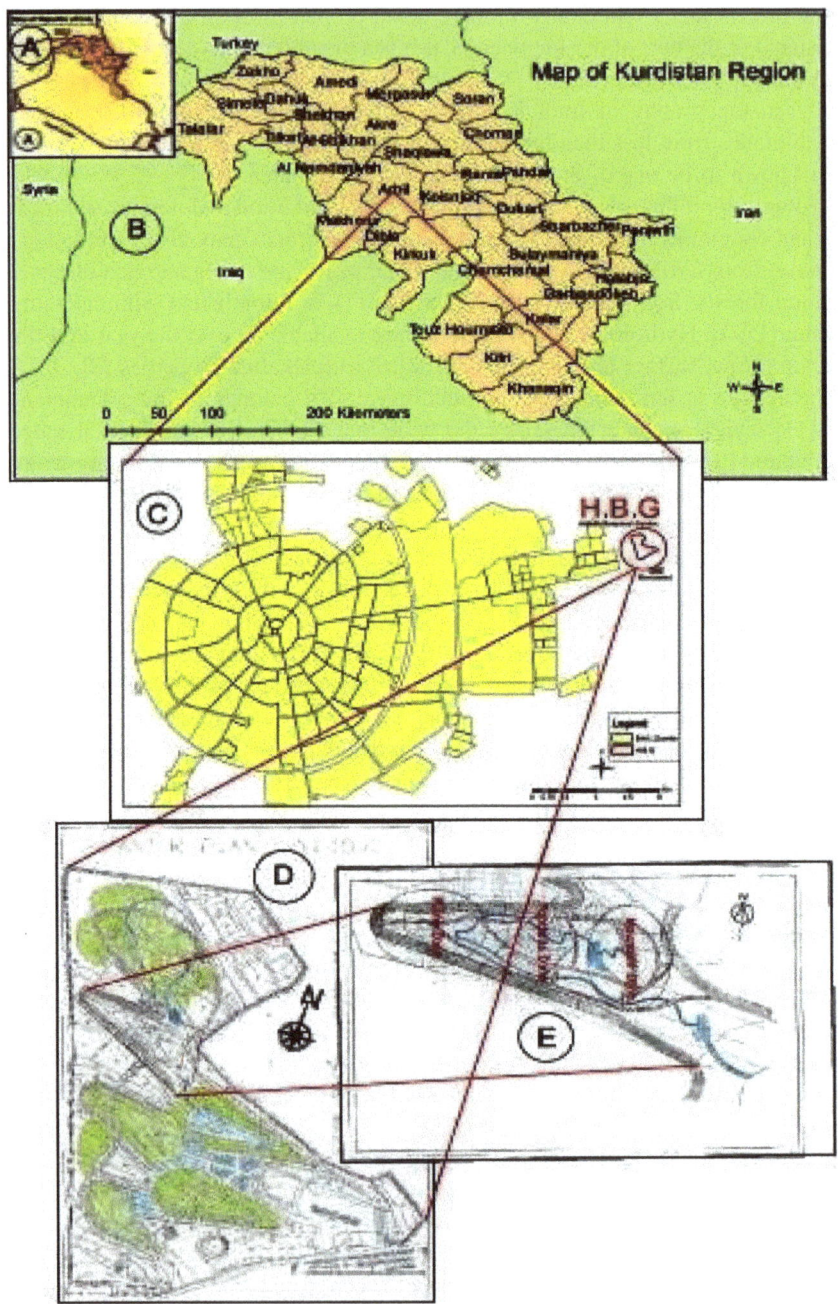

Fig. 1. General map of (A) Iraq. (B) Kurdistan region-Iraq. (C) Location of H.B.G. within Erbil (Hawler) Province. (D) Master Plan of H.B.G. (E) Kurdish Garden. [16].

the proposed Kurdish garden even the water channel and way of irrigation and plantation (Fig. 1A, B, C, D, E).

3   Result

Necessary background knowledge to establish K.G came from all parts of Kurdistan (Kurdistan Iraq, Iran, Turkey and Syria). In fact, the architecture landscape design was performed in such a way to represent natural gardens all parts of Kurdistan with its different topography. In contrast, metrological data and soil analysis were confined to H.B.G only.

Fig. 2. (A1) Pyrus syriaca. (A2) Pistacia terebinthus. (A3) Qurcus aegilops. (B1) Nerium oleander. (B2) Prunus arabica (C1) Fritillaria imperialis. (C2) Calendula arvensis. (C3) Narcissus tazetta.

For the sake of mentioned garden an area of 9 donum within Hawler botanical garden (H.B.G) were allocated and a detail landscape study may be found [16] the proposed area is situated on west part of the main project of H.B.G that was already started since 2008 (Fig. 1D, E).

Table 1. Mean values some climate factors in Erbil.

Years	Some climate factors					
	Air temperature (C°)			Humidity	Rainfalls	Wind speed
	Max. Tem.	Min. Tem.	Av. Tem	(%)	(mm)	(M/Sec)
2014	28.01	16.41	22.21	50.91	32.38	0.52
2015	27.82	16.35	22.09	50.33	35.19	1.008
2016	27.59	16.22	21.9	49.66	31.31	1.49

Landscape design of the spot includes water channel, waterfall, paths, tea house benches, type of rocks etc. as well as plant distribution to involve most recorded flora. Examples are illustrated in (Fig. 2) showing plane land, foothills and mountains.

Metrological data and soil studies of the area are shown in (Table 1) and (Fig. 3) below. Irano-Turanian climate were reflected throughout the result as months of May to beginning of September was dry and hot, whereas rainy season and temperature decline were observed from October on word. Max humidity was recorded during January whereas wind direction was generally South West never exceeds (7 m/s).

Table 2. Some physical and chemical properties of soil samples in H.B.G.

Saturation point %	36.22	37.40	34.50	37.70	36.45	35.90	36.34	35.89	36.77	35.87
Field capacity%	19.50	19.30	18.90	19.80	19.20	18.80	19.90	19.10	20.40	18.90
Soil PH	7.63	7.71	7.60	7.99	7.50	7.61	7.55	7.64	7.29	7.77
Total N %	2.24	2.24	2.28	3.36	2.24	4.48	0.28	3.92	2.20	2.30
Total P %	0.33	0.44	0.39	0.28	0.36	0.41	0.32	0.36	0.33	0.31
Organic matter %	2.45	2.45	2.56	2.34	3.10	2.34	2.44	2.56	2.45	2.22
Ca/mmol/L	4.60	5.10	4.40	3.90	4.50	4.70	4.70	3.60	4.90	3.80
Mg/mmol/L	2.40	3.10	2.90	2.60	2.70	2.40	2.30	2.20	2.80	1.90
So$_4$/mmol/L	8.40	8.80	8.10	7.70	7.90	8.50	8.40	8.60	7.20	9.10
Cl/mmol/L	1.30	1.60	1.60	1.60	1.20	1.30	1.40	1.30	1.10	1.50
K/mmol/L	12.40	13.40	11.70	12.50	12.40	12.70	12.10	11.90	13.20	12.30
Na/mmol/L	7.40	7.30	7.60	7.90	7.80	8.00	7.90	7.70	7.60	7.60
E.C/mmoze/cm	0.67	1.96	0.53	1.14	0.58	1.27	0.22	0.58	0.20	1.67

The soil in this area which is located on upper Bakhtiari region consists of sandstone, metamorphic rocks with low amount of clay. However, (Table 2) shows that the PH never exceeds 7.6 with large pore size. Calcium and potassium was found to be the main dominate bivalent and monovalent cataion respectively in the area.

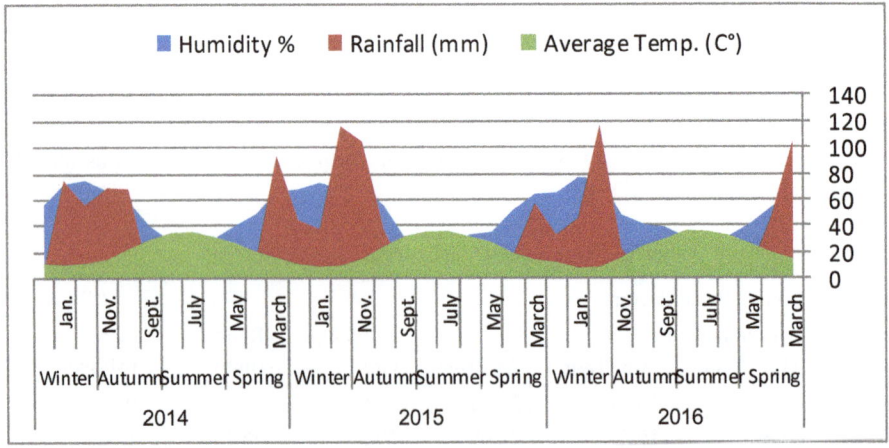

Fig. 3. Average monthly humidity, rainfall and temperature in Erbil, Kurdistan – Iraq (Erbil metrological office).

Forest plant within K.G consisted of about 50 native and naturally distributed trees. All were spotted and planted within the landscape gardening of Kurdish gardens dendrologicaly sketch. Whereas shrubs have been planted and arranged in such away to be as close as to its natural habitation as possible (Fig. 5).

Flowers and grasses contribute the main bulk of Kurdistan flora. The plotted ones make up more than 300 of naturally found plants in different parts of Kurdistan. However, the survey for plant collection is started since last decade in order to produce Kurdistan flora. The survey and plantation extended to threaten plants in the region, as there are quite common in Kurdistan. (List 1) represent local names, common name and taxonomic nomenclature for some representative samples of plant in proposed K.G is listed down below.

List 1. Example of plants in Kurdish garden.

No	Scientific Name	English Name	Kurdish Name	No	Scientific Name	English Name	Kurdish Name
1	*Acer monspesulanum* L.	Montpel-lierMaple	كاتوة ت	11	*Rosa moschata* Herrm.	Musk Rose	طولَة شيلانة
2	*Celtis tournefortii* Lam	Oriental Hackberry	طاوط	12	*Salix purpurea* L.	Purple Willow	بيى
3	*Pinus brutia* Ten.	Calabrian Pine	كاذى زاويّتة	13	*Althaea officinalis* L.	Marsh mallow	طولَة هيّرؤ
4	*Populus euphratica*	Euphra-tesPoplar	دارشيندان ى فوراتى	14	*Bellis perennis* L.	Common daisy	طولَةدوطم ة ،،قؤنضة
5	*Prunus mahaleb* L.	Mahaleb Cherry	دار كنيَز	15	*Fritillaria imperialis* L	Crown imperial	شلتيَر
6	*Quercus aegilops* L.	Tabor Oak,	بةروو	16	*Gladiolus italicus*	Italian gladiolus	طولَة طانزيزة
7	*Anagyris foetida* L.	Mediter-ranean Stinkbush	حافرقترة	17	*Narcissus tazetta* Boiss.&Gail	Narcissus	زطريَن
8	*Cotoneaster racemiflora* Deaf.	Cotoneas-ter	بالووك، كاكيف	18	*Tulipa kurdica*	Tulip	قلايم
9	*Glycyrhiza glabra* L.	Liquorice, licorice	سوو س (ميَكؤك)	19	*Adiantum capillus-veneris* L.	Southern maiden-hair fern	باريَزة ر- خالَةترةشة
10	*Paliurus spina-christi* Mill	Christ's Thorn	دركى قودس, زى, داركةترى	20	*Typha latifolia*	Common bulrush	زقلَ

In fact the final artificial (hard) landscape of K.G. incudes bridges, water channel, different benches, crash cans, illuminate, tea house, pond, water fall and type of rocks in Kurdistan.

Different spots within the 9 donum area of the garden were specified to different habitat and its floral component. The landscape design engineering wise make up in such a way to represent natural Kurdish garden within different altitudes and topography still it designed in such a way to be as close as possible to the natural existing ecosystem. Most landscape architecture design found in Kurdistan, were tended to be included (Fig. 5).

4   Discussion

In fact the earliest gardener of course was neither artist, scientist nor hobbyists they were simply farmers [17]. They learned medicinal and culture power of plant as well as their deadly poisons [18]. In fact agriculture is a prerequisite to civilization [19, 20]. Wild beauty would exist with or without man power [17]. In fact man with his art, brain and labor composed those growing plants to what is known as gardens. The history of garden in our planet may go back to few thousand years B.C. that goes in parallel with different civilization all may be found in [17, 19, 20].

No information for the best of our knowledge exists so far about design, history, shape, architecture component on Kurdish garden [1]. Although it's history may go back to hanging garden of Babylon or Nimrud [21, 22].

Kurdistan is situated along the Hardness zone (7–9) [19] there for a similarity of its flora may be with that of European flora in one side and Arab world and Persian world on the other hand, still it has its Owen specialty and character [1, 16].

Gardens found in Kurdistan along its topography are quite heterogeneous including mountain in high land to fertile plane land in the south region [4]. In fact it was attended to integrate all in one garden. However, the reality of information make one to prepare a more detail study on each type of the garden in feature hens the present work could be regarded as the pioneer for that [16].

Fig. 4. Kurdish garden preparation and design in Kurdistan.

The fence of garden in Kurdistan is quite specific in its structure, component, design and its gates the main ones as well as small ones (Fig. 4). However, the landscape design with water passing and sharing throughout succeeding gardening in such a mechanical engineering water channel design that drive water proportional to the size of the garden, plant type and density also. It is believed that it's a unique way of watering in whole Middle East and may be even in the world.

The cottage map as well as the plantation within the garden in also quite different as fruit trees area designed to be in one side while flower ones plotted always close to gathering space. In contrast to Ossmanian, European and American gardens [23–25] no fountain and water reservoir is exist. In contrast small pond along the running water channel may be the feature of Kurdish gardens. In fact even the proposed paths in K.G are design in such a way that to be reflected the area when the garden is located in (highland in mountain or lowland in plane). All at which are illustrated within landscape design shown below (Fig. 5).

Wild birds of K.G are quite many such as (pigeon, starling, falcon and hawk etc...) many nest, on various trees can be noticed whereas poultry birds such as hens, turkey may be also be seen.

In fact this preliminary study may influence and encourages more detail investigation on K.G. that is not found so far in literature there for the existing gap of knowledge will undoubtedly be reduced.

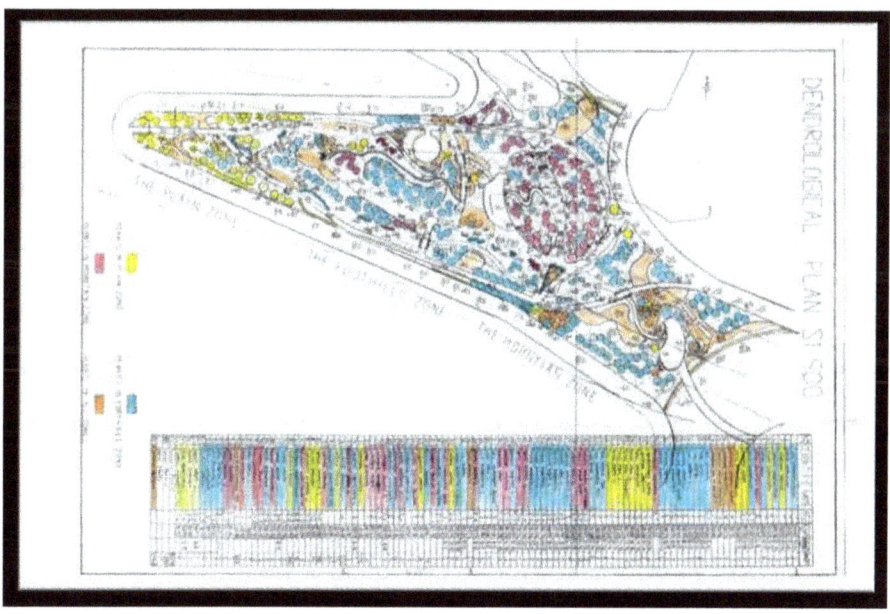

Fig. 5. Dendrological plan in Kurdish garden by diagram plant region [16].

Acknowledgments. We would like to acknowledge and thank Poznan University of Technology in Poland and Hawler Botanical Garden (H.B.G.) as well as the Government of Erbil, Iraqi Kurdistan for their assistance and financial support for this project.

References

1. Kareem, K.N., Bonenberg, W., Maulood, B.K.: Hawler botanical garden (H.B.G) an attempt for scientific greenery and element of Urban Identity Kurdistan – Iraq. Am. Int. J. Res. Form. App. Nat. Sci. **19**(1), 33–40 (2017)
2. Ali, S.S., Culham, A., Maulood, B.K.: A systematic study on the genus erodium in Kurdistan of Iraq. Int. J. Emerg. Technol. Comput. Appl. Sci. **21**(1), 11–17 (2017)
3. Guest, E.: Flora of Iraq. In: Introduction to the Flora. Ministry of Agriculture, Baghdad, Iraq, vol. 1 (1966)
4. FAO: Agro-Ecological Zoning of the three Northern Governorates of Iraq. Erbil-Section, Food Agriculture Organization (2003)
5. Buringh, P.: Soils and Soil Conditions in Iraq. Ministry of Agriculture. Directorate General of Agriculture Research and Projects, Baghdad Government Publication (1960)
6. Jassim, S.Z., Goff, J.C.: Geology of Iraq. Dolin, Prague and Moravian Museum, Berno, Czech Republic (2006). 341p.
7. Anon: Hawler Botanical Garden Report Soil Analysis of Botanical Garden. Erbil, Kurdistan Region-Iraq. (2014)
8. Moulood, B.K.: Feature Aspects of Kurdistan Water Resource. Publication No. 9 of Kurdistan Central Research Office. Ministry Education Press. Erbil, Kurdistan Region-Iraq (1996). (in Arabic)
9. Rzoska, J.: Euphrates and Tigris, Mesopotamian Ecology and Destiny: Monographiae Biologicae. Springer, London (1980)
10. Al- Shalash, A.H.: The Climate of Iraq, Amman, Jordan (1966)
11. Ghazanfar, S.A., McDaniel, T.: Floras of the middle east: a quantitative analysis and biogeography of the Flora of Iraq. Edinburgh J. Bot. **73**(1), 1–24 (2016). Trustees of t he royal botanic garden Edinburgh
12. Al-Saadi, H.A., Al-Mayah A.R.: Aquatic Plants in Iraq. Arabian Gulf Research Center Publication. Basra University (1983)
13. Al-Sahaf, M.: Pollution Control and Water Resources of Iraq. Al-Hurria Printing House, Baghdad (1976)
14. Maulood, B.K., Al-Saadi, H.A., Al-Zubedi, F.S.: Ecology. Ministry of Higher Education and Science Research. Babylon University College of Sciences, Iraq (1992)
15. Weinert, E.: The vegetation zones of Iraq. Bielefelder Okolog. Beitr. **4**, 45–57 (1989)
16. Kareem, K.N.: Ph.D. thesis under preparation. Faculty of architecture. Poznan University of Technology. Poznan, Poland (2018)
17. Miklos, J.V., Fiore E.: The History, the Beauty, the Riches of the Gardeners World (1968). First printing. Published in the United States. Printed in Italy by Mondadori, Verona
18. Moldi-Ravenna, C. and Sammartini, T.: Secret Gardens in Venice (1996). Photographs by Gardin, G.B. translated by Precker, J.A., Sammartini, T.
19. Ullmann, H.F.: Gardening Encyclopedia. Botanica's Pocket (2007). Over 1000 pages and Over 2000 Plants listed. Printed in China
20. Brickell, C.: The Royal Horticultural Society A-Z Encyclopedia of garden plants. Dorling Kindersley Limited, London (2008)

21. Abdullwahed, A.: Planning and Design of Greenery Area and General Land Space in Cities (Landscaping). Knowledge House Library. Al-Azhar University, Egypt (2012). (in Arabic)
22. Noah, A.A.: Gardens. History of Landscape Design. Orchard Knowledge Library. Alexandra University, Egypt (2011). (in Arabic)
23. Calkins, C.C.: Great Gardens of America, 298 p. Coward-McCann, Inc., New York (1969). Printed in U.S.A.
24. Toman, R.: European Garden Design: From classical Antiquity to the Present Day (2005). Translating and editing: translate-A-Book, Oxford, UK. Printed in France
25. Guner, A.: Nezahat Gokyigit Botanical Garden: Guide Book. Istanbul (2011)

The Improvement of the Quality of Public Spaces on the Example of Student Competition Designs: A Case Study

Rafał Radziewicz-Winnicki[(✉)]

Faculty of Architecture, The Silesian University of Technology,
ul. Akademicka 7, 44-100 Gliwice, Poland
rafal.radziewicz-winnicki@polsl.pl

Abstract. The paper discusses the modernisation of public spaces being the subject of conceptual competition designs produced by students of the Faculty of Architecture of the Silesian University of Technology. The design phase was preceded by extensive studies. As a result of these activities, design findings and guidelines were formulated. They were framed in order to improve the discussed public spaces by means of the modernisation of functional, formal and ergonomic aspects. As a result, various architectural designs taking into account contemporary requirements related to this kind of facilities were produced. The paper proves that a competition for an architectural concept enables a broader look at the discussed issues, shows a full spectrum of solutions and, as a consequence, contributes to the improvement of the quality of discussed public spaces. The positive effect is, however, conditioned on the necessity to perform a wide spectrum of analyses ad pre-design studies.

Keywords: Human factors · Public spaces · Sustainable architecture
Urban planning · History of Architecture

1 Introduction

The paper discusses the modernisation of public spaces being the subject of conceptual competition designs produced by students of the Faculty of Architecture of the Silesian University of Technology in Gliwice.

The purpose of the work is to prove that such activities can result in a radical change and improvement of the quality of public spaces.

The study method used is a case study, and design examples are analysed against contemporary movements and trends.

The scope of the discussed issues covers two architectural competitions announced, organised and adjudicated by the Faculty of Architecture together with external institutions in the years 2015–2017. They were related to native areas located in Silesia. The first one involved modernisation of spaces around preserved fragments of historic defensive walls of the City of Żory, and the second one was related to the modernisation of the Steam Engine open-air museum within the Historic Silver Mine in Tarnowskie Góry. Both topics were taken up by the students of the first semester of the

© Springer International Publishing AG, part of Springer Nature 2019
J. Charytonowicz and C. Falcão (Eds.): AHFE 2018, AISC 788, pp. 174–183, 2019.
https://doi.org/10.1007/978-3-319-94199-8_17

master's degree programme. The paper presents selected competition works prepared under the charge of the author of the paper, describing their basis objectives.

2 Contemporary Public Spaces

Public spaces are an important element of harmonious, urbanised space. Their quality is one of the most important elements deciding about the attractiveness of a town both as a place where the residents live and work and as a destination for tourists. Good public space is a peculiar 'identification' and an element of identity expressing the success and aspirations of the city [1].

Unfortunately, we can currently observe numerous negative phenomena in this area. The most important ones include:

- shrinking or dividing public areas,
- privatisation of public space,
- annexation of space by cars,
- residential areas that prevent the access of third parties.

These factors have a negative influence on the quality of life in towns and cities [2].

The Third Congress for the Development of Polish Towns Cities in 2009 has defined public spaces in the social and economic field as areas used and shaped for a particular purpose, being in line with social principles and values, the aim of which is to meet the needs of not only social communities. The demands related to the protection and efficient management of public spaces included:

- maximising the values of the town and its properties by creating high quality public spaces,
- complex local and urban planning of public spaces based on the results of urban architectural competitions,
- social involvement in the creation of public space development and management tools with the active participation of local communities in the process of the preparation of planning documents,
- protection of the cultural heritage and local uniqueness as special qualities of public spaces,
- balanced development of public spaces in relation to the revitalisation of historic spaces,
- access to the public space and the minimisation of conflicts during their development and use,
- shaping public spaces that integrate social groups, respecting their various needs and systems of values.
- active use of public spaces for the purposes of local events [3].

3   Public Spaces in Student Competition Works

In the case of both competition themes taken up at the Faculty of Architecture, the design phase was preceded by extensive studies such as: in-situ studies, urban, architectural or historical analyses. As a result of these activities, design findings and guidelines were formulated. They were formulated in order to improve the discussed public spaces by means of modifications and modernisation of functional, formal and ergonomic aspects as well as accessibility (also for disabled persons) and visual information. As a result of the formulated guidelines, various architectural and urban designs in response to contemporary requirements related to this kind of facilities were produced. Despite similar findings from conducted pre-design studies, the produced designs were very diverse and underlined the extensive potential of the analysed areas.

3.1   The Context of a Medieval City

The first of the discussed student competitions entitled 'The design concept of the reconstruction of historical elements of urban defensive walls in Żory, including a spatial development design covering the area between Bramkowa and Ogrodowa streets was announced at the Faculty of Architecture in the academic year 2015/2016. Substantive aspects were managed by the Team for the History of Architecture and Monument Preservation, and the Municipal Office of Żory was a strategic partner. The design was prepared as part of 'Conservatory Design' subject, under the supervision of Magdalena Żmudzińska-Nowak, Associate Professor, PhD, Eng. Arch. The competition objectives included areas associated with the remains of medieval defensive walls. A 137.4 m fragment of brick defensive walls surrounded the town on the south side. Its height from the ground level ranges from 5.5 to 6.5 m. Despite numerous preservation works and future restructuring, its form is still very attractive and explicitly moves us to medieval times. The curiosities include a joint that is clearly visible in the wall face, which formed as a result of the integration of two of its sections, probably built from opposite directions [4]. Currently, poor exposure is the greatest problem of the discussed area. Moreover, it separates two adjacent plots, one of which is, in fact, a parking lot and the other one consists of unarranged greenery. The area from Bramkowa street comprises, apart from a parking lot for several dozen cars, a small public green area and a public toilet facility. It should be add that there is another, larger parking lot several meters away. Parking along streets is also possible. One can conclude that such a structure satisfies, or even exceeds local parking needs. However, scarcity of urban greenery areas is visible. The fact that the historic defensive wall separates both spaces, hindering the communication in the town, is also problematic. Previous modernisation concepts involved the formation of a hole in the medieval wall in order to solve this problem. In the course of the performed analyses, the student group decided that such a solution would damage the structure of the monument too much, therefore attempts were made to join the two areas in a different way. It was found that it can be made above or under the medieval defensive wall. The elimination of a parking lot from the plot at Bramkowa street was another important issue. It was also found that the opposite plot from Ogrodowa street should be available for the residents and tourists visiting the town. The combination of these two spaces will turn the discussed area into public space. Through these activities, the medieval monument will be properly exposed and can become a tourist attraction of the town as well as a rightful public space for the residents. It was also decided that, in order to make the area attractive, new functions for the residents and tourists should be added. As a result of analyses in the context of the immediate surroundings and the entire town, certain deficiencies in the town infrastructure were observed. They included, among other things: lack of a tourist information point, insufficient supply of catering services as well as few public toilets. It was also found that it would be interesting to look at the city from a higher perspective that is, from the wall top, or even have a bird's eye view.

The above-mentioned problems were tackled by student designs, of which the work entitled an observation tower with a gallery and a multifunctional pavilion (oryg. Wieża widokowa z galerią i pawilon wielofunkcyjny) by the team composed of Monika Grabowska, Eng. Arch., and Adrian Duda, Eng. Arch. is particularly noteworthy. According to the author, it contributes to the improvement of the quality and the creation of attractive public space to the greatest extent. The main objective of the design concept was to link both areas with the use of high quality contemporary architecture and create modern public space having all required functions. The design involves, among other things, the liquidation of a parking lot in favour of a public space including street furniture and recreational space diversified in terms of height. It simultaneously exposes the preserved defensive wall by means of the contrast with new architectural elements, such as: an observation tower and a multifunctional pavilion. It houses the Municipal Information Point, a bistro café and a plumbing system. Vertical communication offers the access to the observation gallery, which in a way symbolises a medieval guard gallery – a hoarding. So we can use a contemporary gallery to pass over the medieval defensive wall to the observation tower, from which we can admire the panorama of the town. As a result of urban analyses, the tower is located within the viewing axis of one of the streets, which enables us to see the tower of the oldest St. Philip and Jacob Church in the town. From Bramkowa street, there is also a modern portal that can enable the communication to the other side of the wall or only serve as a formal or compositional element, highlighting the above-described wall joint (Fig. 1). From Ogrodowa street, there is natural public space also including elements of street furniture, which was intentionally called an urban garden. On this side of the wall, there is an observation tower that dominates in terms of space, the purpose of which is to signal the discussed area and attract residents as well as tourists (Fig. 2). The entire area is available for people with limited mobility and disabled persons thanks to gentle gradients and lifts in cubic facilities [5].

3.2   In the Spirit of the Industrial Revolution

The second discussed student competition entitled *A design concept for the development of the Steam Engine Open-Air Museum within the Historic Silver Mine in Tarnowskie Góry* was organised in the academic year 2016/2017, also as part of the Conservatory Design programme. Substantive supervision was provided by Magdalena Żmudzińska-Nowak, Associate Professor, PhD, Eng. Arch., and Tarnowskie Góry Land Lovers Association acted as a project partner. The organisation of the competition coincided with the efforts of the authorities of the Association to ensure the entry of the

Fig. 1. The design entitled *an observation tower with a gallery and a multifunctional pavilion* by the team composed of Monika Grabowska, Eng. Arch., and Adrian Duda, Eng. Arch. A view from Bramkowa street. Source: study by the authors of the competition design.

Fig. 2. The design entitled *an observation tower with a gallery and a multifunctional pavilion* by the team composed of Monika Grabowska, Eng. Arch., and Adrian Duda, Eng. Arch. A view from Bramkowa street. Source: study by the authors of the competition design.

Historic Mine on UNESCO World Heritage List. The efforts ended in success soon after the adjudication of the competition. However, when working on the theme, the students were aware of the significance of the place and the responsibility associated with design decisions. The area of the open-air museum, apart from the exhibits and the building of the museum, is of hardly any historical value, therefore a wide spectrum of design efforts is possible here. It requires intervention due to the lack of a consistent character, functional problems as well as low aesthetic qualities of certain solutions. The area in question is complicated for several reasons. It results from a combination of numerous elements of different styles.

Most of all, the open-air museum is dominated by the building of the Historic Silver Mine Museum together with a winding machine house and a boiler house. The architectural complex designed in 1967 by Rudolf Witwicki, a prominent architect, is a distinguished example of post-war modernism, yet, its presence is not directly related to silver mining, but with visiting an underground tourist route. It was built over 60 years after the end of the exploitation of the silver mine. Its interesting external form, together with the winding tower, is modelled on coal mine architecture and has such connotations [6]. Apart from architecture, the area of the open-air museum is also full of very valuable 19th and 20th century exhibits: steam engines and locomotives. Unfortunately, most of them are not protected against bad weather, therefore they deteriorate and require frequent preservation works. The layout of the open-air museum is chaotically linked by randomly arranged paths. The whole is completed by greenery, which is the chief asset of the place, but its condition and form requires design intervention. In the immediate vicinity, there are two parking lots for personal cars and coaches, yet, they are not properly communicated with the building of the museum. The users have to walk around the museum building in order to access the parking lot. The presence of a road transport system in the form of the access to back-up facilities and a boiler house within the open-air museum is also problematic. It cuts the area of the open-air museum through into two parts. Despite these drawbacks, the area has a great potential, which is currently not used in full.

These problems were recognised during the analyses and it was found that both the building of the open-air museum and the building of the museum require modernisation. It is mainly about the communication within the open-air museum, offering proper access and possibility to view the exhibits. The current space arrangement is chaotic, full of contrasting buildings and movable elements, and, most of all, there is no overall idea for this unique place. Certain functional and formal issues related to the discussed area are also disputable. Apart from the modernisation of the open space, functional changes in the building of the museum are also necessary. It requires the change of the entry area location in order to enable easy access from the parking lot for the visitors. This solution will also ensure proper control of the access to the facility and within the open-air museum. Some participants of the competitions also found that new elements necessary in a modern public space should be added to the functional programme of the Open-Air Museum.

The most interesting competition works prepared under the supervision of the author include a design concept of the team composed of Zuzanna Kmak, Eng. Arch. and Sonia Machej, Eng. Arch. The authors created a very interesting public space, modernising both the area of the open-air museum and its architecture. The work entitled Green Silver shared the 2nd prize with another design in the competition. The idea of the project is to create a totally new open-air museum, but based on the most valuable elements of the present shape. The design takes a consistent, organic form, where former sheds for exhibits were replaced by pavilion forms 'springing up' from the ground. Modern roofing has a reinforced concrete structure, glass façades and roofs covered with greenery (Fig. 3). They outcrop above the area, resembling survived mine fields with preserved remains of shafts. Seen from above, they can resemble silver nuggets, and the paths that split them are supposed to be modelled on former pavements used by miners (Fig. 4). The new functional programme of the open-air museum covers also, apart from the exhibit protection project, the preparation of new attractions, such as: educational path, entertainment zone with a stage area, external toilets or a catering facility. The entire facility is supposed to be a comfortable area, open for the users of the space. The arrangement of greenery by means of new seedlings, mainly in the form of various kinds of grass, creates an eco area in the spirit of sustainable growth and respect for nature. In terms of functional aspects, the location of the entry to the museum was changed. In the design, it is properly accentuated and communicated with the parking lot and the public space of the open-air museum. It is also associated with the introduction of a unified visual information system that improves the sense of direction within the Open-Air Museum. The entire area is accessible for mobility-impaired persons [7].

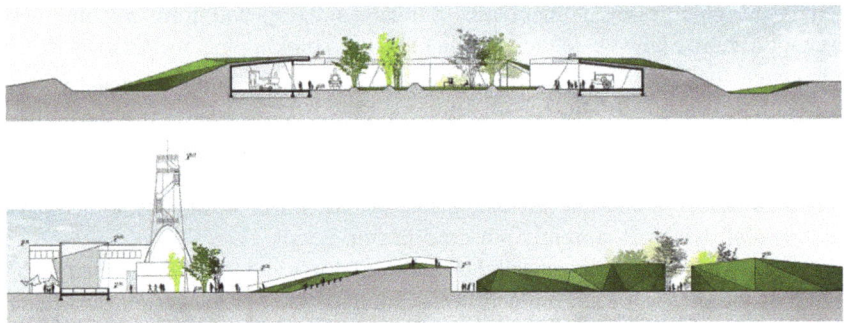

Fig. 3. *Green Silver* competition design by a team composed of Zuzanna Kmak, Eng. Arch. and Sonia Machej, Eng. Arch. Sections of the areas covered by the concept. Source: study by the authors of the competition design.

Fig. 4. *Green Silver* competition design by a team composed of Zuzanna Kmak, Eng. Arch. and Sonia Machej, Eng. Arch. Concept visualisation. Source: study by the authors of the competition design.

4 Findings

Not only the competition for an architectural concept itself, but also the creative 'polemics' with its objectives provide an opportunity to propose development trends and the use of the potential of the area. Such architectural solutions influence the quality of public space, meeting the expectations of the users. As a consequence, they translate into the improvement of the attractiveness and image of the place. The competition works of the students of the Faculty of Architecture are consistent with the trend and have numerous common features.

One can say that this current is reflected in the attempt to create:

– arranged and consciously designed public space,
– contemporary architecture that properly combines the required functions,
– places attractive for users in terms of function and form – a so-called attractor,
– proper exposure of historic objects and culturally important places,
– space meeting the principles of sustainable growth and respect for nature,
– green areas (biologically active),
– accessibility for disabled persons,
– clear visual information system.

Other characteristic features include efforts to change and transform:

– improper functional arrangements,
– unarranged greenery into well-thought-out and carefully designed layouts,
– the road transport system into shared zones with limited vehicle access.

Other important issues include attempts to eliminate:

– low-value architecture that is not associated with the discussed area in formal, functional or historical terms,
– randomly situated parking lots,
– unnecessary road transport system,
– architectural barriers related to the accessibility for disabled persons.

5 Conclusion

The designs presented in the paper offer very diverse solutions consistent with modern trends and requirements related to public spaces. The common feature of the design concepts is to arrange the architecture and urban planning of the discussed spaces and give them a consistent character. Meeting the requirements of the competition and adding new ones resulting from detailed analyses of functions lead to the development of particularly interesting architectural solutions. Public spaces became attractive to the users, meeting their needs in accordance with universal design principles. Objects of historical value were properly exposed by means of appropriate architectural solutions. Effects meeting the principles of sustainable growth and respect for nature were also achieved. Eco-friendliness is one of the most important aspects of the students' works. Green areas that are extremely important for users, together with elements of green architecture, acted as a kind of a carbon dioxide catalyst in the combat with climatic changes.

Such a broad view of the discussed issues would not be possible without the involvement of a large group of students (in this case – Engineers Architects, holders of a bachelor's degree). One can therefore explicitly state that the competition for an architectural concept enables a broader look at the discussed issues, shows a full spectrum of solutions and, as a consequence, contributes to the improvement of the quality of the discussed public spaces. The positive effect is, however, conditioned on the necessity to perform a wide spectrum of analyses and pre-design studies.

References

1. Stangel, M.: Kształtowanie współczesnych obszarów miejskich w kontekście zrównoważonego rozwoju (The development of contemporary urban areas in the context of sustainable growth). Wydawnictwo Politechniki Śląskiej, Gliwice (2013)
2. Palus, K., Zabawa-Krzypkowska, J.: Contemporary public spaces as meeting places. Archit. Civil Eng. Environ. ACEE 9(2), 21–28 (2016)
3. Cichy-Pazder, E., Markowski, T.: Karta Przestrzeni Publicznej. Materiały III Kongresu Urbanistyki Polskiej. Nowa Urbanistyka – nowa jakość życia (Public Space Charter. Materials III Polish Urbanism Congress. New Urbanism – a new quality of life). Biblioteka Urbanisty. Urbanista, no. 14, pp. 234–237 (2009)
4. Żmudzińska-Nowak, M.: Omówienie założeń, problematyki i przebiegu konkursu: "Koncepcja projektowa rekonstrukcji historycznych elementów miejskich murów obronnych w Żorach wraz z projektem zagospodarowania otoczenia pomiędzy ulicami Bramkową i Ogrodową" ("Project concept of the reconstruction of historic elements of the defensive walls in Żory, including a development plan for the area between Bramkowa and Ogrodowa streets" - rules and guidelines for the competition). In: Żmudzińska-Nowak, M., Radziewicz-Winnicki, R. (eds.) Przestrzenie historyczne miasta w projektowaniu konserwatorskim. Historyczne mury miejskie w Żorach jako przedmiot opracowania (Historic town spaces in conservatory design. Historic town walls in Żory as a study subject) Collective work, pp. 25–30, Wydawnictwo Politechniki Śląskiej, Gliwice (2016)
5. Radziewicz-Winnicki, R.: Konkurs architektoniczny jako narzędzie wskazania kierunków rozwoju i potencjału miejsca na podstawie projektów studenckich (Architectural competition as a tool showing the development trends and potential of a place on the basis of student designs). In: Żmudzińska-Nowak, M., Radziewicz-Winnicki, R. (eds.) Przestrzenie historyczne miasta w projektowaniu konserwatorskim. Historyczne mury miejskie w Żorach jako przedmiot opracowania (Historic town spaces in conservatory design. Historic town walls in Żory as a study subject) Collective work, pp. 111–123, Wydaw. Politechniki Śląskiej, Gliwice (2016)

6. Nakonieczny, R.: Muzeum Zabytkowej Kopalni Srebra - dziedzictwo modernizmu i lekcja architektury współczesnej w aspekcie związków z Politechniką Śląską (Historic Silver Mine Museum – modernist heritage and a lesson on contemporary architecture in terms of the relationship with the Silesian University of Technology). In: Żmudzińska-Nowak, M., Radziewicz-Winnicki, R. (eds.) Wokół dziedzictwa poprzemysłowego Ziemi Tarnogórskiej. Skansen Maszyn Parowych przy Zabytkowej Kopalni Srebra w Tarnowskich Górach - koncepcje modernizacji. About industrial heritage of Tarnowskie Góry Land. The Open-Air Steam Engine Museum at the Historic Silver Mine in Tarnowskie Góry - modernisation concepts. Red, pp. 102–124, Wydaw. Politechniki Śląskiej, Gliwice (2017)

7. Radziewicz-Winnicki, R.: Współczesne muzeum techniki na podstawie studenckiego konkursu architektonicznego (A contemporary museum of technology on the basis of a student architectural competition). In: Żmudzińska-Nowak, M., Radziewicz-Winnicki, R. (eds.) Wokół dziedzictwa poprzemysłowego Ziemi Tarnogórskiej. Skansen Maszyn Parowych przy Zabytkowej Kopalni Srebra w Tarnowskich Górach - koncepcje modernizacji. About industrial heritage of Tarnowskie Góry Land. The Open-Air Steam Engine Museum at the Historic Silver Mine in Tarnowskie Góry - modernisation concepts. Red, pp. 224–250, Wydaw. Politechniki Śląskiej, Gliwice (2017)

Structural System for Development of Scenic, Historical, Landscape Parks in China

Teresa Bardzinska-Bonenberg[1]([✉]) and Shoufang Liu[2]

[1] Faculty of Architecture and Design, University of Arts in Poznan, Al.
Marcinkowskiego 29, 60-967 Poznan, Poland
teresa@bardzinska-bonenberg.pl
[2] Liaoning Urban and Rural Construction and Planning Design Institute,
Shenyang, China

Abstract. Although large, beautiful and diversified, Chinese landscape parks
are generally unknown to both the American and European public. Pressure for
economic profit triggered by changes in the country's economy resulted in over
exploitation of some of them. There are many reasons for this situation including
the administrative system, planning, management etc. Analysis of the systems
established in the world, manners and methods of parks' development will help
to ensure successful creation of scenic historical landscape parks and national
parks in China. Furthermore, cooperation with international institutions and
organizations will force Chinese authorities to adhere to the internationally
recognized rules and regulations. Therefore, a model of contemporary Historical
Scenic Area which confirms and supports at the same time Chinese local features and traditions is going to be based upon general experience of parks in the
world supervised by international organizations.

Keywords: Landscape Scenic Historical Area · International regulations
List of World Heritage · Criteria · Management · Tourism · Protection

1 Introduction – Conservation of Nature and Heritage

There are many definitions of a "Landscape Scenic Historical Area" formulated in
different official documents in China, and there exist large discrepancies between them.
In the article titled "Discussion on the Definition of Famous Scenic Sites" that was
published in the April 2007, "Chinese Landscape Journal", Hao Jin who pointed out
that "the Scenic and Historic Landscape Areas are the significant national natural and
cultural heritage. Defining them is essential to the development of landscape enterprise." He studied 292 definitions of "Historic Landscape Areas" by searching and
reviewing the official Chinese documents and relative references, and resolved two
problems caused by current explanation. First of all, the definitions lacked content and
were out of sequence. Secondly, there were no criteria specified. Eventually the author
summarized the final result of the research: "Scenic and Historic Landscape Areas in
China root from the historic landscape and scenic spots of ancient times. The
extraordinary and outstanding cultural, natural scenic resources are the hub of these

© Springer International Publishing AG, part of Springer Nature 2019
J. Charytonowicz and C. Falcão (Eds.): AHFE 2018, AISC 788, pp. 184–193, 2019.
https://doi.org/10.1007/978-3-319-94199-8_18

areas, which should have definite and clearly delimitated boundaries. They contain high scientific, cultural, aesthetic values. In these areas, valuable natural and cultural resources can be conserved well and effectively. Tourism economy should be promoted, as it can contribute to the economic development of local communities. Scenic and Historical Landscape Areas should be designated and managed by the government administration." [1] This problem means also that in present time they should also comply with the international regulations.

2   Conservation of Nature: Comparison of World Heritage Criterions and Chinese Regulations[1]

The International Union for Conservation of Nature (IUCN) an international organization was founded in 1948. This nature monitoring organization has over 1 200 members in the world. [2] In order to accomplish its advisory management function, IUCN maintains a worldwide list of protected areas. Each protected area on the list is ascribed to one of the seven categories based on management goals [3].

IUCN is also one of the three Advisory Bodies to the World Heritage Committee (also including ICCROM and ICOMOS). Its specific roles in relation to the World Heritage Convention include evaluation of properties nominated for inscription on the World Heritage List, monitoring the state of conservation of World Heritage natural properties, reviewing requests for International Assistance submitted by States Parties, and providing input and support for capacity building activities [4].

World Commission on Protected Areas (WCPA) is one of the six voluntary Commissions of the IUCN and is administered by the Programme on Protected Areas from IUCN's headquarters in Gland, Switzerland. WCPA's mission is to promote the establishment and effective management of a worldwide representative network of terrestrial and marine protected areas, as an integral contribution to the mission of the IUCN. With responsibility for the global network of terrestrial and marine protected areas, WCPA also generated and implemented environmental regulations and policies on the use and management of natural resources [5].

In general, a National Park worldwide is understood to be a type of natural or semi-natural protected area, set aside for ecological protection, scientific and education purposes as well as for recreational activity. Because the first idea of a National Park derives from the American Yellowstone National Park, the term "National Park" has existed long before it was defined by the IUCN. Thus, today's definition of a "National Park" varies and is different in countries worldwide. The IUCN defined "National Park" as "Protected Area," and categorized it into seven categories. It embraces nearly almost all types of nature-related National Parks and Reservation Areas throughout the world. The definition of each Protected Area category comprises definition, objectives, distinguishing features, and its roles [5].

---

[1] This chapter is based on the PhD research by Liu [11].

In the recent 2008 version of the "Guidelines for Applying Protected Area Management Categories", Protected Areas are defined as follows: "A clearly defined geographical space, recognized, dedicated and managed through legal or other effective means, to achieve the long-term conservation of nature with associated ecosystem services and cultural values" [5].

The definition emphasizes long-term conservation and effective management of both ecological and cultural values that create clarified three-dimensional spaces out of National Park resources. The definition is a threshold of the World Protected Area dedication, and then a proper Protected Area Management Category can be designated; finally, the appropriate characters, objectives, distinguishing features and roles could be achieved. IUCN also emphasizes that the individual Protected Area should not be isolated; it must be regarded as a part of larger scale system of protected areas.

Most of Chinese Scenic Historical Landscape Areas including Natural and Mix Natural/Cultural Heritage Sites are included in the World Protected Area System. Therefore, studying documentation on the IUCN Protected Areas can help to clarify definitions of nature-based Chinese Scenic Historical Landscape Areas and make them fit into the World Protected Area System so as to strengthen the global environment. Table 1 contains initial comparison of Chinese Scenic Historical Landscape Area Conservation criteria matched with IUCN Protected Area Management Categories.

The comparison proves that the differences between Chinese National Parks and Natural Reserve Areas are that the Natural Reserve Areas are much more strictly looked after than Chinese National Parks [6], and are more relevant to the Protected Area Management Category I. Chinese National Parks can be designated as Category II, III, IV, V, and VI. Therefore, Chinese National Parks (Chinese Scenic Historical Landscape Areas) and their systems should in the future meet the requirements of World Protected Areas, to contribute to the global ecosystems and environment.

Hence, the definition of Chinese Scenic Historical Landscape Areas should adhere to the related contents of the World Protected Areas concept and it can be described in the view of IUCN's Protected Areas as: "A clearly defined geographical space, recognised, dedicated and managed, through legal or other effective means, to achieve the long-term conservation of nature with associated ecosystem services and cultural values" [7].

IUCN Protected Area Definition includes Management Categories and Governance Types. Six management categories are as follows: Ia Strict nature reserve, Ib Wilderness area, II National park, III Natural monument or feature, IV Habitat/species management area, V Protected landscape or seascape, VI Protected areas with sustainable use of natural resources.

Such evaluation varies from the Chinese description, due to the fact, that the areas of protection in China are generally much larger, and contain elements valuable for the whole country culture. History of China in the twentieth century made the problem of conservation and cultivation of local cultures, including architecture, dresses, customs important for the future. Cultural revolution in the sixties on one hand and recent globalization of culture leaves not much time to turn the tendency.

Table 1. Chinese scenic historical landscape area conservation categories matched with IUCN protected area management categories, (by S. Liu). (The data comes from [5, 6].)

Conservation categories of the Chinese Scenic Historical Landscape Area		Matching IUCN Protected Area Management Categories		Remarks
Name	Attributes	Cath	Attributes matched	
Bio-ecological protected area	Special part of the area that is nested inside the Scenic Landscape Area is set aside to protect or preserve the valuable species and their environment for scientific study and research. A strict protected area that prohibits visitation, auto-traffic, and any construction except scientific activity.	Ia, Strict nature reserve	Strictly protected area set aside for biodiversity or geological/geo-morphological feature. The human modification is strictly controlled. It serves as indispensable reference area for scientific research and monitoring.	Bio-ecological protected area should be a nested area of Chinese Scenic Historical Area, that matches Ia of IUCN Protect Area Management category, otherwise the independent area should be designated Chinese Natural Preserved Area
No category match with Ib Wilderness area of IUCN		Ib Wilderness area	No points can match with Chinese Scenic Historical Landscape Area	Usually designated as a Natural Preserved Area of China
Natural landscape protected area	Strictly protected natural area and site that any exploited activities are severely limited. The visitation is controlled, trail and some indispensable security facilities are allowed, auto-traffic is prohibited.	II National park	Large scale, ecological processes are protected within large natural or nearly natural areas. The recreation, education, and scientific research are allowed.	If a Chinese Scenic Historical Landscape Area is designated as IUCN II, the whole area should be managed as II
		III Natural monument or feature	Protected areas contain specific natural feature such as a landform, seamount, submarine cavern, geological feature a living feature such as an ancient grove. Small, protected areas, often of high touristic value.	Outstanding natural values are usually the fundamental cause that the site was designated be a Chinese Scenic Historical Area.
Historical site protected area	Natural area is set to protect historical relics or cultural heritage. Tourism is controlled, trail and some indispensable security facilities are allowed, cars are prohibited.	III Natural monument or feature	Natural resource management objective is similar to III but emphasize to help protecting cultural heritage. The Historical Heritage protection should carry out the cultural heritage management policy.	Outstanding historical values are usually the fundamental cause that the site was designated be a Chinese Scenic Historical Area.
landscape/sea scape recovering area	Area set aside to conserve, recover, foster, nurture, or maintenance by a certain intervention of human being.	IV Habitat/species management area	Priority management is to protect the particular species or habitats. Regular and active management intervention is indispensable to restore the species and habitats in daily life.	Ecological recovering areas within the whole Chinese Scenic Historical Landscape area

| Landscape appreciated area | Attractive attractions concentrated; visitation is gathered for appreciation, recreation, and education purposes. | VI protected area within sustainable use of natural | Large, natural area, where the ecosystem, habitats as well as cultural values and resource management systems are protected. Usually the mutual benefits are the aims of conversation and sustainable use process. | Most of Chinese Scenic Historical Landscape Area fall into VI |
| The development controlled area | Keeping the current traditional landscape management manner, livelihoods of indigenous; the tourist facilities must be limited to a certain reasonable capacity. | V protected landscape /seascape | Protected area is created, conserved and sustained by the interaction between natural forces and human management activities. The natural landscape or seascape is mutually symbiotic between natural resources and native livelihood. | It is a coordinated area or a buffer zone between core zone and the outside boundary within Chinese Scenic Historical Landscape Areas. |

After the comparison of Chinese Scenic Historical Landscape Areas criteria and their World Heritage identification, it is clear that Chinese Scenic Historical Landscape Area is the heritage area, which is established to preserve valuable natural and cultural remains. [8] It was also known from IUCN documented sources that Chinese Scenic Historical Landscape Area is an essential part of the world's natural biological system. In consideration of what was written before, a question arises; how to distinguish the cultural heritage system?

Table 2 refers to criterions that define Chinese Scenic Historical Areas, according to UNESCO World Heritage Criterions discussing a variety of features that undergo different forms of protection.

Protected areas are delimited according to the assets of the area: beautiful and attractive views, historical and ecological resources, protection of which is vital, as well as scientific, educational, recreational activities that are either needed or possible. Such heritage sites should be divided into two groups [9].

- Nature-based Chinese Scenic Historical Landscape Areas – these are protected areas scheduled for ecological protection, scientific and educational opportunities, as well as recreational activity. They should be conserved and managed effectively to achieve long-term preservation and conservation within the clarified space. The Chinese Scenic Historical Landscape Areas are a part of international, national, regional or local Protected Areas' System and fulfill some kind of ecologic demands (ecologic corridors, ecologic stepping-stones and other ecological functions); they should be integrated into these ecosystems and be managed by the Ecosystem Approaches.
- Culture-based Chinese Scenic Historical Landscape Areas – these are protected areas scheduled to protect cultural heritage and conserve cultural diversity for the sake of integrity of present and future generations. They contain one or more Cultural Heritage Sites or Historical Sites that usually integrate with natural beauty. Protection should concentrate on both: natural and cultural quintessence.

Table 2. Main features that define Chinese Scenic Historical Areas, according to UNESCO World Heritage Criterions (by S. Liu)

Main landscape features		Classification of importance		Boundaries
Main landscape features classification	Criterion of importance adopted from the UNESCO World Heritage Criterions			
Natural features — Scenic feature	(VII) contains outstanding natural phenomena or exceptional natural beauty and aesthetic importance; (VIII) is an outstanding example representing major stages of earth's history, including the record of life, significant on-going geological processes in the development of landforms, or significant geomorphic or physiographic features;	parks of national importance	universally outstanding	the area should be of sufficient large size, to keep the natural ecosystem sustainable natural ecosystem in the condition of integrity without or minimum humankind intervention.
Natural features — Biological feature	(IX) is an outstanding example representing significant ongoing ecological and biological processes in the evolution and development of terrestrial, fresh water, coastal and marine ecosystems and communities of plants and animals;		nationally outstanding	
Natural features — Geographic feature	(X) contains important and significant natural habitats for in-situ conservation of biological diversity, including those containing threatened species of outstanding universal value from the point of view of science or conservation	parks of provincial importance	regionally outstanding	
Historical and cultural features — Archeological feature	(I) represents human creative genius; (II) exhibits an important interchange of human values (III) bears testimony to a cultural tradition or to a civilization (IV) is an example of a type of building, architectural or technological ensemble or landscape which illustrates significant stage(s) in human history;			the area should be large enough to maintain the movable and unmovable cultural property in the condition of authenticity. The humankind's intervention and modern civilization should not impair it.
Historical and cultural features — Historical feature	(V) is an example of a traditional human settlement, land-use, or sea-use which is representative of a culture (or cultures), or human interaction with the environment especially when it has become vulnerable under the impact of irreversible change;	parks of local importance	locally outstanding	
Historical and cultural features — Other cultural features	(VI) is directly or tangibly associated with events or living traditions, with ideas, or with beliefs, with artistic and literary works of outstanding universal significance.			

Table 3. Objectives and roles that are specified by the definition of Chinese Scenic Historical Landscape Areas (by S. Liu).

			Characteristics
Objectives	**Primary objective**		The primary objective conditions are to preserve and conserve the national, regional, and local ecological process and natural biodiversity, safeguard the national, regional and local cultural heritage, to promote recreation, education and tourism
	Particular objectives	**National resources conservation**	**a:** to keep the representative formation and typical aesthetic character of regional physiographical feature, **b:** to retain the regional biotical and ecological feature (including biotical character, ecological communities, genetic resource and unimpaired natural processes) in the natural state forever, **c:** to maintain the biodiversity of native species and ecologically functional populations, in order to keep ecosystem integrity and resilience, **d:** to contribute to the regional diversity landscape system in particular by conserving wide-ranging species, regional ecological processes and migration routes
		Cultural resources conservtion	**e:** to preserve the universal, national and regional outstanding cultural heritage site in the state of authenticity, **f:** to reserve the cultural relics for future research, **g:** to keep the culture-diversity by conserving the local movable and unmovable cultural resources
	Usage of park		**h:** keeping the human involvement in the proper capacity not to compromise the significant biological or ecological natural resources and the authenticity of culture heritages, **i:** taking care of the needs and the livelihood of indigenous people and local communities, in the range of not affecting the main management objective adversely, **j:** contributing to local economies through tourism
Roles	**Preservation of valuable heritage**		to meet ecologic services protection function (ecologic corridors, ecologic stepping stones and other ecological functions), and conserve outstanding cultural and historical properties,
	Tourism and recreation		to provide a base for the outdoor recreation and tourism
	Educational and Scientific research		to provide a base for the scientific research and education, as well as a base of valuable treasure for future generations
	Benefit to local community		to provide steppingstones for compatible economic development to contribute to local economies through recreation and tourism

The history of Chinese Scenic and Historic Landscape Areas is relatively short in comparison with the western developed countries. Conservation and appropriate exploitation system should be based on the experience of the parks and protected areas, which were developed over a long period of time. As Chinese Scenic Historical Landscape Areas' System is in the course of integrating with the international natural and cultural protected areas system such as IUCN Protected Area System and World

Heritage Site it is vital to act in the above said manner. The internationalization of the Chinese National Park System is ongoing so the regulations and standards also need adjusting to correspond with the international standards. So far, there have already been 23 Chinese National Historic Landscape Areas inscribed on the List of World Heritage by UNESCO and all the national level Chinese National Parks are designated in the IUCN Protected Areas list [10].

Properties that are defined for each Chinese Scenic Historical Landscape Area include objectives and roles. These are summarized in Table 3.

There are more than 800 Scenic Historical Landscape Areas in China; all of them are inscribed in the Chinese Protected Areas System. Chinese Foreign Affairs Administration became a state member of IUCN in 1996; after that, the Chinese Scenic Historical Landscape Association as a nongovernment organization member was approved by 273rd Conference of Grant Swiss 2003. This means that the Chinese Scenic Historical Landscape Areas System is a part of the IUCN World Protected Area System, and it should be managed by the policies of IUCN Protected Area. From that date, the author found little on the application of IUCN policy in either the official management documents or practice activity.

Therefore, to introduce IUCN Protected Area Management Category into Chinese Scenic Historical Landscape Areas, it is necessary to develop management of the Chinese Scenic Historical Landscape Area system. Efficient managing of the Chinese Scenic Historical Landscape Areas system should follow the standards that are outlined in IUCN Management Category.

- Establishing the protected areas system within ecosystems. Chinese Scenic Historical Landscape Areas are not isolated protected areas; they should become the fundamental element of global protected areas' system, which integrates variety of protected areas into an ecosystem, to contribute to global bio-system, biodiversity and climate change.
- Establishing protected areas in three dimensions. It must be carried out not only on the ground, but under and over the surface as well. Underground mining, and underwater activities such as fishing, dredging for inland water and marine as well as the airspace.
- To protect a Chinese Scenic Historical Landscape Area ecological integrity, number and scope of human activities should be controlled [8].

Category assignment: three stages of assigning a category should be introduced into the Chinese Scenic Historical Landscape Area management process. First of all, a possible category should match initial management objective at the opening stage. Secondly, through assessment and planning processes redefining the primary objective should be carried out followed by designation of the appropriate management category set. the third stage should be to undertake revising and reinforcing the process, which is practiced throughout management activities.

Using the IUCN categories pertaining to the Protected Areas as a tool for conservation planning and policy and while assigning new areas is a core target of the Organization. It is also the aim of the Chinese administration. In the Chinese Scenic Historical Landscape Areas policy, there is also grading referring to "Conservation Category", which should coincide with the IUCN's Protected Area Management Category.

Principles derived from this document should be followed in the lifecycle of management practice within the Chinese Scenic Historical Landscape Areas. They are as follows:

- The overall objective is the fundamental feature; therefore, comprehensive IUCN management category assignment should follow.
- Chinese Scenic Historical Landscape Area is usually a grand, extensively Protected Area. Several exclusive protected zones bear special management primary objective. They are usually nestled within the Chinese Scenic Historical Landscape Area, so assigning IUCN Protected Area Management Category adequately, is crucial for the conservation of the natural or cultural heritage.
- Among the processes of management, the "ecological gaps analysis" is essential, but it can be carried out only on an appropriate analysis basis, which allows to filling the gaps, and strengthen the effects.

The IUCN Protect Area Management Categories constitute a useful tool in the entire natural heritage management process of the Chinese Scenic Historical Areas. It should be introduced into the processes of identifying, designating, and launching the management of a new Scenic Area, as well as designing a Scenic Area system with diversified management purposes and governance types, as well as monitoring the effectiveness of the management.

3 Summary

Incorporating national parks, scenic historical landscape areas and the other discussed in this paper forms that undergo protection in China into worldwide web of protected areas is aimed at widening a scope and range of protected by law heritage. Furthermore, categorization of Chinese resources in a way that complies with the world standardization will enable undertaking global compatible research. Also monitoring of positive and negative occurrences will be comparable.

National parks and protected areas throughout the world provide education, enable development of individual hobbies and interests, encourage tourism which demands physical effort, serve as sports areas.

Contemporary overloaded national parks and scenic historical landscape areas are the sign that a number, scope and area of protected, accessible and properly managed national parks and scenic historical landscape areas should be expanded in order to alleviate congestion in existing spaces. This will expand also rate of protected lands in China. New zones organized as protection for particularly valuable heritage "hubs" will surround them, providing needed recreational space for people. Properly managed they will take over some touristic traffic from existing parks.

There is, however one more theme, that needs research combined with educational effort. Diversified and interesting cultural heritage is locally underestimated. Architecture and customs, that stood the test of really harsh times are disappearing now. Support from local and central administration is indispensable before a network of touristic facilities will encourage city-dwellers to arrive.

References

1. Jin, H.: Discussion on the definition of famous scenic sites. Chin. Landsc. J. **4**, 11 (2007)
2. MacDonald, K.I.: IUCN: A History of Constraint. https://perso.uclouvain.be/marc.maesschalck/MacDonaldInstitutional_Reflexivity_and_IUCN-17.02.03.pdf
3. Weeks. P., Mehta, S.: Managing People and Landscapes: IUCN's Protected Area Categories, 253–263.   www.tandfonline.com/doi/abs/10.1080/09709274.2004.11905747.   Accessed December 2017
4. Operational Guidelines for the Implementation of the World Heritage Convention. WHC, p. 8, 08 January 2008
5. Dudley, N. (ed.): Guidelines for Applying Protected Area Management Categories. IUCN, Gland (2008)
6. Chinese Code for Scenic Area Planning; GB 50298-1999
7. Leung, Y.F., Spenceley, A., Hvenegaard, G., Buckley, R. (eds.): Tourism and Visitor Management in Protected Areas. Guidelines for Sustainability (2014). In collaboration with 54 contributors, Craig Groves Series (ed.), IUCN Review Copy 5
8. The Ordinance of Landscape Historical Scenic Area State Council of the People's Republic of China, September 2006
9. The Green Paper of Chinese Scenic Spot Situation and Prospects: issued by the People's Republic of China, Ministry of Construction (1994)
10. Yangshilong: The Chinese National Park Concept and Development Analysis. In: Eco-civilization and Environmental Law Symposium, Beijing (2009)
11. Liu, S.: Comprehensive layouts of scenic and historical parks exemplified by Sa Er'Hu National Historic Scenic Area in China. Ph.D. thesis, Archives of the Faculty of Architecture, Poznan University of Technology (2007)

Assessment of Parking Demand in the Central Business District of Lahore

Ammad Arshad, Irum Sanaullah$^{(\boxtimes)}$, Amna Chaudhry, Zahara Batool,
and Hina Saleemi

Department of Transportation Engineering and Management,
University of Engineering and Technology, Lahore, Pakistan
ammad31@gmail.com, irum.sanaullah08@gmail.com,
aaminah.ch@gmail.com, zaharabatool14@gmail.com,
hasaleemi@hotmail.com

Abstract. With the increase in motorization, parking problems have become a major issue particularly in urban areas. Lahore is the second most populated city in Pakistan and it is expanding haphazardly. According to the Lahore Urban Transport Master Plan, trip generation would increase up to 48% from the year 2010 to 2030. Road occupation by parking is found to be 25–50%, where there is a commercial activity in Lahore. The increase in traffic volume leads to serious parking problems; hence more parking spaces are required in the city. Currently, it is essential to estimate the parking demand and statistics for the better development of parking facilities. The focus of this study is to calculate the parking statistics for the selected locations in the Central Business District (CBD) of Lahore.

Keywords: CBD · Parking volume · License plate survey · Accumulation
Turnover · Index

1 Introduction

Pakistan is experiencing large scale urbanization due to rural-urban migration [1]. According to the report by Pakistan Strategy Support Program [2], 40% of population in Pakistan lives in urban areas which is the highest in South Asian countries. Lahore is the second most populated city in Pakistan where population has increased up to 46% since 1998. The city is interconnected by a road network of 1266 km and 4.2 million vehicles were registered in the year of 2015 [3]. According to Lahore Urban Transport Master Plan, trip generation in Lahore was 8.2 million per day in 2010 and would increase up to 17 million in 2030.

A modal share of public transport was only 37–40% in 2010. Due to rapid increase in motorization and absence of local transport facilities, the road network is experiencing severe congestion.

The population of CBD area (Fig. 1) of Lahore was 1.2 million in 1998 which would increase up to 2.1 million in 2021 [4]. This area includes historical sites and buildings that attract tourists from all over the world. Per day trip generation and attraction from/to CBD area was 1.12 and 1.13 million respectively in 2010. Currently

J. Charytonowicz and C. Falcão (Eds.): AHFE 2018, AISC 788, pp. 194–202, 2019.
https://doi.org/10.1007/978-3-319-94199-8_19

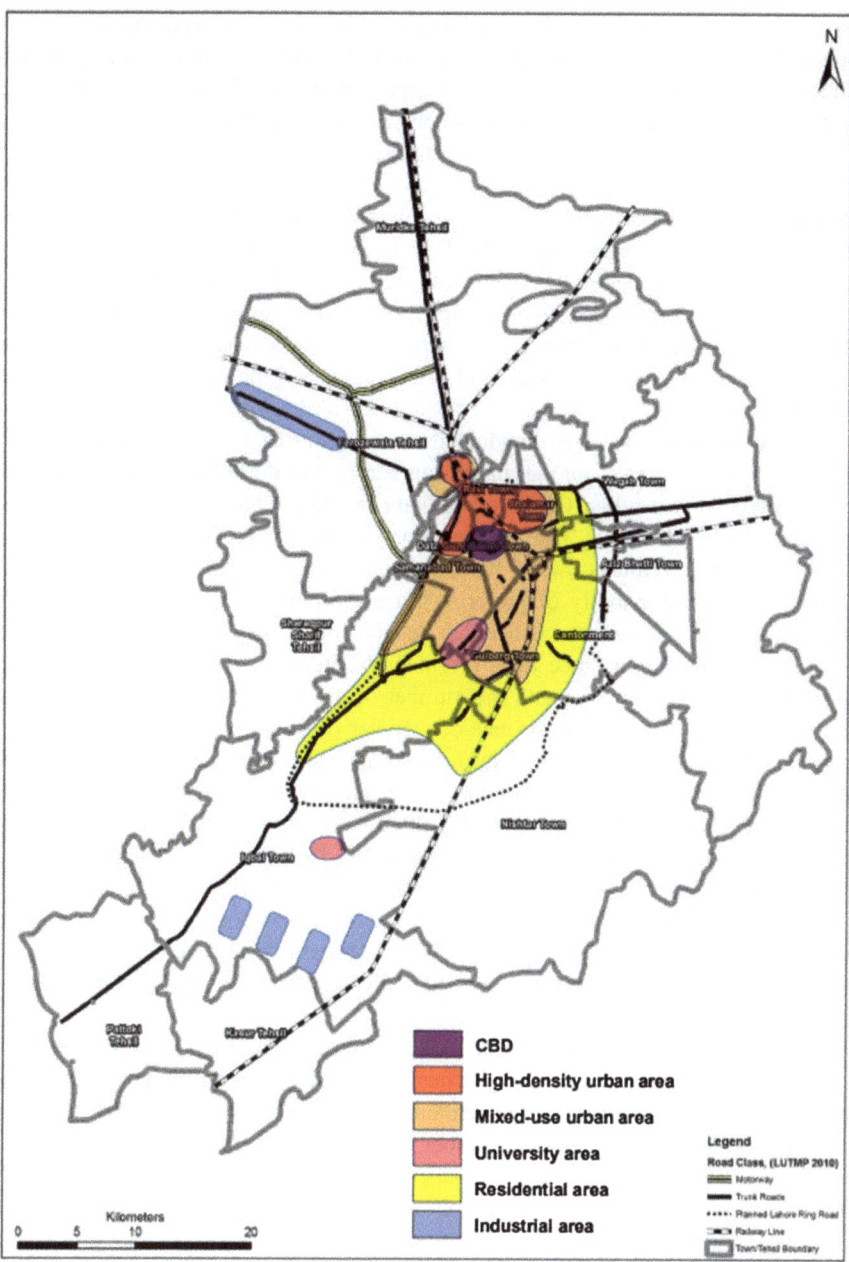

Fig. 1. Map of Lahore showing CBD area.

there are 27 parking sites in CBD of Lahore which are developed, maintained and operated by Lahore parking company known as LePark. On street parking is main member of urban parking system and have great advantages like driver's convenience and short walking distance, however, it directly affects the traffic flows on road network [5]. Similarly, maneuvering due to parking near an intersection has an influence on the level of services of the intersection [6].

Private transport accounts for approximately one third of all the trips in urban areas of metropolitan cities around the globe. Due to lack of parking facilities, demand in urban areas is seldom met and consequently vehicles tend to park illegally on side of the roads. Subsequently, illegal and improper parking results into congestion and bottlenecks in the CBDs of urban areas. Therefore, it is necessary to accurately estimate the parking demand for all types of land uses so that parking supply can be provided accordingly.

Different studies have been conducted to evaluate the parking demand in CBD areas. Sen et al. [7] carried out parking in-out survey to find on-street parking demand for 4-wheelers in two urban CBDs of Kolkata in India. Questionnaire surveys were conducted at two locations. The independent variables used in this study included average number of vehicles owned by household, average parking duration in hours and mode choice (probability of prefering a car over transit). To estimate the mode choice, utility function was established that depends upon the parameters of age of user, annual income and the distance from origin to destination. It was found that parking demand had a high corelation with the number of 4-wheelers owned by a household. Other variables like average hourly rate of vehicles parked, average vehicle ownership per household within catchment (5 km) radius and population within the catchment were considered in a study by Aderamo and Salau [8]. They developed a regression method to calculate the parking demand. Data collection was carried out for 10 major streets on weekdays for the duration of nine hours. Results depicted high correlation between parking demand and the average vehicles parked per hour. Lee [9] proposed parking demand models based on gross floor area of detached housing in Dong-gu, Daegu of Republic Korea. Parking surveys data showed that 11% vehicles were parked in legal parking zone, 61% were parked illegally on streets and 28% were parked in parking prohibited area.

Parking demand models for private hospitals were established by Suthanaya [10]. He developed demand models for 4-wheelers and 2-wheelers by applying simple regression method and using 11 independent variables. These variables include gross floor area, number of VIP rooms, number of class 1 and 2 bed rooms, number of specialists, number of employees and number of doctors in morning and evening shifts. Other parking statistics i.e. parking accumulation, duration, turn over, parking index were also performed. Parking demand for cars was found to be dependent on numbers of class-1 beds in hospital, as it results in high coefficient of determination.

To find out the suitable independent variables for parking demand, parking surveys were conducted at 12 stations of light rail transit system in Kaula Lumpur [11]. In this study, the development of demand models was based on average daily passengers, parking time, parking charges and feeder bus service. The findings indicate that feeder bus service can reduce the parking demand at LRT stations. It was also found that average daily passenger has high degree of association with demand.

Tiexin et al. [12] estimated the parking demand in urban central commercial district of Tianjin city of China. The model was developed using parking generation rates and the types of land uses. The types of land use included residential, shopping centers, office buildings, schools, hospitals, factories and railway stations. Parking in and out survey was conducted for the study. Parking demand was calculated by dividing parking generation rate to gross floor area. Other factors which were considered in establishing demand models include average parking fee, turnover, parking occupancy, and rate of motorization.

In another study [13], parking generation rates were developed for the cities of Palestine. Three land use types selected for this study included residential, office and retail places. Two-day peak volume count was conducted on seventy-three sites and different variables were considered to establish the relationship with parking volume. Simple regression models were used and different parking generation rates were compared with the Institute of Transportation Engineers ITE rates. Parking demand rate was found lower for the selected land uses in Palestine.

Das and Ahmed [14] used fee index model to calculate the parking demand. Parking volume was collected using in-out and license plate surveys at the peak time of the facility. The fee index model was established considering the absence of any transit system and the response on parking prices was evaluated by questionnaire surveys. Results show the decrease in demand by 63% and 59% from original demand on two study locations.

Based on the above studies it can be concluded that it is important to calculate and evaluate the parking demand in order to properly asses and solve the parking problems. With the increase in private vehicles, parking problem is becoming severe particularly in the CBD of Lahore. Currently there is no significant work exists in finding out parking demand and establishing parking generation rates in Lahore. The objective of this study is to calculate and evaluate the parking statistics at selected locations in Lahore.

2 Parking Surveys

Parking in/out and parking license plate surveys were conducted at 8 locations in Lahore. In license plate survey, data related to vehicle type, license plate number, entry exit time of motor bikes and cars were recorded mostly during evening peak of the parking sites. Walking tour were made after every 15 min interval to the parking bays to calculate parking occupancy. In and out surveys were conducted at the sites where the license plate surveys were not possible and where parking volume is on higher side. During In/Out survey, total number of entering and exiting vehicles were recorded. The sites were selected on the basis of major types of land-uses and high trip attraction. Table 1 shows the study locations, evening peak time, and survey methods.

Parking volume is the total number of vehicles parked at a given time interval. The parking volume based on one-hour survey is shown in Table 2. The volume of cars at parking sites ranges from 30 to 347 vehicles/hour in weekdays and 30 to 388 vehicles/hour in weekend. For motor bikes, parking volume ranges from 24 to 1007 vehicles/hour to 26 to 1022 vehicles/hour.

Table 1. Sites inventory and type of surveys conducted

No	Locations	Time evening peak	Time in week	Survey type
1	Mall road	5:00–6:00	Weekend	License plate
		4:30–5:30	Weekday	
		4:40–5:40	Weekday	
2	Hafeez Centre	4:30–5:30	Weekend	In/out
		5:45–6:45	Weekday	
		5:15–6:15	Weekday	
3	Metro park and ride Ichra	5:00–6:00	Weekend	License plate
		5:30–6:30	Weekday	
		5:45–6:45	Weekday	
4	Kareem Block Market	5:15–6:15	Weekday	In/out
		7:00–8:00	Weekend	
		6:00–7:00	Weekday	
5	Minar e Pakistan	6:00–7:00	Weekend	In/out
		4:30–5:30	Weekday	
		5:40–6:40	Weekday	
6	Anarkali	4:15–5:15	Weekday	License plate
		5:00–6:00	Weekend	
		6:15–7:15	Weekday	
7	B.I.S.E Lahore	2:00–3:00	Weekday	License Plate
		1:15–2:15	Weekday	
		2:00–3:00	Weekend	
8	Moon Market	4:15–5:15	Weekday	In/Out
		6:15–7:15	Weekday	
		6:00–7:00	Weekend	

Table 2. Parking volume

No	Locations	Car		Motorbike	
		Weekday	Weekend	Weekday	Weekend
1	Mall road	82	74	354	343
2	Hafeez Centre	79	78	660	678
3	Metro park and ride Ichra	74	73	24	26
4	Kareem Block Market	61	83	288	413
5	Minar e Pakistan	148	167	1007	1022
6	Anarkali	35	42	310	310
7	B.I.S.E Lahore	30	30	119	74
8	Moon Market	347	388	278	330

Accumulation is the summation of all vehicles parked in a parking area at any instant of time. Table 3 shows the parking accumulation for cars and motor bikes. Parking accumulation for cars ranges from 23 to 326 vehicles/hour in week days and 20 to 353 vehicles/hour in weekends. For motor bikes, accumulation ranges from 20 to 826 vehicles/hour in weekdays and 19 to 927 vehicles/hour in weekends.

Table 3. Parking accumulation

No	Locations	Car		Motorbike	
		Weekday	Weekend	Weekday	Weekend
1	Mall road	63	61	259	262
2	Hafeez Centre	75	74	483	615
3	Metro park and ride Ichra	67	68	20	19
4	Kareem Block Market	42	68	163	287
5	Minar e Pakistan	83	125	826	927
6	Anarkali	23	25	278	289
7	B.I.S.E Lahore	26	20	94	56
8	Moon Market	326	353	229	258

Table 4 shows that average parking duration ranges from 0.55 to 0.95 (hours/vehicle) for cars in weekdays and 0.60 to 0.95 (hours/vehicle) in weekends. Average parking duration for motor bikes is in a range of 0.57 to 0.90 in weekdays and 0.69 to 0.93 (hours/vehicle) in weekends.

Table 4. Average parking duration

No	Locations	Car		Motorbike	
		Weekday	Weekend	Weekday	Weekend
1	Mall road	0.76	0.82	0.73	0.76
2	Hafeez Centre	0.95	0.95	0.73	0.91
3	Metro park and ride Ichra	0.91	0.93	0.85	0.71
4	Kareem Block Market	0.69	0.81	0.57	0.69
5	Minar e Pakistan	0.55	0.75	0.83	0.91
6	Anarkali	0.67	0.60	0.90	0.93
7	B.I.S.E Lahore	0.86	0.66	0.80	0.76
8	Moon Market	0.93	0.91	0.82	0.78

Table 5 shows parking turn over for cars ranges from 0.73 to 1.32 vehicles per bay in week days and 0.72 to 1.66 in weekends. Parking turnover for motor bikes ranges from 0.29 to 1.77 vehicles per bay in weekdays and 0.32 to 1.72 vehicles per bay in weekends as shown in Table 6.

Table 5. Parking turnover for cars

No	Locations	Car				
		Volume		Capacity	Turn over	Turn over
		Vehicles		Vehicle	Vehicle/bay	Vehicle/bay
		Weekday	Weekend	bays	Weekday	Weekend
1	Mall road	82	74	80	1.03	0.93
2	Hafeez Centre	79	78	100	0.79	0.78
3	Metro park and ride Ichra	74	73	102	0.73	0.72
4	Kareem Block Market	61	83	50	1.22	1.66
5	Minar e Pakistan	148	167	112	1.32	1.49
6	Anarkali	35	42	35	1.00	1.20
7	B.I.S.E Lahore	30	30	30	1.00	1.00
8	Moon Market	347	388	350	0.99	1.11

Table 6. Parking turnover for motor bikes

No	Locations	Motor bike				
		Volume		Capacity	Turn over	Turn over
		Vehicles		Vehicle	Vehicle/bay	Vehicle/bay
		Weekday	Weekend	bays	Weekday	Weekend
1	Mall road	354	343	200	1.77	1.72
2	Hafeez Centre	660	678	550	1.20	1.23
3	Metro park and ride Ichra	24	26	82	0.29	0.32
4	Kareem Block Market	288	413	400	0.72	1.03
5	Minar e Pakistan	1007	1022	705	1.43	1.45
6	Anarkali	310	310	291	1.07	1.07
7	B.I.S.E Lahore	119	74	80	1.49	0.93
8	Moon Market	278	330	370	0.75	0.89

Parking index shows whether the capacity of existing facility fulfilling the demand or not. Table 7 shows the parking index value for car ranges from 0.66 to 0.93 in weekdays and 0.67 to 1.35 in weekends. Kareem Block and Moon Market shows higher parking demand in weekends.

Table 8 represents the parking index for motor bikes which ranges from 0.24 to 1.30 in weekdays and 0.23 to 1.31 in weekends. The parking places at Mall road, Hafeez Centre and Minar e Pakistan have higher parking demand than parking capacity.

Table 7. Parking index for cars

No	Locations	Car				
		Accumulation (vehicles)		Capacity	Index	Index
		Weekday	Weekend	Vehicle bays	Weekday	Weekend
1	Mall road	63	61	80	0.78	0.76
2	Hafeez Centre	75	74	100	0.75	0.74
3	Metro park and ride Ichra	67	68	102	0.66	0.67
4	Kareem Block Market	42	68	50	0.85	1.35
5	Minar e Pakistan	83	125	112	0.74	1.12
6	Anarkali	23	25	35	0.66	0.71
7	B.I.S.E Lahore	26	20	30	0.87	0.67
8	Moon Market	326	353	350	0.93	1.01

Table 8. Parking index for motor bikes

No	Locations	Motor bike				
		Accumulation (vehicles)		Capacity	Index	Index
		Weekday	Weekend	Vehicle bays	Weekday	Weekend
1	Mall road	259	262	200	1.30	1.31
2	Hafeez Centre	483	615	550	0.88	1.12
3	Metro park and ride Ichra	20	19	82	0.24	0.23
4	Kareem Block Market	163	287	400	0.41	0.72
5	Minar e Pakistan	826	927	705	1.17	1.31
6	Anarkali	278	289	291	0.96	0.99
7	B.I.S.E Lahore	94	56	80	1.17	0.70
8	Moon Market	229	258	370	0.62	0.70

3   Conclusion

This paper develops the parking statistics for the selected parking places in Lahore, which include the estimation of parking volume, accumulation, average parking duration, parking turnover and parking index for cars and motor bikes. Results indicate that parking places such as Kareem Block market and Moon Market have more parking demand for cars in weekend. Mall road and Minar e Pakistan have higher parking demand for motor bikes in weekend. While B.I.S.E Lahore, Minar e Pakistan and Mall road have more parking demand in weekdays. The results of this study can be helpful in evaluation of current parking system in Lahore and for future expansion of parking spaces. Parking data will be collected for the longer duration and for more number of parking places in future to develop the parking generation rates and parking demand models.

Acknowledgments. The authors are grateful to Muhammad Usama Umar, Ali Raza Shah, Majidah Tasneem and Shaukat Mahmood for their participation in parking surveys.

References

1. Jatoo, W.A.K., Fu, C.J., Saengkrod, W., Mastoi, A.G.: Urbanization in Pakistan: challenges and way forward (options) for sustainable urban development, Hongshan District (2016)
2. Kedir, M., Schmidt, E., Waqas, A.: Pakistan's Changing Demography: Urbanization and Peri-Urban Transformation Over Time. International Food Policy Research Institute, Washington, DC (2016)
3. Punjab Bureau of Statistics: Punjab Development Statistics. Bureau of Statistics Planning and Development Department, Government of the Punjab, Lahore (2016)
4. National Engineering Services Pakistan: Lahore Integrated Master Plan. Lahore Development Authority, Lahore (2002)
5. Ye, X., Chen, J.: Traffic delay caused by curb parking set in the influenced area of signalized intersection. In: 2011 International Conference on Chinese Transportation Professionals, Nanjing, China (2011)
6. Cao, J., Vasileios, N., Menendez, M.: On-street parking near the intersections: effects on traffic, Zurich (2013)
7. Sen, S., Ahmed, M.A., Das, D.: A case study on on-street parking demand estimation for 4-wheelers in urban CBD. J. Basic Appl. Eng. Res. **3**, 254–258 (2016)
8. Aderamo, A.J., Salau, K.A.: Parking patterns and problems in developing countries: a case from Ilorin, Nigeria. Afr. J. Eng. Res. **1**, 40–48 (2013)
9. Lee, Y.W.: Study on the variables for on-street parking demand estimation through parking survey. In: Advanced Science and Technology Letters, vol. 100, pp. 43–46 (2015)
10. Suthanaya, P.A.: Development of parking demand model for private hospital in developing country (case study of Denpasar City, Indonesia). J. Sustain. Dev. **10**, 52 (2017)
11. Ng, P.C., Ma'soem, D.M.: The development of model estimation to determine parking needs at LRT stations in suburban area. In: Proceedings of the Eastern Asia Society for Transportation Studies, vol. 5, pp. 877–890 (2005)
12. Tiexin, C., Miaomiao, T., Ze, M.: The model of parking demand forecast for the urban CCD. Energy Procedia **16**, 1393–1400 (2011)
13. Al Sahili, K., Hamadneh, J.: Establishing parking generation rates/models of selected land. Transp. Res. Part A Policy Pract. **91**, 213–222 (2016)
14. Das, D., Ahmed, M.A.: On-street parking demand estimation in urban CBD using FI and CF model: a case study-Kolkata, India. Indian J. Sci. Technol. **10** (2017)

Public Space Projects in the Open Areas

Alicja Maciejko[1](✉) and Roman Czajka[2]

[1] University of Zielona Gora, ul. Prof. Z. Szafrana 1,
65-516 Zielona Gora, Poland
alicjamaciejko@wp.pl
[2] Faculty of Architecture, Wroclaw University of Science and Technology,
Wybrzeze Wyspianskiego 27, 50-370 Wroclaw, Poland

Abstract. The definition of *public space* does not refer to space of highly urbanized city centers or city squares, but to all places, in which people gather and meet with each other. The places in the open landscape may be also called public space. They are usually placed in a difficult location, e.g. near the water, in forests, mountains, tourists paths and in landscape and national parks. The aspects of implementation small wooden architecture in such projects in Poland are discussed in the paper.

Keywords: Small architecture · Landscape architecture · Public space
Park Narodowy Gór Stołowych · Stolowe Mountains National Park

1 Introduction

The paper discusses the problems of designing the small wooden architecture in areas with special landscape and natural requirements and presents project that have been implemented in Poland in Stolowe Mountains National Park. The definition of small architecture objects is not precise, it refers to forms and objects that are not buildings and are not implemented as public in construction areas according a polish building law. This means that building standards do not have to be applied in these projects. These are small elements, information and educational boards, benches, tables, rubbish bins, rest areas, gates at the entrances to protected areas, ticket offices as well as engineering structures implemented in difficult terrain conditions such as shelters, platforms and towers. There are many formal and legal, technical and aesthetic problems connected with designing in a protected and hard-to-access landscape.

There are stereotypes about architectural forms made of wood, which are associated primarily as a traditional polish highlander style and often performed with braces that are not adequate for other regions than the Podhale. (Zakopane style). Stereotypes also refer to the perception of wood as a building material. It is mainly about the conviction that wood is an unstable material, not resistant to fire, atmospheric factors such as moisture and light, which is considered a disadvantage of this type of construction. These issues will be discussed on the basis of the original project designed by arch. Alicja Maciejko, author of this paper and arch. Mirosław Strzelecki, implemented on the tourist trails and routes of the Stolowe Mountains National Park in Poland.

© Springer International Publishing AG, part of Springer Nature 2019
J. Charytonowicz and C. Falcão (Eds.): AHFE 2018, AISC 788, pp. 203–215, 2019.
https://doi.org/10.1007/978-3-319-94199-8_20

2 Designing in Protected Landscape Areas

Areas of protected landscape in Poland such as national parks, reserves and landscape parks, in addition to the main protective function that they perform under the Polish law (the Act of 16 April 2004 on nature protection) are also made available for tourism purposes. This is one of their basic, social tasks, because it enables to meet the elementary needs of human contact with nature, it is a form of acquiring knowledge of nature and sightseeing. Poland currently has 23 national parks. They occupy slightly more than 1% of the country's area. The average size of the national park in Poland is 13,000 ha.

Theoretically, these areas should be free from any forms of human development and interference in the natural environment and all activities should be subordinated to nature protection and take precedence over other activities related to making them available to the public. All nature and specific features of the landscape are subject to protection [4]. Meanwhile, each of them is to some extent managed and covered by the economic activity of man, which causes various changes in the natural environment, including the landscape [5]. As a rule, this activity causes changes in the natural environment, especially in the landscape. Protected areas are a tourist attraction, millions of people visit them annually. Currently, there is a lack of precise data on the number of tourists in Poland in national parks, but, for example, in Stolowe Mountains National Park, newly designed cash registers will monitor the number of people entering.

Tourist walking trails are relatively the simplest form enabling exploring the national park. Usually, they are conducted to show the most interesting and beautiful parts of the parks. They ensure tourist penetration of their areas and should be marked permanently and professionally. Negative effects of sharing the space of national parks are inevitable. However, it should be minimized by appropriate locating of tourist infrastructure, the use of natural materials fully compatible with locally occurring ones For example, the principle in the Table Mountains National Park is using only locally occurring stones for all hardening elements. Negative effects of tourist management of the national parks space are difficult to avoid, because they result from the need to ensure adequate conditions of stay and safety of visitors. The forms of their negative impact on natural and landscape resources are diverse; the most important ones include: urbanization of the environment as a result of the development of tourist and accompanying infrastructure, destruction of the landscape by tourist facilities and objects, their inappropriate location, excessive exploitation, often lack of aesthetic values [5].

Investing in technical infrastructure in the protected landscape is subordinated to nature conservation, therefore, design in these areas largely differs from the design of public space in urbanized areas. In the public space the most important are the preservation of urban, historical and cultural order, accessibility for all users, safety of use and, above all, public good. In the protected mountainous landscape, accessibility is limited, routes are adapted to tourist traffic in a narrow range, interference in natural landscape and nature systems is strictly controlled and monitored. Similar restrictions apply only to the space of historical sites, subject to strict conservation protection. In the situation of activities in the field of tourism and recreation, which is even forced to extraordinary physical effort, and also because the facilities are located in hard-to-reach places, where it is impossible to maintain the standards required for public space, it is reasonable to subject the tourist and recreational infrastructure to the requirements availability and ergonomics? It happens that the designed footbridges are adapted to the movement of wheelchairs for the disabled or for children, (Fig. 2).

Fig. 1. New entrances to the Tatra National Park - 1st prize in the architectural competition; author: 2 pm, (http://m.architektura.info)

National parks are looking for architectural solutions that would change the image of the place and adapt to the requirements of modernity (Fig. 1). Projects are most often selected in competition. One of them was a competition for the Tatra National Park. The recommendation of the competition for small architecture was a simple form, not imitating the Zakopane style, with the use of modern construction and material solutions and the use of wood as the main material. The starting point for the project team was the koliba construction, shape and detail of its gable roof and the gray, wood of larch, naturally silver-plated by sun and rain. In order to avoid direct historical references, the hut shape has been geometrized, which is additionally emphasized by the perpendicular arrangement of decking boards and the detail of the sweep of one roof plane over the other, originally used in shacks. In the opinion of the commission, the work, due to stylistic solutions and small dimensions, fits best in the Tatra landscape, and the large functionality and simplicity of solutions gives the greatest implementation potential.

Fig. 2. Observation platform in Wola Krogulecka near Stary Sącz, photo by A. Klimkowski, (http://www.starostwo.nowy-sacz.pl/pl/wiezewidokowe)

Currently, in many attractive landscape places, in national parks, landscape parks and mountain trails there are not designed information elements (boards, signs) and elements of small architecture (fences, sheds, benches, tables, waste bins) inconsistent stylistically with each other and with the architecture of the region, most often stylized as Zakopianski. On short sections of the routes, stylistically diverse elements meet, for example, as on the tourist route to Śnieżnik in the Klodzko region (Fig. 3).

Fig. 3. A small wooden architecture on the route to Snieznik (Masyw Snieznika, Kotlina Klodzka). Photographs taken on a distance of about 1 km. Alicja Maciejko.

However, the elements of small architecture have a key role in the management of tourist routes and the provision of protected landscape for tourism, because they are, apart from permanent facilities such as hostels and educational centers the only technical elements in the natural landscape. That may affect its devastation, but also a change in the image places and contribute to increasing the attractiveness of tourist routes. Elements such as footbridges, bridges, information signs, field stairs often occur in places that are hard to reach and are the most attractive in nature, properly shaped and protect rock elements and vegetation. Sheds for rest, fireplace shelters set in strategic, analyzed locations limit the number of cases of walking off the routes from the trails.

3   Design Rules

Originally, elements built by man in the wild nature of the mountains served him as a shelter and protection against changing weather conditions and dangerous terrain. Sheds carved in rock or made of wood or stone stairs and paths, elements serving as information. They are still serving as shelters against rain, snow and wind, and they are an extension of civilization in a still unfriendly environment that allows us to hide and survive. However, they are also an element enriching nature. Well-integrated into a landscape with an adapted scale and proportions, adequate materials, very often add beauty to the surroundings. That is why it is so important that all the above-mentioned features are thought out and subordinated to the entire composition of the landscape. These may be features derived from the traditional architectural elements of the region, but they may also be transformed forms of nature, inspired by it and at the same time clearly showing the features of human products.

Infrastructure elements are characteristic "points" scattered in the mountainous environment, they not only affect the feeling of safety and comfort, but also are a determinant of subsequent stages of overcoming space, and they must ensure visual coherence associated with the identification of the place. They are also a place of recreation for all age and social groups; they can be conducive to establishing contacts. They are a social space, an important part of public spaces, in which elements of art, aesthetics and with the help of which you can stimulate intellectual ambitions, in accordance with the values and challenges of the modern world, inspired by natural, often yet virgin environment.

The conditions of the mountain landscape introduces the rigor from which design rules can emerge, aimed at reconciling two mutually exclusive goals, on the one hand protecting the landscape from the maximum extent and, on the other, making protected places available to tourists, which develops intensively in Poland. Visiting protected areas is currently one of the largest tourist attractions. The above issues require the search for universal forms that meet the following assumptions:

- the ability to adapt to various forms of natural routes, assuming the least possible interference in the environment,
- longevity, resistance to difficult climatic conditions,
- regionalism - a sign characteristic for the mountains,
- repeatability of applications, symbol, simplicity,
- expression of a conscious intervening will that signals nature by imposing its rigid foreign form,
- organizing communication and visual system in accordance with the organization of routes.

In addition, structural layers are formed:

- landscape image and background,
- interpretation of the motifs used,
- ordering and marking defining the distance in relation to the landscape,
- a grid of objects with a consistent style.

4 Wood in Traditional and Contemporary Applications

Objects of wooden architecture in an open landscape (especially mountainous) origi- nate from the forms of a mountain hut, shelter for shepherds and hikers, made for obvious reasons from local materials, stones and wood. Therefore, wood is traditionally the most popular material for the construction of small architectural forms on tourist routes, information boards, benches, sign posts, shelters and fences. The reason for the popularity of wood is its easy availability (most often it is forest areas) and the pos- sibility of simple and cheap manual processing, which does not require special equipment. Wooden elements are also much lighter than other materials and - which is an advantage in nature protection areas - they are a natural material, minimally or not at all unprocessed, so they are easily biodegradable. These constructions are connected most often with nails, typical steel joints or carpentry joints. Often these are forms that arise accidentally and are not designed. However, contemporary trends show that making unusual natural places available for tourism, more and more often not only for reasons of attractiveness of landscapes, but also for their protection, requires building spectacular structures, such as post-bridges, stairs, platforms, paths and ramps of several hundred meters in length, constructed over trees or lookout towers. They allow to visit the most attractive places without destroying (trampling on) natural resources. Such buildings are created all over the world. These are engineering objects, con- structions made in unusual conditions requiring creativity not only from the static side but also from the architectural form, because they most often become an icon of a place. And wood is an interesting material, often used in the form of finishing elements (floors, platforms, balustrades, stair steps) but also large-sized construction elements, in addition to steel and reinforced concrete or in the form of hybrid structures, wood steel, using steel elements, e.g. ties. Most often they are made of glued laminated timber, round wood or solid wood construction beams. In Poland, in recent years, in con- nection with the EU subsidies program being implemented to use and make available local natural resources for tourism purposes aimed at reducing pressure on valuable natural areas or projects concerning the reconstruction, extension and retrofitting of ecological education centers, several dozens of interesting structures have been designed and built, using wood as a supporting structure or finishing material. Such buildings were created in response to their popularity with the southern neighbors and in Germany, e.g. in Bavaria. In the Janske Baths, a tower and a path were built among trees, in Lipno nad Vltavou, a wooden walkway over the forest crown, and a few kilometers from the Polish border in the Dolni Morawa resort, a route 750 m long was constructed at a height of 55 meters from larch wood and steel.

Nowadays, the construction of the structure is more and more often used glued laminated timber, due to the fact that it is a much more predictable material, has better strength parameters, possible to obtain larger cross-sections with strictly graded wood class (boards are sorted) and it is dried, which has an effect on the slowdown of biological corrosion. The same is true of the refractory nature of these elements. Well-machined, sheathed elements with increased cross-sections have a much higher fire resistance than solid wood. In the very idea and way of constructing, these con- structions are based on traditional carpentry methods and are their natural continuation.

The combination of horizontal wood layers in a rectangular cross-section imitates the original cross-section of a wooden beam, because glued joints are almost invisible. The natural continuation of the tradition is preserved, which is undoubtedly an advantage in the aesthetic and social sense. Glued laminated timber is one of the best modern construction materials that allows you to create urbanized and open landscaping in a human-friendly way and the use of technical and constructional aspects is useful for creating value-added architecture (multi-layer wood symbolism) and the use of positive material interaction on the broadly understood environment. It is one of the most modern construction materials suitable for constructions with high strength requirements, in addition to steel and reinforced concrete, but also solutions in which wood, characteristic in structure and expression, should be used as a continuation of traditional solutions (Fig. 4).

Fig. 4. Symbolism of wooden construction. Symbolism of forest and trees: strength and longevity, endurance, multiplicity of rhythms, softness and freedom of forms, diversity and random geometry of forms, beautiful space, expression and color, flammability and impermanence, growth and changeability. Symbolism of static work: lightness and mutual coincidence of construction elements, dynamics, movement, technical purity, readability of forms, geometry of forms (creation of spatial systems). The symbolism of technology: eliminating material defects, adapting the material to a significant structural function, tradition [2].

5   The Design of Small Architecture Elements for Stolowe Mountains National Park

Stolowe Mountains National Parkis is an area of legal nature conservation. It was established in 1993, the area is 6,340 ha. It is located in the Central Sudetes in the north-western Kłodzko region. The border of the Park runs partly through the Polish border with the Czech mountains. The highest mountains in the Park are Szczeliniec Wielki (919 m above sea level) and Skalniak (915 m above sea level). Trails and hiking trails of the Stolowe Mountains National Park are characterized by varied difficulty.

The project includes a series of elements of landscaping made in a uniform style and the concepts of ticket office buildings at the entrances to the tourist routes of the Stołowe Mountains - Szczeliniec Wielki and Bledne Skaly. Despite the fact that the project concerns relatively small objects has an impact on the visual, symbolic and cultural perception of the region, it has become a clear image element, in its context it is already referred to as the so-called "new Sudeten style".

The small architecture elements were located throughout the entire area of the Stolowe Mountains National Park, taking into account the requirements of specific tourist routes. In design practice, this means that each place is diverse and requires an individual approach while using mobile, standardized elements. Sometimes these places are uncomfortable for assembly works of infrastructure elements (e.g. very steep slopes, boulders, narrow passages) but strategic for the tourist route, [2].

Fig. 5. Regional architecture of the Sudetes [1].

The ambition of the Stolowe Mountains National Park was to create new design, implementation and utility standards for elements of small architecture in the protected area, emphasizing the natural richness of the area. It was heightened by a strong conviction about responsibility towards nature and the extraordinary natural value of this beautiful place, with a specific character, unprecedented in other National Parks in Poland. This can be seen in the care of every detail, ranging from the choice of places to careful workmanship. Elements of infrastructure are placed in different places, but nevertheless they constitute a coherent and expressive icon, which began to gain the reunification of users and became a role model (Fig. 5).

The aim of the project was to create elements of technical infrastructure on touristic routes such as: a focal world, a rest world, tables, signs and information boards, fences, stairs, footbridges. The idea of the project was to create homogeneous, repeatable, stylistically consistent elements built into various, also hard to reach places, tourist routes. The project is continued, the elements are successively built depending on the needs of the Stolowe Mountains National Park.

In the first phase of design, the investor was offered elements inspired by rock and natural forms, the expression of which directly referred to the surroundings and climate of the Stolowe Mountains. In this way, the authors wanted to gain a perfect harmony with the environment, whose fantastic forms stand out in a unique way from the mountainous landscape in Poland. Lack of examples of existing buildings in this area resulted in the fact that there was no reference as a direct formal inspiration that could be appropriate for the project. On the subject of design with such a strong context of the surrounding, the question always arises as to the path to be followed, whether by traditional way or by modern solutions based on the latest technology and new forms that would express the spirit of the present. According to the authors, this second variant, through references to current technological achievements and the level of civilization, is the only way to create a human environment. Someone may complain about competing project forms with fantastic forms of nature in the Stolowe Mountains, but any form of human creativity resulting from the human intellect will always be a contrast to nature. In design practice, references to historical, local, regional and so-called traditional, must be taken into account by architects in their work, however,

Fig. 6. Stars. Maciejko A, Strzelecki M.: Design of a small technical infrastructure of the Stołowe Mountains National Park, Strzelecki Biuro Architektoniczne, [3].

Fig. 7. Information elements. Maciejko A, Strzelecki M.: Design of a small technical infrastructure of the Stołowe Mountains National Park, Strzelecki Biuro Architektoniczne, [3].

Fig. 8. A building checkout project at Bledne Skaly, shelters for resting anf to make fire inside., [3]. Maciejko A, Strzelecki M.: Design of a small technical infrastructure of the Stołowe Mountains National Park, Strzelecki Biuro Architektoniczne, [3].

the design concept should be so transformed that it can be equated with the modernity in which it is created. Due to the unwilling reception of the investor proposed in the first phase of solutions forms were proposed to a large extent traditional, with the continuing continuation of the Sudeten style, which could become a formal reference of a broad significance, but of a contemporary character. The aim was, on the one hand, to design elements protecting the landscape as much as possible, and on the other hand to create strong, distinctive forms (discouraging from gliding off the trails), remaining in harmony as well as in specific contrast with the surrounding, beautiful and "strong". The traditional forms used were received with applause by the Investor (Fig. 6). However, to modernize the applied tradition of forms, formal modifications of these elements were made by using slightly different proportions of individual fragments of these objects, scaling details, modern materials (Fig. 7). The main design motif is the massive elements of beams made of larch laminated wood, in a natural color, which is a modern modification of the traditional construction material, which is wood (Fig. 8).

Fig. 9. A building details, shelter in winter [3]. Maciejko A, Strzelecki M.: Design of a small technical infrastructure of the Stołowe Mountains National Park, Strzelecki Biuro Architektoniczne, [3].

6 Summary

The project addresses three equally important problem aspects: (1) designing in protected landscape and, consequently, making tourist routes available for tourism, which inevitably connects with possible degradation of natural landscape elements, (2) designing in the existing cultural landscape of the Sudety area in the form of regional architecture with very strong and characteristic stylistic elements and (3) designing contemporary architecture, taking into account strength requirements, durability in a difficult climate, ergonomic, safety for tourists, functional and technical (Fig. 9). Providing protected natural areas to tourists requires compliance with specific procedures that assume the least interference in the existing landscape and the designation of tourist routes, which on the one hand must run through the most attractive places and allow the exposure of nature with protection as much as possible. This makes it necessary to use atypical design rules, which, however, must fit into the canon of requirements for public facilities. The objects themselves must also stand out aesthetically; attract tourists with both functionality and attractiveness of forms.

The project addresses the important issue of designer's responsibility towards nature in the aspect of designing technical infrastructure elements in the protected mountain landscape and designer's responsibility towards the existing cultural heritage of the Sudetenland in the form of regional architecture with very strong and characteristic stylistic elements. The designers tried to draw inspiration from regional architecture largely, without losing the main design assumption that they are contemporary objects and should have a contemporary character. Elements of infrastructure are placed in different places, but they constitute a coherent and expressive icon, which began to gain recognition of users and became a role model. The issue of design in National Parks is the field of many discussions recently, Stolowe Mountains National Park is a precursor of activities aimed at unification of technical infrastructure elements and a good example of the implementation of clear image, and functional assumptions set a few years ago.

In mountain or protected regions, the accessibility is limited, routes are adjusted to narrow tourist's movement, and the interference into natural landscape order is controlled and monitored. There is no possibility to grant access to all the users, because of the landscape conditions (slopes) as well as difficult location conditions. There is also a priority for nature preservation. Does it make it impossible for the disabled to use the space? In fact, there are no ergonomic facilities. Travelling on the routes requires physical strength, but it is also a great stimulant for physical activity. The needs of the disabled should be fulfilled not only through adequate space projects, but also through the modern technological solutions (smart phones, applications, GPS, sounders) and also through the adequate assistive devices.

References

1. Park Narodowy Gor Stolowych (2018). http://www.pngs.com.pl/
2. Maciejko, A.: Analiza przydatności konstrukcji z drewna klejonego o dużych rozpiętosciach do realizacji form architektury współczesnej. Rozprawa doktorska. Wydział Architektury Politechniki Wrocławskiej (2011)
3. Maciejko, A., Strzelecki, M.: Projekt małej infrastruktury technicznej Parku Narodowego Gor Stolowych, Strzelecki Biuro Architektoniczne (2016)
4. Maciejko, A., Strzelecki, M.: Ład natury i architektury. Projekt małej infrastruktury technicznej na obszarze parku narodowego gór stołowych. [w] Zastosowanie Ergonomii. Ergonomia w architekturze i Urbanistyce. Kierunki badańw 2016 roku (2016)
5. Partyka, J.: Udostępnianie turystyczne parków narodowych w Polsce a krajobraz. Krajobraz a turystyka. Prace Komisji Krajobrazu Kulturowego Nr 14 Komisja Krajobrazu Kulturowego PTG, Sosnowiec (2010)
6. Potocki, J.: Kształtowanie sieci turystycznych szlaków pieszych w Sudetach po II wojnie światowej i jego ważniejsze uwarunkowania. [w]: Zarys dziejów turystyki i przewodnictwa w Sudetach, red. Mateusiak, A., Gryszel, P. (red. naukowa części referatowej), Jelenia Góra, s. 23–46 (2013)
7. Stasiak, A.: Produkt turystyczny – szlak. [w]: Turystyka i hotelarstwo, no. 10 (2006)
8. Ustawa z dnia 16 kwietnia 2004 r. o ochronie przyrody (2004)
9. Archive if the company Tatry. http://www.firmatatry.pl/realizacje-i-referencje/budowa-wiez-widokowych-realizacje

Sound and Form in Public Spaces
of Contemporary Hotels

Joanna Jablonska[✉], Elzbieta Trocka-Leszczynska,
and Romuald Tarczewski

Faculty of Architecture, Wroclaw University of Science and Technology,
Prusa St. 53-55, 50-311 Wroclaw, Poland
{joanna.jablonska, elzbieta.trocka-leszczynska,
romuald.tarczewski}@pwr.edu.pl

Abstract. Environmental comfort of public spaces is a result of balanced combination of varied factors, like: architectural geometry of form, space arrangement of equipment, furnishing and finishing, as well as selection of materials, with their texture, surface treatment or color. Full spectrum of these elements allow architects to formulate a proper functionality, usage and safety of interiors. On the other level of perception, this mixture gives a proper character and atmosphere of public areas in a hotel. However, aforementioned collage of factors is used to build proper architectural acoustic parameters of sound filed and may be used to positively influence condition of guests. Thus, this article was devoted to presentation of results of research on sound field connection to architectural design of space.

Keywords: Architectural acoustics · Sound in space
Contemporary architecture · Form and sound

1 Introduction

Public space of a contemporary hotel understood as "hybrid architectural space" [1] with varied functions, is valid element of urban fiber and recognized cities' friendly places. Since huge globalization jump in hospitality business form 80. XX cent., continuing until today, the open-accessed areas have been successively expanded, enriched and developed [2]. This evolvement considers the quality of architectural design, with rising awareness for importance of: functionality, safety and comfort. One of the indicators of such design are optimal parameters of sound field, which are connected to architectural and furnishing forms and materials, used in hotels' public spaces.

In light of this consideration, the issue of presented elaboration was to highlight this connection and mutual influence between sound and form in architectural solutions. And in this way, enrich the discussion of acousticians, architects and ergonomists, bringing this three domains together in practical and scientific approach.

© Springer International Publishing AG, part of Springer Nature 2019
J. Charytonowicz and C. Falcão (Eds.): AHFE 2018, AISC 788, pp. 216–225, 2019.
https://doi.org/10.1007/978-3-319-94199-8_21

2   Aim Method and Scope of Studies

The Aim. The purpose of this elaboration was to show crucial issues of form and sound solutions in the architecture of public spaces in the contemporary hotels. The goal is to demonstrate very close link and mutual influence of two aforementioned domains, in order to encourage interdisciplinary studies and collaboration on the design of such spaces. Also an ergonomic impact of architectural acoustics to the public interiors must be highlighted.

At the same time, it is essential to stress, that selected issues have an universal meaning and can be applied in many other examples, not only for public zones of a nowadays hotels.

The Method. The studies were divided into two parts. First consists of literature review, in order to clear out the definitions and authors' point of view on acoustical phenomenon appearing in architecture. Most of the literature concerning sound is based on practical observations, experiments and has a close connection to the design. Therefore, this part of article is not theoretical, but is continuously linked to the actual solutions in physical space.

Moreover, in this elaboration all occurrences concerning acoustics and architecture are simplified for the sake of considerations. However limitations made do not lessen the converging of studies to reality.

For the literature review critical analysis and synthesis was used. As an outcome of this studies there were made graphical representations, which have been presented further on in text. The case studies were carried out on site, and that continued with the analysis and graphical analysis of gathered material. Conclusions were prepared with the use of comparative synthesis.

The Scope. Case studies examples from Europe and USA selected to this presentation are representative for features of many other reviewed interiors of nowadays hotels. There have been no size limitations to the discussed rooms, in order to present aforementioned universality of issues. The only division was to large scale and minor scale interiors, due to the architectural and acoustical differences in design, being effect of volume differences.

However, it is noted that following paper was limited to the interiors solutions – so architectural acoustics, and few selected case studies, due to a huge wideness of the study field. Yet, all elaboration was kept on a possible most general level, in order to make it as applicable as possible.

3   General Notions and Main Typology

This section is devoted for explanation and refining the authors' point of view of basic architectural and acoustical notions, which will be important in the following passages of text. Firstly, this elaboration is included into the domain of **acoustical architecture**. Here, it will be understood as a scientific and practical area, aiming at creation of the best possible conditions, for the transmission and reception of sound, both speech and music. Of course, creating such parameters is dependent on the function of interiors, and varies clearly between rooms of different usage [3].

In the hotels' public spaces legibility of speech and good conditions for listening music are important. Also reduction of inner noise and creation of spaces providing intimacy of conversation are crucial. Optimal solutions of these factors is positively influencing psychological and physical health of a hotel guest. Depending on the form, related to its dimensions, volume of the room and the method of architectural compartments shaping, each interior will offer different acoustical conditions. This concerns also its' guests capacity, thus, number of people visiting particular space, furniture type and placement, as well as building and finishing materials used.

The simplest method for describing a sound field parameter in the room is to define its **reverberation**. This phenomenon is connected to certain spread of sound waves reflections across the interior from all room compartments, which can be heard by humane ear, after the sound emitted form particular sound source is no longer propagated. This parameter is strictly connected to the human ear sound reception physical capabilities and is defined with the use of reverberation time notion, quoting [3, p. 32]: **"The reverberation time T60** is defined as the time in seconds, necessary for an impulsive or interrupted test sound to drop by 60 dB compared to its maximum level, reduced with 5 dB."

With the use of this three crucial notions: architectural acoustics, reverberation and reverberation time, and following literature definitions [4] there can be distinguished **three types of sound fields types** in the architecture of public use interiors in hotels. These groups can be easily connected to specific subjective humane impressions as well as detailed architectural design parameters. It is as follows:

- **Dry reverberation spaces** – where reverberation time is short (below 1,5 s) and newly emitted acoustical waves do not overlaps with decaying reflections. For listeners speech can be found very clearly heard, however the sound, especially of music or many loud conversations, might be perceived as unpleasant. In a longer exposition dry reverberation interiors may cause feeling of stress and irritation, and in the effect lack of concentration and anxiety among guests. Severe example of such solutions are some recording studios and sound laboratories, where there is no reverberation, and allowed time for humane presence is limited.
- **Short reverberation spaces** – in which the newly emitted acoustical waves do overlap with decaying reflections, however it only enriches the heard sound not masking it. In this spaces listeners will understand both speech and music and sound will be pleasant to the ear. Typical reverberation time for speech is 1,5 s. This kind of spaces are most universal and are characteristic to majority of the rooms for everyday use. Staying in such spaces during standard conditions (i.e. lack of noise) should not expose users to any psychological or physical difficulties.
- **Long reverberation spaces** (so called reverberant) – are reserved for phenomenon in which the reflections mask newly emitted sound. The speech will be illegible, however at many occasions music may sound really well in such spaces. Excessive reverberation also occurs and may lead to unpleasant feeling of lost and disorientation in the interior. These types of halls can be found in old large scale churches, like Christian gothic temples, where reverberation time can endure even up to 12 s.

In the hotel public use spaces maximum long reverberation spaces are not recommended for any interior type (Fig. 1) [5, 6].

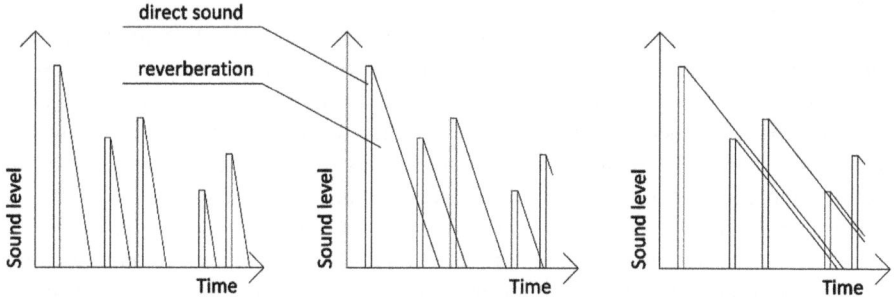

Fig. 1. Scheme for reverberation characteristic in spaces, from left: dry, short, long.

4   Solutions and Materials

4.1   Reflection and Diffusion of Acoustical Sound Waves

In order to combine this three interiors with proper architectural features, it is needed to specify materials characteristics, which influence directly acoustical parameters. As it was discussed, most common sound phenomenon in closed space is **reflection**. Acoustical wave emitted from the sound source at the way of its propagation meats obstacles of substantial size, like: ceilings, floors or walls, and reflects from it in a **mirrored** or **scattered** matter (Fig. 2) [5, 6].

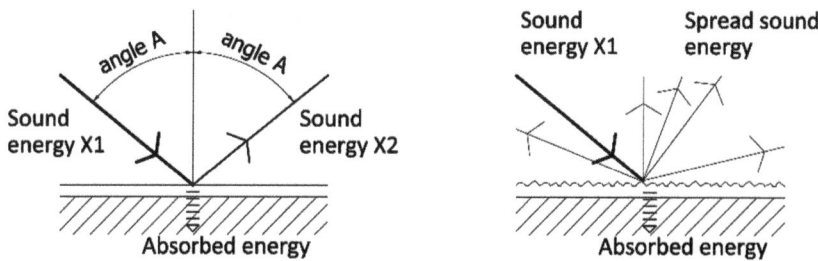

Fig. 2. Scheme for reflection types, from left: mirrored, scattered.

At the same time reflected acoustical wave loses a part of its overall energy to the compartment it bounces from. It propagates as a sound wave in material, transferring its energy into heat, as an effect of friction. In case of mirrored reflection it takes place when reflective surface is hard and even, like: glass, polished wood and concrete, plaster or in large, unstructured boards. Mirrored reflection phenomenon provides a longer reverberation time and impression of 'clear' sound field. Scattered reflection breaks the wave into smaller ones, which go in different directions, caring reduced amounts of acoustical energy. Also some part of energy is propagated into the compartment. It may be provided by the same, as already mentioned materials, however their surfaces must be irregular, uneven, raw. For listeners the sound filed is 'softer' to the ears, more balanced [5, 6].

For public spaces of hotels reflection and diffusion of acoustical sound wave are a proper and needed phenomenon, providing the interiors with impression of liveness and users with spatial orientation and safety. Also materials connected to this phenomenon, like aforementioned: glass, wood, concrete, plaster, are favourable in the hotel public space. The sound wave reflectiveness is especially carefully designed in meetings and conference rooms of a hospitality sector (see: case studies).

However, usually public spaces in contemporary hotels are vast, with excessive volume. It these cases, it is crucial to provide limitation to reflections, to omit too long reverberation time. Same solutions are used, in the places when intimacy is required and this wold be cafes and restaurants, where guests should not hear each other at the separate tables. This is when absorption or isolation are used, which will be discussed in the next paragraph.

4.2   Absorption and Isolation of Acoustical Sound Waves

It also has to be highlighted, that sound arises as a result of periodic vibrations of air molecules, which are not propagating themselves, but only transmit the sound wave. These particles are excited by the vibration of objects, i.e. human vocal cords, the impact on the floor or wall. Only understanding this definition of sound, the phenomenon's of absorption and isolation can be captured. **Absorption** is the friction of air molecules with other materials, i.e. porous concrete, glass and mineral wool, which causes the acoustic energy to turn into heat and thus, the vibrations are lost and the sound weakens. In this way the reverberation is shortened and it may decay faster in the room. If none or highly limited amount of sound energy passes through compartment built form such materials, than **isolation** occurs [3] (Fig. 3).

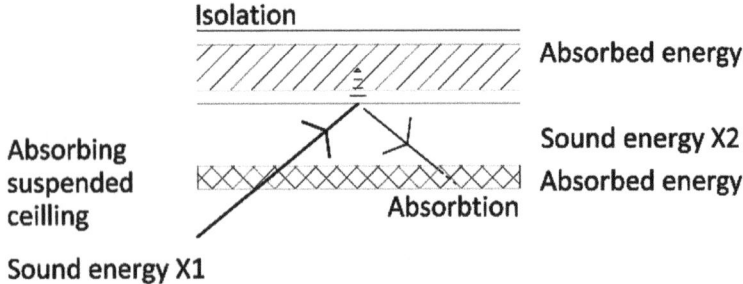

Fig. 3. Scheme for absorption and isolation.

As it was aforementioned in spaces like: vast halls, corridors, bars and cafes, it is especially required to absorb substantial portion of acoustical energy. It is done with use of soft, porous materials, which also can be an important part of an overall hotel design. This can be: carpets, curtains, furniture with soft upholstery, wall textile and so on. If the silence in a space is still not sufficient, additional acoustical solutions are used, like: hard mineral-wool suspended ceilings, sound absorbent acoustical islands also as suspended elements, absorbent paneling for walls. The later have sandwich structure, which combines mineral wool core with thin perforated panel envelopment. Such solutions allow to sufficiently limit the excessive sound energy in public zones.

In contemporary hotel public spaces, there are usually encountered combinations of all four material types. This allows designers to balance the acoustical field parameters, making a hotel a friendly place for visitors.

5 Case Study

This section was devoted to presentation of interesting and good practice study examples, reviling how proper acoustical solutions may benefit architecture and ergonomics of interiors. It is especially important, since hotels' public spaces, became a multifunctional and multicultural centers of hybrid design [1]. There, many activities can take place in a one interior or room, which becomes so called 'shared space' [7]. This process undergoes severe competition in hospitality business [2], therefore all proposed functions must be attractive and encourage wide public to participate. Having so many activities, actions and people in one zone, it seems even more important to properly spread: reflective and diffusive surfaces, as well as provide proper absorbance and isolation.

5.1 Every Day Solutions

Taking into account city and suburban hotels, with moderate conference centers, where larges rooms have the capacity of 50 participants, usually standard acoustical products will be sufficient. The example of such modest solution would be a conference and SPA hotel Sobotel in Sobótka, in Poland. This facility can contain gatherings up to 250 people in four rooms with flexible arrangements. Interiors are serving as meeting rooms, banqueted areas or even one vast ballroom. On the daily basis two connected rooms function as dining spot and publically accessed restaurant. The division of these spaces was carried out with the use of acoustically isolating, movable walls [8]. Each conference room has a rectangular layout, which for small volumes, provides a proper sound reflections from side walls and ceilings. Also flat walls covered with horizontal glazing and plaster, favor mirrored acoustical wave bounce. At the same time the ceiling was treated as suspended with the use of flat gypsum-board surfaces, which reflect, but also absorb some parts of acoustical energy. Soft and heavy flat floor carpeting allows to absorb shock wave coming from the steps, and air-borne sounds of conversations or speech. Additional intimacy of the sound field is provided by the thick upholstery of chairs used in this space. With the application of simple means and clear architectural acoustical solutions, the space is comfortable, functional and friendly to users (Fig. 4).

Similar solutions can be found in a world-known hotel in Chicago the Trump Tower, in USA, which offers 250 guests ballroom and a variety of meetings and

Fig. 4. Multifunctional venue space and its' acoustical solutions in Sobotel in Sobótka, Poland

conference spaces of capacity ranging from 185 people to 20 in the board room. The presented on the photograph Skyline room, which has an flexible agreement, similar to other interiors, is equipped with reflective: ceiling, walls and glazed external façade. All is supplemented with absorbing: curtains, thick carpeting and furniture – chairs with large amount of upholstery. In the described room the structural pillars with a substantial dimensions, are providing noticeable diffusion of acoustical waves, which is very favorable phenomenon in halls of a rectangular plans of layouts. It prevents **fluttering echo** occurrence, which is irritating multiplication of the acoustic wave which, once reflected from the side wall, traverses the inside many times. This results in the overlapping of the same sound reflections on its direct wave, and is perceived as distortion of speech and music [9]. At the same times ceilings are not flat in section, offering a straight angle, which enables obtaining of additional diffusion of sound, and in such way a better balance to the acoustical filed (Fig. 5) [10].

Fig. 5. Multifunctional venue space and its' acoustical solutions in hotel Trump Tower, Chicago (USA).

5.2   Advanced Solutions

For large scale hotels, containing casinos and sport resorts or large conference areas with rooms over 50, 100 and more participants, the especial acoustical solutions are needed. An example of this type solutions would be Ballroom in the Mirage Hotel, Las Vegas (USA), with the capacity between 508 up to even 8650 people. Room can be divided into five smaller meeting spaces, and by such, requires careful acoustical treatment [11]. Thou, the area was planned on a rectangle, the walls are not flat, and received sculptured surfaces, with horizontal and vertical ribs, preventing fluttering echo occurrence. Similar solution was used for the ceiling, where beams of a substantial dimensions, have been designed. They divide whole area into much smaller square fields, filled with sound absorbing tiles. More absorbance was added with the use of thick: carpeting, table cloth and chairs finishing. Due to such solutions, the room provides good conditions for lectures, concerts, free conversations and work (Fig. 6).

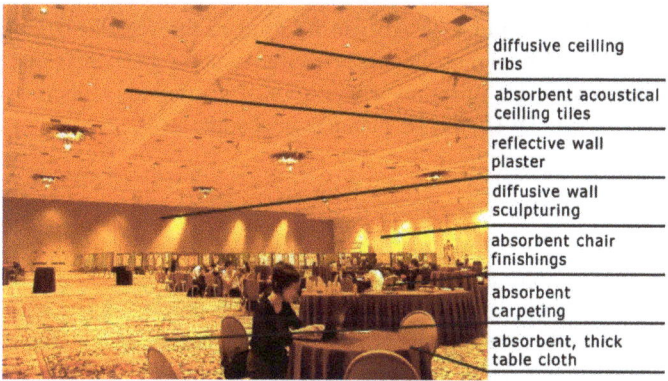

diffusive ceiling ribs

absorbent acoustical ceiling tiles

reflective wall plaster

diffusive wall sculpturing

absorbent chair finishings

absorbent carpeting

absorbent, thick table cloth

Fig. 6. Ballroom in banquet arrangement in the Mirage Hotel, Las Vegas (USA).

Also especial rooms require more advance solutions. The example would be an acoustically absorbing tiled ceiling, suspended in the fitness center of Sofitel Chicago Magnificent Mile (former Sofitel Chicago Water Tower). This element is actually only one absorbent surface in the whole room, due to a sport and sanitary requirements of such areas. Firstly, walls must be flat, thus safe and easy to clean, in a frequent matter. Therefore, they are reflective and do not provide any softening of the acoustical field, because on the sculptures or irregularities absence on their surfaces. Moreover gyms require mirror walls, which are important element of proper training performance. This provides even more hard reflections of the acoustical waves in the interior. Secondly, floor also must be flat and easy to clean. Fortunately, due to the sport requirements, it is elastic, which absorbs perfectly shock sounds from jumping and vibrations produced by sport equipment. Yet, it will not add any absorbance of the reflections from walls. This proves, how important is the proper ceiling solutions for any type of sport center in the contemporary hotel public spaces (Fig. 7).

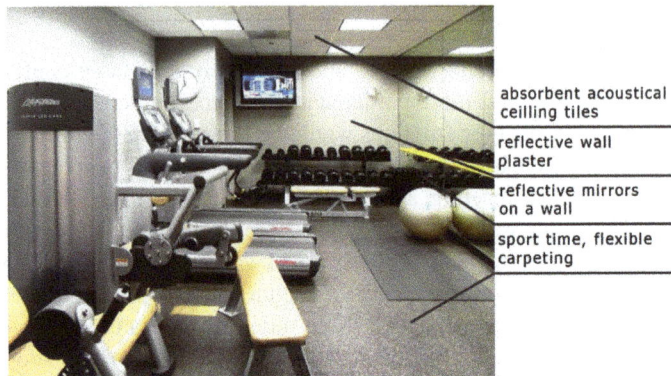

Fig. 7. Fitness center in the Sofitel Chicago Magnificent Mile (former Sofitel Chicago Water Tower), (USA).

It can be added that similar design can be used in the representational halls and areas, where walls and floors must be of a reflective material like stone, due to a certain monumental character. Also in rooms where a lot of glazing an walls is used, and the volume is vast, the ceiling should be planned as absorbent.

6 Summary

The spectrum for architectural acoustical solutions for hotel spaces is vast and complex. Yet, it is advisory to form certain guidelines for venue design, as far as sound field is considered. Taking into account the hotel public spaces typology according to the reverberation time included in point 3 of this elaboration, there were room function and reverberation characteristic formed and presented on the general scheme. For the categories are fluid due to subjective human perception of sound and space, the graph was proposed in a form of continuum. In order to make it more useful also room volume and material indicators were added (Fig. 8).

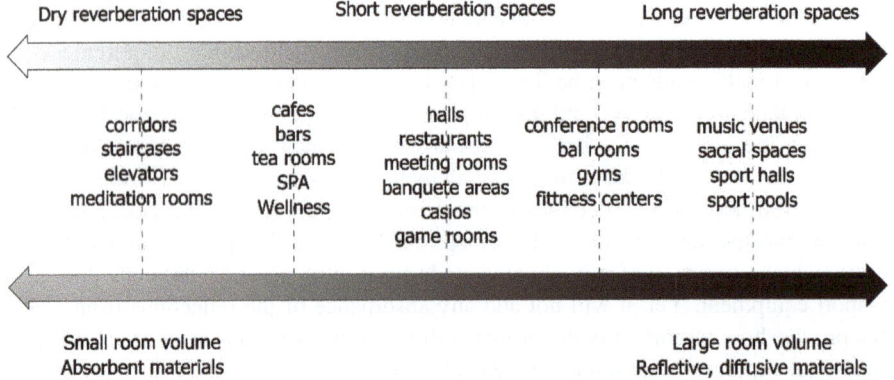

Fig. 8. Reverberation scheme in the architectural acoustics of public use spaces in contemporary hotels.

Architecture and acoustics can be bind together in the public spaces of newly build or refurbished hotels, in order to provide better ergonomics. Advantages to this solutions for humane health are: better speech intelligibility in halls and conference rooms, intimacy in cafes and restaurants and spatial orientations in all zones. Proper solutions in this field will also substantially limit noise exposition of guests in publicly accessed spaces. This will lead to further humanization of our build environment in the hospitality business.

References

1. Hubertus, A.: Strategien und Typologien im Hotelwesen. Staregies and Typologies in Hotel Design: Detail no. 3, pp. 172–181 (2007)
2. Dominik, P., Drogoń, W.: Organizacja przedsiębiorstwa hotelarskiego. Almamer Wyższa Szkoła Ekonomiczna, Warsaw (2009)
3. ArAc Multibook (ed.): International Partnership ArAc Multibook, LLP project of innovation transfer no. 2013-1-PL1-LEO05-37588 (2015). http://arac-multibook.com/www/
4. AV Info: Non-profit Informational Website. http://www.proav.de/index.html? http&&&www.proav.de/acoustic/RT_meetingrooms.html
5. Beranek, L.: Concert and Opera House: How They Sound. Acoustical Society of America, New York (1996)
6. Kulowski, A.: Akustyka sal. Wydawnictwo Politechniki Gdańskiej, Gdańsk (2007)
7. Rutes, A.W., Penner, H.R., Lawrence, A.: Hotel Design planning and Development. Norton and Company, New York, London (2001)
8. http://www.sobotel.pl
9. Barron, M.: Auditorium Acoustics and Architectural Design, London (1993)
10. https://www.trumphotels.com/chicago
11. https://www.mirage.com

Resilience in Housing Regeneration for a Smart City Model

Donatella Radogna[✉] and Manuela Romano

Architecture Department, University "G. D'Annunzio",
Chieti-Pescara, Pescara, Italy
{dradogna,manuela.romano}@unich.it

Abstract. In the European context, the realization of the sustainable city has among its intervention priorities the social housing. The smart city is an important reference point for processes aiming to the optimization of the material and immaterial resources use and management through a concrete involvement of the communities. This work, is founded on the relation between smart city and smart buildings and defines a methodology for the resilience capabilities survey, which can project the housing systems in a sustainable dimension, also thanks to the inhabitants know-how in the resources management. The study tries to define orientations for reaching feasible results and has as experimentation context the city of Pescara (Abruzzi, Italy). Starting from the knowledge of a complex needs framework, given by the inhabitants requests, the management authority needs and the environmental criticalities, we search for a resilient dimension per triggering a regeneration process to be carried step by step.

Keywords: Sustainability · Resilience · Housing quality · Smart city
Building redevelopment

1 Introduction

The challenge of developing degraded quarters and of making efficient settlement systems to guarantee accommodations and basic services for a better life quality represents a preferential intervention setting for the city sustainable development (Carta di Lipsia, 2007; RFCS 2013; Agenda 2030, 2015; UNIHABITAT, 2016). The theme became central as a result of new residential discomfort forms, which were the consequence of social and economic transformation that condition the contemporary housing. The "housing problem", indeed, is still a central question all over the Europe: the inhabit cost represents the families balance.

In Italy, the phenomenon is particularly complex due to the ineffectiveness of the policies supporting the housing sector and the progressive reduction of the invested resources in it. The housing discomfort interests 1,7 millions of families and the social accommodations are about a thousand, therefore insufficient for satisfying a demand in evolution (Federcasa, 2015; Nomisma, 2016; Housing Europe, 2017) and a heavy increase (+62%) of the eviction measures adopted between the 2006 and the 2014.

© Springer International Publishing AG, part of Springer Nature 2019
J. Charytonowicz and C. Falcão (Eds.): AHFE 2018, AISC 788, pp. 226–235, 2019.
https://doi.org/10.1007/978-3-319-94199-8_22

If the realization of new accommodations foresees public-private initiatives, 400.000 are instead the housing units that need urgent performance improvement interventions that is difficult due to social and economic disturbances.

Around the social housing quarters, entire portions of city developed that shows emergency situations of the build systems as well as the urban environment, requiring processes of regeneration and connection with the rest of the city. In this overview, in the claiming back of the common rights, passing from the "house property" to the relationships quality value, using "the housing services", people become protagonist in the quarter life. The concept of "inhabiting as common good" (TAM Associates, 2016) that enhances the needs, to define new intervention ways founded on resilient occupying processes.

Smart and resilient cities and communities have an important role in the multi-disciplinary and international debate about the transition forward living sustainable models and amount themselves as "tools" able to start innovative strategies of spatial-functional configuration and environmental reclamation to make the urban patterns safer, more beautiful and efficient [15]. The co-planning processes offer many opportunities for an efficient and constant of the living quality, intended to answer to the places and inhabitants needs and not only to the market dynamics, rebuilding "people oriented" spaces that encourage the social relations and the common interests.

In these latest years, they defined criteria that in some particularly virtuous and less problematic reality allowed reaching significant results but there are other realities, such as the considered contest, in which the applicability of the best practices is almost impossible. Therefore it is necessary to focus the factors that in realities like that limit the sustainable development and to individuate the "ways" to go forward this development by investigating about the feasibility theme.

2 Smart City and Resilient City

The project of efficient residential systems is connected to the smart city and resilient city concepts, which define innovative processes in the integrated planning ways of the regeneration that are founded on bottom up approaches.

These approaches induce to a deep comprehension of the problems and the consequent individuation of the solutions, starting from the study of specific cases for developing fundamentals lines to be extended and replied.

The smart city concept, based on a systemic organization of the natural resources use, of the communication (smart and green) and safety and well-being monitoring networks, is deeply linked to the "city resilient" concept, that has the capability of absorbing, recovering and preparing respect to future and unforeseeable shocks (OCSE, 2017), educating the community to react against the difficulties.

100 Resilient Cities (international project of the Rockfeller Foundation), on the other hand, defines the urban resilience as people, community and institutions capability of resisting, reacting or adapting to difficulty conditions, coming from chronic stresses and acute shocks. I chronic stresses are those phenomena that daily weaken the city pattern and refer to the social disadvantage, the unemployement, the pollution, the energetic consumption, the services and the public transports inefficiency and the

resources lack. The acute shocks, instead are defined, as the sudden events that threaten the cities and are catastrophic events such as earthquake, floods, epidermis and terroristic events.

The 100ResilientCity activity, involves 100 cities all over the world. Its mission is to a contrive resilience strategies to define a program for the communities self-defense against the risks linked to the climatic changes, the urban environment decay, the resources shortage, the population and poverty growth and the consequent social and housing disadvantage of the urban contests. The key topics of the smart city that is Environment, Energy, ICT, Built Environment, Mobility, Water, Economy, Health & Living e Governance (OICE, 2017), interconnected with the resilience concept seem to give back virtuous processes to face the challenges linked to live in the city of the future.

The comprehension of the problems and of the dynamics that affect the settlement systems operation, is pointed towards the definition of the prerequisites for a resilient behavior and new recovery opportunities [6].

The intervention planning strategies should be developed according to multi-scalar and interdisciplinary levels for a correct management of the environmental system (resources efficient use), of the human system (building and smart control systems efficiency) and of the social system (neighborhood relations, identity and culture). At the urban scale, the intervention planning should aim to the risks elimination (linked to the calamities and the safety) and to the optimization of the energetic flows, also through a "re-naturalization" and re-functionalization of the urban space, and a co-management of the common spaces. At the building scale the strategies should aim to the reduction of the energetic consumption and a functional adaptation, able to improve the performance level linked to the new needs of the contemporary living. The recourse to integrated strategies offers the possibility of reducing the environmental, social and economic vulnerability.

To train communities able to 'regenerate' and 'support' themselves in progress and to asses the 'answer' capabilities of the residential systems, it is necessary a revision of the operative instruments and a definition of a methodological apparatus that gives analytic indicators and design criteria for supporting the decisional processes.

Therefore in this research work, we recognized the utility of defining a framework able to support the authorities and the public administrations, for promoting co-evolution circular processes of the built environment and of the users, starting from the survey of the resilience capabilities and the observed systems vulnerability, at the building and e urban scale.

3 The Resilience Capabilities Survey of Pescara Housing Quarters

The social housing quartiers, can assume a very important role in transforming Pescara in a smart city both because they have significant dimensions and diffusion in the urban territory and because in these quartiers beyond the problems and the disadvantages there are good will and rebirth desires.

For the considered built environment (Fig. 1), the lack of operative tools generated few inefficient interventions. The building systems obsolescence, the settlements marginalization conditions, the high rates of unemployment and the criticality of the migratory flows, are the principal fragility factors of the building sector, that requests a design and procedural approach open forward new answer dynamics in the in progress changing processes.

For improving the applicability of appreciable solutions and make a transition forward sustainable housing systems possible, we should move our actions from the single buildings to the entire settlement systems, measuring against the opportunities given by the technologic innovation and also against new kind of process, which are able to start a dialogue between the demand expressed by the environmental and economic emergency and the inhabitants needs through a better organization and management of the natural, built and social systems.

Fig. 1. The arrangement of social housing quarters in Pescara urban pattern

Our research work inspect about the living quality improvement perspectives in the housing system redeveloping processes, taking as reference the eco-quartier, arisen for promoting a high living quality level through design solutions aiming to combine environmental, social and economic objects e recognized as a model for the European city future. A model able to satisfy the needs coming from the difficulties given by the real estate crisis, the social accommodation and the cultural integration demand, the pollution, the adaptation to the climatic changes necessity, the outskirts de-industrialization, the soil consumption.

The work is focused on the comprehension of the possibilities of transforming the Pescara (Abruzzi, Italy) social housing settlements in sustainable quartiers. These quarters characteristics offer many possibility for a sustainable development, however show a complex situation conditioned by discordant factors (inhabitants needs, authority need, inefficiency of the normative tools).

We tried to understand the applicability of some strategies and best practices and to define a supporting tool for the planning initiatives, good for the evaluation of the requested parameters for reaching a sustainable dimension and a better living quality (Fig. 2).

The research adopts a systemic design approach, based on the need-performance philosophy.

Starting from this observation, the carried out investigation began with the survey of the principal 'disturbing factors' for the environmental, social and economic risks that make "house system" unsustainable.

The cost of living is a crux, on which the performance inefficiency of the settlement systems is significant. In cause-effect logic, indeed, the high building systems operation cost and the absence of services with an efficient maintenance, aggravate the functionality drops of the same systems.

So starting from an organization of the inefficiency causes and aiming to gain a sustainable condition, we defined the indicators able to understand the 'disturbance' factors and the resilience capabilities (sited oriented) through predefined target.

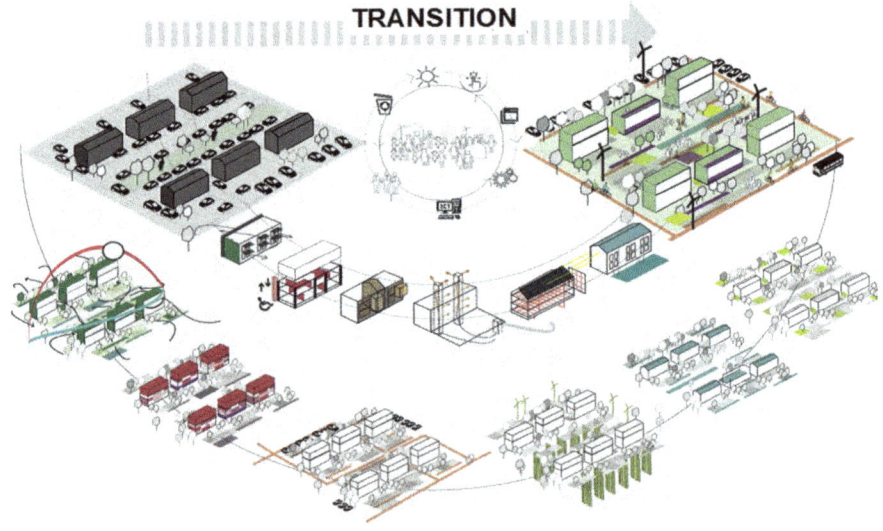

TRANSITION

Fig. 2. The transition process forward sustainability

The indicators are classified in three quality classes, referred to the sustainability dimensions: eco-systemic quality (environmental dimension), use quality (social dimension) and management quality (economic dimension). The assessment is fulfilled by means of organized Check list to identify the performance insufficiency (for the existing buildings), and the performance to reach (for the new buildings), and indicate the transformation and adaptation needs suitable for giving back the percentage values referred to the different target. The selection and the combination of the indicators individuate the specific resilience but the applicability of the strategy is conditioned by an eco-systemic vision that uses the urban metabolism instruments (flows analysis and extended metabolism, Newman, 1999), pointing to the self-sufficiency of the residential systems [16]. The individuated positive factors, answering to the natural and human disturbances (resilience intrinsic capabilities), can concern the redundancy (different functions), the modularity (sub-system independency), the temporal facto (the answers variables recognition), the memory and the knowledge (of the existing disturbance phenomena).

The adopted reference is the eco-quarter that is to say a residential system that support responsible behaviors through the efficient organization of the material and immaterial resources founded on co-design in which the collective action determines high efficiency levels. The developed processes, underline, indeed, as the quarter scale is particularly appropriate to face the challenges tied to the sustainable objective (EcoQuartier in France, Sustainable Quarters in Switzerland, etc.).

The comparison of the collected data wit some experimented systems (Bream Communities, DGNB, GBC, Protocollo ITACA) for the sustainability 'performances' assessment allowed to individuate a rating useful for setting the inefficiency factors and defining the intervention priorities. This passage was fundamental for evaluating the applicability of the systems in the investigated reality, in which the complexity of the needs framework request a more incisive assessment above all in the social and economic dimensions.

As Fig. 3 shows, the methodological procedure includes five phases and considers, both at the quartiers scale and the building scale, environmental, functional and management components in controlling the living quality (environmental quality, economic quality, social quality). The framework of Fig. 3 underlines as, once the intervention priorities have been individuated, the stakeholders matrix (authorities, administrations, inhabitants, associations, etc.) allows to define the intervention scenarios and the design launch (always on the building and quartier double and always in relation to the urban context).

Phase 1: data collection for identifying through co-diagnostics risks and needs. The indicators are 26 aggregated parameters referred to three quality categories (eco-systemic, usability and management). The evaluation assigns a score to each target in a satisfied/not satisfied range aiming to define a percentage value to the quality status.

Phase 2: data processing of the information collected for identifying vulnerabilities (environmental, functional and management), and potentialities (resistance, as a capacity to absorb impact; recovery, as a temporal factor in returning to equilibrium; creativity, as a function of the performances improvement as a result of adversity). The process allows to identify the intervention priorities that represent the risks related to the vulnerabilities and to define the strategic guidelines.

Phase 3: interpretation to individuate the stakeholders, the relationships and the needs framework with respect to the evolutionary potential of the systems are identified. The vulnerabilities and potentialities, after having been defined through co-diagnostics, give rise to alternative intervention scenarios, to be programmed according to the procedural and economic conditions. The evolving needs of the various actors in the process (managing institution, community, public authority, service co-operatives, investors and banks, project teams, external citizens) are revealed through the stakeholder motivation, matrix which allows to build a complex representative needs framework of the requests of each interlocutor. The configuration of the scenarios defines elements of social innovation and neighborhood micro-economies. The new forms of work are linked to the services and to the controlled self-management practices of the efficiency state.

Phase 4: planning to outline, from the understanding of the various stakeholders needs, the planning scenarios aiming to to show project alternatives in relation to the risks and benefits of certain intervention strategies. At this stage, it is desirable the launch of a participation process in which roles, skills and tools for training and information are identified, for the self-management of neighborhood services or self-management of the district and building systems quality.

Phase 5: monitoring of the adopted solutions effectiveness is launched. In this phase the parameters individuated through the three quality categories are monitored to outline a new scenario and to recalibrate the objectives over time.

During the investigation phases, the data collected and processing led to the detection of the main vulnerabilities. From an environmental point of view, the performance shortcomings of technological systems are evident and constitute the main cause of energy high consumption, in addition to the lack of control systems for microclimatic factors. There are also various forms of pollution caused by parking, unused machinery, use of obsolete materials and technologies. From a functional point of view, this forms of degradation influences the accessibility and the possibility of using spaces.

In the interpretation phase, the representative needs framework of the various interlocutors was constructed in the stakeholder matrix, assuming a participatory path aimed at knowing and sharing of the objectives. Once the three 'components' (built, inhabitants, institution) intervention needs were known, design directions for devising an in progress overall quality were prefigured. We refer to a quality viable step by step, in which the design actions aim at eliminating the vulnerability forms.

The priority of the interventions identifies strategies considering different scales, recalibrated during the meta-planning phase and the actual requests of the community, evaluating courageous and shared operations. The scenario outlines a community that takes care of the urban space of relevance mitigating the risks of perturbation with respect to safety (microclimatic, hydraulic and neighbourhood).

4 From Resilience to Smartness for a High Living Quality

The novelty of this work is in the attempt of making, feasible for the considered communities, actions guided by the concept of the 'limit' and coming from the deep knowledge of the contexts and the territorial system reality [11].

The method supporting a collaborative planning, assures the qualitative levels maintenance in progress, demonstrating a better applicability on the existing quarters. Through the participation, the reaction and adaptation capabilities result in organization, enhancement and management of the available material and immaterial resources, developing the circular economy with di green economy logic pointing to people well being and to the eco-system resilience [1].

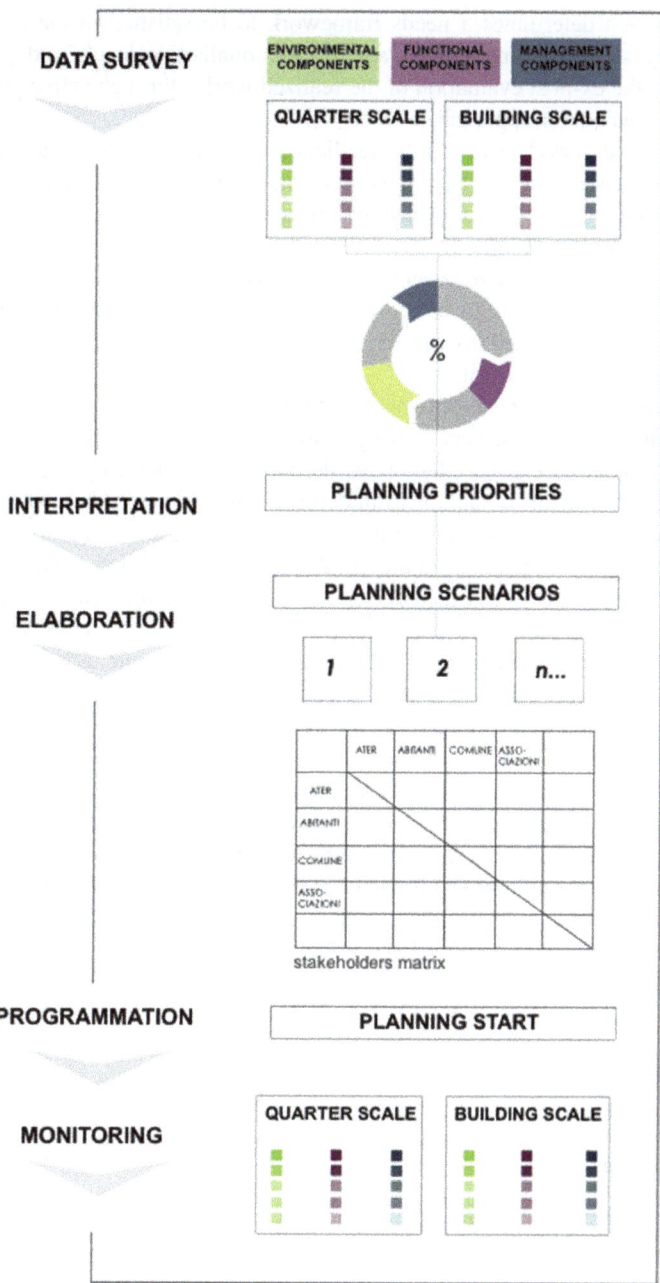

Fig. 3. Framework of the methodology defined for applying the sustainability feasibility strategies

The approach determines a needs framework to be satisfied in the course of the time, step by step, reaching greater and greater qualitative levels and giving back indicators for the ex-post evaluation of the realized works, through active policy maker and a multi stakeholder approach.

In a historical period in which the resilience thinking principles and significances have been discussed and clarified, it is well-timed trying to apply the 'thinking' for facing the principal problems that disturb life quality that is to promote the development of a resilience working.

So the carried out work represent an attempt of translating the resilience concept into operative terms, that is to define utilities and perspectives of application in the intervention logics oriented to the existing buildings regeneration [10] or top the realization of new 'healthy' cities.

We try to stimulate the problems overcoming through the proposal of a design approach that, aware of the very lacking resources availability and of the difficulties due to the social aspects, is founded on the definition of a knowledge framework of the problems to face and of the action to undertake. More exactly, the proposed results, involve the definition of procedural criteria and guide indicators suitable for a gradual and step by step achievement of a sustainable condition, starting from the necessary and sufficient concept. The indicators and the criteria, defined thanks to a critical reading of many existing tools, can permit the management of a complex and diversified framework, due to the considered buildings conditions, the users desires and the limits (above all economic-financial). The needs framework is satisfied in progress, phase after phase that reach qualitative levels greater and greater and give back indications for the interventions ex-ante and in progress control.

The carried out investigations about sustainable living models underlined as through a collaborative approach and an awareness about a responsible use of the material and immaterial resources, we can develop perspectives for reducing the social disadvantage and the environmental impact. The same analysis highlight, indeed, as the realization of sustainable living systems depends deeply on an active participation of the inhabitants and on a co-operation among the different involved persons in the organization and management processes for the correct settlements and city "working".

References

1. Antonini, E., Tucci, F.: Architettura, città e territorio verso la Green economy. Edizioni Ambiente, Milan (2017)
2. Arup & Partners: City Resilience Framework, for "100 Resilient Cities". Project of Rockfeller Foundation (2014)
3. Batty, M., et al.: Smart cities of the future. Eur. Phys. J. Spec. Topics **241**(1) (2012)
4. Bologna, G.: Manuale della sostenibilità. Idee, concetti, nuove discipline capaci di future. Edizioni Ambiente, Milano (2008)
5. Carpenter, S.R., et al.: From metaphor to measurement: resilience of what to what? Ecosystems **4**, 765–781 (2001)
6. Caterina, G.: Introduzione. In: Fabbricatti, K.: Le sfide della città interculturale. La teoria della resilienza per il governo dei cambiamenti, Franco Angeli, Milano (2013)

7. Davoudi, S.: Resilience: a bridging concept or a dead end?' Plan. Theory Practice **13**(2) (2012)
8. Friedman, Y.: L'architettura di sopravvivenza. Bollati Bollinghieri, Torino (2015)
9. Folke, C., et al.: Resilience Thinking: Integrating Resilience, Adaptability and Transformability (2010)
10. Forlani, M.C., Radogna, D.: Sostenibilità e strategie per ricostruire territori in abbandono. TECHNE, J. Technol. Archit. Environ. Beyond Crisis **1** (2011)
11. Forlani, M.C.: Rigenerare, Riqualificare e Valorizzare L'edilizia Sociale. Quodlibet, Macerata (2013)
12. Radogna, D., Romano, M.: Smart city – Eco-quartieri, in Citta sostenibile. In: AAVV, Ve so Pescara 2027. Gangemi, Roma, vol. 02 (2016)
13. Holling, C.S.: Resilience and stability of ecological systems. Annual Reviews, Vancouver, Canada (1973)
14. Lefebvre, H.: Il diritto alla città. Ombre corte, Verona (2014)
15. Lucarelli, M.T., D'Ambrosio, V., Milardi, M.: Resilienza e adattamento dell'ambiente costruito. Architettura, Città e Territorio verso la Green Economy. Edizioni Ambiente, Milano (2017)
16. Mastrolonardo, L., Manigrasso, M.: A.R.M.I: Adattamento, Resilienza, Metabolismo, Intel ligenza. Edicom Edizioni, Monfalcone (2014)
17. Pauli, G.: Blue Economy 2.0. 200 progetti implementati, 4 miliardi di dollari investiti, 3 milioni di nuovi posti di lavoro creati. Edizioni Ambiente, Milano (2015)
18. Settis, S.: Architettura e democrazia. Einaudi Editore, Turin (2017)
19. Ward, C.: Architettura del dissenso. Elèuthera Editore, Milano (2016)

Law in Motion or Passionate Observer on the Shelf? The Ghanaian Disaster Experience

Olivia Anku-Tsede[1]([⊠]), Believe Quaqoo Dedzo[1],
Michael Asiedu Gyensare[2], and Aaron Makafui Ametorwo[1]

[1] Department of Organisation and Human Resource Management,
University of Ghana Business School, Legon-Accra, Ghana
`oanku-tsede@ug.edu.gh`
[2] Department of Business Education, University of Education, Winneba, Ghana

Abstract. The purpose of this paper is to present an overview of the occurrence of disasters and the institutional mechanisms, as well as the legal response to such incidents. The paper provides a viewpoint by using discursive approach on arguments and examination of various legal and documented texts. Findings indicate that although the Ghanaian legal regime on disaster make reasonable provision for various breaches and remedies and establishes internationally acceptable rules in substance and procedure, in certain instances, most regulatory and enforcement agencies are often indisposed to enforcing the rules and fines. Meanwhile, factors contributing to disasters are mainly man-made attributable to non-regulatory compliance, unplanned residential and business communities, and erection of buildings in waterways, among others. It is proposed for an integrative approach and a shift from the traditional response-based thinking to a proactive response in disaster management, taking significant note of the legislations and judicial precedents in the country.

Keywords: Disasters · Disaster risk management · Legal response
Regulatory framework · Ghana

1 Introduction

Disasters are unforeseen destructive events of man-made or natural origins that adversely impact the environment and displace and traumatize human beings and [1]. The UNISDR [2] defines disaster as "a serious disruption of the functioning of a community or a society causing widespread human, material, economic or environmental losses which exceed the ability of the affected community or society to cope using its own resources." (p. 9). Disasters may also be defined as unexpected and harmful occurrences that have undesirable consequences on the natural and physical environment and significantly affects human life and property. Disasters are either natural or as a result of human activities or a combination of both natural and man-made factors and often displace and traumatize human and animal lives. Natural disasters, more than armed conflicts, are regarded as one of the greatest causes of internal displacement in most countries worldwide. Disasters have over the years

© Springer International Publishing AG, part of Springer Nature 2019
J. Charytonowicz and C. Falcão (Eds.): AHFE 2018, AISC 788, pp. 236–247, 2019.
https://doi.org/10.1007/978-3-319-94199-8_23

caused severe and irreparable destruction to communities across the world. Loss of life, injury, disease and other negative effects on physical, mental and social well-being, together with damage to property, destruction of assets, loss of services, social and economic disruption and environmental degradation are some of the consequences of disasters [2].

In Africa, although disaster-associated mortality is decreasing, the economic impact of disasters and the number of people affected on the continent is rising [3]. McClean [4] largely ascribed this to improvements in urbanization and other development efforts and economic activities. While epidemic rates are considered being the highest of all hazards in Africa, the chief hazards affecting people and livelihoods in the continent are hydro-meteorological in nature, including various forms of floods, drought, wild fires, cyclones [5]. Thus, countries over the years have long understood their obligations to work to prevent dislocation in wartime and prevention of industrial accidents and other 'man-made' tragedies that could result in homelessness, loss of lives and property [6], however natural disasters have generally not received this preventive posture. In Asia for example, disasters are as a result of high natural causes and technological factors, whereas other continents are mainly natural causes with little human-induced factors or low-to-medium human causing agents. In Africa, and Ghana in particular, the occurrence is mainly unnatural or human induced causes or less natural caused disasters with underlying human factors. Nevertheless, the repeated or re-occurrence nature of same or similar events, at same or nearby locations, with similar or same impact, attracts same or similar response from the regulatory agencies. In Ghana, the National Disaster Management Organisation [7] has identified hydro-meteorological disasters, pest and insect infestation disasters, geological/nuclear radiological disasters, fires and lightning disasters, disease epidemic disasters (being the highest of all hazards) and man-made disasters as the core disasters/hazards covering wide range of different geographical areas of the country [8] that displace thousands of Ghanaians. The principal focus of this paper is to examine the causes and nature of reported disasters, with particular reference to Ghana, and the response of Ghanaian law to such disasters. The paper is therefore guided by the following queries: What are the main causes of disasters in Ghana? To what extent has the laws of Ghana responded to disasters in the country? Are all reported devastating events resulting in loss of life and property, as well as displacement of communities in Ghana constitute disasters?

2 Reviews

2.1 Existing Institutional and Regulatory Framework for Disasters

Ghana has various policies, legislations, practices and initiatives that have been advanced to regulate the environment and also seek for the protection of life and property during the occurrence of both natural and unnatural crises. Many government policies [9–11] relate directly to unnatural disasters. Although these policies seem to be disaster prevention initiatives, they have the potential to denigrate the occurrence of such hazards. There are also specific legislations regulating certain industries and sectors of the economy such as mining, construction, and manufacturing. A number of international conventions (e.g. the Rio Declaration on Environment and Development, United Nations Conference on Environment and Development, United Nations Framework Convention on Climate Change, United Nations Convention on Biological Diversity, United Nations Kyoto Protocol to the United Nations Framework Convention on Climate Change) that Ghana has ratified have various implications for both natural and unnatural disasters. Nonetheless, in the absence of a clear policy document that could hold individuals, culpable firms and corporations liable to compensate for loss of life and property, some individuals, advocacy groups and public agencies including the Bureau of Public Safety (BPS) seeking to hold corporations responsible for their social and environmental protection responsibilities usually encounter difficulties in doing so. Thus, many of such liable entities go unpunished particularly in the absence of clear parameters or any binding and statutory or contractual obligation to be imposed on such organizations [12]. Organisations across the world are expected to comply with laws and regulations of both local and international governments. These laws attempt to set reasonably minimum principles and responsible conduct in many spheres of organisational operations. Among others, these include such issues as human rights, environmental safety, and health all of which constitute some dimensions of the triple bottom line approach [12].

2.2   Figures Post-legal and Behavioral Responses to the Occurrence of Disasters

Just as other cities and urban areas are growing rapidly across the world particularly in developing countries [13], Ghana is experiencing rapid urbanisation and congregation of population in urban settlements, hence increasingly becoming vulnerable to various threats posed both in natural and anthropogenic nature. Most of the disasters that have occurred in the past, leading to severe destruction and grave losses were due to unplanned or inadequate plans and practices. In fact, poor building construction and unplanned development coupled with lack of enforcement of building regulations have aggravated the consequences of most of the disasters. The demand for land in cities has led to the use of marginal land prone to natural hazards such as floodplains, sloping lands, and reclaimed land, many of which are not habitable [14]. The land and accommodation pressure has further been compounded by limited supplies of essential and basic amenities including water, electricity, transport and inadequate public services [15] hence the urban centres are becoming vulnerable to all forms of hazards. The United Nations attributed the high risk of disasters in the urban towns to frail urban governance, lack of access to available land for low income earning citizens, improper construction, and concentration of economic assets and activities in cities and the decline of eco systems [16]. The situation is even worse in developing countries such as Ghana where many do not have access to basic infrastructural services. Thus, even a small natural hazard such as flooding which could be contained in developed nations result in enormous disaster, killing and exposing many lives and affecting the economic gains of the people. In Ghana, the capital town and the regional capitals where most of the economic activities occur are linked to the survival of the rural dwellers. Solway [15] noted that these cities are deemed important, not only due to their influx of population, but also due to their inherent strategic and economic role. As a result, whenever there is the occurrence of disaster in one such city, the impact is felt across the entire nation. Thus, the costs of disasters are detrimental, especially when they occur in urban environments [17].

2.3 Overview of Public Instruments for Monitoring Systems and Disaster Risk Management

The intervention of the Ghanaian government in the field of disasters includes a variety of public instruments for disaster risk management and goes beyond just civil protection. A public instrument is a legal expression used for any formally executed written document and could be attributed to the public sector that officially articulate a legally enforceable act, process or contractual duty, right or obligations [18]. Risk management on the other hand is a set of pre- and post-disaster measures toward addressing the adverse and harmful consequences of hazards [19]. Whilst pre-disaster efforts may include risk assessment (mapping, monitoring, etc.), risk transfer (i.e. insurance), mitigation efforts (i.e. strengthening infrastructure), and preparedness, post-disaster actions include emergency response (aid, clean-up) and loss sharing for reconstruction and rehabilitation (i.e. public-private and national-local resources) [20]. In Ghana, monitoring, early warning and immediate relief are made by the National Disaster Management Organisation (NADMO) through its various subsidiary and protection systems of the regional, metropolitan/municipal and district offices. Although state and national agencies perform major roles in providing education, warning systems and announcing evacuation strategies, at the pre-disaster phase, the local governments are endowed with the fundamental responsibility for adopting and enforcing building codes, building principles and land regulations to mitigate the impact of disasters [21, 22]. Is Ghana able to follow this process as other developing and advanced nations?

2.4 Governance Structures and Institutional Networks of the Legal Systems

Campanella [23] identified political and economic conditions, city planning and the resilience of its citizens as the three key factors affecting a city's resilience. In effect, a city with a robust, diversified economy will rebound more quickly than a city with a weak economy. This is connected to the issue of planning, as cities with a good economy are not only better to properly design and plan disaster resilient activities but to also allocate resources to disaster resilience. This includes well-rehearsed and functional evacuation and emergency management systems and investment in hazard mitigation planning [24]. In Africa and Ghana in particular, disaster mitigation is rarely integrated in urban planning and development [25] which further increases vulnerability [24]. Disaster risk reduction requires good governance, proper institutional structures, proper planning policies and standards, expertise and skilled workforce, adequate finance for implementation and maintenance [26]. In curbing disasters, Ghana must empower and strengthen the local government system and equip them with the needed logistics and financial resources to ensure compliance with the legal provisions and to promote effective prevention and reduction of disaster incidents, since they have been identified as one of the key stakeholders and the governing body rooted at the local level where disasters occur [16, 24, 27] and their roles and actions become exceedingly crucial [28] although all levels of governments are generally involved in disaster management.

3 Methodology

The study mainly followed a qualitative methodology with the utilization of documented text analysis although quantitative data were also interpreted to address the study objectives. The approach facilitated the analysis of various legal documents, records and other publications involving both numerical and non-numeric data [29]. It utilized secondary data sources such as media publications, government reports, and available literature including scientific articles, research and technical reports on disaster and disaster risk management serving as primary research data. It makes specific reference to contemporary episodes of disaster in Ghana, highlighting the types, number of deaths, human causalities from recorded incidents and the challenges in the legislature in dealing with recorded cases. The goal was to identify selected hazards in Ghana as its basis for analysis and further situate the happenings in the context of the law in understanding how the latter was able to respond to the occurrence of disasters in the country. This methodology has been used by other scholars [30–33] to address set objectives in similar studies.

4 Findings and Discussions

4.1 Findings

Natural disasters have traditionally been considered as acts of nature against which little can be done except to hope for the best and prepare for the worst. The commonest natural disaster in Ghana is flooding which affects significant number of people mainly in the capital and other urban cities. Ghanaian officials always complain of the improper citing of buildings on water courses preventing the flow of rain water into already chocked and destroyed rivers and dams. Meanwhile, the need to safeguard the environment and alleviate the adverse impact of natural disasters on life and property whilst reducing the consequences associated with the occurrence of unnatural disasters are becoming ostensible in the operations of the government and corporate bodies in Ghana. However, such measures and activities are rarely carried out on significant scales and only come up during the period of calamities. The thorough commitment to prosecute such efforts to the letter is always lacking. For instance, in the wake of the June 2015 flood/fire (one of the deadliest disasters ever recorded in history) popularly referred to as the "twin disaster" that destroyed over one hundred and fifty lives and dozens of property including: five properties valued at GH¢1,658,847.00, 17 motor vehicles and a fuel tanker belonging to the Ghana Oil Company (GOIL), a mini market located in the premises, other business establishments and properties [34], many have been concerned about the role of the law in preventing or otherwise ensuring compliance with environmental guidelines in citing structures by individuals and corporate bodies in the country. In fact, the magnitude of damage in terms of death, injuries and loss of property recorded during this June 3, 2015 disaster is the highest ever recorded in Ghana [34]. A government's committee report indicated flooding as the remote cause of the fire, and the underlying and intermediate cause being the displacement of fuel from the GOIL Fuel Station, with the ultimate and immediate cause attributed to a drop of lit cigarette stub by an identified individual into the floating fuel [34, 35]. In reference to this disaster, government quickly announced GH¢ 60 million compensation for affected families with no thorough mechanism for disbursing the funds; also, relief funds by various organisations to support the victims of the disaster were not coordinated [34]. Today, no prosecution has been made with many of the affected families yet to receive their share of the compensation fund. Even when the Committee established to look into the disaster noted with dismay and recommended among others, punitive sanctions against public officers who approved and granted permits for buildings located in inappropriate areas [34], no attempt has yet been made to track the officers responsible for such ineptitude.

In May 2016, about 19 people including children were reported dead when a boat carrying about 60 people capsized on the Volta Lake. Although the boat operator was arrested following reports of negligence [36], prosecution is yet to make any public pronouncement on the matter. Whilst the immediate cause of the incident is unknown, survivors and eyewitnesses have attributed it to overloading of the boat with people and goods [37]. The Ghana Maritime Authority has over the years been blamed for the many disasters on the lake for failing to enforce safety regulations including the need for passengers to put on life jackets [36]. This, unfortunately, is a perennial and common practice considering the prevalence of the underlying causes. Regrettably, whilst in many of such occurrences, culpable persons or operators are not sanctioned or prosecuted, families of victims are often not compensated by government and boat owners.

In February 2016, a deadly Metro Mass Transit (MMT) Bus accident claimed at least 63 lives and injured 25 passengers on the Kintampo-Tamale road. Initial Police investigations indicated faulty brakes leading to loss of control over the vehicle [37, 38]. Meanwhile, internal investigations by Management of MMT and that of the National Road Safety Commission (NRSC) into the circumstances surrounding the carnage are yet to be released. Whilst this may be described as another failure on the part of government and its regulatory agencies in prosecuting culpable persons or perhaps an attempt to shield them, government sadly announced GH¢100,000 to support the victims [38] with no detailed disbursement plans and procedures. The most recent events relating to disasters are the collapse of the Breman Jamra Methodist School block on January 31, 2017; the Kintampo Water Fall disaster on March 19, 2017 and the Atomic Junction gas explosion disaster on October 7, 2017. While the dilapidated school building collapsed and killed 6 Kindergarten school children, the Water Fall disaster took 20 lives of High School students and the gas explosion also resulted in 7 reported deaths. Interestingly, sheer negligence was reported on the part of authorities leading to the collapse of the dilapidated Kindergarten classroom structure and the gas explosion incident attributed to ill-safety measures put in place by the owner of the facility. While the details of such reports (for instance the gas explosion disaster) are kept under wraps, individuals, pressure groups and other civil society organisations are rising to challenge government, with the understanding that doing so enables citizens to fully appreciate the impact of their acts and to hold others accountable for their deeds. To date, no one has been held to account for negligence resulting in any of these incidents although the government, as usual, has announced a number of measures to prevent future occurrences. Meanwhile, a critical review showed that almost all buildings that collapsed in recent years were privately owned properties. Today, not even an individual is known to have been prosecuted for such negligence and ineptitude. For instance, the collapse of the 7-Storey Grand View Hotel at Nii-Boi Town, Accra, in 2014 that resulted in 4 deaths reported no official compensation for the affected persons. Unconfirmed reports indicated among others poor engineering works and lack of adherence to building codes. Surprisingly, there was no prosecution although the owner of the building was reported to have turned a deaf ear to several warnings and notices to stop work by the city authorities as the building had no permit [39]. Meanwhile, the April 24, 2013 incident of the 8-storey Rana-Plaza commercial building collapse in Savar, a sub–district near Dhaka, the capital of Bangladesh, that resulted in the death of 547 people and injured about 2500 caused the suspension of the elected Mayor of the Municipality for alleged negligence in approving the design and layout of the building [40]. In Africa and Ghana in particular, disasters present a significant hindrance to the attempts to attain sustainable development, especially in the light of the continent's insufficient capacities to forecast, monitor and address the impact of hazards [41]. Perhaps, part of the reasons for a lack of progress has been the non-assessment of social vulnerability factors that predispose people to natural disasters [16, 24, 25]. The HFA holds that reducing the vulnerability of the African people to hazards is an indispensable factor or strategy in poverty reduction as well as efforts to preserve developmental gains. Thus, economic and technical support is required to strengthen the capacities of the region and the respective countries, including observation and early warning systems, assessments, prevention, preparedness, response and recovery [41].

4.2   Discussion

In Ghana, Sect. 24 of the National Buildings Regulations, LI 1630 appears to ensure the safety of buildings or building works and that the Authority should also implement regulations to protect safety of the public. It further provides that a qualified building inspector should carry out regular inspections of any project at every stage of the development to ascertain whether the materials as well as proper building safety measures are being employed. Besides, The Environmental Protection Agency Act [42] makes specific provisions for the protection of the environment and prevention of hazards. It also provides that the EPA may also issue and serve enforcement notices, where it appears activities of any undertaking pose serious threats to the environment or to public health. The Act [42] makes it an offence punishable upon summary conviction to a fine and in default to a term of imprisonment for failing to comply with an enforcement notice or permit issued by the EPA. Currently, in Ghana, many private and real estate developers have cultivated an appetite to get returns on their investments in building projects but some are not concerned about the needed protection for the people who occupy such facilities, especially as they believe that natural disasters are unlikely in the country.

In fact, although not all the many incidents of flooding, disease epidemics, building collapse, transport accidents, explosions, industrial and commercial fires resulting in loss of life and property, as well as displacement of communities may constitute or be counted as full-blown disasters, institutional structures of the country are often unable to contain the occurrences. Hence, the resulting impact renders the events as such. In any event, for the severely injured and the people who lost loved ones or property, any occurrence might be considered a disaster. Unfortunately, in many of such happenings, the investigative reports are not made available and no one is held to account for negligence resulting in the disaster. A vivid interrogation of both legislations and judicial precedents in Ghana seems to suggest that even though the law makes pro-vision for various impositions following violations or liabilities in certain instances, most regulatory and enforcement agencies as well as the courts, are often indisposed to imposing such fines. These penalties are more often than not found to be non-punitive enough to prevent the recurrence of such violations. More disturbing is the failure of such regulatory agencies to enforce the payment of the charges by organisations found to have breached the law. Just as the legal system in Ghana enjoins corporate organ-isations to maintain health and safety at the work place, with the provision backed by law; and any breach attracting relevant sanctions, it is essential for the system to also advance such measures and policies to hold to account persons, group of persons and other entities and bodies whose actions negligently result in the occurrence of whichever form of catastrophes in the country. In Ghana, a total number of 1,171,000 people were reported to be affected by all forms of disasters between 1995 and 2004 with 597 people reported killed. The death toll was significantly lesser than 628 people reported killed between 2005 and 2014, considering the marked reduction in the affected population of 686,639 people over the period [43]. With 175 recorded deaths from 5,060 affected people in 2015 [44], perhaps the highest in any singular year, it suggests that whereas the incidents of disaster and affected number of people are decreasing, the death toll is surprisingly on the rise. For instance, whilst the number of affected persons reduced by 80% in Oceania, 66% in Europe, 37% in Asia, it only fell and by 7% in Africa. In general, compared to 1995–2004, the 2005–2014 decade has seen a one-third reduction in the number of disaster deaths and of people reportedly affected across the world. However, the report showed that the number of deaths rather increased by 21% in Africa [43]. In Africa, this may be attributed to many factors including lack of enforcement of regulations and the overly politicized and haphazard nature in which incidents of disaster and related issues are handled in the continent and Ghana in particular. In effect, the socio-economic losses of disasters are rising in Africa, which was a major threat to the continent's ability to achieve the Millennium Development Goals (MDGs) and now the Sustainable Development Goals (SDGs). Besides the social and economic losses, a substantial amount of resources for devel-opment continues to be diverted to relief and rehabilitation assistance to disaster-affected people, which is perhaps beginning to raise its head against the attainment of the (SDGs) [45]. It is mainly argued that although disaster risk reduction (DRR) policies and institutional mechanisms do exist in various forms of completeness in African countries, their effectiveness is however limited and questionable hence a strategic direction to improving and enhancing their efficiency and effectiveness can no longer be delayed. Perhaps in Ghana, time is due for CSOs, NGOs and other human right advocates to begin to fight for the rights of the ordinary Ghanaians who suffer huge losses although the state is enjoined under some form of social obligations to protect the rights and properties of the citizenry. Today, human vulnerability is acknowledged as a key constituent of what results in a natural hazard into a full-fledged disaster including flood-provoked displacement crisis [46]. In complying with the law, the government, the public and civil society must take all necessary precautionary measures to reduce the impact of natural disasters and prevent actions that induce the occurrence of both unnatural and human aided natural disasters in the society as they not only increase the risk and exacerbate the effects of disasters but render the lives and livelihoods of the people and the entire nation to unimaginable consequences [47].

5 Conclusion and Recommendations

This paper has so far examined aspects of the regulatory regime of Ghana relating to the occurrence of disasters and the extent to which all reported devastating events resulting in loss of life and property and displacement of communities in Ghana constitute to disasters. The paper observes that the Ghanaian legal regime on disaster makes reasonable provision for breaches and remedies, for instance, civil and criminal prosecutions, damage repair and restitution of affected communities. It establishes internationally acceptable rules in substance and procedure. Nonetheless, factors leading to disasters in Ghana are largely man-made attributable to non-regulatory compliance, unplanned residential and business communities, and erection of buildings in waterways without alternative paths. It further includes human errors and negligence, open storm drainages, rubbish-filled drainages, poor water and sewerage systems, system inefficiencies and incompetence, intentional breaches (building in waterways and marshy areas), corruption and unresolved conflicts that significantly downplay even the minimal efforts at reducing the occurrence and mitigating the impact of disasters in the country. The paper argues that the legal framework in responding to disaster in Ghana is characterized by poor enforcement mechanisms, lack of respect and recognition of stakeholder rights and absolutely failed or perhaps no prosecution with such actions attributable to human causing agents. Unfortunately, enforcement and compliance has always been a challenge and the main drivers of hindrance, probably due to absence of political will and inadequate logistics. It is argued that poor governance, failed and weak state systems, unsustainable development practices, corruption, lack of accountability and lack of foresight and hope, among others, remain the huge impediments in the path towards development in Africa [33]. It is suggested for government to promote professional training efforts for specific disaster related stakeholders and initiate public education efforts on the responsibilities of individuals, government and corporate entities not only during periods of disasters but as a means of awakening in the citizenry for preventing the human related activities that induce the occurrence of all forms of disasters. Apart from the need for heightened political commitment and attention given to matters of national concern, local authorities and regulatory bodies should be endowed with the needed logistics in detecting and attacking the immediate causes and early onsets of disasters and to ensure stringent compliance with regulations. Indeed, there is the need for rethinking in disaster management strategies and risk reduction initiatives that connect to social and economic vulnerability, with strong recourse to the rule of law, lest with respect to legal reactions to disasters in Ghana, the laws on the shelf are almost perfect but silent and passionate observers in action.

References

1. Ahsan, N., Tullio-Pow, S.: Functional clothing for natural disaster survivors. Disaster Prev. Manag.: Int. J. **24**, 306–319 (2015)
2. UNISDR: UNISDR terminology on disaster risk reduction. UNISDR, Geneva (2009)
3. UN: Global Assessment Report on Disaster Risk Reduction. Information Press, Oxford (2011)
4. McClean, D.: World Disasters Report 2010. IFRC, Geneva (2010)
5. van Niekerk, D., Wisner, B.: Integrating disaster risk management and development planning: experiences from Africa. In: Lopez-Carresi, A., Fordham, M., Wisner, B., Kelman, I., Gaillard, J.C. (eds.) Disaster Management: International Lessons in Risk Reduction, Response and Recovery, 1st edn. Routledge, New York (2014)
6. Fisher, D.: Legal implementation of human rights obligations to prevent displacement due to natural disasters. In: Kalin, W., Williams, R.C., Koser, K., Solomon, A. (eds.) Incorporating the Guiding Principles on Internal Displacement into Domestic Law: Issues and Challenges (2010)
7. NADMO Act 1996 (Act 517)
8. NADMO (2012). http://nadmo.gov.gh
9. Control and Prevention of Bushfires Act, 1990, P.N.D.C.L. 229
10. Ghana National Fire Service Act, 1997 (Act 537)
11. National Fire Service Regulations, 2003 (L.I. 1725)
12. Anku-Tsede, O., Deffor, W.: Corporate responsibility in ghana: an overview of aspects of the regulatory regime. J. Bis. Manag. Res. **3**, 31–41 (2014)
13. United Nations Population Fund: State of World Population 2007. Unleashing the Potential of Urban Growth, New York, USA (2007). http://www.unfpa.org
14. Sinha, A.: Relief administration and capacity building for coping mechanism towards disaster reduction. Shelter, October 9–12 1999
15. Solway, L.: Socio-economic perspectives of developing country megacities vulnerable to flood and landslide hazards. In: Margottini, C. (ed.) Floods and Landslides: Integrated Risk Assessment, pp. 245–277. Springer, Heidelberg (1999). https://doi.org/10.1007/978-3-642-58609-5_15
16. UN-ISDR: Local Governments and Disaster Risk Reduction. United Nations International Strategy for Disaster Reduction–UN-ISDR (2010). http://www.unisdr.org
17. Voogd, H.: Disaster prevention in urban environments. Eur. J. Spat. Dev. (2004). http://www.nordregio.se
18. Grifis, S.H.: Barron's Law Dictionary, 6th edn. Barron's Educational Series, New York (2010)
19. Freeman, P., Mechler, R.: Public Sector Risk Management in Mexico for Natural Disaster Losses. Issues Paper for a Wharton-World Bank Conference on: Innovations in Managing Catastrophic Risks: How Can They Help the Poor? IIASA Laxenburg, Washington, DC (2001)

20. Saldana-Zorrilla, S.O.: Assessment of disaster risk management strategies in Argentina. Disaster Prev. Manag. **24**, 230–248 (2015)
21. Alesch, D.J., Arendt, L.A., Holly, J.N.: Managing for Long-Term Community Recovery in the Aftermath of Disaster. Public Entity Risk Institute, Fairfax (2009)
22. Murphy, B.: Locating social capital in resilient community-level emergency management. Nat. Hazards **41**, 297–315 (2007)
23. Campanella, T.J.: Urban resilience and the recovery of New Orleans. J. Am. Plan. Assoc. **72**, 141–146 (2006)
24. Malalgoda, C., Amaratunga, D., Haigh, R.: Creating a disaster resilient built environment in urban cities: the role of local governments in Sri Lanka. Int. J. Disaster Resil. Built. Env. **4**, 72–94 (2013)
25. Pelling, M.: The Vulnerability of Cities: Natural Disasters and Social Resilience. Earthscan Publications, London (2003)
26. Ginige, K., Amaratunga, D., Haigh, R.: Developing capacities for disaster risk reduction in the built environment: capacity analysis Sri Lanka. Int. J. Strateg. Prop. Manag. **14**, 287–303 (2010)
27. Kusumasari, B., Alam, Q., Siddiqui, K.: Resource capability for local government in making disaster. Disaster Prev. Manag. **19**, 438–451 (2010)
28. Col, J.: Managing disasters: the role of local government. Publ. Adm. Rev. **67**, 114–124 (2007)
29. Saunders, M.N.K., Lewis, P., Thornhill, A.: Research Methods for Business Students, 5th edn. FT Prentice Hall, Harlow (2009)
30. Buhmann, K.: Corporate social responsibility: what role for law? Some aspects of law and CSR. Corp. Gov. **6**, 188–202 (2006)
31. Fedolapo, H.B.: Building collapse: causes and policy direction in Nigeria. Int. J. Sci. Res. Innov. Tech. **2**, 1–8 (2015)
32. McBarnet, D.: Corporate Social Responsibility Beyond Law, Through Law, for Law University of Edinburgh School of Law Working Paper No. 2009/03 (2009). http://dx.doi.org/10.2139/ssrn.1369305
33. Van Niekerk, D.: Disaster risk governance in Africa. Disaster Prev. Manag. **24**, 397–416 (2015)
34. June 3 Report. https://www.dropbox.com/s/o51nw4dred5469m/JUNE_3RD_REPORT.pdf?dl=0
35. JoyFm News (2015). http://m.myjoyonline.com
36. Pulse News (2016). http://pulse.com.gh
37. Modern Ghana News (2016). https://www.modernghana.com
38. JoyFm News (2016). http://www.myjoyonline.com
39. JoyFm News (2014). http://www.myjoyonline.com
40. Corbett, T.J.: Bangladesh and Reasons Buildings Collapse (2015). ermacademy.org/Publication/riskmanagement-article/Bangladesh-and-reasons-buildings-collapse
41. ISDR: Hyogo framework for action 2005–2015: building the resilience of nations and communities to disasters. In: World Conference on Disaster Reduction 18–22 January 2005, Kobe, Hyogo, Japan (2005). https://www.unisdr.org
42. The Environmental Protection Agency Act, 1994 (Act 490)
43. The International Federation of Red Cross and Red Crescent Societies. World Disasters Report (2015)
44. The International Federation of Red Cross and Red Crescent Societies. World Disasters Report (2016)

45. African Union Commission: United Nations International Strategy for Disaster Reduction' Extended Programme of Action for the Implementation of the Africa Regional Strategy for Disaster Risk Reduction (2006–2015) and the Declaration of the 2nd African Ministerial Conference on Disaster Risk Reduction, International Strategy for Disaster Reduction, Nairobi, pp. 1–60 (2010)
46. International Strategy for Disaster Reduction, Living with Risk: A Global Review of Disaster Reduction Initiatives, p. 36 (2004)
47. Malalgoda, C., Amaratunga, D., Haigh, R.: A disaster resilient built environment in urban cities: the need to empower local governments. The "State of DRR at the Local Level". A 2015 Report on the Patterns of Disaster Risk Reduction Actions at Local Level, UNISDR (2015)

The Role of Active Mobility for the Promotion of Urban Health

Cristiana Cellucci[(✉)] and Michele Di Sivo

Department of Architecture, University "G.d' Annunzio" of Chieti and Pescara,
Viale Pindaro 42, 65127 Pescara, Italy
cristiana.cellucci@gmail.com, mdisivo@unich.it

Abstract. In the debate on the issue of designing spaces for 'slow' urban mobility, this paper emphasizes the importance and centrality of the human-centered approach through the observation of the relations that are established among people, technological systems and constructed environments in order to design according to anatomical and metric needs (anthropometric view) as well as to the needs linked with perception and cognitive processes (anthropocentric view). Two levels of interface in the person-system relationship have been identified, the "individual space", where internal variables impact the "user system" (factors related to the psycho-physiological perception of space), and the "prosthetic space", where external variables influence the "environment system" (factors that influence the capability of the architectural space to become physiologically and behaviorally prosthetic).

Keywords: Human factors · Urban health · Human-centered design
Individual space · Prosthetic space · Walkable space

1 Introduction

In recent years, the importance that mass media and popular culture in general attach to the human body and its well-being shows us that they are extremely important to contemporary society; in addition, the awareness that such well-being is indirectly influenced by the morphological-typological features of the places we live, thus negatively or positively affecting the health of the population and the adoption of proper life styles, is constantly increasing.

A growing amount of international research and best practices at a global level (Health for All, the Ottawa Charter for the Promotion of Health, the European Network of "Healthy Cities" within the WHO's Health Cities Project, the "Healthy Cities" project, the WHO's "Age-Friendly Cities" project, The program AHA "Active and Healthy Ageing" etc.) explore the relationships between psycho-physical well-being and urban space, identifying physical activity as a critical factor to protecting against many chronic diseases [1–4]. If, in the past, infective diseases were threats to health, for which urban planning sought to ensure the hygiene and comfort of both cities and individual buildings, now the main killers are the so-called Non-Communicable Diseases (NCD's) – cardiovascular diseases, stroke, cancer, diabetes – whose main risk factors are obesity and physical inactivity. It is therefore necessary to reconnect health

© Springer International Publishing AG, part of Springer Nature 2019
J. Charytonowicz and C. Falcão (Eds.): AHFE 2018, AISC 788, pp. 248–256, 2019.
https://doi.org/10.1007/978-3-319-94199-8_24

and urban planning [5–7]. In many research works, the importance of this can be witnessed by the use of the wording 'urban health' instead of 'public health', as the health of the population refers to the relationships of users with the physical, natural and social setting in which they live [8]. Health, as defined by the WHO, is no longer simply understood as the absence of disease but rather as the overall biological, mental, and social well-being and quality of life [9].

The designing and creation of outdoor spaces and urban pathways (pedestrian and cycling lanes) is an opportunity to create walkable areas that can produce positive effects on environmental health determinants (air, sound and light pollution, waste generation, presence of green areas) with direct consequences on the increase in the "urban heat island" effect, and indirect consequences on so-called "lifestyles" (walking instead of driving, using stairs instead of the elevator, socializing instead of isolation) [10]. The concept of 'walkable' (walking, running, cycling and other forms of physical activity) is a way to promote citizen well-being, and it is also an opportunity to develop and revamp public spaces and architecture to improve life quality. Considering this human – environment – health relationship, the research investigates the main criteria that must be met to achieve the project of active-inclusive mobility that necessarily refers to a series of technological, design and functional aspects directly and indirectly connected to the quality of outdoor areas.

2 The Anthropocentric Approach to Designing a Healthy City

This methodology puts the user and his/her needs at the center of the design process by defining the needs and design criteria that reflect the human-system-environment relationship, wherein for "system" we may mean a cycling path and for "environment" a place or a situation where an activity is carried out. This form of interaction might be physical, through a dimensional interaction (anthropometric approach), or it might be sensorial, through the coherence and appropriateness of stimuli expressed by physical systems, with the physiological structures of individuals (anthropocentric approach). The contribution of the present work consists specifically in refocusing attention on the issue of designing walkable spaces not on the definition of features of the object-space in itself, of services or of environmental situations that one wishes to achieve or adapt, but rather on investigating the relationships and direct or indirect links among the user system, the system utilized and the context of use. The outcomes of this design process are thus aimed at the creation of physical and cognitive interfaces, which are understood as the places where a continuous process of functional interaction among people, systems, and the environment takes place [11, 12].

Design interfaces have an interactive dimension that includes both comparison and contradiction. This dimension advances through subsequent approximations between an external horizon (the relations of the object with its own constituent parts and with its contextual environment) and an internal horizon (the entirety of its definitions in relation to people). It is, therefore, possible to identify certain internal and external variables that design interfaces must encompass (Fig. 1).

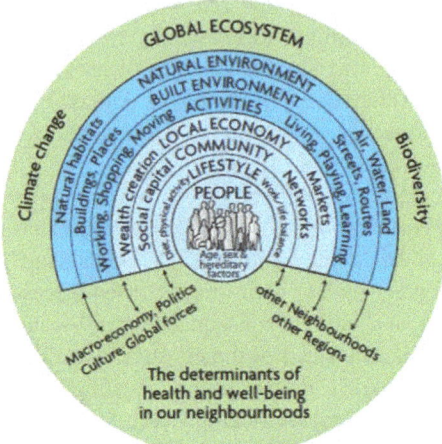

Fig. 1. Determinants of health and well-being in neighborhoods (Source: Barton H, Grant M. A health map for the local human habitat. Journal of the Royal Society for the Promotion of Public Health 2006, 126 (6), pp. 252–261).

2.1 Internal Variables of Individual Space

The interfaces that we design must be derived from an understanding of "spatial modulators" [13], which contribute to the physical, psychological, metabolic and social interactions with designed systems [14]. Considering a person at the center of his/her "individual space" – meaning the physical distance separating us from others (people, objects, design interfaces); in other words, a kind of territorial extension of his/her body in which s/he moves through space – we consider the following spatial modulators:

1. The modulators of the *psycho-sensorial sphere*, which provide for conditions of physical wellbeing and mental balance:
    a. *Sight*. It is the predominant channel of perception for the comprehension of reality. The elements that influence sight include the nature of light (natural or artificial), its distribution, direction, color and personal response of the visual organism.
    b. *Hearing*. It is the channel of perception impacted by auditory stimuli and aural references.
    c. *Smell*. It is the channel of perception impacted by the presence of olfactory stimuli.
    d. *Touch*. It is the channel of perception impacted by the range of information obtainable from the surfaces that surround us.
2. The modulators of the *ergonomic sphere*, which provide for physical movement and the execution of human activities:
    a. *The proprioceptive capability*. A unique sensitivity that endows the organism with a perception of its self in relation to the external world. It is a complex anatomical structure consisting of nerve centers and pathways and structures,

such as muscles, which respond to commands coming from the nervous system. This structure provides for the sense of position and the movement of limbs and the body irrespective from sight.

    b. *The vestibular system.* The ability to move oneself in space against gravity.

3. The modulators of the *social sphere*, which provide for a sense of community and participation:

    a. *Social inhibition.* The ability to share experiences with others and to belong to social groups. The undertaking of an active role in society entails the improvement of psychic well-being in the user.

    b. *Identification*: The possibility to recognize the system of spaces/objects as expressions of one's culture.

2.2 External Variables of the Prosthetic Space

External variables make up the second level of factors that influence the design of interfaces [15, 16]. They stem from the ability of designed systems to correlate at the human scale; in other words, the capability of physical elements to engage with the dimensions that characterize a project centered on the well-being of the user:

- The *physical dimension* – the user's metric relationship with space – meaning the possibility to engage in activities in spaces that are suitably sized and proportioned to users and their diversities. Here we find the emergence of the idea of "inclusive" space (accessible, usable, able to be maintained, safe), meaning a space capable of comfortably and safely "accommodating" people with different specificities and various degrees of freedom. In addition, other meanings of this concept can be considered. These include: structural inclusiveness (dimensional suitability), meaning the possibility for each person to access, move within and utilize a space and its components autonomously, in an equal manner, inclusively, and non-discriminately; perceptive inclusiveness (perceivable information), meaning the capacity of a "product" to be perceivable and understood by all. This is achieved through an ensemble of operations aimed at obtaining environmental communicativeness: allowing the environment to provide information that is perceptible by all, especially individuals with sensory and cognitive difficulties [17, 18].

- The *cognitive dimension* – the user's relational relationship with space – meaning the way in which a space is used that influences physical activity and his/her "active" use. It can be influenced by requisites of proximity, healthiness, accessibility, pedestrian access, density, functional diversity, and the visibility of green infrastructures [19]. If the walkability of a city can be considered the first level of usage for physical activity, from the standpoint of the prevention and the complementary treatment of non-transmissible chronic diseases (obesity, respiratory illnesses, cardiovascular diseases, diabetes) and stress, various case studies have demonstrated how areas of cities designated "open-air gyms" significantly influence healthy lifestyles and social resilience. External fitness facilities can be characterized as stations of exercise circuits that are specialized, developed, and indicated on the basis of age and ability level – defined on the basis of interdisciplinary research in the field of medicine, physical therapy, epidemiology, and psycho-sociology for both specific and wide-ranging targets. This includes, for example, the maintenance of optimum cardiovascular functions, training muscle groups in order to improve balance and resistance with the goal of reducing the risk of injury, the strengthening of muscle extension and the improving of flexibility, and balance coordination in individuals with functional limitations [20].

- The *social dimension* – the user's participatory relationship with space. The capability of the project to define itself as a space of possibility [21]. The ambiguity of the designed "thing" and its outcome must therefore remain within an open process of constant definition, discovery, reflection, and redefinition [22]. Here we find the emergence of the idea of an "adaptive" city (flexible and self-evolving) where the small- and large-scale planning of potential transformations involves a margin of error between the planned solution and the actual future one. This is carried out to include the concept of the casual evolution of the space within the proposed functional and organizational criteria, thus encouraging creative and innovative design. To this effect, participation becomes a driving force in the process. Participation here does not mean only debate and deliberation but also direct action in the "construction" of the city [23]. The space-user relationship is thus translated from a participatory point of view through the fostering of a "Collective Intelligence", which returns the citizen to the center of the processes of transformation and management of the territory in which he/she lives. In this context, citizens reassume the role of protagonist in terms of culture and local identity, and through a process of co-creation transform the space in which they live by adapting it to their own needs [24, 25].

External variables relate to the function of the architectural space to give rise to behaviors. According to this approach, the designed environment is thus considered a prosthetic phenomenon. "It is physiologically prosthetic inasmuch as it favors certain behavioral objectives, maintaining certain physiological demands (behaviorally correlated) and behaviorally prosthetic as it intentionally configures specific behavioral topographies" [26]. This prosthetic view of the designed environment, though it may be translated onto various planning scales (from objects to habitable spaces and open spaces), becomes of particular interest when it is linked to the concept of designing the open spaces of our cities and their ability to foster the vital functions of those who make use of them. In this way, a city's open spaces can be construed to move beyond their current concept as spaces that are technically equipped for movement and instead be reinterpreted as places of movement where virtuous and regenerative relationships among inhabitants moving among the city's resources can be reconstituted and cultivated (Fig. 2).

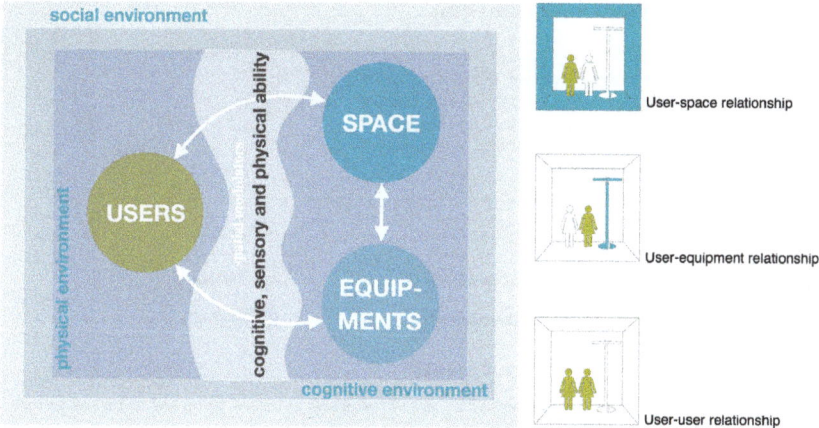

Fig. 2. Relationships between user-space and urban equipment.

3   Strategic Area for Intervention: *roomwalk*

The strategic area for intervention involves urban segments of a city that we have decided to call "roomwalk". A walking or cycle path is often represented in a plan or sectioned in two-dimensional static designs that capture and record key configurations and measurements. While these tools are useful for the dimensioning of these spaces, the intention behind this research is to encourage a more dynamic, territorial approach. Such an approach would have the broad objective to identify the factors that encourage slow mobility and daily physical activity in the urban environment. This is possible by an understanding of the user's experience through his/her senses and the translation of this experience into the physical space. Putting the user at the center of the "roomwalk" encourages certain considerations from the point of view of the user's perspective. The "roomwalk" is not a neutral space, but always an "operating factor" [27]. Pope and Brandt [12] have described the environment as an entity that supports the person, as a sort of mat whose warp is constituted by physical factors, and whose we is constituted by social factors. The capacity of the environment to adequately support the life of people (expressed, in the metaphor, by the solidity of the mat) depends, on the one hand, on its physical features, and on the other on the efficiency of the social support network available in it. The set of relations that develop between these two groups of variables determines the equilibrium of the urban system by favoring or hindering the conditions of anthropo-dimensional and psycho-physical wellbeing, and of anthropo-dynamic and social well-being (Fig. 3).

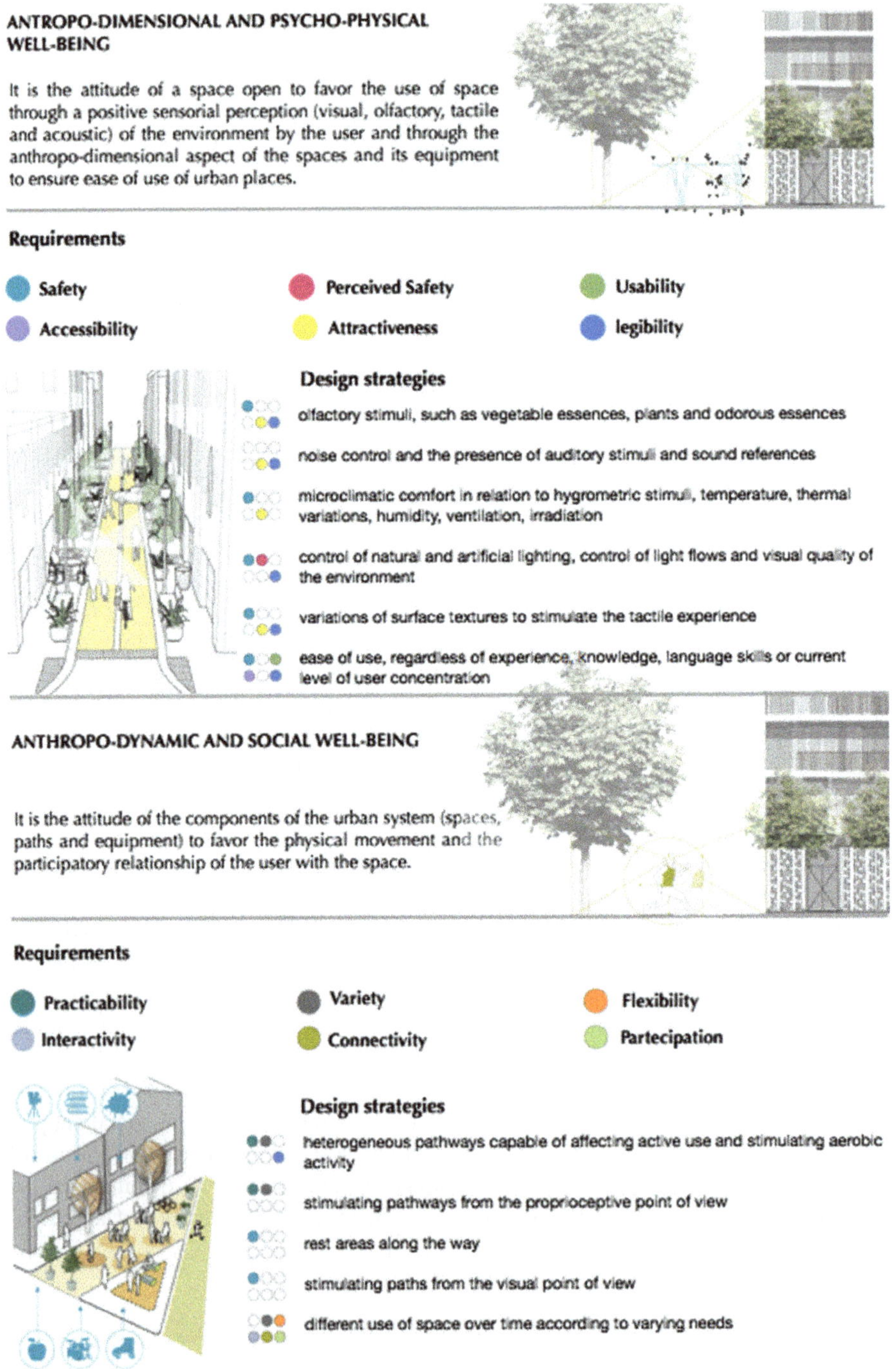

ANTROPO-DIMENSIONAL AND PSYCHO-PHYSICAL WELL-BEING

It is the attitude of a space open to favor the use of space through a positive sensorial perception (visual, olfactory, tactile and acoustic) of the environment by the user and through the anthropo-dimensional aspect of the spaces and its equipment to ensure ease of use of urban places.

Requirements

● Safety   ● Perceived Safety   ● Usability

● Accessibility   ● Attractiveness   ● legibility

Design strategies

olfactory stimuli, such as vegetable essences, plants and odorous essences

noise control and the presence of auditory stimuli and sound references

microclimatic comfort in relation to hygrometric stimuli, temperature, thermal variations, humidity, ventilation, irradiation

control of natural and artificial lighting, control of light flows and visual quality of the environment

variations of surface textures to stimulate the tactile experience

ease of use, regardless of experience, knowledge, language skills or current level of user concentration

ANTHROPO-DYNAMIC AND SOCIAL WELL-BEING

It is the attitude of the components of the urban system (spaces, paths and equipment) to favor the physical movement and the participatory relationship of the user with the space.

Requirements

● Practicability   ● Variety   ● Flexibility

● Interactivity   ● Connectivity   ● Partecipation

Design strategies

heterogeneous pathways capable of affecting active use and stimulating aerobic activity

stimulating pathways from the proprioceptive point of view

rest areas along the way

stimulating paths from the visual point of view

different use of space over time according to varying needs

Fig. 3. Design requirements for the physical-psychological-social well-being of the user and some design strategies.

4   Conclusions

With respect to the issues of individual mobility, we see how design criteria are confronted with the various scales of the project, where the architectural pathway and the perception of the landscape that follows have their roots in the concept of physical space in relation to movement and human senses (which have always been the grammatical elements of spatial appropriation). This involves facing these issues in order to define spaces (cycling-pedestrian space) rather than places (cycling-pedestrian place) [28]. As a consequence, infrastructure turns into space that interrelates with other spaces and is not simply an overlaid and autonomous network in regard to the context through which it moves. Human-object-environment relationships become the focus of design activity (both on a building scale and on an urban scale), where the object may be a technological system, a building component, a domestic tool, or a service. The environment is a place or a situation where activities are carried out. Such interaction can be physical, through manipulative or dimensional contact (linked to the anthropometric approach) or sensorial, through the coherence and appropriateness of stimuli produced from physical systems with the physiological structures of the individual.

References

1. Commissione della Comunità Europea – Libro Verde della Commissione "Verso una nuova cultura della mobilità urbana" (2007)
2. World Health Organization: Physical activity strategy for the WHO European Region 2016 – 2025. Unione Italiana Sport Per tutti, Roma (2016)
3. Ufficio federale dello sport UFSPO, Ufficio federale della sanità pubblica UFSP, Promozione Salute Svizzera, upi – Ufficio prevenzione infortuni, Suva, Rete svizzera Salute e Movimento. Muoversi fa bene alla salute. Macolin: UFSPO (2013)
4. European Innovation Partnership on Active and Healthy Ageing. https://ec.europa.eu/eip/ageing/home_en
5. Lee, I.M., Shiroma, E.J., Lobelo, F., Puska, P., Blair, S.N., Katzmarzyk, P.T.: Effect of physical inactivity on major non-communicable diseases worldwide: an analysis of burden of disease and life expectancy. Lance 219–229 (2012)
6. Spadolini, M.B.: Design for better life, longevità: scenari e strategie. Franco Angeli, Milano (2013)
7. Fries, R.C.: Handbook of Medical Device Design. Marcel Dekker Inc., New York (2000)
8. Galea, S., Vlahov, D.: Urban health: evidence, challenges and directions. In: Annual Review of Public Heath, pp. 341–365 (2005)
9. WHO: ICF, International Classiffication of Functioning Disabilities and Health. Erickson, World Health Organization, Geneve, CH (2006)
10. Cristiana, C.: Inclusiva, attiva e adattiva: la progettazione della città centrata sull'utenza. In Built Environment Technologies and Healthy Architectures. Franco Angeli, Milano (2018)
11. Fitch, M.J.: La progettazione ambientale. Analisi interdisciplinare dei sistemi di controllo dell'ambiente. Franco Muzzio, Padova (1980)
12. Brandt, E., Pope, A.: Models of disability and rehabilitation. In: Brandt, E., Pope, A. (eds.) Enabling America: Assessing the Role of Rehabilitation Science and Engineering, pp. 62–80. National Academy Press, Washington (1997)

13. Righter, C.E., Nowlis D.P., Dunn, V.B., Belton N.J., Wortz E.C.: Habitability guidelines and criteria. NASA n. 8-25100, p. 60 (1971)
14. Gehl, J.: Life Between Buildings, Using Public Space. VNB, New York (1987)
15. Thompson, C.W.: Activity, exercise and the planning and design of outdoor spaces. J. Environ. Psychol. **34**, 79–96 (2013)
16. Canter, D.: Psychology and the Built Environment. Architectural Press, London (1974)
17. Filippo, A., Cellucci, C., Di Sivo, M., Ladiana, D.: The measurable and the real quality of life in the city. Urban regeneration as a technological correlation of resources, spaces and inhabitants. TECHNE, J. Technol. Archit. Environ. **10**, 67–76 (2016). Firenze University Press, Firenze
18. Cellucci, C.: Accessibilità dell'ambiente domestico. In: Cluster in progress. La Tecnologia dell'Architettura in rete per l'innovazione, pp. 53–62. Maggioli Editore, Santarcangelo di Romagna (2016)
19. Duhl, L.J., Sanchez, A.K.: Healthy Cities and the City Planning Process. WHO Regional Office for Europe, Copenhagen (1999)
20. Angelucci, F., Cristiana, C.: The measurable and the real quality of life in the city. Urban regeneration as a technological correlation of resources, spaces and inhabitants. TECHNE, J. Technol. Archit. Environ. **12**, 129–136 (2016). Firenze University Press, Firenze
21. Nardi, G.: Il progetto euristico in architettura. In: L'atto progettuale, struttura e percorsi. Citta Studi, Milano (1991)
22. Lester, R., Piore, M.: Innovation the Missing Dimension. Harvard, Cambridge (2004)
23. Freire, J.: Urbanismo emergente in García-Rosales Mandala, C., Deseo de ciudad: Arquitecturas revolucionarias, Ediciones Peter Walsh (2010)
24. Lévy, P.: L'intelligence collective. Pour une anthropologie du cyberspace. Editions La Découverte, Paris (1994)
25. Cellucci, C., Di Sivo, M.: Shereable City, Regenerated by Making, in Sustainable City 2016. Wessex Press, UK (2016)
26. Studer, R.G.: The dynamics of behavior-contingent physical systems. In: Proceeding of the Symposium on Design Methods. Portsmouth College of Technology (1967)
27. Canter, D.V.: Terence Lee, Psychology and the Built Environment Paperback (1974)
28. Norcliffe, G.: The Ride of Modernity. University of TorontoPress, Toronto (2001)

The Needs of Children and Their Caregivers in New Urban Lifestyles: A Case Study of Playground Facilities in Hong Kong

Kin Wai Michael Siu[1]([⊠]), Yi Lin Wong[1], and Mei Seung Lam[2]

[1] School of Design, The Hong Kong Polytechnic University, Hunghom,
Kowloon, Hong Kong
{m.siu,yi-lin.wong}@polyu.edu.hk
[2] Department of Early Childhood Education,
The Education University of Hong Kong, Tai Po, New Territories, Hong Kong
mlam@eduhk.hk

Abstract. With an increasing number of working couples and a growing elderly population, young children are often taken care of by their grandparents or housemaids from the Philippines and Indonesia. The interaction between children and caregivers has thus changed to accommodate urban life. Yet public facilities for children have not kept pace with contemporary demands. Facilities for children, including playgrounds, have failed to respond to the everyday life of children and caregivers. Taking playground as a case study, this paper examines the mismatch between current playground design and the urban lifestyles of caregivers and children in Hong Kong.

Keywords: Design standards · Playground · Public facility · Urban lifestyle

1 Introduction

Urban lifestyles have transformed tremendously due to rapid changes in social structures and technological advancement. The way children are raised in a family and the nature of the parent-child relationship have also changed. In a traditional Chinese family, a married woman stays at home as a housewife to take care of her children and her or her husband's parents. She also plays with and teaches her children at home or spends a lot of time accompanying them outside of the home. Children who are raised in families of lower socioeconomic status often play with other similar-age peers without being accompanied by adults. However, because of the increased financial demands associated with raising children and taking care of the elderly nowadays, married women often need to enter the workforce to help shoulder the financial burden carried by their husbands. The status of working women in society has also improved and many women have their own careers and are keen to climb the career ladder for reasons beyond financial motivation. Where this is the case, many parents employ housemaids from the Philippines and Indonesia to take care of the elderly and children at home as well as do the housework. Yet, the housemaids are not just caretakers of children and the elderly. They are also, by proxy, the employees of the children and

© Springer International Publishing AG, part of Springer Nature 2019
J. Charytonowicz and C. Falcão (Eds.): AHFE 2018, AISC 788, pp. 257–265, 2019.
https://doi.org/10.1007/978-3-319-94199-8_25

elderly and have to obey them accordingly. Simultaneously, the elderly grandparents are also caretakers of children. The relationship between the elderly grandparents, children, and housemaids may thus be complex and dynamic.

Parents can only play with children on weekends because they are busy during weekdays. The time that parents can spend with their children is limited compared with that in prior decades. The status of children in a family has become more important than ever, and a strong emphasis has been put on the development of children in different respects, including through playing. Many parents and caregivers would not leave children out of sight in public areas. However, the current design of public facilities has often neglected this need. Facilities for children, such as playgrounds, have failed to respond to the everyday life of playground users.

Taking playgrounds as a case study, this paper examines the mismatch between current playground design and the urban lifestyles of caregivers and children in Hong Kong. Play is an important business for children [1], and it is the essence of children's culture to play [2]. Playground is a designed place for children to develop their problem-solving skills and emotional intelligence [3]. As children need to play, and caregivers often bring children to public playgrounds to play and enjoy their leisure time because of limited space at home, public playgrounds for children are an important place where the interaction between caregivers and children occurs. It is thus important to study and understand how mismatches arise vis-à-vis the demands and requirements of users and the actualities of these spaces.

2 Method

2.1 Public Playgrounds in Hong Kong: A Descriptive Overview

Currently, Hong Kong has 634 public playgrounds for children managed by the Leisure and Cultural Services Department (LCSD). These playgrounds are equipped with play facilities including slides, swings, and climbing frames. Around 36% of the playgrounds have swings, which are the most popular play facility [4]. Other play facilities are composite play structures whereby two or more such facilities are attached or functionally linked (Fig. 1). Swings are the only play facility which are not attached to any other play facility [5].

Playground size varies depending on the district. Some large playgrounds, for example, the playground in the Hong Kong Park, are equipped with swings, different types of slides, climbing frames, and different kinds of thematic composite play structures such as a flying saucer, space station, and adventurescape [6]. Some playgrounds are small and only contain a simple slide and climbing staircase. The total area of public playgrounds per child also varies considerably across districts. The average area of public playgrounds per child ranges from 0.16 m^2 in the Kwai Tsing district to 0.55 m^2 in the Central and Western district [5].

Fig. 1. (a-b) Example of a composite play structure (left) and swings (right) in two Hong Kong public playgrounds (photographs by authors)

2.2   Sample Playgrounds

Two typical playgrounds in the Eastern District, Quarry Bay Park Playground and Aldrich Bay Playground, were visited (Fig. 2). The playgrounds occupy 1,535 m^2 and 584 m^2, respectively, and both are located in a coastal area adjacent to residential buildings. The Quarry Bay Park Playground is next to a private residential development, and the Aldrich Bay Playground is next to both private and governmental residential areas. Residents of these areas can enter the playgrounds easily, and both playgrounds are very popular. Playgrounds in the Eastern district were chosen because the average area of public playgrounds per child in the district is 0.21 m^2, thus indicating that the playgrounds would not be too crowded or too empty. Figures 3 and 4 present aerial views of the two sample playgrounds.

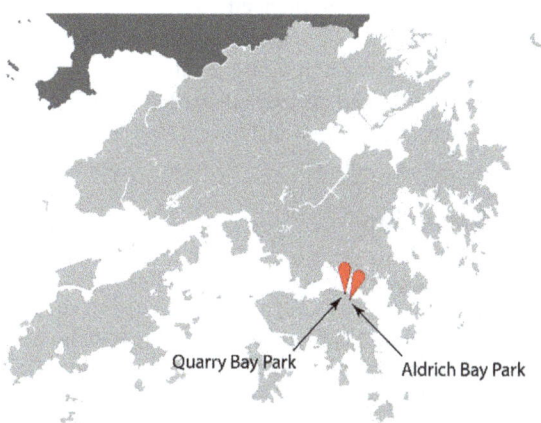

Fig. 2. Location of Quarry Bay Park and Aldrich Bay Park

Fig. 3. Aerial view of Quarry Bay Park (extracted from Google Maps)

Fig. 4. Aerial view of Aldrich Bay Park (extracted from Google Maps)

Two late afternoons in a winter weekend were spent at the playgrounds to study the behaviors and interactions of children and their caregivers. The most popular period, i.e., 4 pm to 6 pm, was chosen for the investigation. Photos were taken at the sites. The observer, i.e., the researcher, did not conduct any interviews so that the caretakers and the children at play were not unduly influenced.

These two playgrounds consist of various composite play structures (including slides and climbing frames), swings, rocking chairs, and sensory play structures. Different play facilities are designed to appeal to 2–5 and 5–12 year old children. Inclusive play facilities were also present at Quarry Bay Park Playground.

3 Findings and Discussion

3.1 Children's Behavior and Facilities Provision

Children were playing happily and actively in the playgrounds. Most of the children played on the composite play structures and the swings. They did not appear to be scared of heights and tried to climb high on the climbing frames. They were also not scared of the planned imbalance created at some play facilities (Fig. 5a). They enjoyed the challenges.

However, some younger children below 5 years old also played on the play structures which were designed for 5–12 year olds. These children feared heights and the imbalance play structure. It can be argued that play facilities for the younger children were inadequate in both playgrounds leading them to use play facilities which were not designed for them. These children were also unable to gather together to play, as there were no empty spaces for them to congregate, and their parents did not sense the need to facilitate play with peers of the same or similar ages.

Some children ran around the play facilities and the limited empty spaces in the playgrounds. Some of them played hide-and-seek and policeman-thief role playing games. A few of them ran to the area containing fitness facilities designed for the elderly (Fig. 5b). This observation also suggested that the available empty spaces for children to run around and congregate in were inadequate. Most of the children were playing together, although they may not have known each other previously.

Fig. 5. (a-b) Children at play in and outside playgrounds (photographs by authors)

3.2 Caregivers' Behavior and Facility Provision

The behaviors of caregivers varied. Some caregivers were standing or sitting nearby watching their children play (Fig. 6a). Some were using their mobile phones and did not watch their children at all. Some of them were leaning against or sitting on the play facilities. Some parents were accompanying both the elderly and children in the playground and had a stroll with the former whilst the latter were playing (Fig. 6b). Unsurprisingly, there were no specific facilities designed for adults in these children's playgrounds.

Fig. 6. (a-b) Activities of caregivers in playgrounds (photographs by authors)

Seating areas were provided next to the playgrounds, but the areas were far away from the play facilities (Fig. 7a and b). Caregivers can only watch the children from a distance. Other caregivers who were looking after infants were also standing or sitting in this area. Some of them left their belongings in the seating areas or next to planted areas and stayed with the children in the playgrounds, while others would leave their belongings with the housemaids.

Fig. 7. (a-b) Seating areas next to the playgrounds (photographs by authors)

In most cases, two caregivers were present to accompany one child in the playgrounds. One of them was responsible for taking care of belongings while the other looked after the children. The number of adults exceeded the number of children in both playgrounds. Both playgrounds were busy and fully occupied.

3.3   Interactions Between Children and Caregivers

It was interesting to investigate the interaction between children and their caregivers. Aside from merely watching their children playing, parents tended to talk to their children very often. For instance, in Fig. 8a, parents are telling their children to be careful on the composite play structure. In Fig. 8b, the parent is encouraging the boy to be braver while walking on the play facility. In general, parents would give instructions to their children on how to play. They would also remind the children to be safe and careful and give encouragement to them. Table 1 summarizes the instructions and guidance that parents gave to children in the playgrounds.

Further, parents also provided physical help to children at play. Some of the help facilitated the development of parent-child relationships, including trust-building. For instance, some parents helped their children to play on the swings, as some were not tall enough to easily master this apparatus because their legs were unable to touch the ground. However, some parents hindered children from enjoying the freedom of play and stopped them from facing developmental challenges at play. For instance, some parents would hold the child's body when they were climbing up or jumping off the ground. Thus, it can be argued that sometimes children were overprotected in the playgrounds.

Fig. 8. (a-b) Interaction between caregivers and children (photographs by authors)

Table 1. Instructions and guidance that parents gave to their children in playgrounds.

Type	Example statements (translated from Cantonese)
Safety issue	"Don't run too fast." "Don't climb too high." "Don't get close to the slides." "Be careful not to knock others off."
Play instruction	"You can slide down now." "Climb on this climbing frame." "You can't walk up the slide." "Don't use your hands."
Encouragement	"Yes! Keep at it!" "You can do it! Keep going!" "Don't be afraid."

3.4 Mismatch Between Playground Facilities and the New Urban Lifestyle

Due to the nature of new urban lifestyles, caregivers, especially parents, only tend to be able to accompany children and play with them in public playgrounds on weekends. Playgrounds have become a place not only for children to play but also for parents and children to spend time together. In other words, play facilities should also cater for the needs of caregivers. Currently available play facilities are unable to cater for these needs and thus do not match the requirements of new urban lifestyles. The mismatch between facilities and lifestyles has caused inappropriate use of existing facilities. Based on the findings and discussion above, the issues associated with current playgrounds are summarized below.

No Facilities Designed for Caregivers. In the context of new urban lifestyles, caregivers have to take care of children in playgrounds and thus need to remain close to them. However, the current design of playgrounds does not provide any spaces or areas for this purpose. There are also no facilities for resting or storing personal belongings. These inadequacies have resulted in caregivers occupying play facilities, and in some cases, leaning on them for physical support, or sitting on them in lieu of alternative places to sit which are close enough to their children. There are also no facilities specifically designed for caregivers to play with children. Caregivers can only assist children while they are at play (hierarchical play relationship) and cannot be involved in the act of play itself. The possibility of caregivers being play companions is neglected.

Inadequate Play Facilities for Communication. It seems that the only play facility that can nurture the caregiver-child relationship is the swings. Other facilities failed to provide this opportunity. Further, given the observation that communication between caregivers and children is frequent in playgrounds, this communication is also an important concern. However, there are inadequate play facilities facilitating communication between caregivers and children. Indeed, it can also be argued that some of the designs and settings of play facilities hindered effective communication. For instance, the design of the composite play structure in Fig. 8a blocks communication between caregivers and children; the former have to talk loudly to be heard and understood by the latter.

Inadequate Spaces for Caregiving. Limited land is a well-known issue in Hong Kong. Play facilities are thus often overly demanded and the playgrounds are crowded in Hong Kong urban areas. Empty spaces between play facilities are very limited, and children have inadequate space to play and run around in the playgrounds. Unsurprisingly, caregivers therefore also have limited space to accompany and assist their children. Caregivers often have to stand closely to each other. With so many children running in playgrounds which cover only a limited amount of space, it can be dangerous for caregivers as well as children.

The current play facilities are based on designs imported from the U.S. and Europe, and most of the designs are standardized. These facilities are unable to cater for the needs of contemporary Hong Kong families, and there are no customized play facilities in Hong Kong public playgrounds. This one-design-fits-all approach does not work effectively in Hong Kong because of the characteristics of new urban lifestyles, the specific caregiver-child relationship, and the limited space. The design of play facilities must be reconsidered so that it better meets the demands of new lifestyles and new needs.

Recently, non-governmental organizations concerned with children's play in Hong Kong such as Playright Children's Play Association have been campaigning to allow caregivers and children to be involved in the playground design process [7]. Caregivers and children have to design and construct their preferred playground using different materials including wooden and paper boards, carton boxes, old tires, gym balls and sticks. Such self-designed, self-constructed playgrounds are perhaps the only customized options which currently fulfill the needs of playground users.

4   Conclusions

Because of different societal changes and financial demands, the interactions between children and their caregivers, i.e., housemaids, elders, and parents, have altered. The urban lifestyle has changed in a way that children have become the center of a family, and significant emphasis has been placed on them. Considering that caregivers and children interact not only at home but also in public areas, it is essential to examine whether contemporary public facilities are able to meet the needs of the new urban lifestyles. This paper takes public playgrounds for children as a case study to investigate this issue. Two children's playgrounds located in Hong Kong's Eastern District were visited. It was found that the current play facilities as well as other facilities in the playgrounds did not provide adequate support to caregivers or their communication with children. Space was also very limited which hindered caregivers from taking care of their children in the playgrounds. We conclude that there is an urgent need to revise the current design and standards of playgrounds so that the needs of children and new urban lifestyles can be fulfilled. Indeed, revising the process of design itself, empowering children and caregivers to design the kinds of playgrounds they want to visit and use, is a promising avenue for future research and development.

Acknowledgments. We are grateful to The Hong Kong Polytechnic University for financial and logistical support. The authors also acknowledge the Eric C. Yim Endowed Professorship in Inclusive Design.

References

1. Bruner, J.: Play is serious business. Psychol. Today **8**, 83 (1975)
2. James, A.: Childhood Identities: Self and Social Relationships in the Experience of the Child. Edinburgh University Press, Edinburgh (1993)
3. Brett, A., Moore, R.C., Provenzo Jr., E.F.: The Complete Playground Book. Syracuse University Press, New York (1993)
4. Legislative Council: Official Record of Proceedings, 7 December 2016. Hong Kong SAR Government, Hong Kong (2016)
5. Legislative   Council.   https://www.legco.gov.hk/research-publications/english/essentials-1718ise04-public-playgrounds-in-hong-kong.htm
6. Leisure and Cultural Services Department. http://www.lcsd.gov.hk/en/facilities/facilitieslist/facilities.php?ftid=55&did=1#1098
7. Playright Children's Play Association. http://www.playright.org.hk/tc/whatshappening.aspx

Problems of Ergonomics in Lecturing History of Architecture and Town Planning Throughout Architectural Studies Course

Teresa Bardzinska-Bonenberg[(✉)]

Faculty of Architecture and Design, University of Arts in Poznan,
Al. Marcinkowskiego 29, 60-967 Poznan, Poland
teresa@bardzinska-bonenberg.pl

Abstract. Issues pertaining to Ergonomics in teaching History of Architecture and Urban Planning in a Faculty of Architecture are discussed in this paper. The analysis is based on a development of the programmes in two architecture faculties in Poznan (Poland): Technical University and Arts University and the subject currently taught within the scope of Architecture and Town Planning History at the University of Arts in Poznan (UAP). Urban landscape, understood in the broad sense, and comfort of using buildings and homes were closely related to to-day's ergonomic problems. Implementation and subsequent changes and adaptations of the programmes for this subject are discussed, with emphasis on ergonomic problems.

Keywords: Architecture and Town Planning · Subject programme
Appropriateness · Suitability · Military purposes · Comfort
Technical specialization · Computer programmes

1 Introduction

The History of the architecture programme and teaching methods: a methodological guidebook [1] book appeared in 2004, written by the author on the basis of the most up to date ministerial guidelines, but still for the five-year programme, which was followed at all faculties offering architecture and urban planning education in Poland.

In accordance with the guidelines and tradition, the programme contained a number of self-complementary subjects. These taught the basic truth: that architects are dealing with a town, a live organism, where each change stimulates even more changes: spatial, social, communication, health. That's what it was like in the past, and that's how it is now, albeit much more intensively and to a larger scope than before [2]. Table 1 is a summary of the first, drafted at the aforementioned stage of initial adaptation of the History of Architecture and Town Planning programme to the Bologna System at the Poznan University of Technology.

© Springer International Publishing AG, part of Springer Nature 2019
J. Charytonowicz and C. Falcão (Eds.): AHFE 2018, AISC 788, pp. 266–276, 2019.
https://doi.org/10.1007/978-3-319-94199-8_26

Such a system aimed to present problems associated with the development of architecture and urban planning in as wide a spectrum as possible, and to furnish future architects with the tools necessary to independently follow further changes in architecture and related disciplines [3, 4].[1]

Table 1. Teaching schedule of History of Architecture and Town Planning, followed by the Faculty of Architecture, Poznan University of Technology between 2000 and 2007 (by the author)

Y	Lectures	Classes	Field classes
Year I	(30 h/term + 30 h/term) HISTORY OF GENERAL ARCHITECTURE Ancient history: Egypt, Mesopotamia, Greece, Rome, early Christian and Romanesque architecture	Introduction to more detailed problems within the scope of the issues discussed at lectures	At the Archaeological Museum
	(30 h/term + 30 h/term) HISTORY OF GENERAL AND POLISH ARCHITECTURE: Romanesque architecture in Poland, Gothic architecture in Europe and Poland, Renaissance architecture in Europe and Poland	Introduction to more detailed problems within the scope of the issues discussed at lectures	In situ analysis of Gothic and Renaissance architecture in Poznan. Field trip opportunity: Romanesque and Gothic architecture
Year II	(30 h/term + 30 h/term) HISTORY OF GENERAL AND POLISH ARCHITECTURE: 18th century architecture: Baroque, Rococo, Classical	Introduction to more detailed problems discussed at lectures	Field analysis: Baroque architecture in Poznan
	(30 h/term + 30 h/term) HISTORY OF GENERAL AND POLISH ARCHITECTURE: technical and technological inventions in late 18th, 19th and early 20th centuries. The gradual changes in architecture of that period.	Introduction to more detailed problems within the scope of the issues discussed at lectures	Field trip opportunity

(continued)

---

[1] This paper is one of a series discussing an array of contemporary problems which should be brought up in the course of architectural studies. The chapter on the Bionics aspects in architectural studies programme was published in 2017, an article on coincidences in forming contemporary architecture and furniture is going to be published later in 2018.

Table 1. (*continued*)

Y	Lectures	Classes	Field classes
Year III	(15 h/term + 15 h/term) HISTORY OF URBAN PLANNING: shaping settlements and towns from ancient times to late 19th century. Theoretical solutions and applications thereof	Literature based work, analysis of selected urban planning systems	Medieval city of Poznan - on foot
	(30 h/term + 30 h/term) POST WORLD WAR II HISTORY OF ARCHITECTURE lectures cover the events of the last 60 years and prepare students for independent observation and assessment of events in the future	Seminars held in five problem groups simultaneously. An option to choose one of five themes	Subject specific trips as part of classes
Year IV	(30 h/term + 60 h/term) HISTORIC MONUMENTS CONSERVATION THEORY Lectures discuss the development of conservation theory from the outset to the present day. The most significant projects are presented, the legal aspects of heritage protection is introduced	Modernisation project of a residential and commercial use building. Drawing up a Monuments Record Card, a building conservation analysis and a conversion design	
	(15 h/term + 15 h/term) HISTORY OF ART: lectures cover basic art and history of art concepts from ancient times	Classes introduce current events in contemporary art and are based on participation in exhibitions	An option to organise a trip outside of Poznan in accordance with the current art events
	(15 h/term) AESTHETICS: lectures cover the evolution of basic aesthetics theories with particular emphasis on the aesthetics and technique relation		
	(15 h/term) TIMBER ARCHITECTURE; lectures on regional forms of timber architecture and traditional building methods. Various types of timber buildings are discussed: houses, churches, farm and production buildings		Trip to the Open-Air Museum. Field classes (two weeks) survey practice
V	MA THESES CONSULTATIONS		

2   Architectural Degree Programme Plans After Changing the Education Model for History of Architecture and Town Planning

The identification of two (essentially three) cycles of higher education had little impact on the History of Architecture and Urban Planning lectures' schedule. Split into two parts, where the first was to provide a comprehensive overview of events which culminate in contemporary projects. The second level of architectural studies was aimed at research connected with contemporary problems of cities and architecture. It provides a facility to expand understanding of the conceptual and structural solutions rooted in functional appropriateness based on a wide interpretation of ergonomic science. Table 2 presents the state of the programme, with modifications made in 2012, following Bologna System regulations.

Within the scope of History of Architecture and Town Planning, the role of the earliest observations of natural phenomena made by man is emphasised, together with understanding the different properties of materials and knowledge on building sites of old.

Ergonomics was introduced in the earliest known building activities and technical solutions: while preparing stone, mud, clay for building purposes. In houses which look different in different parts of the world similar solutions were applied due to similar processes for preparing food. The same wisdom and logics modelled layouts of cities. The paths became streets and streams supplied moats. Finally, tools and equipment adapted to human hands and body were gradually developed.

The knowledge which was accumulated throughout the centuries made gradual modifications possible.

Table 2. History and Theory of Urban Planning and Architecture lectures. Contents of proto-ergonomics solutions in building processes (by the author)

Lectures		Proto-ergonomics solutions used in the arrangements and constructions
Level I		
Year I	(30 h/term) HISTORY OF GENERAL ARCHITECTURE Ancient history: Egypt, Mesopotamia, Greece, Rome, early Christian and Romanesque architecture in Europe (an outing to the Archaeological Museum)	- methods and tools used in extracting stone blocks - equipment used in irrigation systems - different shapes of vases - hollow - bricks
	(30 h/term) HISTORY OF GENERAL AND POLISH ARCHITECTURE: Romanesque architecture in Poland, Gothic architecture in Europe and Poland (in situ analysis of Romanesque and Gothic architecture in Poznan). Field trip opportunity: Romanesque and Gothic architecture in Great Poland	- machinery used in constructing processes - weapons and anti-siege equipment - furniture

(*continued*)

Table 2. (*continued*)

Lectures		Proto-ergonomics solutions used in the arrangements and constructions
Year II	(30 h/term) HISTORY OF GENERAL AND POLISH ARCHITECTURE: Renaissance architecture in Europe and Poland. (analysis of Renaissance architecture in Poznan in situ)	- development of weapons - systems of defensive walls - decisions in the course of rebuilding Louvre's eastern façade - furniture
	(30 h/term) HISTORY OF GENERAL AND POLISH ARCHITECTURE: 17th century architecture: Baroque, Rococo. Organization of a field trip	- function and comfort of interior arrangements - diversification and development of furniture
Year III	(30 h/term) HISTORY OF GENERAL AND POLISH ARCHITECTURE: technical and technological inventions in late 18th, 19th and the first half of the 20th century. Europe and America. Genesis and development of the modern movement	- beginnings of industrial plants and new scope of functional (ergonomic) problems - consequences of industrialisation: workers' housing estates - Bauhaus design - beginnings of ergonomic assessment of tools and machines
	(30 h/term) HISTORY OF ARCHITECTURE TO MODERN TIMES lectures cover the trends and current of the recent decades	- development of ergonomics and its scientific background - Design Council in Britain - design fairs
Year Iv	(30 h/term) POZNAŃ'S SPATIAL GROWTH REGIONAL TIMBER ARCHITECTURE FORMS: Various types of timber buildings are discussed: sacral, residential, farm, production together with the used structural solutions. DEVELOPMENT OF GARDEN FORMS AND PUBLIC SPACES associated with residences and cities	- ergonomics of peasants' tools: similarities throughout the world - interdisciplinary knowledge of vernacular builders
BA Diploma		
Level II		
Year I	(30 h/term) THEORY OF ARCHITECTURE RESEARCH WORKSHOP I Performance of research works associated with early architecture and urban planning problems	Different themes, some topics related to ergonomics

(*continued*)

Table 2. (*continued*)

Lectures		Proto-ergonomics solutions used in the arrangements and constructions
	(30 h/term) THEORY OF ARCHITECTURE RESEARCH WORKSHOP II Performance of research works associated with early architecture and urban planning problems	Different themes, some topics related to ergonomics
II	MA THESES CONSULTATIONS	TERM 9

3 Elements of Ergonomics Throughout History of Architecture and Urban Planning Lectures

Ergonomics, as an interdisciplinary science, is concerned with designing and arranging objects that people use so that they interact most efficiently and safely. Although also called biotechnology, human engineering and even human factors, it is widely applied as a basic element in architecture, under the name of functionality, appropriateness, suitability. The best known, and the oldest written assertation of its importance, is "utilitas" – the first element of *Vitruvius triada*.

Ergonomics covers all the scales of human activity and among them it refers to built structures. Street architecture, details of interiors, furniture, kitchens, cutlery and crockery are now shaped to provide comfort of use. Ergonomic solutions can be a part of financial/scientific success. It is applied to a wide spectrum of disciplines: from medical sciences, structural systems all the way to utility models.

The emergence of computing science, seemed to solve the problem of adherence between needs, dreams and their implementation. Parametric design relieved the process of forming, but at the same time it gave a boost to imagination, opening new possibilities for architects. Frank Gehry's magical exteriors are the result of these new possibilities. The main purpose of History and Theory of Urban Planning and Architecture is to provide students with the knowledge on the reasons why the appearances of cities and buildings is changing. Evolving needs, the meanders of how people experience beauty and ugliness are constantly verified by technical possibilities, particularly from the 20th century onwards.

The problems discussed during subsequent terms of History and Theory of Urban Planning and Architecture emphasise the relation of man/creator and the environment, which he had to, and wanted to change. The imperative to change was at first derived from danger: danger of wild animals, climate, floods. There was also another need: the need for comfort, manifestation of power and pursuit of beauty.

After an earlier period of subordination to the forces of nature, as a result of observations and partial adaptation to everyday phenomena, people have gradually tried to master them. By means of tools adapted to human hands and dimensions, transportation of granite slabs was performed in narrow corridors within the pyramids of the Pharaohs of the IVth dynasty. Sophisticated systems, were constructed

Fig. 1. Edinburgh Castle, outside slits of embrasures, shaped for archer and musket fire (author)

Fig. 2. Poznan, Gunpowder Tower, embrasures inside of the wall (author)

beforehand which could be used to close the corridors after completion of funeral procedures, allowing the last workers leaving the structure to seal them [5].

Many discoveries have been exploited in many ways to the benefit mankind. Tripod support is one of the most multipurpose invention. Although at first it was used as sacral vessels in many European (Greece, Rome) and Asian (China, Japan) cultures, its form as a piece of furniture is one of the most important in vernacular crafts. Today, has been proven to be the safest construction to support industrial loads, as well as a stool to sit on [6]. Throughout the centuries the field on which the rules of ergonomics were the most important, was warfare and military equipment. Starting from Medieval walls, through Renaissance ramparts and the subsequent modern systems – all of them were built with one purpose: effectiveness of defence. Attacking equipment was prepared with the same intention. Studies on cold weapons and early muskets uncover a gradual improvement of their ergonomic qualities (Figs. 1 and 2).

Not many architectural features supporting this thesis survived from the early times. Shooting windows of Medieval fortifications are one such an example. Their embrasures were cut strictly according to the needs of archers standing in them and the size of slits allowed them to look and to fire. Hence the variety of the shapes of slits and gradual introduction of loopholes (keyholes) comfortable for muskets and later even for small cannons.

There was a variety of slits in castles and town walls, depending on time they were built (or modernized) and on the type of weapons held in town armouries. Arbalestinas were advanced forms of cross shaped slits, used in 16th and 17th c. by crossbowmen. They enabled better horizontal and vertical adjustment of arms: for archers and gunmen.

The idea of easier defence in the 20th c. once again employed embrasures, but in an inverted form: stepped outside openings of bunkers contributed to a wider field of fire, while eliminating rebounds [7].

From late Renaissance on, cabinet makers, ebonists and ornament designers of all crafts were the first "designers", who apart from their professional activities popularized master patterns for furnishings for general use. The designs however, were not aimed at comfort, as we see now, but at new ways of exploiting different, sometimes foreign motifs and decorative materials [8].

Monumental palaces and grandiose interiors of Baroque times were furnished with various types of furniture, some of which were equally imposing. On the other hand, boudoir furnishings were minute in a contrasting way. They were comfortable for the ladies, who used them. People in the 17th c were much shorter than today, 168–170 for European men, according to anthropologists research [9]. Despite the notions of this époque, ergonomic approach of the craftsmen and future owners must be noted.

The time of migrations of thousands of European settlers to America in the 19th and 20th centuries marks the moment of change in attitude towards architecture, house equipment and tools on the New Continent [10]. Inaccessibility of the above-mentioned goods on the American market, due the disparity between needs and the possibilities of meeting this demand created a new attitude towards production. This was particularly visible in mass manufacturing of everyday tools and house equipment: catalogues of different types of hammers, pincers, pliers, buckets, pots, provided an opportunity to purchase the ones, that were most suitable for the job that has to be performed or for a new home. The shapes were precisely adapted to the type of work, handles and grips seem to be comfortable to hold. The same pertains to furniture produced *en masse*.

In the USA, ergonomic attitude towards farm arrangements, buildings, houses and their equipment was at its best in Shakers' settlements and houses. Their sternness of discipline derived from religious beliefs resulted in conciseness of solutions, lack of decorations and clear adherence to demands. This made Shakers popular and needed carpenters, joiners, stonemasons, bricklayers and cabinet makers in the wake of America's colonisation [11].

Successive period of development of ergonomic attitude towards design came at the beginning of the 20th c. and there are two important points that should be mentioned.

It is worth underlining, that Antonio Gaudi, whose buildings were a far cry from functional logics and sober economy, is the one, who anticipated the meaning of adjustment to the parameters in architecture. His models conveyed influence of forces of gravity and indicated diameters of structural nodes, according to the material which they were going to be constructed from. Furthermore, while building Park Güell, with the viewing terrace and its top bench/balustrade, he employed some workers to mould the shape of the long and curved seat, by resting for a while on wet concrete. This gave him an idea of how to shape it [12]. Hence, he can be pronounced a forerunner of contemporary parametric design in architecture and an ergonomics-conscient designer (Figs. 3 and 4).

With the formation of the Bauhaus School in Weimar, and then in Dessau, a comprehensive attitude towards design arose. It was based on a combination of a wide variety of arts – architecture, sculpture and painting applied in practice: crafts and industry. The general idea was to create objects that would combine beauty with usefulness. The School's architectural credo was, apart from the ideas mentioned above, triggered by social needs and shortages on the housing market. In this case also mass, factory production was a basis for efforts for delivering goods in different finishes

Fig. 3. Bauhaus designed kettle (one of several versions of the same design) produced by the AEG (author)

Fig. 4. Bauhaus designed hair dryer produced by the AEG (author)

from the same production line. Simplicity and appropriateness for use were the main issues. Goods varied from pre-fabs for houses to kettles, door handles and electric fans [13]. Bauhaus "cool" aesthetics was at first not considered to be beautiful but many of the products developed by Bauhaus have become an integral part of modern households, both the original designs and their derivatives.

Growth in importance of quality of design was marked by constitution of the bodies, that assessed products and gave them their recognition, like the Design Council in Britain in 1944. Similarly, international recognition of the benefits acquired from good design was marked by a growing number of design fairs with Salone di Milano as the flagship of European furniture design fashion in 1961. In architecture, international and local prizes for quality of design multiplied. Architectural journals focused on different aspects of architectonic projects and accomplishments developed and ergonomic design was among many.

End of modernism came gradually with the new generation, seeking universal and recognizable meaning of forms in architecture and design. New trends, that marked their presence from the 1970s in America and Europe, were based on often contradicting theoretical foundations. Mentioning just some of them: post-modernism, deconstructivism, ecologic architecture, new expressionism, contemporary vernacular architecture, new classicism, minimalism, high-tech, rationalism, populistic architecture, new modernism, blob-architecture (*bubble architecture*), parametric architecture makes to possible to realize the spectrum of architectural paths identified at the end of the 20th century [14]. They refer to different philosophical, emotional, historical and functional premises. All of them triggered corresponding tendencies in interior design and design of objects. The "meaning" is the basis for new pursuits. Design fairs organized throughout Europe and world speeded up the process of alleviation of "aesthetic shock" for many non-professional visitors.

In the last decades parametric design has been implemented in all fields of design and production: from shoes and watches to super-structures set in difficult geographic conditions. Parameters, that are imposed on computing programmes opened new possibilities for architects to design buildings, which conform precisely with their surroundings and technical conditions [15]. This is the contemporary way to provide ergonomic shape to extreme structures.

4 Summary

In this paper key moments and turning points in the development of architecture and design of the early centuries were observed from the point of view of application of the rules of contemporary ergonomic science. They led to the conclusion that most of the changes were a result of observation of natural, social, economy and military phenomena. This is strongly underlined in the first semesters of architectural education. The complicated situation in Europe in the 19th and 20th centuries and in particular after the second world war, demands wider discussion, which is in some respects continues during the second level of the studies.

In the first decades of the 20th century, the visions of the avant-garde architects and artists went hand in hand with the social changes, that were not yet widely recognized. Building regulations were subsequently issued by governments and city authorities to ensure safety in buildings. They are clearly visible in urban developments and forms of architecture of that time. WW2 became the turning point, when interdisciplinary cooperation triggered by difficulties of the post-war world, contributed to the spread and gradual acceptance of the new aesthetics among "consumers" in the second part of the 20th century.

The start of the 21st century and the following decades are marked by unparalleled diversity of options and targets in all creative fields of human activity.

Architectural showpieces, a result of parametric design, are the most diversified and imposing status symbols, now also present in the countries formerly unimportant.

Problems of ergonomics in lecturing History of Architecture throughout architectural studies evolve from simple, craft-inspired facts to technically advanced issues, spread between mathematical and aesthetical problems.

References

1. Bardzińska-Bonenberg, T.: Program i metoda nauczania historii architektury, przewodnik metodyczny (History of the architecture programme and teaching methods: A methodological guidebook). Wydawnictwo Politechniki Poznańskiej, Poznań (2004)
2. Bonenberg, A.: Citysscape in the Era of Information and Communicatiom Technologies. Springer International Publishing AG, Cham (2018)
3. Bardzińska-Bonenberg, T.: Environmental protection in lecturing history of architecture and town planning throughout architectural studies. In: Dreszer, M. (ed.) Szacunek dla środowiska przyrodniczego istotnym elementem edukacji projektanta i architekta, pp. 9–13. Uniwersytet Artystyczny w Poznaniu, Poznań (2017)
4. Bardzińska-Bonenberg, T.: Architektura a mebel, krzesła włoskie drugiej połowy XX wieku. In: Charytonowicz, J. (ed.) Wybrane kierunki badań ergonomicznych w 2016 roku. Wydawnictwo Polskiego Towarzystwa Ergonomicznego, Wrocław (2018)

5. Dieter, A.: Building in Egypt. Pharaonic Stone Masonry. Oxford University Press, Oxford (1991)
6. Charytonowicz, J.: Ewolucja form sprzętów do siedzenia od pradziejów do wieku maszyn. Oficyna Wydawnicza Politechniki Wrocławskiej, Wrocław (2007)
7. Bogdanowski, J.: Architektura obronna w krajobrazie Polski. Państwowe Wydawnictwa Naukowe, Warszawa-Kraków (1996)
8. Morley, J.: The History of Furniture: Twenty-Five Centuries of Style and Design in the Western Tradition. Bulfinch Press, Boston (1999)
9. Roser, M.: Human Height. https://ourworldindata.org/human-height. Accessed 10 Feb 2018
10. Szejnert, M.: Wyspa klucz. Społeczny Instytut Wydawniczy. Znak, Kraków (2009)
11. Giedion, S.: Czas, przestrzeń, architektura – narodziny nowej tradycji. Arkady, Warsaw (1968)
12. Moravanszky, A.: Antoni Gaudi. Arkady, Warsaw (1983)
13. Whitford, F.: Bauhaus. Thames & Hudson, London (2003)
14. Jodidio, P.: New forms: Architecture in the 1990s. Taschen, Cologne (1997)
15. Helenowska-Peshke, M.: Parametryczno-algorytmiczne projektowanie architektury. Wydawnictwo Politechniki Gdańskiej, Gdańsk (2014)

Micro-space Planning: Social Action for Popularizing of this Planning Method in Silesian's Case-Study

Katarzyna Ujma-Wąsowicz[1(⊠)], Agnieszka Piórkowska[1],
and Małgorzata Kądziela[2]

[1] Faculty of Architecture, Silesian University of Technology,
ul. Akademicka 7, 44-100 Gliwice, Poland
{katarzyna.ujma-wasowicz,
agnieszka.piorkowska}@polsl.pl
[2] Institute of Cultural and Interdisciplinary Studies, University of Silesia,
plac Sejmu Śląskiego 1, 40-032 Katowice, Poland
malgorzata.kadziela@us.edu.pl

Abstract. Microplanning is addressed to the problems of small spaces in the city structure, which have the potential to be an important place for the local community and to bring about image and pro-social changes. It is not a new issue in the world, but in Poland, for mundane reasons, few local decision-makers, for various reasons, aim to change the image of these small spaces. On the other hand, the microplanning methodology is based on reliable pre-project research conducted with the participation of all stakeholders, including in particular potential users of the developed area.

The article presents the issue of microplanning from a theoretical and practical point of view, as an effect of social involvement of the authors, one of whom is a co-initiator and co-organiser of the action for the benefit of the said microplanning.

Keywords: Microplanning · Public spaces · Pre-project studies
Social competences · Work in interdisciplinary teams · Human factors
Sustainable Development · Homeostasis · Synergy

1 Introduction

The task of modern politics of the countries that regard themselves as the developed ones and thus cities and villages is to be guided by the idea of Sustainable Development. According to its definition drawn up by the UN Commission in 1987, the development of the human environment is to be based on meeting the needs of the contemporary generation without closing the way for meeting the needs of future generations. In other words, there is a concept of socio-economic development that can be achieved and continued for many years. Different stakeholders should be involved in its development and its implementation should take place independently of the changing options, in particular, the political ones.

The essence of the spatial planning process is to keep the environment in balance, i.e. in homeostasis. Most often, especially in practice, the work is based on the

© Springer International Publishing AG, part of Springer Nature 2019
J. Charytonowicz and C. Falcão (Eds.): AHFE 2018, AISC 788, pp. 277–288, 2019.
https://doi.org/10.1007/978-3-319-94199-8_27

distinction between what is related to nature and what is directly dependent on human activity, i.e. the effects of human decisions. On a macro scale, this means above all the infrastructure (road, technical, etc.) which is a major planning issue. On a micro (local) scale, organisation of public, semi-private and private spaces, rotating in the sphere of urban design [1].

Microplanning, which is the subject of the discussion, is addressed precisely to such local "microspaces", to small problems that are escaping because they are simply not the subject of spatial planning. An architectural approach is not applied there which might be the subject of a competition, for example, because there is no cubic capacity here and no "works of art" are expected here.

In Poland (generally) there is no suitable methodology for such small tasks that could correspond to our mentality and therefore no professional competence is observed. These tasks cannot be treated in the same way as the design of building environment (taught at university courses and required from the designing architects) and moreover, this is not the result of functional zone design. This is a completely different problem which requires specific methods of action and the skills and competences of urban planners [2].

Furthermore, in their vast majority the decision makers who manage the spatial development of their cities/municipalities are either focused on multi-million investments or shift these "small topics" away to the undefined future, or delegate tasks of microspace management to non-governmental organisations (whose involvement cannot be denied but are often far from professionalism). Sometimes, however, the same decision-makers do not understand at all how to choose the areas to be micro-planned and what results this action should bring. On the other hand, residents, having no appropriate role models or comparing their problems to the wrong role models lack awareness of what they could expect in such spaces and what actions they could be involved in [2].

Generally speaking, the microplanning methodology is based on reliable pre-design research conducted with the participation of all stakeholders, including in particular potential users of the developed area. It is obvious that in order conduct the said research, designers with the relevant knowledge and skills are needed. Intent observations of the contemporary urban landscape in our country allow us to notice the shortage of professionals as yet (at least in the field of architecture and urban planning, sociology or spatial management) with the developed professional competences in this field; there is also lack of effective, equivalent – as mentioned – to our national mentality, methodology of action.

2 "Microplanning" Methodology as a Planning Rule for Small, Local Areas Within the Cities

There are many methods to search for design solutions. These are proposals developed by psychologists, didactics specialists, praxeologists, and finally creators themselves (i.e. designers). They are rarely used in a "pure" textbook form. Each designer, supported by experience, over time comes to his own often intuitive and thus original method which after analysis turns out to be a compilation (mixture) of various known methods. Intuitive search is for instance sharing, analogy, guessing, associating, jux-taposing images and concepts, searching for similar problems in memory and trans-ferring and improving solutions. The main task of all design methods is to support the intellectual and creative effort [3].

The methods of searching for design solutions known in the world of science include such methods as: morphological (through a new combination of known ele-ments - parts - we obtain a new quality), tree of solutions (we have to approach the subject in total, i.e. create a holistic solution to the problem) or system methods (using abstract models, mathematical, graphical, graphical, analogue, digital model or simply a description) [3].

In contemporary design, the urban planner uses mainly heuristic methods, that is the methods that allow him to discover new solutions by putting appropriate hypotheses. The methods known from the subject literature are for example synectics, brainstorm-ing, morphological analysis, an ideal solution method or superposition method [3, 4]. However, these methods do not guarantee the expected result (particularly in terms of user satisfaction), although they certainly increase its probability. They are not too formalised to leave a certain margin for human intuition. These methods do not require strict adherence to the sequencing of operations or the accuracy of their application. The creator can take full advantage of his knowledge and imagination. However, a drawback of heuristic design methods is that they do not refer to the users of future solutions. Their basis is the practical knowledge and experience of experts (designers).

In turn, referring to the needs, expectations, and limitations of the user, the need/willingness and even the obligation dominates in urban design to use the results of quantitative and statistical surveys, where research results are combined in a descriptive or percentage formula. On their basis, conclusions are drawn about the frequency of phenomena, their intensity and relations between them; users' opinions are examined on a given topic, e.g.: where and how they want to live, what they expect in their work environment, where and how they would like to spend their free time, in what situa-tions they encounter the most spatial barriers, etc.

It is worth asking ourselves here whether sociological research should continue to dominate the assessment of the quality of the environment we live in. Does the fact that a given space corresponds to the majority mean that it is well developed?

The subject matter discussed here revolves around the "micro" scale - a local scale, where the idea of human factors is of key importance. Here, the urban planner's understanding of synergies and homeostasis [1] fits perfectly into contemporary trends in the shaping of space serving human needs. We are reminded of this by such prominent items of literature as Christopher Alexander's "A Pattern Language" [5] or Jan Gehl's "Life between buildings" [6].

In 2016, a pilot edition of the social microplanning campaign was conducted in several cities of the Silesia Region in Poland. This action was initiated and organized in cooperation with Katowice Branch of the Society of Polish Town Planners and the Silesian Association of Municipalities and Poviats, also based in Katowice.

In four different cities, four different project groups undertook to search for a new image of a small, public space area indicated by decision makers to act as a generator of integration of the local community in the future. The final result of each study was the spatial and image vision of a given place; in the first two cases it was based on the expectations of the inhabitants, in the second two cases resulted from the conditions of microplanning, which, as it turned out, were inadequate to the microplanning methodology. The authors of the study wrote:

- Area No. 1: This area will be an accessible, friendly and safe area and a place where residents will be able to spend their free time in accordance with the survey proposals;
- Area No. 2: The assumed features of the concept add up to the organisation of a multifunctional square, whose main feature is the preservation of the historical layout of the square and at the same time transforming its part into a hardened square of an urban character. Its planned elements and new image account for a response to the expressed (in words) expectations of the local community;
- Area 3: A developed urban concept, although having pro-developmental features, was not sufficient to initiate image and pro-social changes in the housing estate. The city has not only handed over too much land to be developed, but also one where socio-economic problems should be solved in the first place;
- Area 4: The authors' ideé fixe is a new quality of the neglected area, developing of a modern design that is a city landmark, creating space for various social groups, as well as activating and encouraging young people to stay in the city. However, the area to be developed could not generate solutions that could be identified with microplanning. There were no voices of residents and other travellers; the city did not exactly know what to expect from designers.

More information about this undertaking is presented in Chapter 4 (case study).

Continuing the above analyses, it is impossible in urban planning to ignore the problem of the Genus Loci (the spirit of the place) and its objective reality. This is an important theoretical concept in the works of the phenomenologists of the place, who try to find out what features of these places correspond to the subjective perception that they have their unique and irresistible specificity. Genius Loci is recognized by a human being, but its source lies in the non-objective features of the place [7]. Obviously, one cannot be sure that the mere fact of shaping microspace through qualitative rather than quantitative analyses will provide the "spirit of space". So how to carry out project work in order to "invite" it? The science is not yet able to answer this question, and it is not yet clear whether it will ever answer.

3 Microplanning Stakeholders: User *(Human Factor)*, Investor (City/Municipality), Designer

Three basic entities are involved in the microplanning process: local and non-local residents (users), designers and investors (cities/municipalities). Their place in the project can be defined at different levels, although their role seems to be equal.

The key to solving microplanning problems is the user and its environmental and phenomenological perception - a human factor. The perception, also called cognizance or impression, refers (similarly as aesthetics) to the perception of human phenomena or processes occurring as a result of the action of specific stimuli on our sensory organs. It includes not only a complex, subjective cognitive process, not only experience and memory but also brings its activities (expectations, values, goals, security, etc.) into the environment. It should also not be forgotten that the perceptual processes are significantly influenced by culture [1]. In the nineteenth century, psychologists studying the problem of perception realized that people are so different that each of us can perceive and describe our own sensory experiences in a different way [8]. Phenomenological experience, including place experience, is sensual and its content depends on the shape and position of our body. It is worth noting that the new paradigm of cognition has introduced the concept of an embodied cognition and has been a basic concept of phenomenological approaches for several decades [9, 10]. Taking up the subject of microplanning taking into account the above assumptions signals that only then will a new space fulfil its role when the designer takes into account as many of the user's expectations as possible. The user is also not an easy partner to cooperate with, for example, due to the frequent inconsistency and chimeric nature of opinions.

The roles and actions of cities represented by decision-makers (officials) can be described from a completely different perspective. Their intentions are often good but the problem to torpedo the development and most of all continuation of project work is the atomisation of competences and tasks in the structure of offices, including difficulties in undertaking integrated actions. Often, they are used to either predefined or known solutions, thus expecting the only technical implementation of specific (already established) ideas from the project teams. And last but not least, the difficulty of communicating with the project team - the tradition of "order-acceptance" [2].

And the last group of stakeholders i.e. the designers. They are expected to have the knowledge, i.e. a command and understanding in an in-depth way of the theoretical and applied basics of methods and technologies, phenomena and processes as well as organisational solutions. And in terms of skills, the skill of anticipating the development of a situation for example, or a well-targeted modification of methods of action depending on the situation. It is hard to imagine a well-designed space. Here, however, experience with the conducted research shows such shortcomings as lack of interdisciplinary cooperation skills; still prevailing belief in the environment that the role of the urban planner/architect is to play the role of the space creator; lack of professional identity in the environment of sociologists in design and dominance of the research attitude towards engineering (i.e. the activities of designing, constructing, modifying and maintaining economically sound solutions for practical issues); and in the environment of economists, difficulties in moving within the microscale of the economy.

4   The Concept of Social Actions Popularizing the Micro-planning – *Case Study* – Poland, Silesia Region

As it was mentioned before, in 2016 the Society of Polish Town Planners Katowice branch and the Silesian Union of Municipalities and Districts took the initiative to develop the character of a social project, called "Urban micro-planning". The project aims to promote the idea of micro-planning in relation to the public space of local importance while taking into consideration that the project is not oriented at commercial activities and does not force the execution of the initial assumptions. Two editions of the project have already taken place.

The project works take place in Southern Poland, in the cities of Silesian Metropolis. This region is an area of around 2.5 thousand km^2, inhabited by more than 2.2 mln of people. It is one of the most densely populated cities in Poland, with the highest degree of urbanization (above 77%) and rich industrial tradition.

4.1   The Example of the Performed Actions - the Assumptions

The methodology of the project works, conducted by the authors of the article for one of the given areas, will be discussed as a case study.

The representatives of the local authorities took part in the project "Urban micro-planning" in the role of the investors. They proposed some undeveloped and neglected areas of public space which, in their opinion, fit the idea of micro-planning. The estimated time for the project was more than 2 months and it required close cooperation between interdisciplinary project team, local society and the city in the role of the investor.

The interdisciplinary project team consisted of the specialists who hold the graduate degree in architecture and urban development, spatial economy, geography, law, philosophy and cultural studies (other teams included the sociologists as well). Moreover, the project team involved both experienced professionals and students. As a result, a few interdisciplinary teams were created, in total 9 participants in each group. The team together with the local authorities selected the area which complied with the requirements.

4.2   The Example of the Performed Actions - Pre-project Examination

During the project works numerous consultations took place. In the first phase, it involved the local authorities as the ones responsible for the area and in the role of the investor. At this phase, the project teams received first information about the area, local society, problems and spatial conflicts which are relevant for the city. Throughout the project duration, it was possible to obtain the necessary materials, information but also the support of the representatives of local authorities as well as the consultation in the direction of the actions. At the same time, some local inspections were taking place which included: inventory of the current state, spatial analyses concerning the type and height of the buildings, functional analyses, analyses of transportation, the ownership analyses, problems and spatial conflicts, land-use, determining the users groups, availability barriers and other regarding the given area. The local inspections took place in different parts of the days and on different days of the week, taking into account the weekends in order to understand the use of the space in the best way possible.

The next phase consisted of the meetings with the representatives of the local authorities and the influential stakeholders of the chosen project area, including local school management, cultural center, county council, leaders of the local church or local entrepreneurs directly connected with the project area.

In the process of the pre-project works, the consultations the with urban preservationist were of high significance. Those discussions helped to preserve the identity of the place and the urban arrangements which are valuable in the cultural and historical sense.

The following phase of the works relied on the direct examinations, including unstructured interviews with the inhabitants and users of the project area, in order to identify the needs and expectations towards the public space in their neighborhood. Apart from the group discussion, there was a possibility of expressing an anonymous opinion by the use of special cards placed on the model prepared by the project team. (Fig. 1).

Fig. 1. Model prepared by the project team (photo: A. Piórkowska)

Fig. 2. Meeting with users – cards placed on the model (photo: A. Piórkowska)

Besides the offered cards with observed and suggested problems during the interview (such as vandalism, lack of the places to sit, poor lighting, lack of dustbins, heavy car traffic, dangerous pedestrian crossing, hampered pedestrian traffic, the need of car parks, the need of a place destined for meetings, the need of a park, the necessary facilities for the elderly people and for the disabled, and many more), the empty cards were prepared to allow for raising other problems and needs. During and after the discussion, the participants of the meetings placed the cards in the areas of the spacial conflicts (Fig. 2).

4.3   The Example of the Performed Actions - Urban Planning Concept

The identification of the problems and the needs enabled the preparation of the project concept which referred to forming, organizing, choosing the elements for reconstruction and the target functioning of small public spaces in the city (covered by the project). Those spaces constitute or are to constitute an important place of social activity and at the same time should be discriminant of the identity on the scale of the city or the district.

The project took into consideration the influence on the social development of the area, engagement of the local society, reinforcing a sense of identification and a sense of responsibility for the surrounding area as well as commercialization of the space which will allow for involving private fundings at the building phase, but above all, it will reduce the operating costs. The result of that will be also the fact that the constant public fundings in this area will not be necessary. Another significant issue was maintaining coherence with other activities led or planned by the investor (the city).

The area destined for the development is situated in the center of the district, characterized by the typical for Upper Silesia historical housing development (known as company town) which is protected by preservation activity. Inside the district, there can be found a library, some services, primary school, fire department and the church. The area contains numerous availability barriers both in the form of plentiful fences and the difference in levels. It does not contain the clear space of public nature.

After the consultation with the local users, the concept included cleaning the area and developing the public space by creating an area of urban nature (Fig. 3), which was adjusted to the needs of all age groups, including the elderly people and the disabled people. The project assumes elimination of the availability barriers and offers the solutions which enable engagement of the users in order to commercialize the space and counteract low economic activity.

According to the above, the following objectives for the project were set:

- Creation of the spaces with the new quality, non-existing spaces or the spaces with previously limited use: recreational, commercial, parking spaces and spaces for winter sports;
- Organization of the space which would satisfy the needs of each group of the residents;
- Facilitating the availability of the space (public use);
- Retaining the original urban planning assumptions;
- Non-interference with the current lifestyle of the residents;

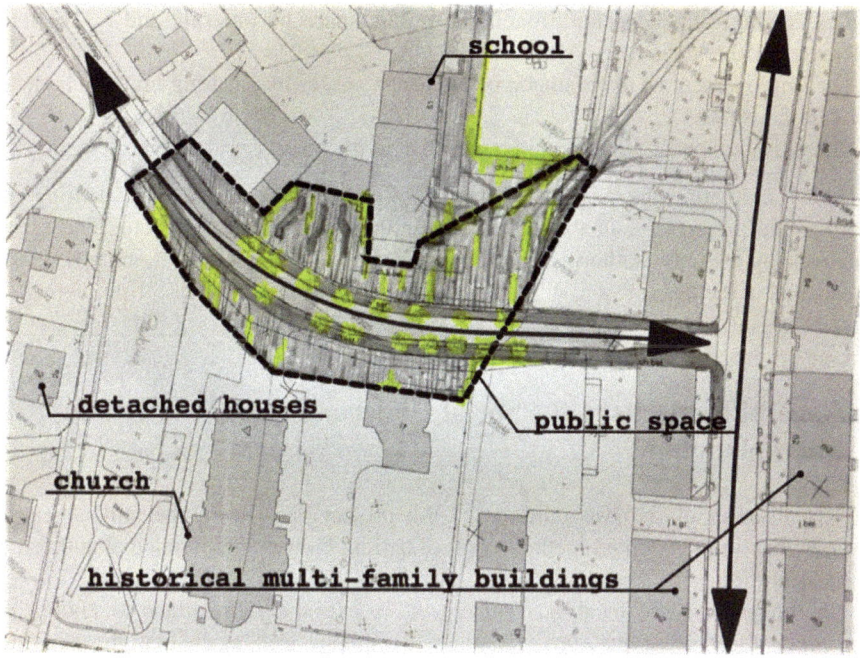

Fig. 3. Sketch of project concept (draw.: A piórkowska)

- Retaining the current way in which the area is used and the addition of the options of the new activities;
- The increase of users' safety;
- Availability in many seasons;
- 24-h access to use;
- The residents' acceptance of the management practices and financing the maintenance of the space.

For the purpose of the discussed example, the following consequences of social, economic and spatial nature were assumed:

- The improvement of infrastructural conditions for the cooperation of public institutions and the residents;
- The increase of the quality of the residence conditions by increasing the availability of commercial services and creation of new places and forms of free time activities;
- Greater openness to the residents - making the space public will allow for greater availability for the events initiated by the local institutions;
- The improvement of the public safety as a result of reorganization of pedestrian and car traffic solutions and the enhanced lightening of the space;
- The organization and equipping the space for continuing, seasonal and periodic commercial activities; allowing for gaining financial resources for maintaining the investments made within small operations;

- The creation of a new, multifunctional public space by ordering and by the integration of the area;
- The improvement of the aesthetic of the urban spaces by ensuring unified character of traffic routes;
- The introduction of small architecture objects for passive and active recreation.

After completing the project phase, the official meeting took place. During the meeting, the projects of all the groups (including ours) were presented. The representatives of the city authorities, who were involved in this social initiative, also participated in the meeting and discussed the initiative in the group of the policymakers in their cities.

5   General Remarks to the Social Undertaking Called "Urban Microplanning"

The organisation of the two editions of the project has shown the still appearing shortcomings in the adopted methodology of action, both by the municipal authorities and project groups.

On the part of the local government, it is extremely important to choose an appropriate location of the area. It should be emphasized again that the activities connected with microplanning are not the same as the classic understanding of revitalization, although in a certain extent resemble them. As mentioned above, the aim of urban microplanning is to carry out activities that are characterised by "micro intervention", i.e. stimulate the activity of the local community in the area under development, but not only – the advantage of the solutions is also a skilful encouragement to visit it, or even to stop for a proverbial moment the people from the outside, who are temporarily staying at a given place. This should therefore not be strategic areas or areas that accumulate excessive problems, for example in the social sphere. The classic space for microplanning procedures and methodology are small, i.e. of a public character (owner is the commune), open to everyone, areas in residential districts. The essence of these places is the lack of their potential spectra-latency. The executed projects do not have to, or maybe should not even bear the names of the icon/ "monument" of the investor; they should, however, contribute to equating the inhabitants with a given place in order to make them feel like their hosts.

Another issue is the appropriate selection of members of the project group, which in a pro-professional way is able to conduct consultations with residents and other potential users of a given place, is able to predict the social and financial impact of the generated idea, which finally (in the final phase) is able to propose solutions not only practical, but also aesthetic. It is also here that there are some non-defects. In the idea of the originators, the groups were supposed to be multidisciplinary and even-balanced, which was not always successful - in two mention cases the architects had a definite majority, who, despite their willingness, put more emphasis on the aesthetics of solutions than on other aspects, equally important in the case of microplanning.

The problem of communication between the members of the group turned out to be a separate issue: it seemed that people of different professions were unable to communicate with each other because they speak (not through their own fault) "different languages" spoken in their profession.

6 Summary

Despite these shortcomings, the social project undertaken by the initiators to promote microplanning and disseminate this idea brings about a lot of positive effects. Participants gain knowledge and skills of conducting consultation and interprofessional cooperation. The social competences achieved relate to taking up professional responsibility and making decisions in difficult situations; relate to cooperation and responsibility based on communication, interprofessional relations and ethical standards. Furthermore, the willingness to take into account the quality, spatial, social and economic aspects, to promote a pro-quality culture and other important effects of professional activity is shaped.

Modern urban planners/architects in Poland, when shaping a modern built environment, are able without any difficulty, to meet the requirements of maintaining spatial order, designing in the context of their surroundings and in accordance with universal architecture (without spatial barriers) or modelling the visualization of a place. What they lack, however, is that they have developed methodological instruments to solve seemingly trivial problems such as the public microspace "between buildings". However, bringing about positive changes requires a lot of effort, education and social action.

References

1. Kądziela, M., Ujma-Wasowicz, K.: The homeostasis and the synergy in the contemporary and future landscaping. In: Bijedić, D., Krstić-Furundzić, A., Zecević, M. (eds.) Book of Conference Proceedings. Keeping Up with Technologies in the Context of Urban and Rural Synergy. 4th International Academic Conference. Places and Technologies 2017, Sarajevo, 08–09 June 2017, pp. 38–47, Arhitektonski fakultet Univerziteta u Sarajevu (2017)
2. Borsa M., Ujma-Wąsowicz K.: Planowanie mikroprzestrzeni. Praktyczne doświadczenia współpracy wielodyscyplinarnych zespołów z gminami (Micro-space Planning. Practical Experience of Cooperation of Multidisciplinary Teams with Municipalities). In: Problemy planistyczne – jesień 2017, pp. 179–192. Stowarzyszenie Urbanistów ZOIU, Wrocław (2017)
3. Ujma-Wasowicz, K., Fross, K.: Beauty - aesthetics - senses research of attractiveness and magic of the built environment. In: Charytonowicz, J. (ed.) Advances in Human Factors, Sustainable Urban Planning and Infrastructure. Proceedings of the AHFE 2017 International Conference on Human Factors, Sustainable Urban Planning and Infrastructure, 17–21 July 2017, Los Angeles, California, USA, pp. 22–32. Springer International Publishing (2017)
4. Niezabitowska, E.: Metody i techniki badawcze w architekturze (Research Methods and Techniques in Architecture). Wydawnictwo Politechniki Śląskiej, Gliwice (2014)

5. Alexander, C.: A Pattern Language. Towns – Buildings – Construction. Oxford University Press, New York (1997)
6. Gehl, J.: Life Between Buildings. Using Public Space. Island Press, London (2011)
7. Ujma-Wasowicz, K., Sulimowska-Ociepka, A.: Genius Loci - examples of changes of the image of post-industrial areas in Poland in the region of the Upper Silesian Conurbation. In: Charytonowicz, J. (ed.) Advances in Human Factors, Sustainable Urban Planning and Infrastructure. Proceedings of the AHFE 2017 International Conference on Human Factors, Sustainable Urban Planning and Infrastructure, 17–21 July 2017, Los Angeles, California, USA, pp. 53–63. Springer, Cham (2018)
8. Bell, P.A., Greene, T.C., Fisher, J.D., Baum, A.: Environmental Psychology (polish version). Gdańskie Wydawnictwo Psychologiczne, Gdańsk (2004)
9. Merleau-Ponty, M.: The Phenomenology of Perception. Routledge, London (2001)
10. Tuan, Y.: Space and Place. The Perspective of Experience (polish version). Państwowy Instytut Wydawniczy PIW, Warszawa (1987)

Urban Informal Settlement and Infrastructure for Sustainable Urban Design: Investigating the Correlates and Mitigation Strategy

Oluwole Soyinka[1,2(✉)], Ben Spencer[2], Kin Wai Michael Siu[1,3],
Jeff Hou[2], and Laure Heland[2]

[1] School of Design, The Hong Kong Polytechnic University,
Hong Kong Room V616, Public Design Lab, Jockey Club Innovation Tower,
Kowloon, Hong Kong
oluwole.a.soyinka@connect.polyu.hk,
m.siu@polyu.edu.hk
[2] Department of Landscape Architecture, College of Built Environment,
University of Washington, Seattle, USA
{bspen,jhou}@uw.edu, heland@u.washington.edu
[3] School of Design, Wuhan University of Technology, Wuhan, China

Abstract. Adopting a case study methodology with mixed methods data collection and analysis, this study investigates causes and mitigation strategies related to urban informal settlement and infrastructure (UISI) in Hong Kong Special Administrative Region (SAR), Seattle, Washington, United States and Lagos metropolis, Nigeria. The study indicates that informal urban settlement is associated with factors including socio-economics, environmental conditions, and governance. In Lagos metropolis, UISI stems from inadequate infrastructural development, economic challenges, governance and other factors. In Seattle and Hong Kong, the escalating cost of living (housing) and a widening gap between the rich and the poor create homelessness and street sleepers which contributes to UISI. In Seattle, significant efforts and resources are directed towards this urban threat. However, administrative bureaucracy, inadequate coordination of organizations and issues of equality, equity, and race hinder significant advancements. The study recommends policy reforms, community-based participatory design, and inclusive socio-economic and environmental planning design as vehicles for sustainable development.

Keywords: Equity · Equality · Homelessness · Inclusive design
Participatory design · Sustainable development

1 Introduction

Urban informal settlements and infrastructure are complex and diverse urban challenges involving social, economic, environmental, and political variables. They pose threats to urban livelihoods, urban landscapes, and governance at different magnitudes all over the world. Developed countries such as Hong Kong and the United States (US) are not exceptions. The challenge of UISI is associated with several other urban challenges.

© Springer International Publishing AG, part of Springer Nature 2019
J. Charytonowicz and C. Falcão (Eds.): AHFE 2018, AISC 788, pp. 289–302, 2019.
https://doi.org/10.1007/978-3-319-94199-8_28

For example, the UN-Habitat world urban dialogue, UN-Habitat III Urban Agenda, and other research organizations have identified these issues as a critical obstacle to sustainable urban development [1]. Sustainable urban development is the ability to adequately use the urban resource without depleting its capacity for future use. Urban informal settlement (UIS) is described in different ways. This study adopts the UN-Habitat world urban dialogue's definition of informal urban settlements as residential areas with one or more of the following characteristics: (1) lack of security of tenure vis-à-vis the land use and/or the dwelling structures, (2) lack of basic facilities, services and city infrastructure; and (3) non-compliance with the current planning and/or building regulations and/or location within a geographically hazardous area. Informal urban settlements encompass homeless, street sleepers, squatter settlements and permanent structures of different kinds with these characteristics. Similarly, infrastructure is the network of facilities and services required for social, economic and environmental development. Considering these basic definitions, UISI is argued to interrelate with urban sustainability concept (social, economic, environmental and administrative) differently across the globe.

Globally, the challenges of UISI are significant. Several studies have attempted to address them from different perspectives with different conclusions. This study investigates the relationship between urban informal settlement and infrastructure in different contexts (developed cities and a developing city), as a means of gaining insights into sustainable urban design strategies with implications in different contexts.

2 Literature Review

Urban challenges are enormous, severe and multi-faceted. They differ from region to region. UISI has been identified as significant in the challenge of urban areas globally. Different factors and causes are associated with UISI in different places and a wide range of mitigation strategies have been adopted to confront UIS [1–3]. The work of Hernández et al. [3], Schneider [4], Deden [5] and other literature establishes the existence of UISI in different countries (developed and developing) with different qualities.

The study of Schneider [4] in 110 countries identified the existence of UIS globally and associated it with economic and other causal factors:

> "The average size of the informal economy as a percentage of the official GNI in the year 2000 in developing countries is 41%, in transition countries 38% and in OECD countries 18%. A large burden of taxation and social security contributions combined with government regulations are the main determinants of the size and causes of informal economy" [4, p. 1]

Several studies also describe UISI in terms of its social, physical, environmental and infrastructural characteristics. Majale [6] characterizes UISI as lacking:

> "adequate privacy, adequate space, physical accessibility; adequate security, security of tenure, structural stability and durability, adequate lighting, heating and ventilation, adequate basic infrastructures such as water supply, sanitation and waste management facilities, suitable environmental quality and health-related factors, adequate and accessible location with regards to work and basic facilities: all of which should be available at an affordable cost" [6, p. 1].

Table 1 presents the selected literature review for this study and describes UISI as it relates to sustainability (social, economic, environmental) as well as political and/or administrative issues.

Table 1. Selected research on urban informal settlement and infrastructure planning

Authors/Tile	Country	Significant summary/findings
Urban informality		
Alter Chen, M. (2005). Rethinking the informal economy: linkages with the formal economy and the formal regulatory environment	Global Perspective	The study suggests why and how more equitable linkages between informal and formal economy should be promoted through appropriate inclusive policy and regulated environment
Roy, A. (2012). Urban informality: the production of space and practice of planning	Global South	Urban informality is perceived as a 'planet of slum' with 'advanced marginality' and as a city 'embodiment of entrepreneurial energies'. It is a way of life; the characteristics and condition of living which is associated with several factors
Douglas, G.C.C. (2016). The formalities of informal improvement: technical and scholarly knowledge at work in do-it-yourself urban design	USA	The study argues the formalities of informal improvement through technical and scholarly knowledge of Do-It-Yourself (DIY) urban design
Hernández, F., Kellett, P.W., & Allen, L.K. (2010). Rethinking the informal city: critical perspectives from Latin America	Latin America	The book attempt to generalize the perspective of informal cities for spatial rethinking and development strategies. It, however, argues that the formal and the informal have become not only inseparable and interdependent but also indefinable. "Formal and informal are best thought of as part of a continuum: a few activities are wholly formal, a few wholly informal, but most home and household are some combination of both" (p. 795)
Elsheshtawy, Y. (2013). Where the sidewalk ends: informal street corner encounters in Dubai	Dubai	The overall perception of Hor Al Anz is that informal settlement is a menace that needs to be contained. The role of supportive environment facilities and services act as a system composed setting which supports each other. There is a sense of anomie and alienation due to the absence of a strong community sustained social, economic and environmental network. The area experiences a strong degree of isolation from the city

(continued)

Table 1. (*continued*)

Authors/Tile	Country	Significant summary/findings
Jabareen, Y. (2014). "Do it yourself" as an informal mode of space production: conceptualizing informality	Israel	The study addresses the ideal right to the city, resolving injustice, insecurity, poverty, and inequality. Informality is social, economically and culturally constructed and it has its own unique structural element
Marjit, S., & Kar, S. (2009). The Urban Informal Sector and Poverty: Effects of Trade Reform and Capital Mobility in India	India	The study reflects that a large number of the urban poor in India work and live in the informal arrangements. The improved conditions of the informal workers leave a significant and sustained impact on the incidence of poverty in urban areas of the country
Ali, M.H., & Sulaiman, M.S. (2006). 'Shaping the Change-The Causes and Consequences of the Informal Settlements in Zanzibar'	Zanziba Tanzania	The basic problems with this urban development pattern have been the inappropriate conception of space as somehow separable from other dimensions of the society such as the economy, social, policy, and environment. Added to the inadequate political will to restructure and improve the area
Hegazy, I.R. (2016). Informal settlement upgrading policies in Egypt: towards improvement in the upgrading process	Egypt	The findings reflect that despite the continuous attempt to develop informal settlements and improve their conditions in Egypt, it still faces intractable challenges that affect sustainable development. It also identifies the significance of upgrading through infrastructure "It is essential to develop affordable and participatory measures for upgrading housing conditions with related infrastructure support in informal areas to achieve improvement" (p. 272)
Brown-Luthango, M., Reyes, E., & Gubevu, M. (2016). Informal settlement upgrading and safety: experiences from Cape Town, South Africa	South Africa	One of the significant findings and conclusions of the study states that "Physical improvements in the built environment are of absolute importance. However, without accompanying social and economic programmes, they will not bring about the transformation of settlements (p. 491)

(*continued*)

Table 1. (*continued*)

Authors/Tile	Country	Significant summary/findings
Olajide, O. (2010). Confronting the Lagos Informal Land Use: Issues and Challenges	Lagos, Nigeria	The land defines the social, economic, environmental, and political relations in a society. The findings further show that the challenge confronting informal land development in the area is multi-dimensional ranging from social, economic, cultural, environmental and physical. The paper thus suggests confronting the challenge with a sustainable urban land use management system
Wu, R., & Canham, S. (2009). Portraits from Above-Hong Kong's Informal Rooftop Communities	Hong Kong	The study present pictorial analysis of the types and characteristics of urban informal settlement in Hong Kong. It describes the condition of living of the people and the state of the facilities and services available in these areas. It also argues that the neoliberal system of government with the socio-economic and housing challenge in the region contributes to the issue of urban informality in the area
Infrastructure planning		
Goodman, A., & Hastak, M. (2015). Infrastructure Planning, Engineering, and Economics	Global perspectives	The study discusses diverse perspectives towards infrastructure planning; the concepts, model, and methodologies of infrastructure planning. It concludes that the sustainability (social, economic, physical and environmental) factors is significant towards achieving sustainable infrastructure planning
Uddin, W., Hudson, W., & Haas, R. (2013). Public infrastructure asset management	Global perspective	The study describes the design of significant infrastructure in relation to planning needs assessment and performance indicators
Verheijen, M. (2016). Infrastructure: Infrastructure by Design	Global perspectives	Including 15 research perspectives and over 30 international best practices this book presents design perspectives for infrastructure, the study states that infrastructure is more than a functional necessity. Intelligent infrastructure design can influence the social, economic and physical value of living conditions

(*continued*)

Table 1. (*continued*)

Authors/Tile	Country	Significant summary/findings
Timmermans, J., & Beroggi, G. (2000). Conflict resolution in sustainable infrastructure management	Netherlands	The study opines that developing effective infrastructure involves compromising between sustainability, safety, economic and environmental factors and that different organizations must agree on a course of action
Olaseni, M., & Alade, W. (2012). Vision 20: 2020 and the challenges of infrastructural development in Nigeria	Nigeria	Nigeria's sustainable economic and infrastructural development are related to the social, economic and physical infrastructure of the metropolis. The quantity and quality of infrastructure needed to propel, economic, social, physical and environmental development are absent
Ng, S.T., Wong, J.M., & Wong, K.K. (2013). A public-private people partnerships (P4) process framework for infrastructure development in Hong Kong	Hong Kong	The study advocates re-strategizing the pragmatic approach underpinning public engagement to enhance social, economic and physical infrastructure and services. Adopting P4 framework it envisages that the framework will change the public aspiration, and involvement in infrastructure planning and policy formulation

Source: Author 2017

The literature evidence of this study established that UISI is a global challenge, it occurs differently in different regions with different severity and there are limited literature on it comparison across the developed and developing cities with sustainability perspectives.

2.1   Research Objective and Framework

Considering the gap identified, the argument on the interrelationship between UISI and sustainability concept towards sustainable urban development. This study investigates UISI across developed cities and developing city through the lens of sustainability and sustainable urban design strategies. Adopting the case study areas of the Hong Kong Special Administrative Region (SAR) of China, Seattle, Washington, United States and Lagos metropolis, Nigeria, the objective of this study is to integrate literature-based and field-based findings to generate strategies to address the challenges of UISI. These areas are adopted to foster an understanding of UISI in developed and developing cities. Figure 1 illustrates the research framework adopted to achieve this objective in the selected study area.

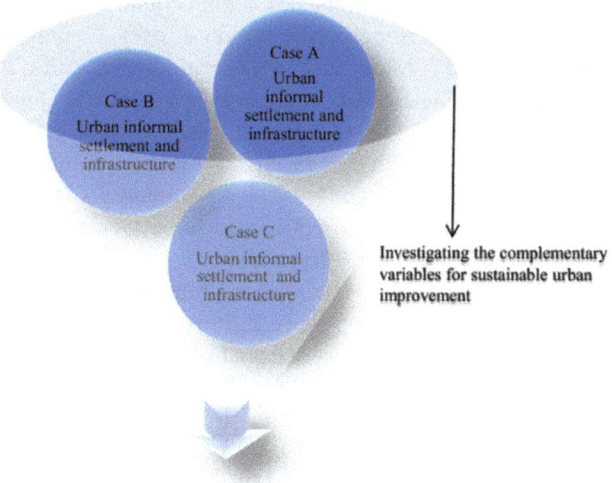

Comprehensive sustainable urban design strategy

Fig. 1. Research objective and conceptual framework. Source: Authors 2017

3   Methodology

3.1   Case Study

Hong Kong, one of the study areas of this study, is located at 22o19'42" N and 114o11'30" E on the southern coast of China. The city has an estimated population of 7,152,000 and a land area of 1,110 km^2. This study investigates the three main geographical regions of the city - namely New Territories, Kowloon Peninsula, and Hong Kong Island - which accommodate eighteen districts [7]. Hong Kong has experienced rapid growth in urban prosperity and challenges. It is adopted for this study considering the theoretical and practical evidence of this phenomenon in the SAR [8]. The city is one of the world's fastest growing regions with highly developed services industries, tourist activities, effective and efficient transportation systems (MTR, road, and airport). However, poor urban communities exist amidst Hong Kong's prosperity [9–11]. The region's neoliberal policies create a wide gap between the rich and the poor and contribute to the persistence of street sleepers, homelessness, UISI and squatter settlements, among other urban challenges [9, 11, 12].

The seaport city Seattle in Washington on the west coast of United States is located at 47o36'35" N and 122o19'59" W. It has an estimated population of 713,700 and a land area of 217 km^2. The city is one of the fastest growing cities in America and is home to leading company's such as Microsoft, Amazon, and Boeing. It is a city blessed with social, economic, environmental and natural resources. However, there is growing concern that Seattle's cost of housing, street sleepers and homelessness are increasing. UISI represents a significant challenge for urban management. The government, citizens, and professionals are working to improve the conditions of the city [13–15].

Lagos metropolis is a coastal city in the developing country of Nigeria. It is located at 6o23'N and 2o42'E. The metropolis' population is over 17 million and its geographical land area occupies 3577 km^2. One of the world largest cities, it is home to an urbanizing economy and significant urban challenges [16–18]. It has huge latent natural resources and is the commercial hub of Nigeria. Despite social, economic, environmental and natural wealth, it is a city characterized by rapid population growth with inadequate infrastructure facilities and services to support its growth. There is a growing concern that Lagos metropolis will experience a breakdown of urban settlement, infrastructure, and administrative structures.

This study considers the existence of urban informal settlement and infrastructure challenges in developed and developing cities and investigates their common characteristics as means of gaining insights into sustainable urban design.

3.2   Methods

This study adopts a case study method, mixed method data collection with triangulation, and mixed method data analysis. The criteria for selecting the case study areas include literature and practical evidence of this issue with a pilot study conducted in the study areas. Triangulated data collection techniques ensure the validity and reliability of the study. This involves the use of more than two data collection techniques to obtain data from both primary and secondary sources.

Subsequent to the identification of the case study areas as discussed above, the study triangulates complete observation (site visit and reconnaissance study), interviews (professionals and residents) and questionnaires to extract primary and secondary data. Professional interviews engaged built environment professionals with work experience in UISI and homelessness. Respondents held a minimum of a Master's degree in a related field and had at least 5 years of experience in public or private enterprise. Resident interviews involved residents with a minimum of 5 years of UIS residency or homelessness and an understanding of the environment. Sampling techniques involved the concept of saturation. There are no specific benchmarks that determine how many interviews are necessary but the saturation (repetition) of similar feedback is important [19]. A minimum of twelve interviews with a saturation of responses has been established as sufficient. This study adopts 12, 12 and 18 responses in Hong Kong, Seattle, and Lagos metropolis respectively [7, 19, 20]. The approach follows the snowball sampling approach (a chain-referral sampling).

The data analysis adopts mixed method data analysis of exploratory design (qualitative data analysis results building into quantitative data analysis and results; then followed by the interpretation).

4   Discussion

The findings outlined in Table 2 illustrate that the UISI is a significant challenge in the study areas and that current approaches are inadequate to ameliorate the UISI challenge.

The interviews conducted indicate that the government and citizens of Seattle and Hong Kong are working to improve UISI but that their efforts are insufficient to improve conditions. Table 2 presents the detailed frequency distribution of the response across the study areas.

Table 2. Urban informal settlement and infrastructure characteristics in the study areas.

Response	Strongly disagree	Disagree	Neutral	Agree	Strongly agree	%
UISI is significant in your area						
Seattle, US	0.00%	0.00%	0.00%	14.30%	85.70%	100.00%
Hong Kong	6.00%	21.05%	15.05%	10.53%	47.37%	100.00%
Lagos metropolis	0.00%	0.00%	5.00%	15.05%	80.05%	100.0%
Current approaches to addressing UISI in your area are effective						
Seattle, US	14.20%	42.90%	28.60%	0.00%	14.30%	100.00%
Hong Kong	48.00%	35.00%	5.00%	12.00%	0.00%	100.00%
Lagos metropolis	21.74%	60.87%	13.04%	0.00%	4.35	100.00%
The scope of efforts to address UISI in your area sufficient						
Seattle, US	0.00%	42.80%	28.57%	14.30%	14..30%	100.00%
Hong Kong	35.00%	45.00%	20.00%	0.00%	0.00%	100.00%
Lagos metropolis	63.00%	30.00%	7.00%	0.00%	0.00%	100.00%

Source: Authors' Fieldwork (2017)

In order to establish the relationship between the urban challenges in the study areas, this study utilizes literature review, questionnaires, and interviews. It investigates the significant causes of UISI and classifies them within the context of sustainable development concepts and infrastructure (social, economic, environmental, institutional/administrative).

In Seattle the causes of UISI are attributed as follows; 28.57% high cost of living, 20.40% drug abuse, 18.36% inadequate/non-affordable housing, 16.32% unemployment and 8.16% lack of infrastructure and migrations. In Hong Kong the causes of UISI are attributed as follows; 33.00% inadequate/non-affordable housing, 25.00% high cost of living, 20.00% unemployment, 17.00% migrations and 5.00% lack of infrastructure. In Lagos metropolis, the results are similar to Hong Kong. The causes of UISI are attributed as follows; 37.00% high-cost living, 18.00% unemployment, 25.00% lack of infrastructure, 15.00% inadequate/non-affordable housing and 5.00% migration effects with no response on drug abuse. Table 3 presents the detailed results of the cause ranking and provides sample descriptions of the causes of UISI in the study areas.

The significant correlation of this study finding in the study areas reflects that the challenge of UISI exists in developed and developing cities and is associated with sustainable development factors based on the investigation of sustainability indices such as social, economic, and environmental variables. Results indicate that socio-economic variables - high cost of living and inadequate/non-affordable housing - are the most significant causes of UISI in the study areas. Figure 2 shows the correlates of UISI and sustainability concept identified for sustainable urban design strategies in the study areas.

Table 3. Ranking the causes of UISI and descriptions of UISI in the study areas

Kindly rank the causes of UISI in Seattle		%	How would you describe UISI in your area
Seattle, US	Drug abuse	20.40%	"… there are many different types of people that fit into this IUSI category. We often refer to them as people experiencing homelessness because many are only temporarily without a home…., I would also add that we have rolling homeless, or people who sleep in their cars"
	Inadequate/Non-affordable Housing	18.36%	
	Lack of infrastructure	8.16%	
	Unemployment	16.32%	
	Migrations	8.16%	
	High cost of living	28.57%	
Total		100.00%	
Hong Kong	Drug abuse	0.00%	"…. It is very small in Hong Kong and it mostly occurs in the old apartment. It is caused because of money, living with families and getting government house or a lower rent is not easy. So, it is because of financial problem, staying with families and there are no enoughd public houses."
	Inadequate/Non-affordable Housing	33.00%	
	Lack of infrastructure	5.00%	
	Unemployment	20.00%	
	Migrations	17.00%	
	High cost of living	25.00%	
Total		100.00%	
Lagos metropolis	Drug abuse	0.00%	"…informal settlement environment in Lagos is not supposed to be a severe urban challenge the way it is now… a lot has however gone wrong and they are characterized with shark's structures, and all inhuman condition you can call it….no open space, facilities like sports ground, no countryside, no open views, no community facilities…"
	Inadequate/Non-affordable Housing	15.00%	
	Lack of infrastructure	25.00%	
	Unemployment	18.00%	
	Migrations	5.00%	
	High cost of living	37.00%	
Total		100.00%	

Source: Authors' Fieldwork (2017)

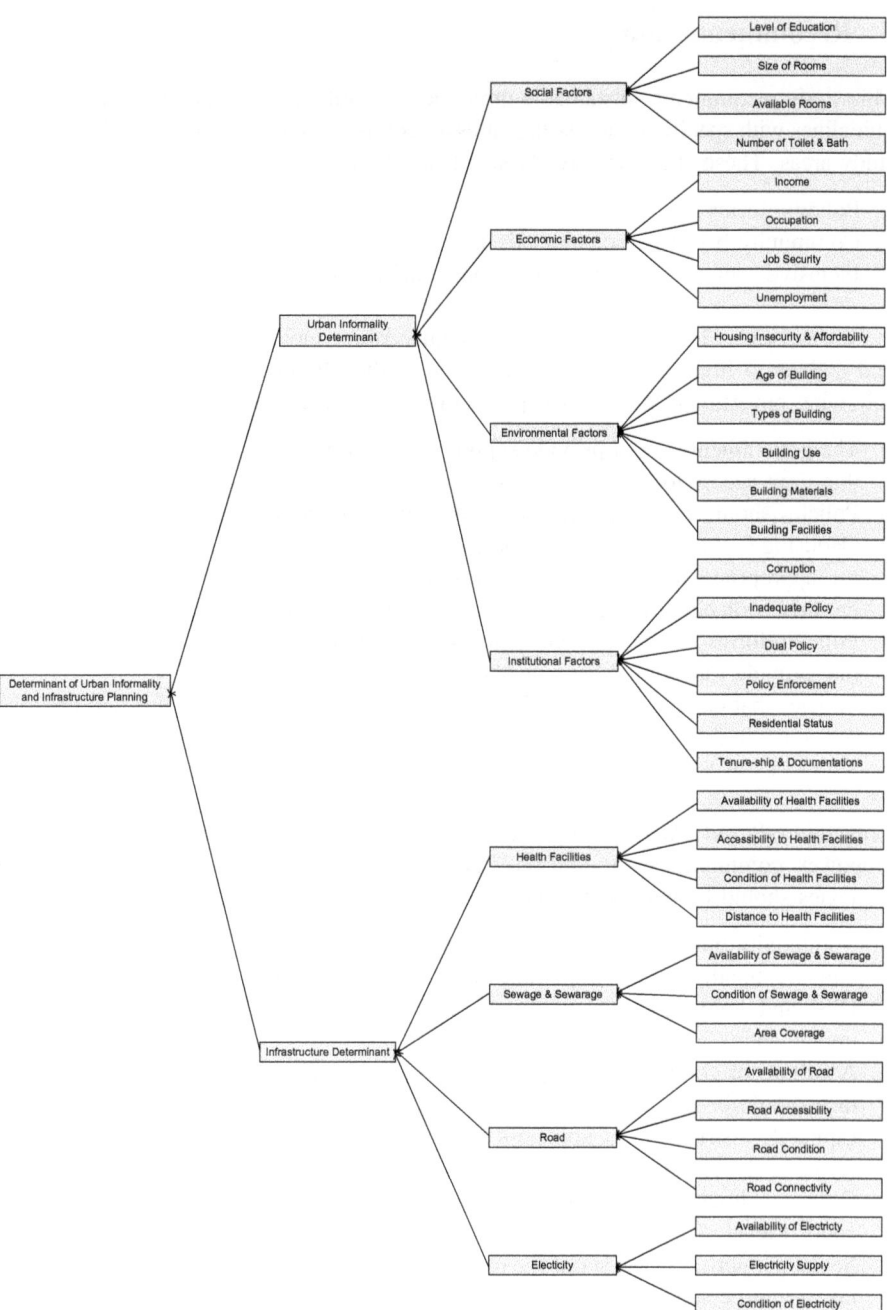

Fig. 2. Significant correlations and factors of urban informality and infrastructure with sustainability concept. Source: Authors 2017

5 Recommendation

This study recommends sustainable urban design strategies that should be adopted as guidelines with specificity across the cities and/or as vehicles for improving UISI in the study areas. These strategies are classified as follows;

1. Policy reforms
2. Community-based participatory design
3. Inclusive socio-economic and environmental planning design.

Policy Reforms: This aspect of the recommendation includes government programs and actions to improve UISI. Based on the study's findings in the study areas, the following government actions and programs are recommended:

a. The government should provide support for UISI programs without long processes that limit access to government aid.
b. Policies should ensure government aid is given without prejudice.
c. "Housing first for all" policies should be introduced to minimize long waiting times for appropriate housing, and the resulting exposure of the homeless to health hazards, violence, human trafficking, prostitution and even death.
d. Policy reforms that benefit tent-cities, tiny structures, and other temporary shelters that provide transitional housing should be considered.
e. Policies that engage homeless citizens in income generating activities will improve the situation faster, help the homeless integrate into society and help them manage high costs of living.

Community-Based Participatory Design: This involves activities and actions that empower communities to confront UISI. Community-based participatory design strategies should be tailored to local contexts:

a. Community outreach to create awareness (about causes and dangers of UISI) and human empathy is very important to fostering community involvement.
b. UISI and homeless citizens should be involved in the act of formulating and implementing solutions.
c. A culture of community acceptance and lack of prejudice to ethnicity, color, and race should be promoted.
d. Equitable distribution of homeless facilities are important to avoid some residents rejecting the location or development of these facilities in their neighborhood.
e. Outreach should be community-based and coordinated at the municipal and/or the council level.
f. Social support systems should avoid prejudice, accept the lifestyle decisions, mistakes, and personal choices of UIS residents and encourage reinstatement or integration into the community.

Inclusive Socio-Economic and Environmental Planning design: This involves the integration of policy reform and community-based participatory design with technical strategies, sustainable practices, and citizen engagement (Fig. 3).

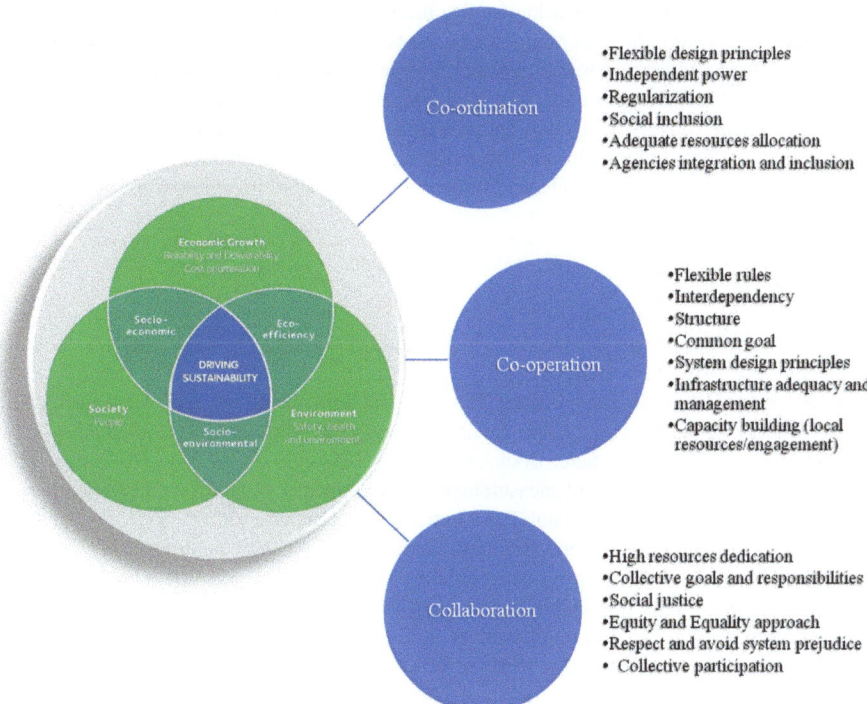

Fig. 3. Inclusive socio-economic and environmental planning design. Source: Authors 2017

More interdisciplinary work to address UISI in a more holistic and sustainable way, unified efforts and interconnected solutions should be encouraged.

Inter-agency collaboration and coordination is also important to address UISI. For instance, hospitals, police, and fire departments should work together to lessen addiction, mental illness, and systemic prejudice.

6 Conclusion

This study concludes that UISI is a global challenge associated with several factors, among which sustainable development factors are particularly significant. The most significant challenges of UISI in the study areas are socio-economic and environmental. High cost of living, inadequate and/or non-affordable housing are more significant Seattle and Hong Kong than in Lagos metropolis. Drug abuse was identified as an issue peculiar to Seattle and inadequate infrastructure was identified as an issue particular to Lagos metropolis. The study recommends policy reforms, community-based participatory design, and inclusive socio-economic and environmental planning design approaches as means of addressing the challenges of UISI.

Acknowledgments. The authors acknowledge the Hong Kong Research Grant Council for the Ph.D. Fellowship award, the opportunity for overseas study and the School of Design, The Hong Kong Polytechnic University for their research support to carry out this research. The support of the Department of Landscape Architecture, Department of Urban Design and Planning, College of Built Environment, University of Washington, Seattle, United States was also appreciated for their research support. The authors also acknowledge the support of all the public and private professionals (urban planners, architects, and other environmentalist) interviewed for this research.

References

1. UN-Habitat: Habitat III Issue Papers, 18 - Urban Infrastructure and Basic Services, Including Energy, pp. 1–10, New York (2015)
2. UN-Habitat: Habitat III Issue Papers, 22 - Informal Settlements, pp. 1–9, New York (2015)
3. Hernández, F., Kellett, P.W., Allen, L.K.: Rethinking the Informal City: Critical Perspectives from Latin America. Berghahn Books, Oxford (2010)
4. Schneider, F., (ed.): Size and measurement of the informal economy in 110 countries. In: Workshop of Australian National Tax Centre, ANU, Canberra (2002)
5. Deden, R.: Urban Planning and Informal Sector in Developing Countries (2007)
6. Majale, M., (ed.): Towards pro-poor regulatory guidelines for urban upgrading. a review of papers. Presented at the International Workshop on Regulatory Guidelines for Urban Upgrading Held at Bourton-on-Dunsmore, UK (2001)
7. Soyinka, O., Siu, K.W.M., (ed.): Urban informality and infrastructure planning in hong kong and lagos metropolis: professionals perspectives. In: International Conference on Applied Human Factors and Ergonomics. Springer (2017)
8. Soyinka, O., Siu, K.W.M.: Investigating informal settlement and infrastructure adequacy for future resilient urban center in Hong Kong. SAR Procedia Eng. **198**, 84–98 (2017)
9. Tam, I.Y.S.: Hidden Slum-Poor People in Rich Hong Kong (2012)
10. Rufina, Wu, Canham, S.: Portraits from Above-Hong Kong's Informal Rooftop Communities. MCCM Creations, Pepperoni Books, Berlin (2009)
11. Chui, E.: Rooftop Housing in Hong Kong: Hong Kong's Informal Rooftop Communities. Pepperoni Books, Berlin (2008)
12. Kennett, P., Mizuuchi, T.: Homelessness, housing insecurity and social exclusion in China, Hong Kong, and Japan. City, Cult. Soc. **1**(3), 111–118 (2010)
13. Hopper, K.: Reckoning with Homelessness. Cornell University Press, Ithaca (2003)
14. Ellen, I.G., O'Flaherty, B.: How to House the Homeless. Russell Sage Foundation, New York (2010)
15. Heben, A.: Tent City Urbanism: from Self-Organized Camps to Tiny House Villages. Village Collaborative, Eugene (2014)
16. Olajide, O.: Confronting the Lagos Informal Land Use: Issues and Challenges (2010)
17. Lukeman, Y., Bako, A., Omole, F., Nwokoro, I., Akinbogun, S.: Socio-economic attributes of residents of slum and shanty areas of Lagos State, Nigeria. Mediterr. J. Soc. Sci. **5**(9), 656 (2014)
18. Olajide, O.A.: Understanding the complexity of factors which influence livelihoods of the urban poor in Lagos' informal settlements (2015)
19. Baker, S.E., Edwards, R., Doidge, M.: How many qualitative interviews are enough?: Expert voices and early career reflections on sampling and cases in qualitative research (2012)
20. Mason, M., (ed.): Sample size and saturation in Ph.D. studies using qualitative interviews. Forum Qualitative Sozialforschung/Forum: qualitative social research (2010)

School Architecture: Components to Improve Quality and Sociability in a City in the Northeast of Brazil

Maria Juliana Morais[✉] and Terezinha Silva

Universidade Federal de Pernambuco,
Av. Professor Moraes Rego, 1235, Recife, PE 50670-901, Brazil
ju_morais86@hotmail.com, terezinhapsilva@hotmail.com

Abstract. This article brings a brief analysis of the educational world within the city of Macaparana, located in the Northeast of Brazil. Based on data collected in this county about the interaction between the community and the school equipment and its usage, this study looked for a better understanding required to create a set of guidelines in order to improve the architectonic model of the schools in the county, seeking to redesign its structure and make it more suitable to its broad service to the community: the learning and sociability of the space.

Keywords: Brasil · School · Community

1   Introduction

Largely discussed among pedagogics and the own learning composition process, the learning technique reaches a conclusion that goes beyond the basic formula: student; teacher; school structure. The relationship between pupils and the comfort of the physical space is increasingly intimate, guided by a playful and comfortable environment.

It was not a surprise to notice this additional usage of the space in the public schools of Macaparana, a county between the inlands and the northeaster coast of the state of Pernambuco, Brazil. Not because it presented examples of it, but for quite the opposite: from an architectonic point of view, the schools do not have an adequate structure to benefit the learning process. Many of them resemble houses or simply look inhospitable, with cold and unwelcoming colours, depreciating the potential of playfulness for learning. The social function of the school space can and should transcend the merely pedagogical education and become education and social inclusion, as an equipment for reference and interaction within the community where it exists.

This project analysis is based on architecture and urbanism criteria, with technical and influential rigor for the implementation of a pleasant school environment, suggesting changes to specific points based on guidelines that should humanize and socialize the space, bearing in mind the relationship between school spaces and the community, where the latter make use of it for gatherings and events on a regular basis.

© Springer International Publishing AG, part of Springer Nature 2019
J. Charytonowicz and C. Falcão (Eds.): AHFE 2018, AISC 788, pp. 303–311, 2019.
https://doi.org/10.1007/978-3-319-94199-8_29

Therefore, this project aims to contribute, under the architecture thinking, to the learning development and to a better reception for the ones who experience it.

Considering the important roles that schools play and that in the county of Macaparana they do not serve educational purposes only, but also as a venue for gatherings and events for the local community, this research aims to find a more suitable standard to meet social and educational needs.

2 Methodology

This research project intends to offer components to improve the quality and sociability in the county of Macaparana.

To compose this project, bibliographic research will be used, ranging from scientific papers, books and official data on the subject; as well as field research to understand the county needs. To do so, qualitative research aspects should be analysed and interpreted, since it involves the social scope.

Further on, the research will make use of case studies to bring project references, in accordance to the goals of a descriptive methodology. "The descriptive research consists of the study, the analysis, the register and the interpretation of the facts in the physical world without the interference of the researcher" (Barros e Lehfeld) [1].

As proposed by Lakatos and Marconi [2] in Fundamentals of Scientific Methodology, the outlines of a monographic elaboration base the research in intensive investigation to analyse the "facts that influenced it and the complete analysis of its aspects". From this point of view, the project should then make use of the inductive method, as described on its objective, leading to broader conclusions than the presented data.

As result, the material used for theoretical reference and the analysis of the county of Macaparana will point, among other aspects, to the guidelines which will lead the process which this research aims for.

3 The County of Macaparana and Its Schools

A 2010 research from Brazilian Institute of Geography and Statistics, IBGE [3], Macaparana is a city in the state of Pernambuco, that is located in the mesoregion of Mata Pernambucana and in the microregion Mata Setentrional Pernambucana. Distant 84 km from Recife, having as coordinates latitude 07°33'17" South and longitude 35°27'11" West, and is located at 350 m above sea level. The county area occupies 108.05 km^2, representing 0.10% of the State of Pernambuco, and has a population of 25,114 inhabitants and density of 221,43 hab./km^2.

In the of education field, the county has 31 elementary schools with 6230 students enrolled, and 04 high schools with 866 students. It comes to 121 classrooms: 04 state run schools, 77 are public operate schools at the city level and 04 private schools.

To better understand the dynamics of local education, the agent representative of the Education Department of Macaparana, Coutinho [4] and the technician in charge of the school buildings Cavalcanti [5], besides some school principals and students, were interviewed. Visits to the sights were also undertaken to comprehend how school structures work.

Even though the city has a club and a municipal gymnasium, during the weekends the schools have its recreational areas reserved for meetings, talks and local gatherings. The city hall runs a schedule to organise the events, by date and time of occurrence, and the busiest school is the Moura Cavalcanti, followed by other requested schools. To Coutinho [4], the preference for the schools lies on the fact that they have accessible furniture, while the Club and Gymnasium do not contain any of it and people would hence have to rent it, adding extra costs.

Currently, the public schools of the county of Macaparana present some issues, the main one being the acquisition of buildings designed for other usages to adequate to the architectonic program of a school. At many times, this process is made as an emergency, leaving a whole list of demands unattended. The new schools are built by Wladimir Cavalcanti, a building technician who works in the city hall and follows the guideline established by the program FUNDESCOLA and some requisites from the Secretary of Education. According to Cavalcanti [5], there is not a fixed guideline, not for choosing the lands nor the school projects, which are built and expanded according to its momentary needs.

Regarding school structure, the largest urban school of the county has 20 classrooms, the Governador Moura Cavalcanti school, and the smallest, 04 classrooms, the Antônio de M. Andrade school. However, it is relevant to mention the irregularities in these school spaces, often undersized designed and not meeting the needs for a comfortable environment.

To Cavalcanti [5], the local school structure present two major problems. The first one is the lack of organisation for expanding the schools, which happen randomly. Another issue is that the lands originally destined to schools, when not completely used, end up being donated to the construction of housing for the community. He claims that this affects the proper development and expansion of future school establishments.

Among students, the complaint about the lack of recreational areas was unanimous, since none of the schools have any playgrounds, as it has been mentioned, or playing fields. Besides, the patios, when present, virtually have no vegetation and relaxing areas (Figs. 1, 2 and 3).

Fig. 1. Patio from the municipal Antonio Coutinho school. Source: original.

Fig. 2. Patio from the municipal Governador Moura Cavalcanti school. Source: original

Fig. 3. Patio from the municipal Anexo school. Source: original

4   School Architecture and Its Social Function

In county of Macaparana, schools do not play only na educational role, but also work as a venue for gathering and events for the local community. That being said, following the urbanist idea that the city is a place to inhabit, work, circulate and entertain, it should also be a place to learn. Therefore, it is crucial the existence of a school structure which is compatible with the real needs of the local population.

Duarte [6], states that new schools, apart from seeking to answer a pedagogical idea, should work from the premise that the solutions interact with the place. "When we embody the organism, we find physical incidences that lead us to a range of solutions looking for harmonizing them with the given plan".

One of the trends discussed regarding school project and architecture is humanization, related to providing happiness through a quality experience of the space. Humanized architecture should have qualities that put emphasis on human needs, with low-rise buildings, tones of vegetation, variation and spatial order, possibility of manipulation by users, harmony of colours and ornamentation, usage of softer materials, low abrasion and proper care and maintenance [7].

In the social perspective, Funari and Kawaltowski [8], state that school building is an equipment of significant relevance within the social cultural and economic context of a country, since it is an important tool for the population education. Furthermore, when mentioning a developing country, such is the case of Brazil, which presents huge economic and social gaps that can be minimised by education. The physical configuration of the school environment and the settling of the pupil to it constitute a massive influence on the evolution of learning. Within this context, it is possible to say that the quality of school buildings play a relevant role in the social and economical development of a country.

Thus, it is crucial that the government take action in building public equipment which are relevant not only to the city, but mainly to the community where they are inserted. From this work of action it also becomes possible to guarantee access to education without leaving quality architecture aside, both visual and spatial, that can also be able to meet the needs and desires of the people of Macaparana.

5   Analysis and Guidelines to the School Architecture Project Aiming for Sociability in Macaparana

5.1   Analysis of the School Structure in Macaparana

Displayed equations or formulas are centered and set on a separate line (with an extra line or halfline space above and below). Displayed expressions should be numbered for reference. The numbers should be consecutive within each section or within the contribution, with numbers enclosed in parentheses and set on the right margin.

The municipal school chain of Macaparana uses the colours and the symbol of the City Hall. The colours chosen to represent the school are grayish blue-green and brown, cold and dull palettes which do not reflect the pleasant and relaxing environment of a school. The choice was made by the workers without any previous study, with a solo argument – the convenience of its maintenance.

The schools fronts are high-rise walls and some windows and doors, which means a total lack of relation with the surroundings and no visual permeability between the inside and the outside of the school. Not to mention that school fronts sometimes resemble housing structures, as a result of the adaptation of old houses to attend the educational program.

It is also important to highlight the undersized and rather dull recreational patios. In some schools it is limited to a big area of circulation, as seen in the Antônio Coutinho school (Fig. 1) Anexo School (Fig. 2). And when they do exist, the patios are completely paved, with nearly no vegetation and relaxing (Figs. 1, 2 and 3). The schools have recently undergone makeovers in order to adapt to the NBR 9050.

None of the schools, state or municipal, have any sports fields. It is notorious the lack of recreational areas in the schools. Its dry patios do not have a proper structure to benefit the interaction of the students with the environment. The dull colours, lack of vegetation, furniture and playgrounds cannot provide a playful environment to students, let alone comfort and leisure.

Despite all the poor school architecture space of the county, schools are used for weekend events (Fig. 4). This happens as a result of the basic structure provided – kitchen and furniture, even though it is not suitable for this purpose. The gratuity of the space is also factor.

Fig. 4. Event organised in the Municipal Severino Francisco school. Source: original.

We can, thus, conclude that it is notorious the vocation of the space for social interaction. However, for this aptitude to be cultivated, a series of makeovers would be necessary in order to humanise and facilitates the spatial interaction between the city and the school equipment.

5.2 Guideline for Humanization of School Architecture of Macaparana

To Funari and Kawaltowski [8], the six needs of the physical school environment are: comfort, to meet the sensorial needs of heat, light, sound and smell; territoriality and privacy; safety; spatial orientation and constancy; aesthetic visual stimulus and beauty; a range of sensorial stimulus.

Using the topics mentioned by Funari and Kawaltowski [8], the reality of the municipal education, its characteristics and needs, guidelines were created to improve the quality of the space in the municipal schools. Furthermore, the guidelines proposal seeks to be economically sustainable by making small changes of aesthetic and functional, without the need of major makeovers.

1. Change of the colour palette to a more suitable tone serving its purpose;
2. Implementation of gardens and green areas as a playful element, isolating private areas;
3. Create greater interaction between internal and external areas;
4. Installing a more attractive set of furniture in the patios and social areas;
5. Installing playful elements, made with recycled material for the recreational areas (e.g. a tire swing);

6. Favour the usage of the social spaces of the school, such as the library and patio for events, by the local community;
7. Accessibility should be guaranteed as in the NRB – 9050;
8. Schools should be in accordance with the noise restrictions as in the NBR – 10152 and lightning NBR – 5413;
9. The project should contemplate the rationalization of expenses through the proper choice of a building system, aiming for the control of budget planning and cost reduction;
10. Include the community in the process of planning and developing the new identity of the schools and, if possible, their participation in the execution of the project.

In collective spaces, such is the case of schools, these standards can create the image of a building associated with the local culture and the surrounding residents, something essential when the school community gets involved in the development of the architecture project of the new school. A feeling of belonging is created and it helps to value the resources the community invest in the school, avoiding possible correction costs once the construction has started, or after the beginning of the usage of the spaces [8].

5.3   The Project

The guidelines mentioned serve as guidance to improve the school architecture of the county and make the community own the structure, establishing a relation between the place and outer world instead of isolation, making it more flexible and possible to always reinvent itself.

In the project developed (Fig. 5), the schools has virtually no walls, and its outline delimitates outside space in a dynamic way, breaking orthogonality and rigidity, typical of large school groups, its patios communicate with the outside, favouring the interaction with the external area.

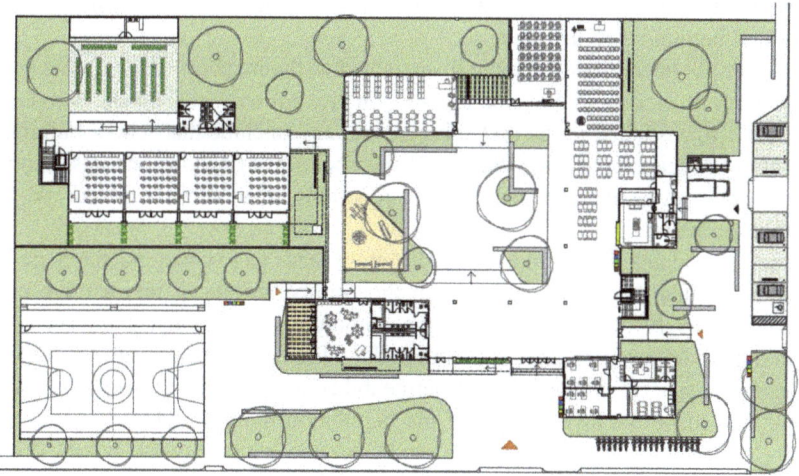

Fig. 5. The school project floorplan.

The central patio was used as a reference for the positioning of the blocks and will work as square for the students to gather in, besides being a venue for the community to get together in festivities. The patio has a free center to allow for a variety of activities, and on its vertices were created spaces that benefit the interaction between people. The building received large setbacks from the paths. These setbacks were made to create squares surrounding the school, turning it into an element of reference and sociability among residents. Two big squares were created, the Access Square and the Sports Square, where the sports court and other spaces are located.

Regarding the classrooms, its square shape allows different configurations. It is important that this space can adapt to different pedagogical activities, making it possible to set a variety of internal space organisations, that is, in groups, circles, lines and others.

Wishing to achieve a better structure, modernization and urban interaction, to para "[...] Reinforce the social function of space in the city as a gathering space that contributes to the goals of social sustainability and for a democratic and open society [9]. Thus, with the model developed it was possible to see improvements in the school space not only for the pedagogical process, but also for the interaction with the community that tries to benefit more from its structure, bringing vitality and a feeling of belonging.

6   Conclusion

Based on what has been collected as data, the schools of the county share a mixture of learning center and the feeling of belonging to the space, since its physical spaces are also used for gatherings and events: a way to humanise and facilitate social interaction.

From this understanding, changes have been suggested. In fact, the guidelines for the improvement of the space. Lightning, landscaping, colours and alterations in the furniture intend to improve sociability and human relation. This way we have an image associated with the local culture from those who use it, hoping to establish a feeling of identity and belonging to the place.

For what has been proposed to be investigated as the main objective of this project, with the analysis seeking to question its functionality not only in the usual and physical aspect, but also aesthetic, social and cultural, we can conclude that school buildings of the county of Macaparana follow the established concept of making schools, that is, an equipment that serves its basic pedagogical functions, even if has a certain use by its population in events.

Therefore, the analysis conducted on this work leads us to conclude that there is the need to strengthen the school structure in the county, turning it into a more playful space wich favours interaction within the community. This objective was reached through the guidelines proposed, which intends to humanise the educational buildings, creating green spaces, dynamism and permeability to the form and establishment of elements that make the involvement of the population more present.

References

1. Barros, A.J.P., Lehfeld, N.A.S.: Fundamentos da metodologia científica, 3ª edn. Makron, São Paulo (2007)
2. Lakatos, E.M., Marconi, M.A.: Fundamentos de metodologia científica, 5th edn. Atlas, São Paulo (2003)
3. IBGE. http://www.ibge.gov.br/cidadesat/painel/painel.php?codmun=260900#topo. Acessado 21 Oct 2016
4. Coutinho, M.: Agent representative of the Education Department of Macaparana. Interviewed on 13th September 2016
5. Cavalcanti, W.: Building Technician. Interviewed on 13th September 2016
6. Duarte, H.: O problema escolar e a arquitetura, no 4. Revista Habitat, São Paulo (1951)
7. Kowaltowski, D.C.C.K.: Arquitetura escolar: o projeto do ambiente de ensino. Oficina de textos, São Paulo (2011)
8. Funari, T.B.S., Kowaltowski, D.C.C.K.: Arquitetura escolar e avaliação pós- ocupação. In: VIII Encontro Nacional de Conforto no Ambiente Construído/IV Encontro Latino-Americano de Conforto no Ambiente Construído, Maceió. Associação Nacional de Tecnologia do Ambiente Construído, Maceió (ANTAC) (2005)
9. Gehl, J.: Cities for people, 3rd edn. Perspectiva, São Paulo (2015)

Barriers of Building Maintenance in Private Tertiary Institutions in Nigeria

Douglas Aghimien[1](✉), Ayodeji Oke[2], and Clinton Aigbavboa[1]

[1] Sustainable Human Settlement and Construction Research Centre,
Faculty of Engineering and the Built Environment, University of Johannesburg,
Johannesburg, South Africa
aghimiendouglas@yahoo.com, ciagbavboa@uj.ac.za
[2] Department of Construction Management and Quantity Surveying,
University of Johannesburg, Johannesburg, South Africa
aoke@uj.ac.za

Abstract. Maintenance of educational buildings is crucial if quality and sustainable education is to be delivered. However, while studies have emanated as regards maintenance issues of public educational buildings in Nigeria, there seem to be a deficiency in literature as regards maintenance happenings within the private tertiary institutions. This paper therefore presents the findings of the assessment of the barriers of building maintenance in private tertiary institutions in the country. The study adopted a quantitative approach and information were harnessed from maintenance officers and construction professionals within two private tertiary institutions. Data gathered were analyzed using percentage, mean score, standard deviation, and factor analysis. Findings revealed that the maintenance units of these institutions are characterized by inexperienced staffs, while the most used maintenance approach is the corrective maintenance. The factors affecting building maintenance are mostly; building design and usage factors, policy and management factors, and human resources and funding factors.

Keywords: Building design · Building elements · Maintenance management
Higher education institutions · Tertiary institutions

1 Introduction

Just as humans need care, buildings require proper maintenance for them to survive their expected life span and serve their intended purpose effectively. The regular and correct maintenance of a building is critical to their conservation. This assertion is affirmed by the increasing number of researches being carried out in the area of building management and maintenance [1–4] just to mention a few. It has been established that buildings play a vital role in the delivery of quality education, and their function goes beyond just serving the present generation [5, 6]. Its importance in the education sector and the condition of the indoor, in providing comfort and satisfaction for its users have been pointed out in previous researches [7–9]. It is believed that to continually achieve this stated comfort, maintenance of the building is crucial. According to Abdul Lateef [10] educational buildings, especially higher institution buildings, requires maintenance if a

© Springer International Publishing AG, part of Springer Nature 2019
J. Charytonowicz and C. Falcão (Eds.): AHFE 2018, AISC 788, pp. 312–323, 2019.
https://doi.org/10.1007/978-3-319-94199-8_30

conducive environment that supports and stimulates learning, teaching, innovation, and research is to be achieved. BSI [11] stated that this building maintenance is a combination of any actions carried out to retain the building in, or restore it to an acceptable condition in which it can perform its required function. In the view of Abisuga *et al.* [12] this definition clearly points to the fact that maintenance is a crucial part of any organization's existence in relation to its asset management.

Edmond *et al.* [13] in a study on "benchmarking success of building maintenance projects" observed that maintenance of existing building assets is a top priority in most client organisations. Most organisations have realized the need for maintenance in their daily activities. However, the same is not the case with government organizations in Nigeria. Abigo *et al.* [14] noted that the government seems to be more interested in setting up buildings than maintaining the old ones, thus leaving the old buildings to rot and decay. Despite the importance of maintenance in achieving sustainable educational buildings - buildings that serve the present and future generations - the case of maintenance in most government owned higher institutions in the country is appalling. According to Adenuga *et al.* [15], public buildings in the country, including higher institution buildings, are in very poor and deplorable conditions. Although millions of Naira is spent to erect these buildings, they are left, immediately after commissioning to face steady and rapid deterioration, decay and dilapidation.

The poor state of most government owned higher institution buildings in the country and dissatisfaction among users of these buildings have been pointed out in past studies [16–20]. In similar vein, studies on the maintenance management of these institutional buildings are equally evident [1, 12, 21]. However, these studies placed focus on government owned institutions. Understanding issues surrounding building maintenance in private institutions is also important, since the growth of private institutions in the country is becoming overwhelming. As at 2012, Nigeria had one hundred and twenty-two universities, and 43% of them were private owned [22]. Adama *et al.* [23] believe that this number today would have increased with more privately owned institutions springing forth in the country. Thus, understanding the factors affecting building maintenance practices in these private institutions also is necessary, since quality education depends on having conducive and well-maintained building facilities. Based on this background, this study assessed the barriers to proper building maintenance within private tertiary institutions in Nigeria, with a view to proffering possible solutions towards increasing maintenance activities in tertiary institutions in the country.

2 Literature Review

BSI [11] defined maintenance as "the work or a combination of actions associated with initiation, organisation and implementation carried out to retain an item in or restore it to an acceptable standard". Smith [24] stated that the maintenance of the built environment has a continuous effect on everyone, since we depend on the state of our homes, offices and factories not only for our comfort, but also for our economic survival. Building maintenance could be classified into reactive and proactive maintenance; however, there have been cases of the combination of these two approaches in practice. While the reactive maintenance represents the response to the identified problems or work only when requested, proactive maintenance is carried out to avoid the occurrence of any problem in the nearest future. The reactive approach is advantageous as it reduces the total time and cost to a minimum, and it can be integrated with other maintenance. The proactive maintenance is gradually becoming a vital part of building management. Its adoption can assist in the production of mechanisms that can help avoid and/or monitor these components' processes of becoming progressively worse by adopting some of the maintenance actions that are both practical and cost-effective [25]. Unfortunately, Mukelasi et al. [26] carried out a review of critical success factors in building maintenance management of local authority in Malaysia and observed that maintenance approach adopted is more of reactive rather than proactive in nature. A similar scenario was observed in Nigeria and Ghana [15, 27, 28].

Horner et al. [29] in the study of building maintenance strategies observed three different strategic approaches of maintenance, which are, corrective, preventive and condition-based maintenance. Lee and Scott [30] made a similar observation in the study of acceptable maintenance standard and resources from a building maintenance operation perspective. It was noted that maintenance strategy in general includes corrective, preventive or maintenance based on condition. The preventive maintenance which is a planned maintenance approach and involves a time-based maintenance, requires regular task of maintenance irrespective of the condition of the structure. The corrective maintenance is an unplanned maintenance approach, and it is a failure-driven maintenance referring to running building and its associated facilities, until unexpected event breakdown of these facilities or malfunctioning. The condition-based maintenance also entails periodic inspection of building facilities to check and replace when a faulty condition is observed before breakdown can occur.

The poor conditions of buildings and auxiliary facilities in most countries around the world most especially the developing ones has been related to maintenance issues. Several factors relating to these poor maintenance issues have been proposed over the years. Iyagba [31] noted that the sufficiency of the design, constructional details and the methods of building construction determines the usefulness of buildings. It also determines the way that the building is used and the maintenance policies and practice undertaken during its life. In Iraq, Dakhil et al. [25] assessed the barriers to the implementation of proactive building maintenance. It was observed that the major barriers can be categorized into - technical, management and administration, financial, human behaviour and attitudes, spare parts, and lack of institutional and training facilities. More specifically, factors such as corruption, lack of standards, quality control, and misuse of the building were dominant in none implementation of proactive maintenance in the country. Issahaku [28] noted that the factors affecting building maintenance in Ghana include; insufficient fund for maintenance job, persistent breakdown through indiscipline and ignorance of building users, attitude of users and misuse of facilities, lack of skilled manpower to maintain work in buildings.

In Nigeria, Asiabaka [32] noted the issue of lack of policy guidelines for infrastructural development and maintenance in buildings in Nigeria, is a major problem affecting building management and maintenance. Mohammed and Hassanain [33] submitted that most buildings, especially the residential buildings, house more than the specified number of users than the building had been designed for. This puts a strain on the building and its services, thus making maintenance a bit difficult. Ogunmakinde et al. [34] carried out an analysis of the factors affecting building maintenance in government residential estates in Akure, Ondo state, and observed that funding of maintenance work is a crucial factor affecting this process. It study discovered that maintenance is only carried out whenever a fault is detected. This implies that corrective maintenance is majorly adopted in the area. While maintenance management of a building is the duty of the owner, users also have a duty to care for these structures. It was observed that maintenance of building has received little attention from their users. These users make poor use of these properties, which they believe that its maintenance is up to the owner. Olarenwaju et al. [35] carried out an assessment of the challenges of building maintenance in some public and private buildings in Nigeria and found that the use of untested or inferior materials was the major factor causing building deterioration. The availability of qualified and competent construction professionals was discovered to be the most significant factor that would affect the drive to achieve quality of maintenance operations in the country.

3 Research Methodology

The study assessed the barriers of building maintenance in private tertiary institution in Nigeria. The study employed a quantitative survey approach and data were gathered through questionnaire administered on maintenance officers within two (one polytechnic and one university) private tertiary institutions in Osun State. Data were gathered through a questionnaire survey conducted among members of the maintenance unit, and Works and Physical Planning Department of these institutions. The research instrument used was structured questionnaire designed based on information gathered from the review of related literatures. The questionnaire was designed in two parts. Part A dwelt on the background information of respondent. Information gotten from this section provides quality check to the data gotten from the other section of the research instrument. Part B dwelt on questions geared towards identifying the major maintenance approaches adopted in these institutions, and the barriers of building maintenance in the institutions. A 5-point Likert scale was adopted in the study. A total of 54 questionnaires were distributed, and they were all retrieved and ascertained fit for analysis. The reason for the 100% response rate is because the total population of the study falls within a manageable and reachable size.

The reliability of the instrument was tested using Cronbach's alpha test. This method is used to measure the reliability of the questionnaire between each field and the mean of the whole fields of the questionnaire. The normal range of Cronbach alpha value is between 0.0 and +1.0, and the higher value, the higher degree of internal consistency. The Cronbach alpha value of 0.801 was derived for the barriers of building maintenance in the identified institution. This result shows that the instrument is reliable since the degree of reliability of an instrument is more perfect as the value tends towards 1.0 [36]. Data analyses was done using percentile in analyzing the background information of the respondents, while mean score was used to rank the identified building maintenance approaches. Factors Analysis was employed in analyzing the barriers of building maintenance.

4 Findings and Discussions

4.1 Background Information of Respondents

Result on the background information of the respondents shows more male (71.9%) than female (28.1%). Majorities (59.4%) of the respondents are not construction professionals, nor do they have prior knowledge of building maintenance before being drafted into the maintenance unit of these institutions. About 69% stated that they have not been involved in any building maintenance related activities prior to their engagement as maintenance officers. In terms of number of years working with the maintenance unit, most of these respondents (87.5%) have spent just between 1 to 2 years in the unit. The average years of working in the maintenance unit is 1.8 years. This result points to the fact that the maintenance unit of these institutions is filled with "greenhorns". This situation is most like to affect the manner at which maintenance work will be carried out by these individuals, as they are not vast in the act of building maintenance nor do they have reasonable prior knowledge of it. This result is similar to the submission of [21] who observed that in public tertiary institutions in Nigeria, there is a structure in place to carryout maintenance work but it is characterized by inexperienced staff. Thus, there is no difference in the maintenance structure of educational buildings in both private and public institutions in the country. There is therefore the need for critical evaluation of those charged with the maintenance of educational buildings it in the country.

4.2 Approach of Maintenance Adopted in Private Tertiary Institutions

As observed from literature, maintenance approach has been categorized in different ways by several authors. However, this study employs the classification of [29, 30] which state that the approaches to maintenance can be preventive, corrective or condition-based. These three approaches were presented to the respondents and they were asked to rank them based on their level of usage. For clarity, a brief definition of these three approaches was given in the questionnaire so as to enable the respondents adequately answer the stated question.

Result in Table 1 shows that there is a considerable amount of consistency in the view of the respondents as regards the level of usage of these maintenance approaches, as a Standard Deviation (SD) of below 1.0 was derived for all the approaches assessed. The most used approach is the corrective maintenance with a mean value of 3.78 and a SD of 0.975. This is followed by condition-based maintenance with a mean value of 3.28 and SD of 0.991. Preventive maintenance is the least used approach with a mean value of 2.97 and SD of 0.782. This result implies that maintenance work in the institution are mostly reactive than proactive. This finding is in line with [34] submission in public buildings in Nigeria that maintenance is carried out only when a fault is detected. The finding also agrees with [17] submission that building maintenance management practices of higher education institutions in the country is mostly corrective in nature. This situation is not peculiar to public and educational buildings in Nigeria alone as [25], [28] and [26] has observed a similar situation in Ghana, Iraq and Malaysia.

Table 1. Level of usage of maintenance approaches

Approach	Mean	Rank	SD
Corrective	3.78	1	0.975
Condition-based	3.28	2	0.991
Preventive	2.97	3	0.782

4.3 Factors Affecting Building Maintenance in Private Tertiary Institutions

In assessing the key factors affecting building maintenance in the institution, some factors were identified from the review of literature and respondents were asked to rank them based on their level of significance. The data gathered was subjected to factor analysis (FA) in order to analyze and group the identified factors into more manageable and significant size. For this to be done adequately, the suitability of the gathered data was first determined. Firstly, the sample size and number of variables under study was considered. Zhao [38] conducted a study for the minimum sample size in FA and discovered that several authors have proposed several sample sizes. Preacher and MacCullum [39] suggested that as long as the communalities are high, the number of expected factors is relatively small, and model error is low, researchers and reviewers should not be concerned about small sample size. Zhao [38] went further to suggest a communalities figure of above 0.6 as being suitable irrespective of the sample size being adopted. Result from the communalities test conducted revealed that eleven out of the thirteen assessed factors have communalities figure of 0.6 and above. Another suitability factor considered in this study is the number of variables being assessed. Hair et al. [40] suggested that FA is most suitable for 20 to 50 variables; however, there has been evidence of FA being conducted on lesser number of variables [41]. Thus, it can arguably be stated that the data gathered in this study is suitable for FA, based on the number of variables, in relation to the sample size, and the communalities figure obtained.

Kaiser–Meyer–Olkin measure of sampling adequacy (KMO) and Bartlett's test of sphericity conducted gave a 0.684 value for KMO and significant level of 0.000 for Bartletts test. Tabachnick and Fidell [42] stated that the KMO index ranges from 0 to 1, with 0.6 suggested as the minimum value for a good factor analysis. Also Pallant [43] submitted that Bartlett's test of sphericity shows whether the correlation matrix is an identity matrix and this should be significant ($p < 0.05$) for FA to be considered appropriate. The result of the KMO and Bartletts test derived in this study, coupled with the 0.801 value obtained from the reliability test carried out, also proves that the use of FA for the data gathered is appropriate.

Having established that the data gathered met all the necessary requirement, FA was conducted using principal component analysis (PCA) with varimax rotation. Result revealed that three components with eigenvalues greater than 1 were extracted with 0.50 set as the cut-off point for the factor loading. The final statistics of the PCA and the components extracted accounted for approximately 59% of the total cumulative variance. This fulfils the criterion of factors explaining at least 50% of the variation [44]. Result in Table 2 shows the three extracted components and the factors loading on them. According to Spector [45] a clear component structure is present when a variable has significant factor loading (loading > 0.50) on one component only. Hence, only factors with 0.5 and above are deemed significant and are considered for discussion under each principal component.

Table 2. Rotated component matrix

Factors	Component		
	1	2	3
Poor initial planning	0.844		
Faulty design and construction	0.796		
Lack of communication	0.677		
Poor maintenance culture of the building users	0.641		
Inadequate facilities usage information	0.572		
Age of the building	0.437		
Problem of policy implementation		0.784	
Lack of support from institution management		0.737	
Corruption		0.663	
Insufficient maintenance personnel			0.762
Poor technical knowhow of maintenance personnel			0.729
Lack of funds			0.530
Lack of incentives for maintenance personnel			0.505

4.4   Discussion of Extracted Factors

Building Design and Usage Barriers

The first principal component has the highest factor loading of five factors and it account for 32.3% of the total variance explained. These factors include; poor initial planning, faulty design and construction, lack of communication between users and maintenance personnel, poor maintenance culture of the building users, and inadequate facilities usage information. Considering the nature of these factors, the component was therefore named "building design and usage barriers".

Iyagba [31] noted that the sufficiency of the design, constructional details and the methods of building construction determines the usefulness of buildings. It also determines the way that the building is used and the maintenance policies and practice undertaken during its life. Waziri [46] made a similar observation by stating that the use of defective construction materials, poor supervision, non-compliance with specifications and poor quality control on site during construction are the most significant factors affecting building maintenance in Nigeria. It has been observed that designers in most cases are unaware of the consequences of their designs until during post-occupancy survey [47]. It is therefore important to consider maintenance at both design and construction stages by incorporating maintenance variables in order to reduce subsequent maintenance effort during occupancy.

Aside these design issues, the usage of buildings has also been noted to have significant effect on their maintenance. Issahaku [28] observed that in Ghana, building maintenance is affected by persistent breakdown of building services due to indiscipline and ignorance of building users, attitude of users and misuse of facilities. Dakhil *et al.* [25] also mentioned that technical issues relating to building designs, and human behaviour and attitudes in terms of usage of the facilities are affecting proper maintenance in Iraq. A similar observation was made by [33, 34] in Nigeria.

Policy and Management Barriers
The second principal component accounts for 13.9% of the total variance explained. The factors under this component includes; problem of policy implementation, lack of support from institution management, and corruption among maintenance personnel and management. This component was further named "policy and management barriers" based on the characteristics of its factors. Lee [48] sees maintenance policy as the ground rules for allocation of resources between the alternative types of maintenance actions that are available to management. According to Seeley [49] the benefits of a maintenance policy is the ability to formulate a long term maintenance strategy and prepare budgetary forecasts. However, Asiabaka [32] noted the issue of lack of policy guidelines for infrastructural development and maintenance in buildings in Nigeria, is a major problem affecting building management and maintenance. Dakhil *et al.* [25] also found management and administration issues such as corruption and lack of standards to be a major challenge to effective maintenance in Iraq. Therefore, if building maintenance is to increase within the education sector and indeed all aspect of the nation's economy where construction products are required, the Government must be proactive in championing its course. In terms of the education sector, institution management must also be ready and willing to give special attention to maintenance of their institutional facilities.

Human Resources and Funding Barriers
The last principal component extracted accounts for 12.6% of the variance explained and has four factors loading on it. These factors are insufficient maintenance personnel, poor technical knowhow of maintenance personnel, lack of funds, and lack of incentives for maintenance personnel. Considering the latent characteristics of these factors, this component was subsequently named "human resources and funding barriers".

Abdul Lateef [10] carried out an appraisal of the building maintenance management practices of Malaysian universities and observed that expenditure on maintenance is inadequate. Issahaku [28] also observed that issues surrounding funding have significant effect on building maintenance in Ghana. Dakhil *et al.* [25] in Iraq made a similar observation. Ogunmakinde *et al.* [34] also observed the issue of lack or insufficient fund allocation allowed for maintenance is a major factor affecting effective maintenance of buildings in Nigeria. The finding of this present study is in-line with these submissions as lack of funds is seen as a major factor deterring proper maintenance practice within the identified institution. Aside financial issues [28] also noted that lack of skilled work force to maintain work in buildings is also a major problem affecting maintenance of buildings. Ofide *et al.* [37] also noted maintenance staff shortage as a major factor affecting building maintenance. If poor maintenance issue is to reduce, proper funding is therefore necessary. Also Olarenwaju *et al.* [35] believed that the availability of qualified and competent construction and maintenance professionals is a significant factor that would affect the drive towards achieving quality maintenance operations.

5 Conclusion and Recommendations

This study set out to assess the barriers of building maintenance in private tertiary institutions in Nigeria. Using a survey approach with quantitative data gathered from maintenance officers within private tertiary institutions, the study has been able to determine the characteristics of the maintenance unit in the institution, the maintenance approaches mostly adopted, and the factors affecting building maintenance in the institution.

Based on the findings, the study concludes that although there is a maintenance unit within private institutions, they are characterized by inexperienced staffs, while corrective maintenance is the most common approach being adopted. The barriers of building maintenance can be categorized under building design and usage factors, policy and management factors, and human resources and funding factors.

If building maintenance is to increase in terms of performance within the education sector and indeed the entire country, then there is the need for more capable hands. Only maintenance professionals with considerable experience should be employed within maintenance units. In addition, there is the need for a radical shift from corrective maintenance to a more proactive approach. This will help improve the life span of structures within the institutions and help reduce cost spent on total replacement of building facilities, as faults would have been detected and corrected earlier. Although private institutions are privately operated, Government can still help in ensuring standard and quality education is delivered through adequate maintenance, by creating policies that will encourage the adoption of proactive maintenance culture and creating avenues for the enforcing of it. Building designers should also painstakingly design buildings and give thorough details and specifications, while during construction, proper monitoring should be done to ensure buildings are constructed to specification. In addition, users should understand that they have a duty to care for these structures. In the case of tertiary institutions, sensitizing the students and staffs as regards the need for proper usage of school structures is necessary.

This study has been able to contribute to the body of knowledge as it has bring to light the maintenance management of buildings within private tertiary institutions in Nigeria; an aspect deficient in the discussions of building maintenance within the country. It is believed that the findings of the study will assist top management of tertiary institutions within the country in making crucial decisions regarding the maintenance of buildings within their institutions. Findings of the study provide direction for future research as it was limited to two private tertiary institutions. Similar studies can be conducted in other private tertiary institutions around the country in order to compare results.

References

1. Abisuga, A.O., Famakin, I.O., Oshodi, S.O.: Educational building conditions and the health of users. Constr. Econ. Build. **16**(4), 19–34 (2016)
2. Falorca, J., Rodrigues, R.C., Da Silva, R.M.: Research measurement of knowledge advances in building maintenance issues. Struct. Surv. **32**(1), 61–71 (2014)
3. Leung, M.Y., Fung, I.: Enhancement of classroom facilities of primary schools and its impact on learning behaviors of students. Facilities **23**(13/14), 585–594 (2005)
4. Ogungbile, A.J., Oke, A.E.: Assessment of facility management practices in public and private buildings in Akure and Ibadan cities, south-western Nigeria. J. Facil. Manag. **13**(4), 366–390 (2015)
5. Abdul Lateef, O.A., Khamidi, M.F., Idrus, A.: Appraisal of the building maintenance management practices of Malaysian universities. J. Build. Apprais. **6**, 261–275 (2011)
6. Aghimien, D.O., Awodele, O.A., Aghimien, E.I.: Providing sustainability in educational buildings through the use of compressed stabilized interlocking earth blocks. J. Constr. Eng. Technol. Manag. **6**(2), 130–140 (2016)
7. Kok, H.B., Mobach, M.P., Omta, S.W.F.: The added value of facility management in the educational environment. J. Facil. Manag. **9**(4), 249–265 (2011)
8. Lavy, S.: Facility management practice in higher education buildings: a case study. J. Facil. Manag. **6**(4), 303–315 (2008)
9. Uline, C., Tschannen-Moran, M.: The walls speak: the interplay of quality facilities, school climate, and student achievement. J. Educ. Adm. **46**(1), 55–73 (2008)
10. Abdul Lateef, O.A.: Quantitative analysis of criteria in university building maintenance in Malaysia. Australas. J. Constr. Econ. Build. **10**(3), 51–61 (2010)
11. British Standard Institution: BS 3811 - Glossary of maintenance terms in terotechnology. British Standard Institution (BSI), London (1984)
12. Abisuga, A.O., Ogungbemi, A.O., Akinpelu, A.A., Oshodi, O.S.: Assessment of building maintenance projects success factors in developing countries. J. Constr. Bus. Manag. **1**(1), 29–38 (2017)
13. Edmond, W.M., Lam Albert, P.C., Chan Daniel, W.M.: Benchmarking success of building maintenance projects. Facilities **28**(516), 260–305 (2010)
14. Abigo, A., Madgwick, D., Gidado, K., Okonji, S.: Embedding Sustainable Facilities Management in the Management of Public Buildings in Nigeria (2012)
15. Adenuga, O.A., Olufowobi, M.B., Raheem, A.A.: Effective maintenance policy as a tool for sustaining housing stock in downturn economy. J. Build. Perform. **1**(1), 93–109 (2010)
16. Ajayi, M., Nwosu, A., Ajani, Y.: Students' satisfaction with hostel facilities in federal university of technology, Akure, Nigeria. Eur. Sci. J. **11**(34), 402–415 (2015)
17. Akinpelu, O.P.: Students' assessment of hostel facilities in the polytechnic Ibadan, Ibadan, Nigeria: realities and challenges. Res. Humanit. Soc. Sci. **5**(17), 74–81 (2015)
18. Amole, D.: Residential satisfaction in students' housing. J. Environ. Psychol. **29**(1), 76–85 (2009)
19. Edukugho, E.: Education sector stinks! Infrastructure bad, now worse. Vanguard Newspaper (2013). http://www.vanguardngr.com/2013/12/education-sector-stinks-infrastructure-bad-nowworse/
20. Sawyerr, P.T., Yusof, N.: Student satisfaction with hostel facilities in Nigerian polytechnics. J. Facil. Manag. **11**(4), 306–322 (2013)
21. Olatunji, S.O., Aghimien, D.O., Oke, A.E., Akinkunmi, T.: Assessment of maintenance management culture of tertiary institutions in Nigeria. Civil Environ. Res. **8**(6), 98–105 (2016)

22. Okojie, J.: The Punch Newspaper, Wednesday, February 29th, p. 9 (2012)
23. Adama, J.U., Aghimien, D.O., Fabunmi, C.O.: Students' housing in private universities in Nigeria: influencing factors and effect on academic performance. Int. J. Built Environ. Sustain. **5**(1), 12–20 (2018)
24. Smith, R.: Best maintenance practices. J. Maint. Maint. Manag. **16**(1), 9–15 (2003)
25. Dakhil, A., Qasim, I., Chinan, J.: Barriers for implementation of proactive building maintenance in Iraq: Basra City as case study. Kufa J. Eng. **8**(1), 26–36 (2016)
26. Mukelasi, M.F.M., Zawawii, E.M.A., Kamaruzzaman, S.N., Ithnin, Z., Zulkaranain, S.H.: A review of critical success factors in building maintenance management of local authority in Malaysia. In: IEEE Symposium on Business, Engineering and Industrial Applications, pp. 653–657 (2012)
27. Adenuga, O.A.: Evaluation of maintenance management practice in public hospital buildings in south west, Nigeria. Ph.D. thesis, Department of Building, University of Lagos, Nigeria (2008)
28. Issahaku, M.I.: Evaluation of maintenance management practices in Ghana highway authority's bungalows in greater Accra region. Masters dissertation submitted to the Department of Building Technology, Kwame Nkrumah University of Science and Technology, Ghana (2013)
29. Horner, R.M.W., El-Haram, M.A., Munns, A.K.: Building maintenance strategy: a new management approach. J. Qual. Maint. Eng. **3**(4), 273–280 (1997)
30. Lee, H.H., Scott, D.: Overview of maintenance strategy, acceptable maintenance standard and resources from a building maintenance operation perspective. J. Build. Apprais. **4**(4), 269–278 (2009)
31. Iyagba, R.O.A.: The menace of sick buildings – a challenge to all for its prevention and treatment. An Inaugural Lecture Delivered at University of Lagos, Lagos (2005)
32. Asiabaka, I.P.: The need for effective facility management in schools in Nigeria. N. Y. Sci. J. **1**(2), 10–21 (2008)
33. Mohammed, M.A., Hassanain, M.A.: Towards improvement in facilities operation and maintenance through feedback to the design team. Built Hum. Environ. Rev. **3**, 72–87 (2010)
34. Ogunmakinde, O.E., Akinola, A.A., Siyanbola, A.B.: Analysis of the factors affecting building maintenance in Government residential estates in Akure, Ondo state, Nigeria. J. Environ. Sci. Resour. Manag. **5**(2), 89–103 (2013)
35. Olanrewaju, S., Babatunde, O., Anifowose, O.: The challenges of building maintenance in Nigeria: (A case study of Ekiti state). Eur. J. Educ. Dev. Psychol. **3**(2), 30–39 (2015)
36. Moser, C.A., Kalton, G.: Survey Methods in Social Investigation, 2nd edn. Gower Publishing Company Ltd., Aldershot (1999)
37. Ofide, B., Jimoh, R., Achuenu, E.: Assessment of building maintenance management practices of higher education institutions in Niger State – Nigeria. J. Des. Built Environ. **15**(2), 1–14 (2015)
38. Zhao, N.: The Minimum Sample Size in Factor Analysis (2008). https://www.encorewiki.org/plugins/servlet/mobile#content/view/25657
39. Preacher, K.J., MacCallum, R.C.: Exploratory factor analysis in behaviour genetics research: factor recovery with small sample sizes. Behav. Genet. **32**, 153–161 (2002)
40. Hair, J.F., Anderson, R.E., Tathan, R.L., Black, W.C.: Multivariate Data Analysis. Prentice Hall, Upper Saddle River (1998)
41. Ahadzie, D.K., Proverbs, D.G., Olomolaiye, P.O.: Critical success criteria for mass house building projects in developing countries. Int. J. Project Manag. **26**, 675–687 (2008)
42. Tabachnick, B.G., Fidell, L.S.: Using Multivariate Statistics, 5th edn. Pearson Education, Boston (2007)

43. Pallant, J.: SPSS Survival Manual: A Step by Step Guide to Data Analysis Using SPSS for Windows (Version 12), 2nd edn. Allen and Unwin, Crows Nest (2005)
44. Stern, L.: A Visual Approach to SPSS for Windows: A Guide to SPSS 17.0, 2nd edn. Allyn and Bacon, Boston (2010)
45. Spector, P.: Summated Rating Scale Construction: An Introduction. Sage Publications, Newbaury Park (1992)
46. Waziri, B.S.: Design and construction defects influencing residential building maintenance in Nigeria. Jordan J. Civil Eng. **10**(3), 313–323 (2016)
47. Chohan, A.H., Che-Ani, A.R., Memon, Z., Tahir, M.M., Abdullah, N.K.G., Ishak, N.H.: Evaluation of user satisfaction towards construction faults in medium-cost housing of under-developing metropolis. Am. J. Sci. Res. **13**(1), 6–17 (2011)
48. Lee, R.: Building Maintenance Management. Blackwell Oxford Ltd., Oxford (1987)
49. Seeley, I.H.: Building Maintenance. Macmillan Press Ltd., London (1987)

Creative Cluster and Urban Rehabilitation: Case Study in Northeast Brazil

Christianne Soares Falcao[(⊠)] and Alberico Paes Barreto Barros

Creative Industries, Catholic University of Pernambuco,
Rua do Principe, 526, Boa Vista, Recife, PE, Brazil
christiannefalcao.arq@gmail.com

Abstract. Nowadays, terms such as *Creative Economy* and *Creative City* get criticized in the planning of cities in the face of the emergence of a new economy based on the valuation of knowledge, innovation and creativity as economic assets. Based on this understanding, urbanism concepts, and economic and human relations are interrelated. This paper aims to help better understand the relationship between the implantation of a *Creative Cluster* and the development of the urban space, based on the conceptual analysis of Creative City and the observation of a specific case in the city of Recife, Brazil, the Porto Digital, envision the possibility of transforming this city into "creative environment" together with the formation of innovation core, and the creation and diffusion of knowledge and supports answering questions related to urban rehabilitation as a way of transforming existing urban environments.

Keywords: Porto digital · Creative cities · Creative economy

1 Introduction

Faced with the convergence of several factors such as globalization, there is growing presence of communication and information technologies, the possibility of capital transfers between countries, among others, the discussion on "creative economics" began in countries such as Australia, the United Kingdom, the United States and Canada, by professionals related to economics, sociology and urbanism. In Brazil, this debate began only in the first decade of the 21st century, when the economic and cultural benefits generated by this new global economy were already emerging.

The term *creative economy* covers the entire business environment that exists around the *creative industry*, based on goods and *creative services*. In the current context, creativity becomes the most active component for innovation, the center of economic policies, assuming prime importance as the main basis for competitiveness and growth of new economy.

Based on this, creative economy establishes the connection between urban discussions and their historical, economic and social context in contemporary times, the formation of the concept of creative cities emerges, emphasizing initially the importance of creativity for the city and its economy. Related studies started from a different paradigm, but are converging to complementary influence. Jane Jacobs, an American journalist, already pointed out in 1961 that urban centers, as places of concentration of

© Springer International Publishing AG, part of Springer Nature 2019
J. Charytonowicz and C. Falcão (Eds.): AHFE 2018, AISC 788, pp. 324–330, 2019.
https://doi.org/10.1007/978-3-319-94199-8_31

people as well as cultural and population diversity, favor the emergence of ideas and the processes of innovation [1]; John Howkins, an English journalist, points out that creativity is a fundamental part of economic competitiveness and that the individuals involved should have a fair return on their ideas [2]; Richard Florida, an American economist, says cities that claim to be "creative" need to attract a new class of knowledge workers to drive innovation and economic growth [3]; the British architect Charles Landry, extends the notion of creative city beyond the focus on artistic activities or creative economics, bringing themes such as dynamic social policies to foster creativity, creative environments, and the importance of history and tradition in creativity [4].

Landry [4], states that cities are living beings with phases of growth, stagnation and decline, and based on this, it is necessary to rethink the role of cities and their available resources, as well as the functioning of their urban planning at the moment the world goes through dramatic changes. This urban viability is precisely "the ability of cities to adapt and respond to changing circumstances" [4].

Jaime Lerner, Brazilian architect and urbanist, points out that any city can be "creative" and that this process, based on the construction of a "collective dream", is what will induce a collective aspiration and consequent mobilization of efforts so that this "dream" realize itself. For him, urban quality of life is expressed in three concepts: sustainability, mobility and solidarity..." the integration is the key word for characterizing a 'creative city' [5] stimulating the coexistence of diversity (ages, incomes, uses, functions and typologies).

In fact, with the imminent transformation of the economy, cities' need to create conditions for people to think, plan and act with imagination. The built environment - the stage, the scenario, the container - are crucial to create a social environment" [6]. Among numerous activities and factors that make a "creative" city, the formation of "creative clusters" is at the genesis of this concept.

This paper therefore aims to investigate the relationship between the establishments of a "creative industry" center (cluster) called Porto Digital, and urban development generated from its deployment in the city of Recife, Brazil, witnessing the possibility of transformation of this city to creative environment.

2 Creative Cluster

Michael Porter defines cluster as a geographic concentration of interconnected companies, specialized suppliers, service providers, institutions and associated companies in related industries [7].

Williamson [8] emphasized the network economies and argues that the process of corporate re-alignment exists due to the need to minimize transaction costs. In this new paradigm, the local proximity of innovative companies is necessary to facilitate multiple and complex interactions. In this way, the networks provide an extremely favorable, creative and innovative environment for the constant flow of information related to new technologies, new markets and products.

This interdependence between companies should also seek relations with entities supported by the government, professional associations and the scientific community.

Therefore, relations have been created in the clusters that do not normally exist between geographically dispersed companies: cooperative relations.

In addition to these concepts, this geographical concentration of institutions and producers of knowledge and interconnected activities (linked to technology, design, architecture, fashion, literature, gastronomy and the visual arts, among others), which offer or provide complementary services between itself, provoke positive changes, both economic - transforming the region into reference in specific subjects, as well as social - improving the quality of life of the area where it is implanted.

Therefore, as an instrument of local development and as inductors of innovation processes coupled with new technological and economic demands, clusters are provoking transformations in the urban space. The territory needs to reorganize itself in a new way, with new functions and demands, based on the implementation of these new productive arrangements based on creativity and innovation.

The implantation of *creative cluster* in historical value in the city has provided the rehabilitation of old industrial spaces, such as old warehouses, factories, markets, military barracks, among others. Without underestimating the importance of the recovery of outer space, the main concern is in the interior space where rehabilitation is done with greater rigor and care for the implantation of the new spaces and its new public [9]. In this context, these creative poles are a favorable environment for development: on it, people exchange ideas, develop projects, work and market their products together. These sites are cozy and offer a wide range of experiences (food, arts, production, heritage, nature, etc.) in reset, inspiring and alternative scenarios.

For the city citizens, these poles of creativity begin to be visible as references of culture, technology, academic formation [10]. In this context, it is not difficult to imagine that this "creative spirit" brings about positive changes in the urban environment in which these new poles take place, and, in the long term, have repercussions on the city as a whole. However, the question arises: Are these initiatives enough to make the city "creative"?

3 Urban Rehabilitation

Urban rehabilitation mainly consists of the regeneration and conservation of the built heritage or the urban environment, in order to improve and adapt its conditions of use. This concept aims to maintain the architectural character of the objects being built and includes the modernization and improvement of historical equipment and respect for norms and environmental and safety rules [9].

By observing "creative clusters" implanted in different locations and situations, such as LxFactory in the city of Lisbon in Portugal [9] and the Bhering Factory in the city of Rio de Janeiro in Brazil [11], the following questions were highlighted:

(a) How can the concept of "creative industry" subsidize urban rehabilitation?
(b) How can "creative activities" (re) qualify urban spaces?
(c) Are innovations or innovative environments coming first?

(d) What are the types of materializations from principles of the "creative economy" that can value and add to the preexisting specificities of the place where they are implanted?

(e) Do these experiences have similar levels of complexity or are just concentrations of companies in the same geographic space?

As a contribution to the understanding of these issues, researchers studied the case of the Digital Port in the city of Recife, located in the northeastern region of Brazil.

4 The Creative Cluster Porto Digital

The center of Recife, today corresponds to a site of various cultural activities such as museums, places for exhibitions, educational institutions, festivals, associations, and the cluster Porto Digital.

Porto Digital was founded at the turn of the 21st century in the emblematic year 2000 by a group of businessmen, members of the academy and representatives of the public sector who decided to bring to the state government the idea of creating a public policy for the technology sector of Information and Communication (ICT) that interlinked actors, companies and organizations that at that time worked independently and in isolation. Two goals were very clear: to prevent the young professionals from the IT area, graduated from the prestigious undergraduate and postgraduate courses (mainly from the Federal University of Pernambuco - UFPE), leaving the city attracted by the great centers promoting state-of-the-art technology; and to revitalize the Recife neighborhood, a historical region of the city that in the 1990s had an ephemeral "re-discovery" due to investments in the areas of culture and entertainment, but which in the following decade already presented a path of degradation of its built heritage.

Initially designated as a "technological cluster", Porto Digital operates in the areas of software and services in Information and Communication Technology (ICT) and Creative Economy, with emphasis in the multimedia, audiovisual, music, photography, design, and more recently, urban technologies [12].

The Porto Digital already attracted dozens of companies from other regions of Brazil, besides several multinationals and technology centers, has been expanding its territory and has already crossed the Capibaribe River incorporating the "Quadrilátero de Santo Amaro" (Fig. 1), besides to create the Creativity Warehouse, at "Polo da Moda", in the interior of the state of Pernambuco in the city of Caruaru (Fig. 2). Currently there are more than 260 companies in the technology park, research institutes, incubators, accelerators and investment funds, employing around 8,500 people [12].

Porto Digital complements its activities with revitalization actions, in an effort to demonstrate that it is possible to combine technological development with the preservation of history and culture. In this way, it has recovered several prominent buildings (since its founding, managed to recover more than 80 thousand square meters of historical buildings), in order to adapt the infrastructure of the neighborhood to receive modern companies, while maintaining its architectural characteristics.

Fig. 1. Area of expansion of Porto Digital to the Quadrilatero de Santo Amaro.

Fig. 2. Armazem da Criatividade located in the city of Caruaru, Pernambuco.

As a result of the implementation of Porto Digital, the entire area belonging to the former Port of Recife has been receiving government investments in partnership with private initiative since 2012, through projects such as Porto Novo (public initiative) and Porto Novo Recife (private initiative). With are great initiatives that seek to return to the city spaces previously dedicated to the port operation. These projects aim at the redevelopment and re-urbanization of noble areas in order to establish entertainment, culture, commerce, archeology and tourism options in the Recife neighborhood [13]. Reformed warehouses and new constructions already house the Pernambuco Handicraft Center (the first, inaugurated in 2012), the museum Cais do Sertão Luiz Gonzaga, the Passenger Terminal, offices, bars and restaurants, representing an economic and urban impact in this region.

5   Conclusion

A cluster, by definition, encompasses, from the economic point of view, structures much more complex than just a concentration of certain sectors in certain localities. Although local development cannot be reduced exclusively to the promotion of clusters, it is possible to consider them as an important alternative for the renewal of the territory of certain cities, constituting a new productive strategy in the "new" economy of the 21st century [14].

According to Leite and Awad [14], the understanding of clusters as instruments of urban rehabilitation, drives to some common points in the strategy of recovery of productive areas in transformation in certain urban poles. The first point is the need for interaction between public management and private initiative, delimiting for each a specific role within an integrated planning process. The two main requirements that are made to the public power are the implementation of quality infrastructure and adaptation to the current technological requirements and the creation of an articulation body between the various agents. The issue of infrastructure, which also involves accessibility and mobility, is fundamental to enable large private-sector investments linked to high technology. The second is private investment, such as the purchase of land or rehabilitation and restoration of buildings with historical, cultural or architectural value.

In this scenario, as in the case of Porto Digital, the implementation of clusters has a local productive arrangement format with immense potential for development as a strategy in the restructuring of old degraded industrial areas, providing the possibility of working creatively to construct an ideal identity. Urbanists and economists can speculate on these forms of arrangements where the big bottlenecks of today's cities (such as lack of mobility and adequate infrastructure) are gradually overcome and new models emerge through new technologies. One can imagine the creative expansion that the city would achieve if a massive investment were made in a system of advanced teaching and professionalization, with curricular adaptation to new professions of the creative sectors; in long-range public policies aimed at the creative economy; in social development policies; in preserving its historical and cultural heritage; in an integrated public transport system; among many other factors that foster development.

Following the example of what has been presented in this paper, a city to become "creative" must, first of all, seek out and recognize the wealth of its heritage, its creative potential, and the talent of its people.

References

1. Jane, J.: The Death and Life of Great American Cities. Vintage, New York (1961)
2. Howkins, J.: The Creative Economy: How People Make Money from Ideas. Penguin Books, London (2001). 246 p.
3. Florida, R.: The Rise of the Creative Class. Basic Books, New York (2002). 434 p.

4. Landry, C., Franco, B.: The Creative City, vol. 12. Demos, London (1995)
5. Lerner, J.: Every city can be a creative city. Creative City Perspectivas, pp. 31–36(2009)
6. Landry, C.: The Creative City: A Toolkit for Urban Innovators. Earthscan, London (2012)
7. Porter, M.: Clusters and the New Economics of Competition. Harvard Business Review, Rio de Janeiro (1998). https://hbr.org/1990/03/the-competitive-advantage-of-nations
8. Williamson, O.E.: Market and Hierarchies: Analysis and Antitrust Implications. The Free Press, New York (1975)
9. Ribeiro, C.A.D.S.: Reabilitação Urbana e Sustentabilidade – LxFactory: um exemplo de reabilitação sustentável na cidade de Lisboa. 2012. 91f. Dissertação (Mestrado em Arquitetura) - Universidade Técnica de Lisboa – Faculdade de Arquitetura, Lisboa (2012)
10. Reis, A.C.F.: Cidades Criativas – análise de um conceito em formação e da pertinência de sua aplicação à cidade de São Paulo. 2011. 312 f. Tese (Doutorado em Arquitetura) - Universidade de São Paulo – Faculdade de Arquitetura e Urbanismo, São Paulo (2011)
11. http://www.fabricabhering.com.br
12. http://www.portodigital.org
13. http://www.portodorecife.pe.gov.br
14. Leite, C., Awad, J.D.C.M.: Cidades sustentáveis, cidades inteligentes: desenvolvimento sustentável num planeta urbano, p. 127. Bookman (2012)

Ergonomics in Building and Architecture

The Impact of Ergonomic Guiding Principles on the Formation of Modern Monumental Art

Agnieszka Gębczyńska-Janowicz[✉]

Faculty of Architecture, Gdańsk University of Technology,
ul. Narutowicza 11/12, 80-233 Gdańsk, Poland
agnjanow1@pg.edu.pl

Abstract. Commemoration of historic events is now a commonplace phenomenon both in culture and in academic research. The phenomenon is linked to an increase in the representation of the visualisation of the past in public sphere, which in turn leads to reformulation of monumental art. Modern monuments are often spatial forms consisting both of sculpture installation and bigger constructions. In case of these realisations, the designers, wishing to create certain atmosphere among the users, have undertaken certain steps that are contrary to the guidelines set out in the norms concerning ergonomic use of space. Architecture, whose basic purpose is to provide a sense of security, implements construction solutions which make the user disoriented and feel uneasy. The purpose of this article is an analysis of intentional use of design practices related to non-ergonomic organisation of public spaces, creation of typologies of realised solutions and an attempt to present the consequences of undertaking such steps.

Keywords: Human factors · Architecture · Ergonomic disorders
Memorial · Museum

1 Introduction

This article presents how the principles of conceptual ergonomics participate in the process of forming the commemorative guidelines. In several realizations of modern commemorative art, ergonomic disorders of space have been used to provoke strong emotional experience among the visitors of the sites of commemoration. The research presented in the article focuses on the analysis of monumental guidelines in the context of the use of means of artistic interpretation related to ergonomics.

The first part of the work presents characteristic features of architecture and ergonomics, which make both disciplines participate in artistic creations used in creating commemorative sites, such as monuments or museums.

The second part of the article presents the most representative realizations of commemorative art, in which ergonomic disorders were applied through the architectural syntax. The presented case studies concern monumental guidelines and museum buildings that illustrate the past, influencing the user's feelings and simulating experiences related metaphorically to the past. The sites visualize events of World

© Springer International Publishing AG, part of Springer Nature 2019
J. Charytonowicz and C. Falcão (Eds.): AHFE 2018, AISC 788, pp. 333–341, 2019.
https://doi.org/10.1007/978-3-319-94199-8_32

War II and are located in Europe and North America. A typology of the analysed implementations was proposed and a critical evaluation of their impact on the development of monumental art was presented.

2 Relations Between Ergonomics and Architecture

Ergonomics is a field that supports research into the safe use of space. According to the majority of definitions given by researchers related to ergonomics, its activities are mainly related to the human – machine (building) relationship within the working environment. As a scientific discipline, it deals first of all with adjusting the working conditions to the psychophysical capabilities of a human being [1]. The modern approach to this discipline also leads to research that adapts ergonomic assumptions in facilities related to other human activities, such as place of residence, facilities for recreation, or the development of public spaces on an urban scale. The results of research on the functioning of the workplace space are successfully applied in studies on spatial solutions of buildings with a different function [2]. Modern ergonomics should therefore cover in the area of its interest all spheres of human life [3].

Ergonomics is an interdisciplinary and transdisciplinary science, which allows the application of its achievements to other fields, including architecture. What links ergonomics and architecture is primarily a concentration around human needs in shaping the physical space intended for him. The subject of interest in both fields is the adaptation of the entire material environment to the needs and abilities of a human being. The parameters of dependence between human and the environment elaborated by ergonomists not only support the comfort of using building objects, but also minimize the level of potential threats. Architects must consider the most comfortable and efficient use of space while taking care of aesthetic values. The primary, basic purpose of architecture is to provide shelter and security to people in their everyday environment. Contemporary perspective focused on people causes that the architectural theory significantly examines the influence of architecture on the physical condition and mental state of the user of the developed space. Understanding the basic rights that the human psyche is subject to has become a significant element in the work of architects. The space of building objects has begun to be created in such a way, so as to affect the user also by extra-visual means. This tendency got adopted also in the monumental art. Some studies referring to the competence of architecture to act as a memory carrier begin with rhetorical questions, whether the genesis of creating architecture based on securing basic human needs can transfer metaphorical functions [4], and whether tools that create a framework for a sense of security can deliberately instil anxiety and steer towards reflection [5]. The first ergonomic disorders began to be used to evoke anxiety. Architecture has the power to affect the senses of the recipient. The building constructed with the intentional negation of the bases of the architectural syntax (which is manifested, for example, by: uncomfortable stairs, disturbance in the floor level, claustrophobic rooms) triggers a feeling of threat, and physiological reactions increase the sensitivity to the process of commemoration.

3 The Role of Ergonomic Disorders in Creating Monumental Guidelines

Human factors and ergonomics are based on anthropometry, a research method based on comparative measurements of parts of the human body. The practice of designing products, systems or processes to properly consider the interaction between them and their users, works with the physical parameters characteristic to the human body. With the data obtained this way, the environment can be created not only in an optimal but also safe manner [6].

The antithesis of ergonomic sciences inspired artists designing from the middle of the 20th century the sites commemorating events of the World War II. In particular, the holocaust, or history of the programmatic racial extermination of the Jewish population by the regime of the Third Reich, has become a frequent topic of sculptural and architectural installations, where an important element is to show the fate of man facing terror. The places of extermination were created for industrialized processes. In addition, according to modern philosophers, the rationalization, industrialization and bureaucratization which are characteristic for modernity led to mechanisms that German national socialism used during the Second World War to spread terror [7]. Labour camps, concentration camps and other places serving the Nazi regime were designed as an environment in which human needs and constraints were placed at the lowest priority compared to other purposes. The drama associated with it became an inspiration for designers of monumental guidelines for the next post-war decades.

Characteristic features through which the physical environment is perceived by the user as carrying potential threats have been applied to artistic solutions in the monumental art. The first intuitive attempts in this direction were made in the 1960s An example of this type of implementation is the Monument to the Victims of the Extermination Camp in Treblinka, unveiled in 1964. A large-spatial monumental guidelines was established in a place where in 1942–44 functioned the Nazi extermination camp, in which nearly 800,000 Jews from various European countries were murdered. In the area of seventeen hectares, a moving monumental guideline was planned by Franciszek Duszeńko and Adam Haupt, which also serves as a necropolis for mass graves. This heartrending place of memory is opened by a gate with a non-traditional form of two concrete cuboids, and then two routes lead into the guideline: a symbolic railway line formed of rhythmically arranged concrete blocks and a path made of stone pavement uncomfortable for pedestrians. The entire monument guideline was carefully thought out so that each point on the way of commemoration would introduce new sensations in the reception of the work. The stone route makes the observer cease to be a passive recipient of the tragedy and begin to participate in a journey through the space, in which several decades earlier thousands of people were led to death.

Similar influence of the commemorating place on the recipient was used in the monumental structure erected in 1969 at the site of the former German concentration camp KL Lublin (now known as Majdanek). There are elements here to prepare arriving people to come into contact with the space where the tragic events took place before in the past. In the foreground of the monumental sculptural installation functioning as a symbolic gate leading to the post-camp area, the designer Wiktor Tołkin placed a few meters long track located three meters below the ground level. On the outskirts of this channel there are boulders that hang on the heads of passers-by. By their location they arouse a sense of danger. The author of the monument recalls the time of the monument installation and the reaction to the idea of the first users: "I decided to hang these blocks over the road, which symbolizes danger. Already at the beginning it worked so strongly on the imagination, that the organizers wanted the area under the blocks to be separated for security reasons [8]". The anxiety that arouses in the visitors because of the boulders is compounded when they walk out of the channel to the steep, unergonomically designed stairs. After passing through the gate, the visitors see the view of the remains of the camp infrastructure and the mausoleum, in which the ashes of the prisoners of the camp were placed (Fig. 1).

Fig. 1. The monument to struggle and martyrdom in Majdanek on the grounds of the former German concentration camp KL Lublin, 2008, Photography: Gębczyńska-Janowicz A.

The two above-mentioned projects point to a tendency in contemporary commemorative art to treat recipients of monuments or museums as active participants who are to receive the visualization of the past with all their senses. In the last decade of the 20th century, commemorative arts have undergone another metamorphosis. It saw the birth of a typically modern phenomenon called by the scientists *dark turism*. This concept, originally defined by Lennon and Foley [9], recognizes places commemorating the tragic historical events as a tourist attraction. Traveling people want to experience the reality of the event in isolation from the interpretation of media revelations, verify in situ the images shaped by contemporary culture. Therefore, in the commemorative realizations, there has been a gradual change in the interest of visitors from passive sightseeing to the experiencing the space which visualizes the past. The metaphorical approach of a person visiting a physically determined place of remembrance to the experiences of victims of the World War II persecution has become a frequent tool for visualizing the past in the 1990s This is particularly the case for museum buildings. There traditional exhibition practices were extended to experience scenarios. In this concept, the physical shaped place of memory is getting close to the reconstruction of dramatic events, and the users participate in an empathetic journey. Learning historical facts is accompanied by kinaesthetic, haptic and other individual experiences based on non-visual senses [10]. In the United States Holocaust Memorial Museum in Washington, opened in 1993, visitors pass through claustrophobic tunnels and replicas of railway cattle cars, in which the Germans transported people to places of extermination. Similar solutions were applied in the Warsaw Uprising Museum established in Warsaw in 2004, where visitors can go through a tunnel arranged to look like the drain channels used by the insurgents in 1944.

The most controversial implementation in which the means of physical simulation of experience was applied is the Beit Hashoah-Museum of Tolerance opened in 1993 in Los Angeles. It differs from other conventional museums by presenting historical facts and interpreting them in the context of contemporary social realities. The method of creating the exhibition openly declares the intention to use history to teach tolerance and social responsibility. The latest solutions related to multimedia technologies have been used to attract young people and manipulate their emotions. The museum uses dioramas, films, interactive installations and computer-controlled exhibits. The aspect related to evoking emotions in the recipient caused that the way of conducting the educational activities of the Museum of Tolerance is often criticized by the researchers of the commemorative art [11]. The main doubts concern the use by the designers of exhibition of the television formats to communicate and educate. Creating dramatic performances is primarily to shape the way in which visitors will feel the visualizations of the past presented to them. According to Nicola Lisus and Richard V. Ericson this concept of representing historical facts turns a museum into a factory of emotions and functions as a *format of control* [12].

Despite the many concerns about the ethical aspect of making educational activities based on the simulation of experiences, monuments and museums in the 21st century were still formulated as objects, in which you can enter, touch and experience the space also in the physical way. This concept, popularized by the *countermemory* [13] installations, has been developed in a direction that made the cubature architecture enter the monumental art. Some new solutions were also leading to *embodied experiences* [14].

At the beginning of the 21st century, two realizations created in Berlin became icons of world commemorative art. These are: the expansion of the Jewish Museum building and the Memorial to the Murdered Jews of Europe. Both sites significantly used ergonomic disorders in architectural solutions leading to the creation of a place of memory.

The expansion of the Berlin Jewish Museum was established in 2001 according to the concept of architect Daniel Libeskind. The author titled his project *Between the Lines*. On the basis of multi-aspect semantics, a building was created, the convention of which deviates from the traditional museum architecture in the direction of creating the object as a place giving an impulse to discover the memory of the past. The space created by Libeskind requires from the recipient both intellectual and emotional involvement. The architectural layout affects the human psyche, arousing anxiety, insecurity and loss. It stands structurally in opposition to the paradigm of a building object: there is no entrance zone indicated from the outside and the method of material solutions on the façades makes it impossible to understand the internal structure of the building. The layout of communication links between individual functional zones causes a sense of confusion for the visitors. The levelness of floors and angle of walls inside the structure are far from generally accepted ergonomic standards. This is not an environment in which a person feels comfortable. This way the recipients are forced to be more sensitive to the exhibits they are looking at. Ergonomic disorders are applied not only on the main routes, but also in objects that constitute their metaphorical summary. The three main elements building this concept are the corridors of impressions, the Holocaust Tower and the Garden of Exile. The visiting routes symbolize the three realities form the history of German Jews – the history of their functioning in the German society throughout history, the process of migration and the Holocaust. The longest route - the continuation axis leads from the old building to the staircase linking the exhibition spaces of the museum. Another route - the emigration axis leads outside to the Garden of Exile. Along the way, the walls are slightly inclined, the floor is uneven, which introduces visitors to the anxiety resulting from the disturbance of the perception of the surrounding. Going outside to a symbolic garden is uncomfortable because of the heavy door wings. The Garden of Exile is located outside the main building. The 49 concrete, perpendicular pillars were set on a square surface, where the floor is tilted 12° from the level to create a sense of destabilization. The third route of visiting, the Holocaust axis is like a blind street - getting narrower and darker - which leads to the Tower of the Holocaust. The 24-m high tower is a building standing outside the main body of the building. Empty, not fitted with air conditioning and unprotected against the influence of external weather conditions, this premise is terrifying with cold alienness. The senses here receive first of all quiet, distant sounds from the outside world and a small amount of light, which is passed by the aperture located at the top. This striking aura is meant to remind you of emptiness, loneliness and danger - feelings that Jews could experience during the World War II persecution (Fig. 2).

Fig. 2. The corridors of impressions, the garden of exile and the holocaust tower in Berlin jewish museum, 2011, Photography: Gębczyńska-Janowicz A.

The success of the Libeskind concept, which has been confirmed by numerous publications of art theoreticians and public interest, paved the way for the next implementation, deviating from the traditional definition of a monument. In the center of the German capital, in a quarter with an area of 19,000 m², in the symbolic location between the Reichstag and the ruins of the former Chancellery of the Reich, the Memorial to the Murdered Jews of Europe was created. Its official unveiling in 2005 preceded the years of social disputes about its form, purpose and location. Several competitions have been unsuccessful. Finally, the implementation of such a sensitive monument was made according to a project proposed by Peter Eisenman and Richard Sierra. The designers of the winning concept proposed a formula of the monument in which the user is engaged in the memory of the Holocaust through somatic and kinaesthetic experiences [14]. Users captured as pilgrims move along narrow corridors created by cold grey steles. The pavements of this labyrinth have uneven surface and they lead to unidentified destination. Visitors to the monument are symbolically cut off from the noise of the city and left in spatial disorientation (Fig. 3).

Fig. 3. The memorial to the murdered Jews of Europe, 2011, Photography: Gębczyńska-Janowicz A.

Visitors of the Memorial Site in Bełżec are subjected to similar experiences. It was created in the place where, in 1942, the Germans established on the surface of approx. 6 hectares an extermination camp created as part of Operation Reinhardt. About 600,000 people, mainly of Jewish descent, were brought here to death. The monumental premise established in 2004 according to the concept of Andrzej Sołyga, Zdzisław Pidek and Marcin Roszczyk is commemorating these dramatic events based on the experience of the monumental art from the 1960s. The non-ergonomically selected parameters of architectural structures such as the fissure sliding into a historical hill, or stairs leading out of the depths are solutions that bring no comfort to the user. Similar impressions are caused by the sterile interiors of the museum located near the monument. The Contemplation Hall located at the end of the building is an empty, dark room, where every murmur, as a result of acoustic treatments, transforms into a moving rumble.

4 Conclusions

The presented analysis of several implementations of commemorative art devoted to the events of World War II allows the separation of stages in which artistic creations applied ergonomic disorders to monumental guidelines and museum objects.

The first realizations used simple solutions related to the unergonomically prepared building elements included in the spatial composition having a commemorative function. It consisted primarily of an intentional induction of discomfort in users moving through the routes leading through the monuments. Unevenness of the floors was used or a sense of danger was evoked by the monumental form of sculptural installations and elements positioned in a manner suggesting the risk of an accident. The first realizations in this discourse were directed more towards creating the atmosphere of sacred, calm and being an impulse for reflection, rather than the manipulation of feelings. Ergonomic disruption practices were used in the entrance zones, at the point of entering the commemoration zone, so to metaphorically recall the passage to the sacred space.

The next stage was associated with techniques simulating physical experience in order to manipulate the reception of broadcasted content relating to the past. Multimedia technologies and solutions used in television scripts play a greater role here. The user directly enters into the replicas of historical objects designed to trigger transient trauma. The controversial pedagogical tool aims to temporarily get close to real experience [15].

The last of the discussed stages mitigates the impact of the commemorative implementation on the users' senses. There is no such a strong process of simulating experiments as in the previous implementations. However, compared to the projects from the 1960s, commemorative objects implemented in the new century benefit to a greater extent from architectural syntax and ergonomic disorders. Moving through the objects exposes the user to a physical sense of being trapped, lost, and isolated. There is a possibility of disturbance of balance and spatial disorientation. Visitors follow the directions imposed by architecture. The commemoration order adopts a literal, behavioural formula. Users of memorial sites are asked to identify with the pain of people from the past and reconstruct the past through the experience of superficial trauma [16].

All of the three cited stages of using ergonomic disorders in commemorative art evoke critical opinions. Simulating the experience of museum and monumental guidelines users has led to suspicions about the possibility of a dangerous phenomenon of identification with victims of traumatic events from the past. In addition, the aesthetization of evil and the presentation of cruelty in a visually attractive way are also a concern [17].

What initially was a slight arousal of discomfort in the 1960s was deepened by the realizations of the 21st century. The transition stage consisted of exceed the application of experimental simulation techniques in museum implementations. Despite the controversy that arouses with using the ergonomic disorders in the architecture of monumental guidelines and museum objects, there is no indication that the monumental art will give up such techniques of user manipulation.

References

1. Górska, E.: Ergonomia. Projektowanie, diagnoza, eksperymenty. Politechnika Warszawska, Warszawa (2007)
2. Charytonowicz, J. (eds.): Zastosowania ergonomii. Wybrane kierunki badań ergonomicznych w roku 2015, Wrocław (2015)
3. Charytonowicz, J.: Architecture and ergonomics. In: Proceedings of the Human Factors and Ergonomics Society Annual Meeting, vol. 44, pp. 6-103–6-106 (2000)
4. Wierzbicka, A.M.: Architektura jako narracja znaczeniowa, Politechnika Warszawska, Warszawa (2013)
5. Young, J.E.: Daniel Liebeskind's Jewish Museum in Berlin: the uncanny arts of memorial architecture. Jewish Soc. Stud. 6(2), 1–23 (2000)
6. Giacomin, J.: What is human centred design? Des. J. 17(4), 606–623 (2014)
7. Bauman, Z.: Nowoczesność i Zagłada, Wydawnictwo Literackie, Kraków (2009)
8. Gębczyńska-Janowicz, A.: Polskie założenia pomnikowe: Rola architektury w tworzeniu miejsc pamięci od połowy XX wieku, Wydawnictwo Neriton, Warszawa (2010)
9. Lennon, J.J., Foley, M.: Dark tourism. Continuum, London (2000)
10. Williams, P.H.: Memorial Museums: The Global Rush to Commemorate Atrocities. Berg, Oxford - New York (2007)
11. Marcuse, H.: Experiencing the Jewish Holocaust in Los Angeles: The Beit Hashoah—Museum of Tolerance. Other Voices 2 (2000). http://www.othervoices.org/2.1/marcuse/tolerance.php
12. Lisus, N., Ericson, R.V.: Misplacing memory: the effects of television format on holocaust remembrance. Br. J. Sociol. 46(1), 1–19 (1995)
13. Young, J.E.: At Memory's Edge: After-Images of the Holocaust in Contemporary Art and Architecture. Yale University Press, New Haven - London (2000)
14. Sion, B.: Affective memory, ineffective functionality: experiencing Berlin's memorial to the murdered Jews of Europe. In: Niven, B., Paver, C. (eds.) Memorialization in Germany since 1945, pp. 243–252. Springer, London (2009)
15. van Alphen, E.: Zabawa w Holocaust. Literatura na świecie 1–2, 217–244 (2004)
16. Arnold-de Simine, S.: Mediating Memory in the Museum: Trauma, Empathy, Nostalgia. Palgrave Macmillan, Basingstoke, New York (2013)
17. Chrudzimska-Uhera, K.: Kamienne Piekło. Projekty Mirosława Nizio upamiętniające miejsca kaźni i zagłady: były niemiecki obóz koncentracyjny w Gross Rosen oraz Mauzoleum Martyrologii Wsi Polskiej w Michniowie. In: Chrudzimska-Uhera K., Gutowski B. (eds.) Rzeźba w Polsce (1945–2008), pp. 115–120. Centrum Rzeźby Polskiej, Orońsko (2008)

Space Architecture

Klaudiusz Fross[(⊠)] and Maria Bielak-Zasadzka

Faculty of Architecture, Silesian University of Technology, Gliwice, Poland
{klaudiusz.fross,maria.bielak-zasadzka}@polsl.pl

Abstract. We find ourselves at a moment where it is quite realistic to develop space tourism and hotels, e.g. on the Moon. Is an architect needed in space? Is the unavailability of the terrain and difficult conditions justifying the container placement in a free manner, without thinking about the landscape, about aesthetics. As is currently the case in Antarctica. Without the application of developed principles of architectural planning, urban planning and spatial planning. The authors hope that, with the development of space engineering, astronauts-architects will fly into space. At the Faculty of Architecture of the Silesian University of Technology in 2017, the "Space Architecture Research Program" was inaugurated. The program aims to support scientific work and design concepts in the field of space architecture and cooperation in the field of virtual reality. The article will present two MA theses: "City on Mars" and "Spaceport Abu-Dhabi".

Keywords: Architecture · Cosmos · Spaceport · Mars · Design by research

1 Introduction

Interest in space architecture, shown by students from the Faculty of Architecture at the Silesian University of Technology in their diploma papers, could be observed for more than ten years. However, colonization of Mars has never been as realistic as it is today. These works are not submitted very often as they constitute quite a challenge. Moreover, they require a great deal of knowledge, a lot of passion and rich imagination. Recently, there have been two really intriguing works submitted as MA Papers and these were: "*City on Mars – Jednostki kolonizacyjne na innych planetach z możliwością wykorzystania ich w ekstremalnych warunkach*" [Colonization Units Designed for Other Planets with the Option of Being Used in Extreme Conditions] (with a Virtual Reality presentation) author: Kamil Lis, MSc, Eng. of Architecture, supervised by: Maria Bielak-Zasadzka, PhD, Eng. of Architecture, "Spaceport Abu-dhabi – A concept project for a spaceport in Abu-dhabi" author: Roman Wala, MSc, Eng. of Architecture, supervised by: Klaudiusz Fross, PhD, Eng. of Architecture. These works contributed to the initiation of the "Research program for space architecture". It was started in 2017 at the Faculty of Architecture of Silesian University of Technology - the first faculty in Poland. Klaudiusz Fross, PhD, Eng. of Architecture was the initiator of the program. The program is aimed at promoting and supporting scientific papers as well as design concepts in the scope of space architecture. The faculty is equipped with a test facility as well as resources in the scope of Virtual Reality (VR). Combining problems and

© Springer International Publishing AG, part of Springer Nature 2019
J. Charytonowicz and C. Falcão (Eds.): AHFE 2018, AISC 788, pp. 342–353, 2019.
https://doi.org/10.1007/978-3-319-94199-8_33

issues connected with space with virtual reality open up a vista of new possibilities. The Dean would like to take the opportunity to invite all space architecture enthusiasts who are interested in cooperation (www.polsl.pl). The faculty has been in talks with the Polish Space Agency about a potential scientific and didactic cooperation.

The scientific staff from Department of Design and Quality Assessment in Architecture are specialist in "design by research", for examples publications: [1–5].

Does the Space Need Architects?

We find ourselves at a moment where it is quite realistic to develop space tourism and hotels, e.g. on the Moon. Does the space need architects? Does the inaccessibility of the area and adverse conditions justify the placement of containers in a random way, regardless of the landscape, regardless of esthetics? Such approach is visible only on Antarctica. It is devoid of any principles of architectonic, urban or spatial planning. The authors are hoping that together with the development of cosmic engineering, space architects will be able to go into space. Just like it is done on Earth, on Mars or in the outer space, where we envisage building structures, one should specify binding principles of the spatial order, adjusted to the specificity of the conditions. Unfortunately, for the last few years the space has not only been conquered and explored, it has also been littered. That is what happens when there is an uncontrolled expansion, in a place where there are no principles, rules or responsibilities. Perhaps, before the process of building things in space begins, we should first clean up all cosmic trash which remains on the orbit. This trash includes objects which were created by people and are no longer of any use. Space trash is estimated at weighing 5 tons. Interestingly, PW-Sat2, a Polish noteworthy satellite, should be praised for not leaving any trash in space (Inna Uwarowa - project coordinator).

2   Background of the Issue

April 1961, the first man ever travels into space. It was none other than Major Jurij Gagarin, an astronaut from the Soviet Union. This space flight lasted only 1 h and 48 min. It was not until 8 years later, 1969, that for the first time a man landed on a celestial body - the Moon. The crew of Apollo-11, commanded by Neil Armstrong, spend one night on the Moon, only to return to Earth by way of a splashdown. The space conquest has been in progress for decades now. At present, there are several global institutions which are involved in works on manned travels onto other celestial bodies in our Solar Panel. These institutions include: NASA - National Aeronautics and Space Administration, Mars Society, ESA - European Space Agency, Mars Underground as well as many other agencies and groups that do research in space. In recent years, more and more is being said about the construction of extraterrestrialbases. There are, of course, orbit stations so it is assumed that the next planet where a man will be sent is going to be Mars - the first flight, with crew, is being planned for 2024, and is supposed to last approximately 180 days. Before Mars is colonized or before resources can be extracted so that infrastructure can be built there, several pioneer expeditions will have to take place. The first unmanned expedition to the Red Planet, with resources necessary to build the first base, is planned for this year. Onboard there are going to be vehicles that will search for the proper location for the construction of that base. On the 6th of February 2018, Falcon Heavy was successfully launched thanks to Elton Musk. That mission may change all future opinions concerning space. That rocket is able to transport, onto the circum terrestrial orbit, a total load of approximately 63 tons and, what is more, it is three times cheaper.

Concerns

At first, some basic questions should be raised:

- What is the reason for the colonization of other planets?
- What makes us think about life on Mars?

Mars is the fourth planet from the Sun. It is two times smaller than Earth and is ¼ of the area of our planet. There are no seas or oceans on the Red Planet. Its total area is equivalent to the area of Earth's continent. Unlike Earth, Mars is tectonically and geologically inactive.[1] As a planet, it is really attractive for "us" as it possesses a significant amount of natural resources and water, while its soil is highly fertile. In the future, Mars may become an invaluable storage room and a source of natural resources. The second reason why this new planet should be colonized is the fact that testing and implementation of innovative technical solutions can be done in its extreme conditions. The "Martian" solution will make it possible to look at Architecture on Earth from a different perspective. Development in space technology may positively influence people's life on Earth, especially when it comes to dealing with multiple problems. Colonization of Mars is a fantastic opportunity for contemporary architects who could soon play the key role in the shaping of the "cosmic" built environment. The basic assumption will be to make it safe in both physical and mental terms.

Before we start our ruminations connected with the creation and development of a base on planet Mars, or even before we start thinking about residential units set in extreme situation, one should take into consideration the following issues:

- Does the space need architects?
- What role does he or she play in creating a permanent place of residence on Mars?
- What knowledge must he or she have in order to proceed with creating *The Space Architecture*?
- What are the optimal tools, strategy and technology which would allow a human being to take up residence on the surface of the Red Planet?
- In what way can the development of "Martian" technology positively influence and help people's life in extreme conditions on Earth?

---

[1] Its red color comes from the fact that there is a lot of iron oxide contained in minerals such as "which can be found on the surface or deep underground. Mars has two natural satelites - Phobos and Deimos. At present, there are three artificial satelites circulating Mars: 2001 Mars Odyssey, Mars Express and Mars Reconnaissance Orbiter. Mars is visible from plant Earth. There is a large number of impact craters, canions, valleys, deserts and polar icecaps. Mars features the highest mountain in the Solar System - Olympus Mons and the largest canyon - Valles Marineris.

Complexity of the Issue

Construction of a permanent base on Mars, with conditions more similar to those on Earth rather than those at the orbit stations, constitutes a unique task of high complexity. Mental health of the colonizers is one of the most important factors that might influence the success of space missions. It is closely connected with the creation of a safe life, residential environment for the future users and with the creation of the "due" behavioral quality. One should bear in mind that the quality and functionality must be kept at a high level. Creating perfect life and development conditions for future occupants of the Mars base is a highly complex problem. Therefore, one should take into consideration different factors and issues from various branches of knowledge. One should definitely analyze the following factors: geophysical conditions present on Mars; issues connected with physiological and psychological aspects of human life on Mars, including nutritional problems; scientific projects and agencies doing research on the universe; available technology which enables the construction and operation of the Martian base; theoretical projects which have been created to date, concerning The Space Architecture.

The first step before proceeding with pioneer research works, with proper and reliable operation of the space structure in mind, would be to perform a detailed analysis of the conditions present on the Red Planet. One should make an analysis of the following aspects: temperature, atmosphere, water, solar cycles and radiation, transportation means, existing resources, energy, geology and gravity. Having analyzed these issues, one can make a realistic list of challenges which the unknown environment puts in front of the architects and scientists. On planet Mars, one can find resources such as iron, found in the clay dust, abundant amounts of carbon, hydrogen as well as silicon dioxide. This could pave the way for the production of construction and finishing materials as well as water synthesis.

1. The terrain is highly rough and uneven. The elevation difference in the crater can reach up to 8 km. Due to that, an adequately flat terrain should be found so that the base could be built.
2. Mean temperature on Mars is about $-63\ °C$. The yearly temperature amplitude inside the crater amounts to 80 °C (from $-8°$ to 0 °C). This means that the structure to be designed should be well insulated.
3. Possible sources of energy on Mars will include wind, photovoltaic and geothermal energy. The most desired source of energy would be a wind power plant or, alternatively, photovoltaic panels brought from Earth, which could later be produced on Mars.
4. The atmosphere on Mars is 100 times thinner than the atmosphere on Earth. The amount of the oxygen is not sufficient for the needs of human livelihood, so it is necessary to use spacesuits when staying outside. High content of hydrogen within the atmosphere allows for the synthesis of water from air.
5. Water is present in significant amounts, in the frozen underground lakes and craters, therefore its extraction is not so problematic and may be delivered to the base by means of a space exploration vehicles, also known as rovers.
6. Soil on Mars is highly fertile. In closed conditions it might be used for farming.

7. Surface of Mars receives a lot more UV radiation than Earth so it is of essential importance to design the base in such a way that the inflow of the hazardous radiation to the interior of the base is limited.
8. Gravitation on the surface of Mars amounts to … of Earth's gravitation, which makes it possible to decrease the adopted dimensions of constructional elements by half.

Issues which scientists and engineers need to address first are connected with the general safety of the astronauts (future residents), protection against space radiation and totally different gravitation present on Mars. One should be aware that one of the key factors which can influence the outcome of the mission is the human being as well as his/her mental and physical condition. The second important issue includes problems connected with technological possibilities of civil engineering on Mars as well as problems with materials. In order to receive necessary information, one should analyze scientific projects together with theoretical architectonic projects connected with *Space Architecture*.

3 Inspirational Examples of Space Architecture

Base on the Moon - Foster + Partners
Base on the Moon proposed by Foster + Partners in 2013 constitutes a theoretical prototype designed by the European Space Agency (ESA). The project combines inflatable architecture with 3D print in order to create a residential space on the surface of the Moon for astronauts. The biggest advantage of such solution is the thorough analysis of the structure of the Moon with a view to making use of its resources to print the habitats. Both agencies - NASA and ESA are running a policy which consists in taking advantage of resources available on that planet. Application of modern 3D technology as well as the technology based on inflating structures may also be introduced on Mars, which unlike the Moon is rich in various resources.

ZA Architects - Colonization of Mars[2]
ZA Architects laboratory has devised a conception which is connected with the construction of solid settlements on Mars, using robots and materials available on the spot. The project is aimed at sending drilling robots in order to locate basalt columns which were found in the crater, inside its walls, close to Marte Vallis by Mars Reconnaissance (NASA).

Andreas Vogler - Moon Capital[3]
Moon Capital, proposed by Andreas Voglers, constitutes a project of a residential base on the Moon. This proposal would be located on the outer edge of the Shackleton crater, on the south pole of the Moon. Voglers envisions that up to 69 people will be able to live on the surface of the Moon by 2069. The project will combine inflatable,

---

[2] Source: http://www.gizmag.com/martianarchitecture/28999.
[3] Source: Andreas Vogler. "Moon Capital" – Life on the Moon 100 Years after Apollo.

self-supporting concrete domes together with the technology of small robots which resemble bugs.

Mars One - Successive colonization of Mars[4]

The mission will commence in 2018, as soon as the first landing module has landed on the surface of Mars. Its task will be to examine the area in the Gale crater, which has already been visited earlier, in order to indicate key elements that will make life and general functioning on the "Red Planet" possible. Two solutions are going to be tested. They will include the placement of a thin surface with solar energy generating cells as well as a regolith heating oven on top, in order to examine the presence of water in that region. At the same time, the orbital satellite will try to create a system of communication between Earth and Mars, which will facilitate future manned missions.

Monika Lipińska - Test Lab[5]

Test Lab constitutes the future of lunar life as well as the exploration of the cosmic space. The concept of the base is simple - fill out the area on the Moon step by step. The whole structure is supposed to take a couple of years, starting with a simple settlement which will be inhabited solely by a small group of astronauts. As soon as the settlement starts to develop and evolve, it will be possible to send a larger number of astronauts. Once the Test Lab becomes autonomous, it will be possible to send ordinary men so that they can participate in the development of the station on the Moon. They will be able to experiment with fruit and plants growing in space and experience 3D and 4D prints. The whole idea consists in forming a cooperation between experienced astronauts and ordinary people, passionate about cosmic space, who wish to find new and fascinating ways to experience live.

Spaceport Abu-Dhabi - author Roman Wala, supervised by: Klaudiusz Fross (Fig. 2)

MA diploma project submitted in 2016, at the Faculty of Architecture of the Silesian University of Silesia. "*Spaceport Abu-dhabi – A concept project for the Abu-dhabi spaceport*" author: Roman Wala, MSc, Eng. of Architecture, supervised by: Klaudiusz Fross, PhD, Eng. of Architecture.

The subject of this work is the conceptual design of a commercial spaceport, located off the coast of Abu Dhabi. The complex is an innovative investment, the form and nature of the related infrastructure, is a direct result of functional requirements, which are: service suborbital flights - stay in the space around Earth by means of specially created for the manned aircraft (spacecraft); complementary functions: taking out the cargo into space, staying in the hotel space - a commercial alternative to existing spaceports. The resort provides the opportunity of a 4-day mandatory training and testing for space tourists (including accommodation in isolation, along with pilots and instructors), attractive stay for flight, a wide range of activities related to the theme of space exploration. Spaceport is designed as a set of buildings erected on artificially banked up island in the Persian Gulf, west of the capital of the emirate, is connected to the mainland by a two-story bridge (the main assumption is a formal continuation of the

---

[4] Source: www.mars-one.com/mission/roadmap.

[5] Source: WWW.eleven-magazine.com/?entrants=test-lab-ec2323.

trend connected with BANKING UP artificial islands along the coast of the United Arab Emirates). The complex of buildings consists of three main buildings spread radially around the centrally located entrance area of the railway station and car park: Segment 1: hangar with administrative part and traffic control; Segment 2: medium education and training, along with habitat for astronauts; Segment 3: Public hotel with a leisure area. Streamlined, modern forms of the buildings, as well as their concentric arrangement, inspired by the motives of well-known sci-fi elements characteristic of space exploration, the traditional architecture of the local Arab Emirates (Sheikh Zayed Mosque); biological forms reflect the local natural environment. The basis of the main structure of objects are massive reinforced concrete foundations on the embankment of the precious metal-reinforced edges. Curtain walls are a lightweight glass facade of high performance acoustic and thermal properties. Designed runway measuring 3800 x 95 m will be able to accept each aircraft (and space using the runway), including prototype Stratolaunch Carrier Aircraft with a wingspan of approx. 117 m. Determination of classification code airports (based on the guidelines of the International Civil Aviation Organization - ICAO Aerodrome Design Manual) 4 F. Apron adapted to the adoption of the space fleet and chartered planes port, also designed two helipads. Spaceport communicated to the mainland by road and rail.

City on Mars - Kamil Lis, supervised by: Maria Bielak-Zasadzka (Fig. 1)

MA diploma project submitted in 2017, at the Faculty of Architecture of the Silesian University of Technology. "*City on Mars – Jednostki kolonizacyjne na innych planetach z możliwością wykorzystania ich w ekstremalnych warunkach*" [Colonization Units on Other Panets with the Option of Being Used in Extreme Conditions]. (With a Virtual Reality presentation) author: Kamil Lis, MSc, Eng. of Architecture, supervised by: Maria Bielak-Zasadzka.

It is an innovative example of indicating lines of solutions in the scope of space architecture. Completion of the analysis and research on problems related to Mars (and the existing conditions), the synthesis of modern technology and engineering materials which would enable the construction of the base and examples of architectonic projects connected with Space Architecture, would make it possible to create a conceptual model adjusted to the conditions found on that planet. At the initial - predesign stage, the following factors were analyzed: location, atmospheric conditions, geology of the terrain, resources, including water, cost of the investment, psychological aspects. Area within the Gale crater was chosen for the construction of the base. The location guarantees that the building is near a potential source of water, is located under the shade of a massif and makes for a relatively easy construction, considering there are significant differences in the area in other parts of the planet. The crater offers a gentler climate in comparison to other climate zones of Mars. Placing the structure in relation to the directions of the world is optional due to the adopted architectural solutions. Direct influx of light into the inside of the building will be avoided due to strong radiation. The light which will appear will be reflected light. As a result, the structure may be oriented regardless of the directions of the world.

According to the project assumptions, the base will meet the following criteria:

- Intended number of users - 50 to ensure the most optimal environment for life, work and normal functioning.

- The base will be a totally independent unit, independent of Earth (capable of only using the resources available on Mars)
- The base will be located in natural depressions such as craters or caves - in order to minimize the effect of hazardous UV radiation (gentler climate).
- It will meet the conditions of sustainable development.
- It will provide the possibility of dividing the zones of different purposes, i.e. the residential zone, working zone, farming and cultivation zone, technical zone and the common and recreation zone.
- It will be adjusted to handicapped people and people who wish to stat family.
- It will be ready for the possibility of migration as well as multiple usage of engineering and finishing elements.
- It will anticipate cultivation of plants and in later stages, also the breeding of animals inside the complex.
- It will make use of modular elements for a quick, cheap and easy modification and expansion of the complex.

The base on Mars was designed on the basis of the principles connected with the staging of the investment (the project includes the first three stages - it also plans for a future development)

Stage I – includes a plan for 3 people to take up residence there. With the help of robots these individuals will erect the structure. All elements necessary for the assembly must be transported from Earth due to the lack of possibility of creating them on the spot. The project includes: 3 single residential cabins with annexed garden modules, laboratory module, dining and kitchen annex, storerooms, control room, waste disposal room, 3 farming modules and a decompression chamber - all arranged around the central atrium. The garage for the space exploration vehicle will be a separate structure, accessible from the outside only. At that stage, it will only be necessary to depend on provisions and materials delivered from Earth.

Stage II – plans to accommodate 12 users. Number of floors is larger by two, the garage for the space exploration vehicle would be connected with the building by means of a workshop. In the vicinity of the garage, there will be a second entrance. The cultivation part's clear height will be increased two times, which will increase the total area of the hydroponic cultivation. An apiary will be added. The residential and recreational function (additional 9 residential rooms with gardens) will be located on the second floor. Due to the increase in the number of floors it will be necessary to include the vertical passage way staircase. Passageways and corridors on the first floor will be arranged on the basis of a gallery along the central atrium, from where one will be able to reach residential rooms. Apart from that a viewing balcony was added from where one can observe the cultivation part located below.

Stage III – includes a plan for further development of the base; 50 users. It is independent (large arable areas and large scale animal breeding). The surface of the laboratory is significantly bigger due to the larger number of users, mainly scientists. The residential part will see the introduction of two-room modules due to the fact that there will be a possibility of starting a family on the premises of the structure. The proposed solution for stage III is a complete solution as further expansion, especially in terms of a larger distance off the atrium, will increase the passageways, which is

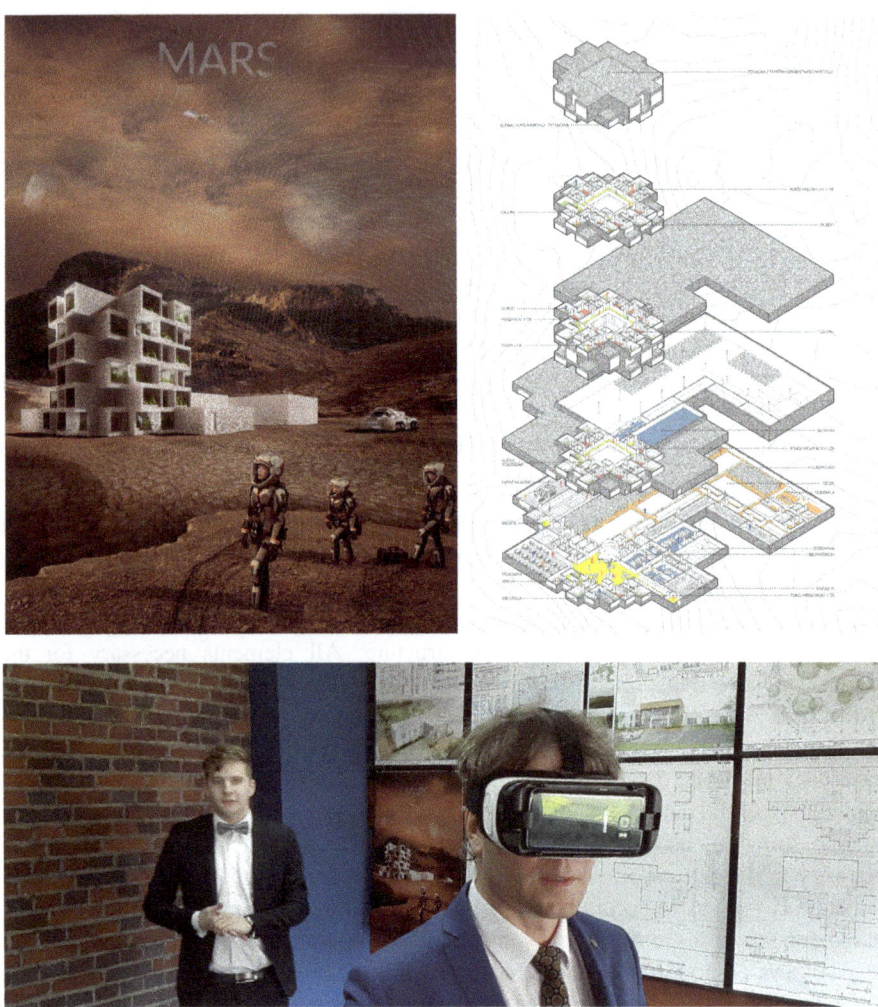

Fig. 1. Project: City on Mars – colonization units designed for other planets with the option of being used in extreme conditions (with a virtual reality presentation) author: Kamil Lis, supervised by: Maria Bielak-Zasadzka.

unnecessary. As a result, in subsequent stages duplication of the plan of the third stage is proposed, together with a connection by means of the cultivable part with a con-current introduction of an air bridge which will connect single entrance cabins.

In the first three stages of the base's development, the following infrastructure is planned for: main building of the base, emergency center, communication tower, storehouse for samples, waste disposal and a garage for vehicles.

The application of innovative technical solutions constitutes an important part in the project. Starting from building materials applied (such as: steel, glass, polymers, aerogel, conductive polymers) through to construction solutions and installations,

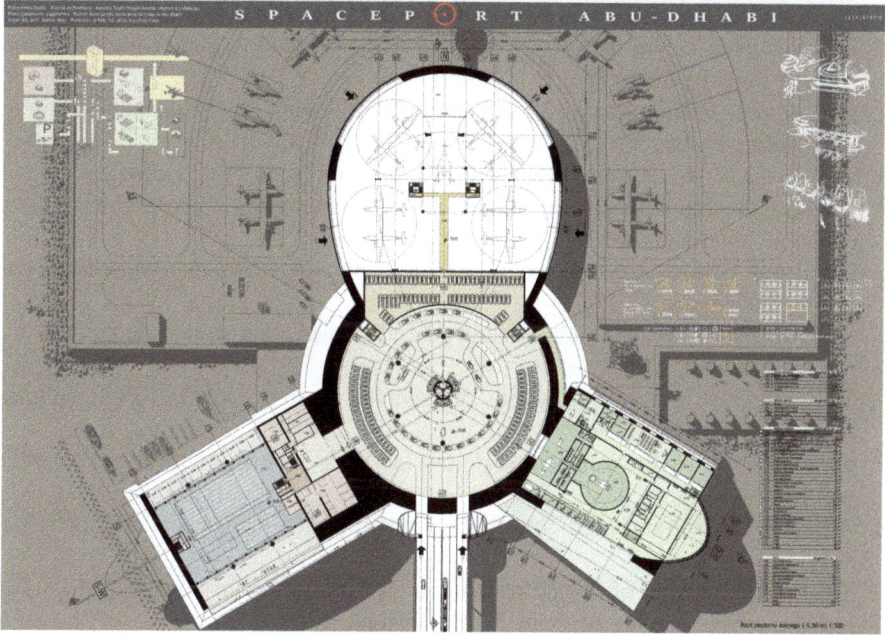

Fig. 2. Project: "Spaceport Abu-dhabi – a concept project for a spaceport in Abu-dhabi" author: Roman Wala, supervised by: Klaudiusz Fross.

ending with issues connected with sources of energy (solar, wind, geothermal and energy obtained from biogas). The perfect example of innovative solutions are the external walls which are made of a 20 cm 3D printed polymer, thermally insulated with a double layer or aerogel with a total thickness of 20 cm and covered externally with a facade made of triangular elements made of plastic combined with titanium, with a total thickness of 5 cm. These walls come in white or light-gray colors. These boards offer a double feature - they protect the building and the users against hazardous conditions (especially the radiation) and they are capable of producing water through subjecting the hydrogen present in the Martian air to chemical reactions. In the event of strong winds and due to the risk of a meteorite strike, all the panels will be closed to create a full and uniform protective layer. The glass facade (located on the garden modules and within the dining part) will be made of a double glazed panes based on the "heat mirror" technology. Between the layers a system of pipes supplied with warm air had been planned for. This will ensure proper thermal insulation within the division. Additionally, it will be covered with foil with an UV filter. Interestingly, the proposed architecture is similar to the one found on Earth. It does not resemble containers nor capsules. This aspect seems to be important due to the fact that it forms a bond between the residents of Mars and planet Earth.

4 Conclusions and Guidelines for Space Architecture

1. The most recommended solution is to make use of the highest possible number of materials available on the spot as this will allow people to significantly lower the costs of the investment.
2. Lack of precipitation and lower gravity make it possible to significantly decrease the building elements.
3. Two levels of the housing ensure proper protection against UV radiation.
4. Locating the base under the shade of the elevations ensures protections against radiation, wind and dust.
5. Production of own food will allow people, who will reside there, to render the colony independent.
6. In the first stage of the base's construction, all construction works can be done by means of robots.
7. The project is based on a system of modules. It constitutes the most optimal strategy of building and developing the base.
8. Creation of virtual environment may turn out to be beneficial for the people who will live on Mars.
9. A large number of divisions constitutes good protection in the event of a meteorite strike.
10. The optimal number of people residing at the base in the first stages of the colonization process should be from 25 to 50 people.

Assuming that one of the key factors that can influence the outcome of the space mission is the human being itself, one should really consider all the problems connected with his or her physical and mental condition and also consider all the issues connected with nutrition or medical health care available on Mars. At present NASA is conducting detailed series of research into the effects of a long-term life in space where the gravity is similar to the one on Mars. The task is connected with finding answers to the following question: "Do space travels influence the human body at the genetic level. If so, do these changes help protect or select the future astronauts?" As we can see, what we are facing at this point is a completely new, and multi-faceted, series of problems connected with atmospheric and geological conditions in an unknown environment where human beings plan on interfering. We are well aware of how extensive knowledge we must dispose of and how complex problems scientists, architects and constructors must face. These individuals wish to create innovative guidelines, principles of architectonic, urban and spatial planning for the new planet. These problems will include numerous interdisciplinary issues and well as different forms of cooperation between specialists from multiple disciplines.

At the end of the day, we should once again raise a very important questions: does space need architects? The answer is YES. Who else will design structures in space?

References

1. Fross, K., Sempruch, A.: The qualitative research for the architectural design and evaluation of completed buildings – part 1 – basic principles and methodology. ACEE Archit. Civil Eng. Environ. **8**(3), 13–19 (2015). Silesian University of Technology
2. Fross, K., Sempruch, A.: The qualitative research for the architectural design and evaluation of completed buildings – part 2 – examples of accomplished research. ACEE Archit. Civil Eng. Environ. **8**(3), 21–28 (2015). Silesian University of Technology
3. Fross, K., Winnicka-Jasłowska, D., Gumińska, A., Masły, D., Sitek, M.: Use of qualitative research in architectural design and evaluation of the built environment. In: Ahram, T., Karwowski, W., Schmorrow, D. (eds.) Proceedings of the 6th International Conference on Applied Human Factors and Ergonomics 2015 and the Affiliated Conferences, AHFE 2015, Las Vegas, USA, 26–30 July 2015, pp. 1625–1632. Elsevier (2015)
4. Tymkiewicz, J., Winnicka-Jasłowska, D., Jastrzębska, M.: Pre-design studies on the example of modernization project of geotechnical laboratories. ACEE Archit. Civ. Eng. Environ. **10**(2), 43–52 (2017)
5. Tymkiewicz, J., Winnicka-Jasłowska, D., Jastrzębska, M.: Ergonomics of laboratory rooms - case studies based on the geotechnical laboratories at the Silesian University of Technology. ACEE Archit. Civ. Eng. Environ. **10**(2), 35–41 (2017)

Ergonomic Aspects of Development of Architecture in the Context of Sanitary and Hygiene Safety

Rafał Janowicz[(⊠)]

Faculty of Architecture, Gdańsk University of Technology,
ul. Narutowicza 11/12, 80-233 Gdańsk, Poland
rafjanow@pg.edu.pl

Abstract. Ergonomics answers the need of safe development of space by creating spatial forms which help to implement the safety procedures and limit the threats involved both in ordinary use of the spaces and in case of unforeseen events. Using the knowledge of ergonomics and architecture on the basis of defining the routes of germ transmission, allows to limit the spread of those organisms. Ergonomics of developing architectural spaces allows to improve safety of all users. Changing reality, including the risks and the level of knowledge about those risks results in the fact that development of architecture with the use of ergonomics is not only a multi-dimensional issue, but also requiring constant analysis and validation of solutions due to a changing character of those risks. The work is an attempt to present how the deficit of detailed ergonomic guiding principles in the process of programming and designing of architectural objects experienced in Poland influences the sanitary and hygiene safety.

Keywords: Architecture · Ergonomics · Sanitary and hygiene safety

1 Introduction

Man cannot eliminate infectious agents from the environment in which he/she is currently completely eliminated. Monitoring the level of infections and taking action to control and fight them is one of the basic goals of the World Health Organization (WHO). This organization recommends the use of disease control programs built into a comprehensive health care system [1].

In an environment in which humans reside on a daily basis there are microbial reservoirs, some of which have a detrimental effect on human health and can be a source of infection. In some cases, as a result of diagnostic processes, the presence of infectious agents is probable, and thus the introduction of epidemiological risk management procedures becomes possible and necessary. The infectious process is widely described in the literature, nevertheless, from the point of view of the proper shaping of architectural structures and ergonomic safeguards, it is important to correctly define the ways of spreading the infectious agent, which passes from the source to the susceptible individuals.

© Springer International Publishing AG, part of Springer Nature 2019
J. Charytonowicz and C. Falcão (Eds.): AHFE 2018, AISC 788, pp. 354–363, 2019.
https://doi.org/10.1007/978-3-319-94199-8_34

The ways of spreading infectious agents can be divided into the following main groups:

- Direct transfer of the infectious agent to the appropriate entry gates through which it goes to the host organism. This can happen through contact with an infected person or be referred to as the so-called droplet;
- Indirect transfer by objects that were previously contaminated. The microbe is then transferred passively, and the carrier is a biological medium. In this case, the microorganism can multiply;
- Transmission by air through fine aerosol particles that may persist in the air and by inhalation not being delivered to the lungs. Dry fine particle dust can thus transmit fungal spores to a minimum extent;
- Horizontal transmission, i.e. transmission of infections between representatives of the same species with the exception of transmission from mother to child in the period from conception to delivery inclusive;
- Transmission by vectors such as insects (e.g. Ticks, mosquitoes) that are living carriers [2].

In addition, hazards at workstations can also be divided into groups with different levels of harmfulness to human health and can be attributed to harmful biological agents, which in this case are cellular microbes (also genetically modified), non-cellular units capable of replication or transmission of genetic material, including genetically modified, cell cultures and internal parasites.

Polish legislation for the above classification, the division into four groups of the hazard level for human health is introduced and they are assigned to disease entities:

- Group 1 hazards includes the least dangerous biological factors for human life. In this case, it is unlikely that human disease will occur. Prevention is based in this case on the obligation to comply with the basic rules of personal hygiene.
- Group 2 hazards are factors that can cause diseases for which effective treatments have already been developed. A further spread of this type of threat in the human population is unlikely.
- Group 3 hazards - factors that can cause severe illness in people, are dangerous for employees, and their spread in the human population is very likely. There are usually effective methods of prevention or treatment for them.
- Group 4 hazards requires the utmost attention, because the factors mentioned in this group cause diseases that pose a direct threat to human life and are not always subjected to simple therapeutic methods. The risk of the biological agent spreading in this case is very high [3].

Knowledge about the pathways of pathogenic agents should be considered in the context of the frequency of occurrence of such factors, as well as the degree of risk they pose to humans. Such an analysis allows the introduction of architectural and ergonomic solutions as elements of the occupational health and safety management system, which should be formed based on risk assessment [4]. The element of risk analysis is the indication of actions eliminating or limiting the occurrence of risk. Among the elements limiting the risk of infection, spatial organizational measures, trainings and the application of collective and individual protection measures can be mentioned.

Architectural and ergonomic solutions created as a result of risk analysis should be adapted to the type of hazards occurring in rooms and other areas and intensity of their occurrence.

2 Primary Safeguard of Epidemiological Risk

Minimizing the occurrence of epidemiological risk by extorting or promoting behaviour concerns a number of facilities with various functions. Not all hazard groups associated with a biological agent are used in each function of the use of architectural space. Dining premises are an example of the widespread implementation of solutions affecting safety. It is usually required here to undertake hermetic safeguards that are characteristic of Group 4 hazards due to the estimated risk and for organizational and functional reasons (for catering facilities located outside infectious hospitals). This is because adapting public spaces to the requirements of airtightness for Group 4 hazards is not rational. They require restricting access to rooms only for authorized persons and isolating these zones from the remaining functional system in the building. In addition, there is an obligation to enter and extract air through HEPA filters [3]. In the context of a publicly accessible room in a dining area, where any user of public space can have access, implementing such restrictions is not possible.

In practice, this means the need to adopt and apply individual solutions adopted based on the standards resulting from the applicable law, standards, and technical knowledge from broadly understood good practice and procedures implemented by organizational units.

Architectural and ergonomic solutions should refer to both risk analysis as well as to take into account adequacy and rationality of applied security in the context of both threats and possible transmission routes, thus causing a wide interdisciplinary scope of the issue.

The basis for implementation of architectural and ergonomic solutions aimed at minimizing the occurrence of infection risk in gastronomic premises is the FAO/WHO Codex Alimentarius recommended by the Commission to implement the Hazard Analysis and Critical Control Points (HACCP). It is a methodology of proceedings aimed at ensuring food safety by identifying and estimating the scale of hazards from the point of view of health food requirements and the risk of hazards during the course of all stages of food production and marketing [5]. The HACCP management system is implemented by two programs created by the Codex Alimentarius Commission: Good Hygienic Practice (GHP) and Good Manufacturing Practice (GMP). In Europe, this requirement introduces as a mandatory principle: Regulation of the European Parliament and Council No. 178/2002 of 28 January 2002 setting general principles and requirements of food law [6], and Regulation No. 852/2004 of the European Parliament and of the Council of 29 April 2004 on the hygiene of foodstuffs [7].

From the point of view of this article, the most important are recommendations regarding the correct functionality of the spatial layout required for the appropriate organization of food production plants, defined as Good Hygienic Practice. Actions that must be taken and hygiene conditions that must be met and controlled at all stages of production or marketing in order to ensure food safety [5]. Good Hygienic Practice refers to various areas of activity within gastronomic establishments, among which, for example, spatial activities enabling the implementation of correct procedures can be mentioned. Reducing the risk of infection is possible by shaping the environment that eliminates the possibility of infection, or its reduction through architectural and ergonomic solutions enabling the implementation of activities in the field of infection prevention, including the correct attitudes of users.

The correctness of this thesis can be traced on the example of shaping a dishwashing room in a catering establishment. In today's cities, mass caterers, among others restaurants, are commonly found. In most of them dishes are served on reusable utensils, which directly introduces the epidemiological risk. Among consumers, there are people who are sick or who carry biological agents During consumption, they infect tableware and post-consumption remains. For these reasons, dishwashing should be positioned so that dirty dishes do not come into contact with ready meals and clean tableware [8].

This relatively simple principle of separating "dirty" from "clean" processes is not, however, a sufficient answer to the risk of contamination of products. In small gastronomic premises, transient dishwashers are rarely used, which allowed the reception of clean dishes in the clean area as a separate room. Thus, from an architectural point of view, it is important to clearly distinguish zones in which cleaning activities are performed and to separate them from dirty activities. The basic source of the disinfection process in such a functional system is a dishwasher with a steaming function as a source of disinfection. This solution reduces the hazards resulting from the use of meals from reusable containers, but does not completely eliminate them. The steaming standard commonly adopted in Polish gastronomy does not lead to 100% removal of hazards despite its high efficiency and effectiveness, due to the fact that sterilization would be necessary to sterilize vessels and utensils.

The presented figure visualizes the possibility of ergonomic arrangement of architectural space in a way that ensures progressive movement, that is, the introduction of space organization in a way limiting the possibility of infections (Fig. 1).

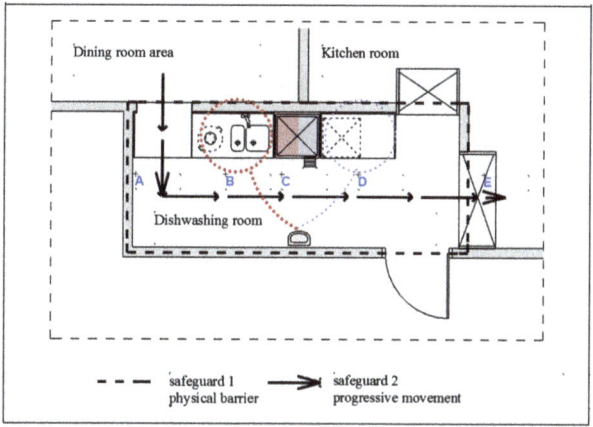

Fig. 1. Organizational diagram of the dishwashing room. Author: Rafał Janowicz.

Adopted solutions limiting the risk of food contamination by an infected client in the case of reusable vessels as a transmission path through:

Selected elements of risk analysis for the functional system:
The risk of contamination of clean products prepared for consumption by post-consumption vessels and utensils

Safeguard 1

Physical barrier between the clean part (kitchen) and the dirty part (dish washer) carried out by separating the rooms with walls and crossing cabinets.

The risk of contamination of clean dishes by contact with dirty dishes

Safeguard 2

Progressive movement:

A - Post-consumption return of vessels, dishes and utensils (risk of biological contamination)
B - Post-consumption waste removal
C - washing station using a dishwasher with a steaming function - a technological process at a temperature of 80–90 °C
D - unpacking position for dishwasher basket
E - passage cabinet serving the function of storing dishes and separating them from clean rooms of the kitchen

The risk of transmission of infections by droplets

Safeguard 3

The use of mechanical ventilation as well as walls and passage cabinets as solutions limiting the possibility of migration of pollutants through the air as the transmission path.

The risk of transmission of infections through reusable utensils

Safeguard 4

The use of dishwashers with the steaming function as a solution limiting the possibility of infections.

The risk of transmission of infections through personnel resulting from contact with both "clean materials" and "dirty materials"

Safeguard 5

Introduction of organizational solutions in the field of ergonomics of workplaces that allow for maintaining the correct procedure. Allowing washing and disinfection of hands after carrying out dirty activities and before performing "cleaning" activities.

The scope of activities resulting from the introduction of the above safeguards includes, among others:

- Ergonomic elements - organization of the workstation in a way that allows proper execution of the procedure;
- architectural elements - separation of rooms/space;
- technological elements - organization of the flow of vessels/dishes/utensils;
- epidemiological elements - hazard identification;
- ventilation elements of the room - separate ventilation system;
- elements of risk management - creating procedures of conduct.

The above ergonomic and architectural solution allows to reduce the risk of transmission of infections due to secondary vessel infection. The presented diagram based on the risk analysis in the room of the dishwashing machine allows to conclude that the ergonomic shape of the rooms may be of key importance for introducing solutions that improve hygienic sanitary safety, but it requires cooperation of specialists from several fields of science.

The analysis of sanitary and hygienic security in terms of architectural and ergonomic solutions is important due to the lack of information and research on non-legal conditions for safe shaping of usable space. For cases not covered by detailed legal regulations, there is a need to individually assess the design solution. This requires a sanitary-hygienic risk assessment in the context of possible infections, their potential effects and the possibility of occurrence. Thus, in such a case, the primary difficulty for the discussed subject is its interdisciplinary character including both architecture, ergonomics as well as epidemiology.

3 Complex Ergonomic and Architectural Epidemiological Risk Constraints

Minimizing epidemiological risk by forcing or promoting behaviours from the use of ergonomic and architectural solutions in particular for medical facilities. Over the last decade, patient safety has become one of the main elements of healthcare management. Hospitals identify threats, examine errors, implement changes to reduce risks and train staff in patient safety [14]. One of the effects of this approach is the implementation of complex functional systems, whose task is to reduce the epidemiological threat in individual organizational units.

Caring for a sanitary regime in the hospital, proper disinfection and sterilization of equipment, the proper way of preparing meals, utilization of dirty material and procedures in accordance with the procedures of washing hospital underwear is the minimum necessary to maintain the conditions limiting the occurrence of epidemiological threats. In order to ensure sanitary and hygienic safety in medical units, however, the functional system that enforces specific behaviour, e.g. the passage through the apron-washbasin sluice, as well as personnel education is of key importance. It is assumed that despite the introduced safeguards, one of the key elements of ensuring safety is proper hand hygiene. Hands and aprons are the main source of infections. Undoubtedly, hospital-acquired infections testify to the quality of medical services [10]. In this context, the creation of safe use of medical rooms is related to the implementation of ergonomic solutions that enable implementation of safety procedures limiting epidemiological risks.

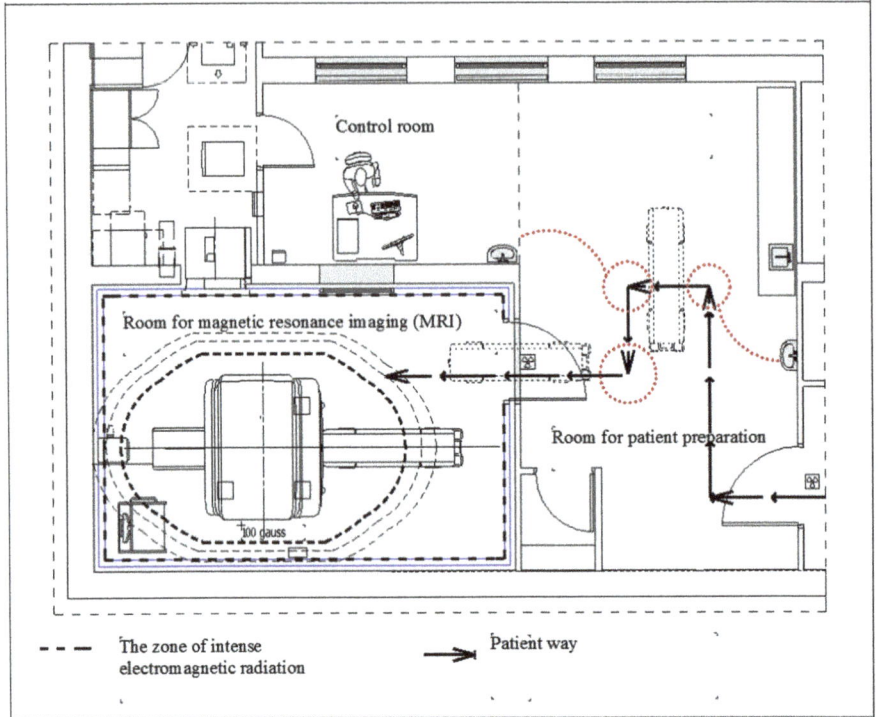

Fig. 2. Organization diagram of the areas for magnetic resonance imaging (MRI). Author: Rafał Janowicz.

This issue is included in the attached drawing that presents the analysis of the room architecture of the magnetic resonance room in the context of ergonomic solutions (Fig. 2).

Selected elements of risk analysis for the functional system
The risk of transmission of infections between patients by personnel due to the lack of proper individual hygiene

Safeguard 1

Ensuring the possibility of individual hand hygiene.

The equipment of work stations with wash basins or stands with disinfectant is one of the elements, whose placement in space requires analysis of functional ergonomic as well as sanitary and hygienic conditions.

"Hand hygiene is widely regarded as a fundamental element of infection prevention and control programs. (...) The reason for insufficient compliance with hand hygiene is the uncomfortable placement of the above-mentioned dispensers and/or washbasins overloaded by work, which in turn is associated with staff shortages, or dermatological problems (skin irritation) and cultural issues, such as the lack of specimens and forgetting about the applicable procedures. (...) Despite hundreds of years of research on the effectiveness of different hand hygiene methods, many health care workers are still not convinced about the superiority of hands disinfection over their washing [9]".

In the case presented below, individual hand hygiene should be performed each time after contact with the patient as a potential source of infections. Due to the magnetic resonance in the room, an intense magnetic field are limited in introducing water points to such a room. A place with a hand-wash basin should be provided as close to the place of patient preparation as possible, and a place with disinfecting liquid within the MRI room, so that the individual hand hygiene procedure can be carried out as soon as possible after direct contact with the patient - regardless of personal protective equipment such as disposable gloves. The solution should first of all be convenient for the staff.

The risk of exposure of unauthorized persons to the strong electromagnetic field

Safeguard 2

Due to the high level of the electromagnetic field, it is necessary to additionally introduce barriers limiting the risk of being in the vicinity of the field of unauthorized persons. The development of this barrier requires the simultaneous introduction of solutions allowing for the free movement of staff members. With a patient lying on a mobile table between rooms. (Due to the construction with the use of metal elements, no standard hospital beds or chairs for the disabled should be placed in the resonance room).

Such conditions make it necessary to provide a relatively large maneuvering area.

The risk of taking undesirable actions by the patient

Safeguard 3

Ensuring safety through the direct appearance of personnel to the test room, for example through an observation window. In some cases, the patient undergoes severe stress during the tests, which results in the need for a quick reaction of the staff.

The risk of unauthorized personnel entering the magnetic resonance room

Safeguard 4

Ergonomic solutions should reduce the possibility of adverse reactions occurring for example, due to the potential negative effect of an intense magnetic field on the well-being of a patient with a pacemaker. Each time the patient should be checked for having metal parts before entering the room. Therefore, the ergonomic solution should prevent the resonance from hanging directly from the general communication, omitting the personnel performing the check. On the presented solution, the laboratory consists of a complex of three rooms, where the role of avenging the patient's preparation is, among other things, limiting the risk of intrusion into the magnetic resonance room.

The risk of improper disinfection of surfaces with which the patient has contact

Safeguard 5

Other solutions due to the necessity of cleaning and disinfecting the rooms, ergonomic solutions should ensure the possibility of proper implementation of this process by limiting the number of hard-to-reach places.

The scope of activities resulting from the introduction of the above safeguards includes, among others:

- Ergonomic elements - organization of the workstation in a way that allows proper execution of the procedure
- architectural elements - separation of rooms/space
- elements of interior architecture - material solutions
- psychological elements - in relation to the patient being examined, in the device, in relation to the staff.
- technological elements - organization of a research laboratory
- epidemiological elements - hazard identification
- ventilation elements of the room - separate ventilation system
- elements of process and risk management - creation of procedures
- elements of radiological protection - protection against the action of the electro-magnetic field

The presented ergonomic and architectural solution reduces the sanitary and hygienic risks in the magnetic resonance laboratory. However, the correct solution of all elements of the project requires work based on an interdisciplinary team of specialists, and individual optimization should result from reliable research.

4 Summary

Efforts to improve epidemiological security require cooperation of specialists from many fields: architects; ergonomists; psychologists of doctors, specialists in the field of management in order to create interdisciplinary research teams. Effective implementation of solutions requires the integration of solutions from the theory, carried out research, but also the verification of the effects of work in the form of specific implementations of the work, which creates the opportunity to reliably evaluate them.

There are many sanitary and hygienic risks in the work environment. The source of these threats are bacteria, viruses, fungi, parasitic diseases and parasites as well as radiological, electromagnetic and other harmful substances occurring in the work environment. Issues related to solving them are the subject of both legal regulations as well as specialist studies. The combination of ergonomics and architecture does not usually indicate one optimal solution to a given problem, including sanitary and hygienic safety, it creates a framework for solutions considered acceptable. In this context, architecture is an area that supports sanitary and hygienic security.

Dynamic changes concerning technological conditions in the design of health care facilities cause the necessity of introducing changes to medical facilities based on expert opinions that currently provide a medical, economic and ergonomic context.

This is facilitated by the search for optimal solutions for medical units in support of scientific research. In accordance with the method that has been widespread in Europe for more than a decade, Evidence-Based Design in the architect's workshop are increasingly used in practice scientific analyzes regarding the contemporary needs of health care facilities [10].

References

1. World Health Statistics 2017: monitoring health for the SDGs, Sustainable Development Goals. World Health Organization, Geneva (2017). http://www.who.int/
2. Sadkowska, M., Todys, A.: Epidemiology of Infectious Diseases in Hospital Infections. PZWL, Warsaw (2016)
3. Regulation of the Minister of Health of 22 April 2005 on harmful biological factors for health in the work environment and protection of health of workers professionally exposed to these factors (Journal of laws, item no. 81 pos. 716 as amended)
4. Polish Standard; Occupational health and safety management systems. General guidelines for occupational risk assessment ICS 03.100.01; 13.100; PN-N-18002; (2011)
5. Act of 25 August 2006 on food and nutrition safety (Journal of Laws 2006 No. 171 item 1225 as amended)
6. Regulation of the European Parliament and of the Council No. 178/2002 of 28 January 2002 establishing general principles and requirements of food law
7. Regulation of the European Parliament and Council No. 852/2004 of 29 April 2004 on the hygiene of foodstuffs
8. Żabicki, W.: Organization, Health and Safety at Work. WSIP, Warsaw (2013)
9. Bulanda, M.: Hospital Infections. PZWL, Warsaw (2016)
10. Denys, A.: Hospital Infections. ABC, Warsaw (2012)
11. Cama, R.: Evidence-Based Healthcare Design. Wiley, Hoboken (2009)
12. Regulation (EC) No. 1935/2004 of the European Parliament and of the Council of 27 October 2004 on materials and articles intended for contact with food
13. Regulation (EC) No. 852/2004 of the European Parliament and of the Council of 29 April 2004 on the hygiene of foodstuffs
14. Carayon, P.: Human Factors and Ergonomics in Health Care and Patient Safety. CRC Press, Boca Raton (2017)

Architecture of Public Toilets in the Landscape – Disorder or Integration

Anna Jaglarz[(✉)]

Faculty of Architecture, Wroclaw University of Science and Technology,
Prusa Street 53/55, 50-317 Wroclaw, Poland
anna.jaglarz@pwr.edu.pl

Abstract. A toilet in the public space does not have to be bashfully hidden as a unaesthetic and unattractive building. Public toilets can also be a place for good design and prestige, especially with the willingness and commitment of the architects. The hygienic and sanitary facilities located and functioning in the vicinity of the natural landscape, in close proximity to nature, are a particular challenge in this respect. Green and recreational areas, landscape parks, mountain and seaside areas, marinas and others are equipped with public toilets to provide basic amenities to the users. Public toilets are also located on the tourist routes. Keeping in mind that these facilities are often the permanent elements of the landscape, designers should take care of the formal and material aesthetics of toilet buildings and propose solutions that do not interfere with the surroundings, but rather provide a harmonious and consistent complement to the landscape. By entering into the surrounding environment, public toilets should not disfigure, disturb and deter, but on the contrary, through interesting look, they may even become a tourist attraction. Assuming that the design of public toilets is not a necessary evil, architects can also present surprising ideas and feel satisfaction. The purpose of this study is to identify and present contemporary trends in this area. An important issue in the research is the analysis and evaluation of existing examples of public toilets located in the natural landscape. Conclusions will be the basis for developing design guidelines.

Keywords: Architectural design · Public toilets design · Landscape

1 Introduction

Providing a range of sanitary facilities in areas that are directly adjacent to the natural landscape may be necessary. The scope and type of facilities depend on the character of the place, the type of users for whom they are intended. It can also be related to the regulations regarding public health, hygiene, safety, as well as the demand for various user activities. Most green and recreational areas, including city parks, should be equipped with public toilets in accordance with legal requirements. Sanitary facilities are also provided in landscape parks, mountain and seaside areas, such as marinas and beaches. Public toilets are also located near tourist routes and roads. Their quantity, size, type, organization and appearance are different. Hygienic and sanitary facilities located and functioning near the natural landscape, in close proximity to nature, are a

© Springer International Publishing AG, part of Springer Nature 2019
J. Charytonowicz and C. Falcão (Eds.): AHFE 2018, AISC 788, pp. 364–376, 2019.
https://doi.org/10.1007/978-3-319-94199-8_35

special design challenge that skillfully combines function and aesthetics. These are objects that on one hand must respond to the basic needs and functional requirements of users, on the other hand they should not disturb their aesthetic impressions.

Aesthetic appreciation of the surroundings can be primarily visual and kinesthetic (through movement) [3]. However, in many cases experiencing surroundings is multisensory - and sometimes hearing, smell and sensitivity may be more important than vision [3]. Nevertheless, it is assumed that the vision is about 87% of the human perception of the environment, so it is proportionately more important [2].

Given that sanitary facilities are often permanent elements of the natural landscape, designers should take care of the formal and material aesthetics of toilet buildings and propose solutions that do not disturb the environment, but rather provide a harmonious and consistent complement to the landscape.

2   The Need of Providing Public Toilets in the Landscape Places

The decision to provide public toilet facilities depends on several factors:

- **Location and nature of the place**
  Remotely located areas with a wild, natural character suggest that the impact of human activities in this area should be kept to a minimum. Toilet buildings, although well designed, may spoil the atmosphere of remoteness and isolation associated with such areas. On the other hand, if there is a large number of visitors, the health risk and the problem of pollution caused by too many people resorting to "walking behind the bushes" may be definitely worse. Urban, rural or other locations involving more buildings are not such a issue [1].
- **The number of people visiting the area, especially on weekends and during holiday periods**
  In places where there are fewer guests, it may be difficult to justify the need and capital expenditures associated with the construction and maintenance of a sanitary facility, especially if there are no visible problems related to visual and olfactory pollution. In areas with limited space, unpleasant smell and view of bushes used by previous visitors may be troublesome, especially in hot weather conditions. This situation requires the provision of at least a small toilet, a overwhelm regime, covering or removing. In larger areas, visited by a larger number of people, the need for sanitary facilities becomes evident [1].
- **The distance traveled to a given place**
  A short trip to the visited area reduces the need to use the toilet, while a long journey usually makes use unavoidable. This is especially important if there are no alternative sanitary facilities along the route [1].
- **The duration of the visit in a given place**
  If most of the visitors come to a given place for a short time, for example for a walk with a dog in a suburban park, the use of sanitary facilities is not necessary. If guests spend most of the day in the area, then they need toilet facilities, especially if they consume food [1].

- **Presence of gastronomic objects at the area**
  In some places catering facilities and food objects are provided, which function especially during periods of intensive use of the area. The presence of catering facilities attracts a larger number of visitors and may affect the need for a toilet, although in principle a similar situation arises when people bring their own snacks [1].
- **The possibility of water recreation at the site**
  The health risk resulting from the use of contaminated water exists always. If water is used for bathing, swimming, use of water equipment or similar activities, toilet facilities are necessary for sanitary purposes, but also for changing, washing and other hygienic actions, although usually suitable facilities for these purposes are usually provided [1].
- **Winter use of the place**
  If the area is regularly used throughout the year, this may determine a balance in terms of providing comfort to tourists in very cold conditions, including sanitary needs [1].

3   The Scope of Providing Public Toilet in the Landscape Places

Along with the decision to provide public toilets in a given area, the extent to which they are provided, their quantity and scope, are partly determined by the factors described in the previous chapter. A remote location where a small or medium number of visitors spends some time may justify a single, universal, unisex object, while a large area with many visitors staying there for a long time requires many facilities divided into parts for women, men and disabled. The effective scope of provision may partly depend on:

- the legal regulations regarding hygiene and public health,
- the social acceptance of unisex facilities,
- the right balance between facilities for men and women (equal access),
- the need to adapt the facilities to the requirements of people with disabilities,
- the need to adapt the facilities for small children use (mothers with small children),
- the possibility of using the area by larger tour groups, including school groups [1].

Hygiene and public health regulations, as well as other arrangements, may help to establish a minimum level of sanitary facilities provision in the area. Because these laws are different in different countries, there is no point in citing them, but it is worth stressing that they need to be analyzed. If workers are employed in a given area, for example as rangers, then there may be completely different provisions related to labor law. Also, the social acceptance related to the use of devices intended for both sexes (without gender segregation) may vary. In many cases, it is assumed that facilities for men and women should be located separately, but universal (unisex) are often used. If a sense of security, privacy and intimacy is assured and a number of users is small, then such an approach may be completely acceptable. This solution avoids unnecessary duplication of objects and reduces the costs associated with the construction and maintenance of facilities. The balance between men's and women's facilities is often difficult to predict and estimate because the needs and requirements of users vary. Generally, men spend less time in the toilet than women. Women, due to anatomical and biological reasons and due to the way they dress, need more time to use the toilet than men. In addition, women often use toilets together with babies and small children. For the above reasons, they have much more needs in terms of using public toilets. They use them more often than men. They also require more space inside the toilet stall [4]. Ways to plan, design and organize public toilets are developed to allow equal access for both women and men. Usually this is regulated by laws and standards. However, sometimes other factors may also influence these proportions. For example, some areas are more often used by men because of certain actions or sports activities. People with disabilities need in most cases separate rooms, equipped with additional space for wheelchair access, special handles, handrails, specialized devices mounted at appropriate heights and other amenities. In a large building it is worth providing toilets for the disabled both in the women's and men's zone. While in a smaller object, one toilet for disabled women and men is probably enough. Another reason for favoring a unisex toilet is the fact that a person with a disability can be accompanied by a person of the opposite sex. In many cases, there is nothing to prevent the able-bodied people from using a special toilet, thus enabling the functioning of a universal sanitary room in the smallest buildings. If the area is used by tour groups, including school groups, it is probably necessary to increase in toilet provision due to the large number of people who need to use the toilet at the same time. In such cases, it is usually expected to use devices for equal numbers of girls and boys [1].

4   Types of Toilet Facilities

The choice of toilet type to be used in a given location is based on the method of the sewage disposal and related construction and installation requirements. Other technical elements are also important, such as the type of ventilation, the type of sanitary equipment, etc.

- **Composting toilet**
  This solution requires a toilet seat, which is located above the container or opening to which all sewage is collected. If the moisture content in toilet can be kept reasonably low, for example by adding soil, ash or bark, the impurities break down into relatively odorless compost, which can be emptied from containers located at the back of the toilet building and then safely removed. The use of plastic containers prevents contamination of groundwater. A special bark placed in the toilet after use minimizes the unpleasant odor and accelerates the composting process. This type of toilet does not require water. The device can be constructed above the ground. This is particularly important in areas where groundwater is susceptible to contamination, where the soil is rocky or susceptible to freezing in the winter, making it difficult to dig a hole in it. This type of dry toilet is common in Scandinavia and increasingly found in North America. The interest and popularity of this low-budget, sustainable sanitary technique is gradually increasing [1].

- **Vault toilet (pit toilet, pit latrine)**

  This type of a dry toilet is similar to a composting toilet. It differs from it with greater capacity and tendency to accumulate more moisture. The main disadvantage of such a toilet is the unpleasant odor, intense especially in hot weather. Some users,

Fig. 1. Example of composting toilet, Skuleskogen National Park, Umea, Sweden. (Source: own work)

Fig. 2. Example of vault toilet, Balos Lagoon, Kissamos, Crete, Greece. (Source: own work)

especially children, may be afraid of a possible falling into it. The use of adequate ventilation in this solution is not always possible. Some of the toilets can be ventilated by fans powered by solar or rechargeable batteries. Pumping out impurities is done using special trucks. Such toilets can also be equipped with chemicals that help to break down impurities and mask unpleasant odors. Both this type of toilet and composting toilet can be used in places where there is no access to water and power. Composting toilets may only be used with low consumption. The vault toilets also provides intensive use. Next types of toilets require access to water and power supply to be able to operate [1] (Figs. 1 and 2).

- **Flush toilet with cesspit or septic tank**
  This type of toilet requires a constant supply of water in order to operate the flushing system. It can be obtained by means of connection to the water supply service or water intake from a nearby lake or river. In some cases, rainwater can be collected and stored in special tanks. Flush toilets and wash basins with running water help to maintain hygiene in public toilets. They also enable regular and easy cleaning. Sewage is removed to a septic tank, which can be connected to a soak-away field or pumped out [1].

- **Chemical toilet**
  This type of toilet operates with a special chemical that neutralizes the smell of sewage. The toilet does not have to use water, but it needs some power in order to the substance to be recirculated and to work as a flushing agent. This system is most often used in temporary objects, portable, mobile, tourist toilets, but it can also be used in permanent places. The chemical causes the sewage to be sterile and, if necessary, it can be stored for a long time. They must be pumped out periodically in order to be removed. Considering sustainable design, this type of toilet is not recommended, due to the need to use and remove large amounts of sterilizing chemicals affecting the environment [1].

- **Flush toilet with access to service (water, drainage, electricity)**
  In some landscape areas, full media services are available at a reasonable distance from the toilet place. Usually, they are located along a public road and can be connected to facilities with short links. They allow to provide the highest standard of sanitary facility, with full flushing, cold and hot running water for washing hands, with lighting, heating, ventilation, electric equipment, for example, hand dryers. Most users expect such a standard everywhere. This is due to some concerns related to the use of toilets that are not flushed. However, well-equipped, clean and properly maintained toilets, no matter if flushed or dry, usually do not cause fears and complaints [1].

5 Designing Public Toilet Facilities in the Landscape Areas

When designing public toilets in landscape areas, solutions chosen from a wide range of examples of sanitary objects that are widely used throughout the world can be used. It might be anticipated that some standard systems, forms, technical and construction solutions might have evolved over the years based on practices and experiences, but it happens that some mistakes are still duplicated in new toilet objects. The basic principle of a good toilet building design is a skillful combination of the internal functional and spatial layout and sanitary equipment with the form of the building, which in turn fits into the surrounding landscape without intrusion. The solution to this problem should be balanced with the considered location of the toilet facility, preferred as inviting and friendly, and not hidden in the out of reach or camouflaged out of sight. It happens that the toilet is located away from exposed places, parking lots, open ground, in the shade of trees. Such location of the object causes that it is hidden, camouflaged, and its unfriendly setting may cause anxiety among some users, for example women and children. It is also not the right way to harmoniously connect with the landscape. Many used forms of toilet buildings have inappropriate proportions in relation to the landscape, which negatively affects the visual perception of them by observers. Sometimes their appearance resembles backyard latrines. Due to the limited horizontal plan, their vertical proportions are strongly emphasized and stand out in the open landscape. If there are two identical objects next to each other, the result is stronger. Often the scale of these buildings seems to be too small compared to standard materials used on external walls and roofs (shingles, roof tiles, cladding, elevation boards, etc.). Every place located in the natural landscape in which the toilet object is to be placed should be considered separately, and the type of building, its size, form and material solutions should be selected to match the essential character of the place [1].

When considering the location of a toilet facility in a woody area, the specificity of the place should be taken into account. The forest has many vertical forms and lines in its structure. The scale of the space depends on the degree of tree density. Subdued, neutral, earthy colors and coarse, raw textures are specific to the forest landscape. These features can be expressed in the designed object through a raw, rough, minimalistic form. Expressive, coarse textures of natural materials can additionally increase the sense of reference to the "forest" style. Local stone, wood, roof covered with slate or grass can enable harmonious inclusion in the forest surroundings. The location of the object hidden in a hollow or "nestled" into a hill can emphasize the effect of the building "growing out" from the landscape. In open spaces, glades, it is worth considering the use of horizontal forms and lines that can give the effect of a building that "hugs" the ground [1] (Fig. 3).

Similar assumptions may be associated with the design of public toilets in parks. If it is a city park, it is worth considering the application of noble, elegant material solutions. Smooth, polished, reflective materials such as glass, mirror surfaces, polished steel cause reflection of the surroundings, and thus a clear reference to the landscape, without disturbing it. Correlation with the park surroundings can be achieved by breaking up the building block into smaller objects scattered in the existing greenery [9]. However, in many cases such a solution is not recommended and it is preferable to group the facilities in one place (Fig. 4).

Fig. 3. Example of toilet with trunks in the forest. (Source: student project depicted by M. Pilarczyk, tutor - A. Jaglarz)

Fig. 4. Example of friendly toilet in the park. (Source: student project depicted by D. Malinowska, tutor - A. Jaglarz)

The location of a sanitary facility in a rural setting may suggest forms based on vernacular architecture, referring to local farmsteads, huts, barns, small farm buildings. These references can be strengthened with the help of characteristic architectural, structural and material elements that correspond to the existing rural landscape. The use

Fig. 5. Sharp form of toilet in the mountains. (Source: student project depicted by K. Kucharzyszyn, tutor - A. Jaglarz)

of traditional colors of doors, window frames, details and fencing elements may also be helpful [1].

If the toilet object is to be built in the seaside area, it may be worth considering the opening of the landscape coming out of its interior, for example by the form of a periscope, which will allow users to have direct eye contact with the marine environment. The building - periscope can emphasize the tourist character and seaside theme of the place and at the same time affect the undisturbed integration with the landscape. The shape of the monolithic, concrete building block may result from the hidden construction of the periscope, while the rounded facade and arcades as well as raw materials will allow it to become similar to the breakwater and sea waves. The wooden trim will gently warm the raw material and minimalist form of the building, and pebbles will suggest a seaside atmosphere [5, 8].

When designing a toilet building in high mountain terrain, it is worth referring to a location that suggests angular, sharp forms and raw, natural materials. The use of a stone coating make the object look like a piece of rock. The glass reflects the surrounding landscape, and at the same time allows users to observe a nearby cliff or waterfall. "Hugging" to the slope allows maximum integration with the mountains. In the context of the raw, rough exterior walls of the building, its interior can be finished with wood, that harmonizes with the greenery growing on the slope [6].

An alternative to all locations is to design a neutral, simple form of sanitary object that does not borrow anything from the environment. Through the simplicity and neutrality of its shape and material solutions it fits the landscape and is not an intruder, because takes second place in relation to the rest of the surroundings [1] (Fig. 5).

It is important that the form of the building is shaped based on the needs and requirements of the toilet facilities. This may allow to obtain a proper functional and spatial solution. For this reason, it is necessary to analyze several problems that affect the design assumptions [1].

- **The need for a lobby or vestibule**
  When designing single sanitary cabins without hand-washing facilities, no vestibules are planned - the cabins are opened directly into the surroundings. In larger objects where toilets for women and men are separated and equipped with hygienic hand washing appliances, there is usually a lobby area from which users can access each toilet. An additional reason for designing the lobby is to provide a covered shelter in the case of queues. Separate lobbies in front of toilets for men and women are more preferred than the shared lobby in front of the shared toilet. Another solution involves the location of entrances to the appropriate toilet units at the opposite ends of the building, so that they are out of sight. An appropriately extended roof can provide sufficient shelter and help arrange the entrance zone - vestibule, without the need to separate the lobby from the part of the internal layout [1].

- **Providing hygienic devices for hand washing**
  For this purpose, additional space should be provided in the toilet cabin, which will allow the installation of a washbasin. If there is more than one toilet in each unit, the vestibule (lobby) should be used for the location of the washbasins. Even if there are more sanitary cabins, one or two washbasins with accessories (soap dispensers, paper towel dispensers or electric dryers, mirrors) should be enough. This zone should accommodate two or three people at the same time [1] (Fig. 6).

- **The way of lighting the interior**
  Windows located in the lower part of the toilet building prove to be impractical because they do not ensure privacy for users. High-level windows (located above eye level), fanlights or skylights are recommended. The bright interiors of toilets provide a sense of friendliness and security. The roof can be perforated or moved away from the top of the walls so that natural light can reach the internal space. This solution also provides natural ventilation. Artificial lighting may be necessary at certain times of the year, but should not be considered as the main form of lighting objects [1].

- **The season of operation of the facility during the year**
  If the toilet is expected to be available throughout the year, provision should be made for heating, protection against freezing, maintenance and other protection. This is also related to the type of materials, equipment and installations [1].

- **The risk of vandalism**
  In places where collective antisocial groups can be expected, a certain degree of vandalism should be anticipated. This affects the type of materials used both construction and finishing, as well as related to equipment and installation. The use of wood may be a fire hazard, interior finishes encourage graffiti, ceramic devices are easily smashed. For this reason, it is preferable to choose materials that help reduce the possibility and effects of vandalism, such as resistant, durable sanitary devices, high-level windows, stainless steel, non-solid surface and finishing materials that facilitate the removal of graffiti, appropriate anti-graffiti preparations and other protection [1].

Fig. 6. The toilet entrance zone design – as a lobby, shelter, tourist information and relax place, as a bicycle rack. (Source: student project depicted by P. Cygnar, tutor - A. Jaglarz)

6 Final Conclusions

Sanitary objects located in landscape places provide a range of facilities that support the needs and thus the comfort of users. The design, deployment and organization of these objects is crucial because they have a visual and smelling effect on the surroundings and may be elements that disturb the perception of the landscape. The guarantee that the designed buildings will be sensitive to the surrounding landscape is the main goal. Their form, scale, material solutions and choice and location method have a significant impact on the natural, subtle features that define landscape areas. The shaping of these buildings should be a positive contribution and a harmonious complement to the landscape while meeting the needs and functional requirements of a diverse group of users [7].

Criteria for the design of hygienic and sanitary objects surrounded by natural landscapes:

- **Selection of the appropriate location of the object in the context of the site**
  - taking into account how an object can be adapted for possible increase of use and supporting infrastructure, such as paths, access to vehicles, parking lots (objects may affect the increase of pedestrian and vehicular traffic in a given area, which is a strain for a sensitive environment),
  - locating objects into the landscape avoiding open spaces, on the outskirts, in order to minimize negative visual impacts,
  - combining or grouping structures as much as possible to minimize the overall visual impact,
  - locating objects on the background of greenery to reduce the visual impact,
  - locating objects near the area with the highest activity to optimize their use,
  - providing safe and legible access, without confusing camouflage,
  - analysis of the time and manner of using facilities by users in order to provide additional means of transport or necessary parking spaces [7].
- **Designing the appropriate form of the object and the internal functional and spatial arrangement**
  - ensuring that the building corresponds with the particular history of the place, culture, landscape and local community,
  - the use of forms and materials that can help in the interpretation and reference,
  - careful selection of the cladding of buildings, bearing in mind their integration with the surrounding landscape, and at the same time resistance to graffiti,
  - defining the range of user groups and their individual needs and requirements,
  - ensuring accessibility for users of all ages and physical abilities,
  - if possible, shaping multifunctional structures to ensure the best practicable use,
  - if possible, incorporating flexibility into the building design so that it meets a number of current or future user needs,
  - orienting views from objects towards the landscape to emphasize important view lines,
  - providing basic amenities and toiletries, including soap, toilet paper and hand dryers or paper towel, to promote hygiene and public health [7].
- **Achieving a sustainable building architecture**
  - the use of energy efficient design principles, the use of energy efficient light sources, the use of solar energy, the use of renewable materials,
  - the use of on-site generated energy (if possible) and including it within the implementation costs (e.g. taking into account the costs of solar panels),
  - the use of the sustainable water management function, the use of water efficient appliances, for example low-flush toilets, and water-saving taps,
  - collection and use of rainwater for flushing toilets,
  - the use of green roofs for rainwater detention and filtration,
  - providing external shady places for shelter [7].

- **Ensuring well maintenance and management of the facility**
  - allowing easy regular maintenance,
  - ensuring easy access for service vehicles,
  - a selection of building materials that can be easily transported to a location, preferably local materials,
  - the use of resistant materials and strategic solutions aimed at counteracting acts of vandalism [7].

References

1. Bell, S.: Design for Outdoor Recreation, pp. 72–84. Spon Press, Taylor & Francis Group, London (2005)
2. Bell, S.: Elements of Visual Design in the Landscape, p. 2. Spon Press, Taylor & Francis Group, London, New York (2004)
3. Carmona, M., Tiesdell, S., Heath, T., Oc, T.: Public Places – Urban Spaces: The Dimensions of Urban Design, 2nd edn, p. 169. Routledge, Taylor & Francis Group, London, New York (2010)
4. Molotch, H., Noren, L.: Toilet: Public Restrooms and the Politics of Sharing, pp. 118–119. New York University Press, New York, London (2010)
5. Nadmorski Peryskop – Projekt Toalety Publicznej w Gdyni. http://www.archinea.pl
6. Takie Rzeczy Tylko w Norwegii: Toaleta z Widokiem na Wodospad. https://www.mojanorwegia.pl
7. The importance of buildings, shelters and toilets. http://www.aucklanddesignmanual.co.nz
8. Toaleta z Pieknym Widokiem, http://www.sztuka-krajobrazu.pl
9. Wasilkowska, A.: Shadow Architecture – Architektura Cienia, p. 231. Fundacja Inna Przestrzen, Warszawa (2014)

Human Scale in Architecture of Schools Located in Dense Urban Fabric

Andrzej Dudzinski[✉]

Faculty of Architecture, Bialystok University of Technology,
ul. Oskara Sosnowskiego 11, 15-893 Bialystok, Poland
a.dudzinski@pb.edu.pl

Abstract. Designing new schools located in high density cities usually need to face complex, and sometimes contradicting requirements. With problems finding proper building lot in already built environment to begin with, and much higher cost of land comparing to the suburbs, in many cases it is a huge challenge for architects to balance the needs of users against feasibility of the project. Dense urban fabric is usually associated with high population density and overcrowded school districts and classrooms. On the other hand there are existing neighborhoods with its unique residential atmosphere, which should determine the scale of new public buildings located there. Therefore the planning and design of new schools should be performed with great attention to the requirements and needs of users, but with respect to already existing environment. Well-designed schools can influence children in a positive way. Architecture of schools plays significant role in child development, therefore its human scale should be one of its most important factors. This paper presents some of the latest school projects completed in New York as representation of educational facilities located in dense urban fabric. Author shows results of the analysis of scale of given examples in relations with the adjacent space. The subject of Author's research is based on his active participation in the design and implementation of schools in New York.

Keywords: School architecture · Schools in New York · Design of school

1 Introduction

Urbanization processes in recent decades have been strongly accelerated, and forecasts indicate further development of this trend. More than half of the world's population lives in cities today. According to the UN forecasts, in the next twenty years, 90% of the total population growth in the world will reach urban residents. The most urbanized continent is North America (without Latin and Caribbean) where 82% of the population lives in cities [1]. In 2014, urban areas in the world inhabited 3.9 billion people, or 54% of the population. By 2050, it is to be 66%, or 2.5 billion more [2]. Figure 1 shows these changes in relations between the population of cities and villages in the world. We are the generation that witnesses those demographic changes, and we can notice that in some parts of the World actual numbers are already higher than estimated in the projections.

© Springer International Publishing AG, part of Springer Nature 2019
J. Charytonowicz and C. Falcão (Eds.): AHFE 2018, AISC 788, pp. 377–386, 2019.
https://doi.org/10.1007/978-3-319-94199-8_36

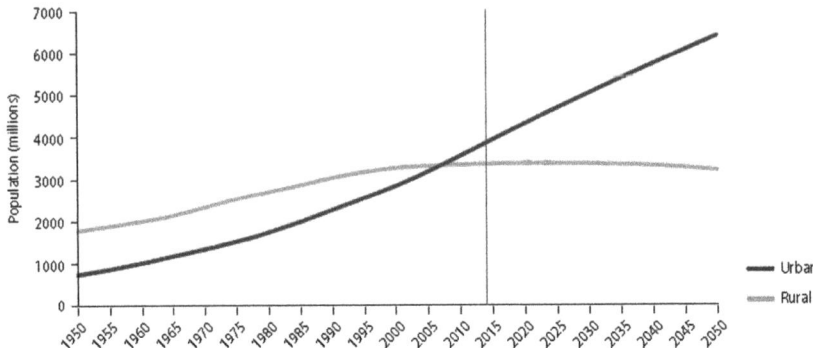

Fig. 1. The hitherto and projected diversification of the population of cities and villages in the World. It shows that in 2007, the urban population predominated for the first time over the rural population [2].

This trend seems to be irreversible therefore urban planners and architects need to adapt quickly to rapidly changing circumstances.

One of the megacities that like a magnet attract mostly new immigrants, but also young Americans looking for great opportunities it offers is New York. This city is inhabited by about 40% of the New York state population [3]. It is estimated that by 2030, the city's population will range from 9.2 to 9.5 million people (Table 1).

Table 1. The forecast population size of New York in 2000–2030 divided into boroughs [4].

	2000	2010	2020	2030	Change 2000–2030
NYC	8,008,278	8,402,213	8,692,564	9,119,811	13,9%
Bronx	1,332,650	1,401,194	1,420,277	1,457,039	9,3%
Brooklyn	2,465,326	2,566,836	2,628,211	2,718,967	10,3%
Manhattan	1,537,195	1,662,701	1,729,530	1,826,547	18,8%
Queens	2,229,379	2,279,949	2,396,949	2,565,352	15,1%
Staten Island	443,728	491,808	517,597	551,906	24,4%

In view of the above trends, it is obvious that with the increase in the population of cities, there are needs whose satisfaction is the basis for their proper functioning. Ensuring the efficient organization of schools, education should be one of the priority tasks of the local communities.

The choice of New York for the analysis being the subject of this article is primarily due to the fact that it is the largest school district in the USA, serving 1.1 million students in over 1800 schools [5].

In the case of new design and implementation of new school buildings the main determinants, often very individual and specific architectural solutions, are the conditions resulting from the location in the existing, intense and dense urban fabric. One of the questions that remain open is how to find and maintain human scale in architecture of schools located in dense urban fabric. Is it possible in a megacity like New York?

2 New York – Magic of the City

It is difficult to find an unambiguous answer to the questions related to proper scale of the city and its elements for many reasons, mostly because architectural design is a field infused with subjectivity. Subjective is also human perception where we touch concept of perceptual psychology, which is a subfield of cognitive psychology that is concerned specifically with the pre-conscious innate aspects of the human cognitive system [6]. Human perception of scale is always affected by cultural background, individual experiences, personal preferences, expectations, and many other factors. For some people the space, scale, and energy of New York would be unbearable, but there are people that cannot imagine living anywhere else. The skyline of Manhattan could be described as unhuman space by some, but many of us would admire the beauty of its skyscrapers, and find it acceptable living in such environment.

Christopher Alexander in his "A Pattern Language" book in chapter "Magic of the city" admits that "only a city such as New York can support a restaurant where you can eat chocolate-covered ants, or buy three-hundred-year-old book of poems, or find Caribbean steel band playing with American folk singers". That perfectly describes diversity and energy of the city. But some pages later of the same book we can read "There is abundant evidence to show that high buildings make people crazy (…). In any urban area, no matter how dense, keep the majority of buildings four stories high or less. It is possible that certain buildings should exceed this limit, but they should never be buildings for human habitation" [7]. This type of postulate is not possible in most modern cities, but it is worth taking into account the impact of high buildings on the human psyche at the stage of spatial planning (Fig. 2).

Fig. 2. Views of downtown Manhattan, phot. A. Dudzinski

New York – as many other big cities – is not an easy place to live for families bringing up children, the elderly, needing care etc. In most cases young New Yorkers as soon as they think about starting a family they plan moving to suburbs. But on the other hand we have to admit that living in such a large city brings undeniably a huge amount of attractions, such as access to theaters, cinemas, museums, sports centers, etc. In recent years New York with its lower crime rate, green and sustainable initiative, access to parks and recreational centers, bike lines, pedestrian zones etc. became much more family friendly than in the past. Quality of life of families living in a city like New York is also dependent on the quality of schools available (Fig. 3).

Fig. 3. Family scenes in New York. Skyscrapers of Manhattan visible in the background, phot. A. Dudzinski

Contemporary solutions being implemented in modern city planning, and architectural designs make cities more livable. Concepts of "living cities" are not new, and theories have been already tested and put into practice, for example in work of Danish architect Gehl [8]. He carefully observes how people behave in public spaces, then analyzes on how and where they stand, sit, talk, walk, how they use urban space. As a conclusion comes with ideas on how to develop space that favors its social use. Also in the documentary "The Human Scale" - which is above all a great lecture by Jan Gehl – we can see change in the way people think about a big city [9]. The example of New York, which director Andreas M. Dalsgaard relies on in his document, testifies to the way of thinking about the city and the lifestyle of its inhabitants.

The dynamic development concerning both the quantity and quality of public school infrastructure in New York, created in recent years, made the quality of education of children and youth, and hence the quality of life and prospects of the residents of this city significantly improved.

3   New Public Schools in New York

The beginning of the 21st century is a special period for the development of infrastructure related to education in New York, mainly due to broadly understood policy at both national and local level. According to data published by the Department of Education of New York, since 2002 - that is, since the beginning of the first term of Michael Bloomberg as mayor, – 656 new schools have been opened [10]. This is an impressive number, even for this city. The number of school investments is due to enormous demographic needs.

Designing new schools located in high density cities usually need to face complex, and sometimes contradicting requirements. The location of schools in intensive urban tissue often causes nuisance related to communication noise, air pollution, and increased risk of accidents. Lack of recreational areas is often the reason for using for this purpose any terraces, roofs or even parts of underground buildings. The insufficiency of investment areas often results in the location of schools in a nuisance neighborhood.

The urban environment also has its advantages. School buildings are usually easily accessible, both in terms of public transport and existing walking routes. A city such as New York has to offer a huge number of cultural facilities, museums, theaters, cinemas, libraries, sports facilities etc. which pupils can use both as part of school programs and as additional classes. Living in a multicultural environment, they have a greater opportunity to interact with peers with a different skin color, religion and country of origin. Undoubtedly, these children are well adapted to living in a modern, globalized world [11]. Well-designed schools can influence children in a positive way. Architecture of schools plays significant role in child development, therefore its human scale should be one of its most important factors. But what is the human scale in a city like New York?

The human scale as a term does not have an unambiguous definition and is used intuitively, usually when describing architectural objects that do not have this feature.

As explains architect Christian Sottile "...human-based design principles could provide part of this answer. They cross the boundaries of time, style, and history because they are derived from human scale, form, and psychology. Understanding and using them does not limit the creativity of architects; in fact, quite the opposite. Architectural styles can continue evolving while a common link is maintained between buildings within the city. This link is based on an approach that puts the primary visual purpose of the building first: human perception..." [12]. Then he lists eight principles that can be observed and measured within the built environment: Materials, Composition, Scale, Proportion, Rhythm, Transparency, Articulation, Expression.

Selected examples of representative school buildings built in the years 2000–2015 in New York were created as a result of the activities of the SCA agency, which determined the entire investment process. Despite the standardization of individual solutions imposed by the investor, as well as similar functional requirements, various spatial solutions were obtained. Most of the locations of these objects should be described as very difficult. Individual design solutions are determined mainly by limitations resulting from the existing, intense, compact urban development. Often, in order to carry out the investment, it was necessary to demolish existing buildings.

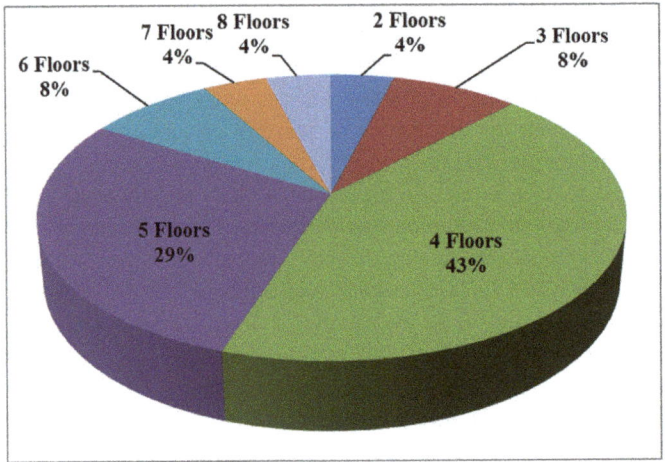

Fig. 4. Number of floors in new public elementary schools built in New York in the years 2000–2015. The diagram is the result of the author's own research.

Fig. 5. Front elevation of PS 244Q- The Active Learning Elementary School. Four floors primary school surrounded by much higher residential towers. Distinguished main entrance feature at the right scale. phot. A. Dudzinski

Fig. 6. Left – Main entrance at PS244Q. Right - Aerial view of PS 244Q. phot. A. Dudzinski

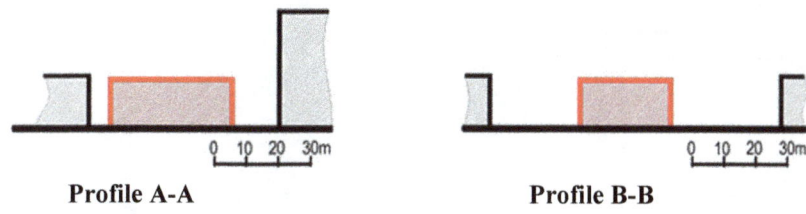

Profile A-A   **Profile B-B**

Fig. 7. Profiles of PS 244Q (indicated on Fig. 6) with its surroundings. Source: Drawings are the result of the author's own research.

The author of this article tries to focus on the positive aspects of contemporary public schools in New York, being aware of the disadvantages associated with a specific location.

Bearing in mind the issues of the scale of the building, which is primarily adapted to the users - that is, to children of school age - but also implemented in the existing city fabric, it seems appropriate that most new schools do not extend beyond the four stories. Anything beyond that limit seems to be excessive, even in a city of that scale (Figs. 4, 5 and 7).

The limited area of the plot – which is the usual deficiency in schools located in dense urban fabric - affects the reduction of outside areas designated for recreation, sports and spaces such as the pre-entrance space, which is a kind of link between the outside world and the school building interior, but also often performs more complex functions than just a communication buffer. The external space of the main entrance to the school building is formally a very important part of it. It should be its showcase, but also meet specific functional requirements related to security, compositional, aesthetic, etc. The entrance area to the school building, as a keystone for external and external space, requires a detailed study not only for functional and aesthetic reasons. "The main entrance is a showcase of the school, the most memorable of its fragment, as it is the first impression in the meeting with the building. It should be distinguished from other inputs in a determined way" [13].

Another aspect of the proper functioning of the educational facility in the context of the attempt to shape a city with a human scale worth mentioning is the way the student reaches his school every day. Architects rarely have an impact on this issue, because it results from the size of school districts that the given branch handles. This one is a derivative of the settlement structure and demographic conditions. However, despite the American traditions associated with both private and public transport as well as organized by schools (yellow busses) it is worth remembering that the most appropriate solution is the possibility of safe walking to school. An important element (although not less difficult to meet) is the attractiveness of the way to school, due to its cyclical overcoming.

Fig. 8. Main entrance at PS 273Q. Clearly articulated part of the building. On the left masonry pier with the artwork displayed. phot. A. Dudzinski

Another interesting feature of Public Schools in New York is participation of the investor (City represented by School Construction Authority-SCA) in program called Public Art for Public Schools (PAPS). This unit within the New York City School Construction Authority, was established in 1989 to oversee the Department of Education's collection of permanent artwork in the city's schools. The program's mission is to commission new projects for new schools, and to maintain and preserve artwork in existing schools. The ultimate goal is for every school in the city to possess at least one

Fig. 9. Terracotta panels insets in masonry piers of the fence around school PS273Q in Richmond Hill-Queens, artist: Janet Morgan, "The Circus Comes to Town" phot. A. Dudzinski

Fig. 10. Left - New Settlement Community Campus, artist: Scott Trimble, "Paths of Discovery". Right - IS/HS 362, artist: Amy Yoes, "Spectrum" Source: [14]

professional artwork, thereby visually, artistically, and educationally enhancing school learning environments [13]. That makes every school in New York unique. Pupils can easily identify their school, not only by regular features of the building, but also something more personal – artwork, that can be interpreted differently by every spectator.

One example of artwork displayed are colorful terracotta panels inset in masonry piers of the fence around school PS273Q in Richmond Hill-Queens shown in Figs. 8, 9 and 10.

4 Conclusion

Designing educational facilities in the dense urban fabric of modern cities is not a simple task. The correct formation, planning, should be the subject of careful analysis already at the pre-design stage.

Too often, poor spatial solutions are noticed today, which, especially in primary level schools, may result in not only lowering the quality of space around the school building, but, what is worse, it may negatively affect the level of user safety.

The building's design should respond to its urban setting in a densely populated neighborhood, and implement contemporary urban planning theories and approaches raised in the 21st century in order to provide a high and sustainable quality of life. City planners and architects need to take into account psychological and social aspects of the design of buildings or public spaces in order to find, balance and maintain human scale of the architecture they create.

Contemporary public schools implementations located in the environment of intensive urban development in most cases have a significant impact on their neighborhoods not only enriching existing public spaces, but also in the social sense by integrating both existing and new communities living around them.

References

1. World Urbanization Prospects. The 2014 Revision. United Nations, New York (2014). https://esa.un.org/unpd/wup/publications/files/wup2014-highlights.Pdf. Accessed Dec 2017
2. https://esa.un.org/unpd/wup/publications/files/wup2014-highlights.Pdf. Accessed Dec 2017
3. http://www.nyc.gov/html/dcp/html/census/popcur.shtml
4. New York City Population Projections by Age/Sex & Borough 2000–2030 REPORT, THE CITY OF NEW YORK DEPARTMENT OF CITY PLANNING, December 2006. https://www1.nyc.gov/assets/planning/download/pdf/data-maps/nyc-population/projections_report.pdf. Accessed Dec 2017
5. http://schools.nyc.gov/AboutUs/default.htm. Accessed Dec 2017
6. https://en.wikipedia.org/wiki/Perceptual_psychology. Accessed Dec 2017
7. Alexander, Ch., Ishikawa, S., Silverstein, M.: A Pattern Language. Oxford University Press, New York (1977)
8. Gehl, J.: Life Between Buildings, Using Public Space. Island Press, Washington DC (2011)
9. The Human Scale – documentary movie, dir. Andreas M. Dalsgaard, Denmark (2012)
10. http://schools.nyc.gov/Offices/mediarelations/NewsandSpeeches/2012-2013/040213_administrationincityhistory.htm
11. Dudzinski, A.: The development of New York public school architecture in the first decade of the 21st century, Architecturae et Artibus, Oficyna Wydawnicza Politechniki Bialostockiej, Bialystok (2013)
12. Sottile, Ch.: The Humane Principles of Good Buildings. https://www.preservationsociety.org/blog/2017/02/14/the-humane-principles-of-good-buildings/. Accessed Dec 2017
13. Włodarczyk, J.: Architektura Szkoły Arkady, Warszawa (1992)
14. http://www.nycsca.org/Community/Public-Art-for-Public-Schools/About-Us

Ergonomics as the Common Denominator and Vital Condition to Achieve Sustainability of Buildings of Different Types on Example of Two Built Projects

Pawel Horn[(✉)]

Department of Housing, Industrial Architecture and Interiors,
Faculty of Architecture, Wrocław University of Science and Technology,
B. Prusa Street 53/55, 50-317 Wroclaw, Poland
horn@hornarchitekci.pl

Abstract. The author searches for universal aspects in design of buildings of different types to show how ergonomics becomes not only common denominator but also indispensable condition for sustainability of a building independent of its type. Ergonomics starts from human needs, to assure best working and living conditions in terms of health, comfort and efficiency. Buildings are designed for humans, and without this important human measure may become unfriendly or even harmful. Sustainability is overall human approach to protect our own kind at present with consequences for future. From the point of view of the architect it means to make our built environment not only best quality for ourselves, but harmonized with natural as well. To achieve this, our human measure has to be applied properly at the stage of design of our built environment. This reference plane remains constant independent of the type of human activities which architecture is designed to shelter. In the article the author as the practicing architect shows general or repetitive aspects in ergonomic design as universal conditions of sustainability, taking as the examples housing estate and 3D printing factory, recently completed (2016–2017).

Keywords: Sustainable architecture · Ergonomics in architectural design
Sustainable ergonomics · Sustainable architectural design

1 Introduction

Architecture is one of the domains, where sustainable development[1] theory is visible in terms of influence on methodology of design and consequently in spatial character, functionality and technical parameters of new built objects. After the formulation of

---

[1] The author assumes that the term 'sustainable development' is nowadays already clear and popular enough so as not to be explained. However, for better understanding of sustainability in built environment: "In 1987, the Bruntland Commission published its report, Our Common Future, in an effort to link the issues of economic development and environmental stability. In doing so, this report provided the oft-cited definition of sustainable development as "development that meets the needs of the present without compromising the ability of future generations to meet their own needs" (United Nations General Assembly, 1987, p. 43). Albeit somewhat vague, this concept of sustainable development aims to maintain economic advancement and progress while protecting the long-term value of the environment" [1].

© Springer International Publishing AG, part of Springer Nature 2019
J. Charytonowicz and C. Falcão (Eds.): AHFE 2018, AISC 788, pp. 387–399, 2019.
https://doi.org/10.1007/978-3-319-94199-8_37

objectives and principles of sustainable development, countries are introducing new policies and regulations according to guiding decisions made on global scale by agreements of their representatives. Creators of the built environment, and architects among them are faced with the need to adapt their philosophy and know-how to the demands of this direction, with discussion of the topics in professional circles as natural consequence. This trend can be observed especially since the nineties of XX century[2], when architectural domain in Poland witnessed first attempts to define precisely sustainability in reference to design and buildings with sustainable design prerequisites. Architectural practice absorbs these trends and indications from the two sources. First is top-down approach - country's governmental formal and branch regulations (building standards) and the other is professional environment adopting to the new demands in the bottom-up approach, which is more time-consuming in the process of academic and self-education, expanding know-how and implementing science in practice.

In one of the Polish flagship studies on this topics from nineties, sustainable design prerequisites were described as following contexts: "ecological consciousness and thinking, eco philosophy and environmental ethics, new paradigm of science, holism, inclusion of evaluation in cognitive and design processes, new social paradigm, revival of the concept of social bonds, common ecological education, environmental-behaviour-design, concept of ecological costs, green economy, alternative/environment friendly technologies based on renewable resources, resource-time-energy spatial concepts, searching for a new aesthetics, meaning as a tool for shaping of users' behaviours, spatial structure as a communique" [2]. These contexts were followed by implications for sustainable design, which lead to formulation of the rules for practice including: deriving values from environmental ethics, minimalizing pressure on environment in spatial planning and transformation, respect for resources with special consideration of space as a limited welfare, consideration for the space-time character of spatial structures and their complete life-cycle, highest possible integration of social, cultural, economic, ecologic and spatial aspects, allowing effect of synergy [2]. Literature referring to programming and design methodologies shows adapting design in architecture to sustainable approach by incorporating research methodology (examples: [3–5]). In practice, sustainability is actually a new dimension in design, while retaining its fundamental principles. This article is a search for place of ergonomics in architectural design in the context of sustainable development. It is based on the analysis of studies (publications) referring to discussion of ergonomics against sustainable development. Conclusions from the analysis are then referred to architectural practice on examples of built objects. The article follows questions raised by specialists of both domains regarding mutual dependences and influences, to find the level of convergence.

---

[2] It is the reference to timing in Poland, incorporating sustainability in policy, economy, science and in development in general since nineties; as the declared rule of statutory tasks sustainable development is stated in constitution of the Republic of Poland 1997, article 5 [6]. It can be also observed in scientific publications, with a start of defining sustainable buildings and design.

2   Relationship of Ergonomics and Sustainable Development

Ergonomics is established and indispensable part of the architectural design if it is to respect human as a centre of reference and measures. Thus ergonomics as an approach in design process needs the same shift in the light of sustainable development. From the point of view of architectural practice, with its interdisciplinary character and holistic view, adopting ergonomic design to new demands first of all needs extending the scope of issues considered. The same process is happening in the domain of ergonomics, where scientists involved in discussion make efforts to define the mutual relationship and influence with sustainable development, and similarly to architecture, it takes place at the level of general foundations and detailed scopes of interest. Extensive study on research trends, present in the publications on ergonomics relating to sustainability [7] reported that "since the perspective on sustainable development is relatively new for ergonomists, very few publications are available on where and how ergonomics can contribute well to the sustainable development". This vast search through literature in last two decades (1992–2011) allowed the authors of this study to formulate observations and conclusions: "sustainability issues are currently attracting more and more attention from the scientific community", and in terms of classification method they used, the main research areas in ergonomics are 'methods and techniques', 'human characteristics', 'work design and organization', 'health and safety', the last two steadily increasing (2007–2011). Simultaneously the two researchers specified the most popular research areas in the field of sustainable development: 'environment and social', 'industrial and product design', 'education', 'health and safety', 'renewable energy and technology', and 'architecture'. The aim of their scientific project was to find convergent areas in both disciplines, what turned out to be difficult due to intertwining of categories of ergonomics studies (publications) with more than one categories of sustainable development; "in the field of design for sustainability, ergonomic design (User-Centred Design) is becoming an important design strategy for design culture innovation, providing designers with the necessary knowledge regarding human characteristics (…) 'health and safety' should be important for both ergonomics and sustainable development, since it is the common classifying category in both fields. 'Workspace and equipment design' and 'environmental ergonomics' in ergonomics field can be connected to sustainability features of 'architecture' that promote the wellbeing of the occupants and protect the environment through energy conservation and green buildings", 'health and safety' and 'work design and organization' were indicated as promising also from the point of view of business sustainability in context of economy. General improvements, that ergonomics can bring to the global society through collaborative approach of ergonomics societies, enterprises and governmental authorities, were also identified in the study as possible upon an interdisciplinary design-driven and systems approach.

The authors of above study admitted its limitations, in both fields influencing the search and classification criteria for related publications. Another author investigating links between ergonomics and sustainability on examples present in literature (mainly regarding consequences of failing to apply ergonomic principles to systems designs) discussed the future of ergonomics as 'a missing part of sustainability' [8]. Starting with the examples of ergonomic issues, mainly occupational safety and health, the author pointed out their impact on the health, well-being and welfare of both individuals and society, linking them to environmental implications. In scope of human reliability and errors, environmental effects of neglecting the ergonomic aspects were shown on drastic examples. These lead to important conclusion the author made in terms of human factor: designs should be sustainable not only as green buildings, but also being safe to build, maintain and use, and that "a building that fails to promote the health and welfare of its occupants cannot be considered sustainable". Further examples of health, safety behaviours and inclusion of disabled and elderly issues showed social and economic aspects of sustainable design that was discussed in the context of useful tools, which ergonomics can provide to help solving these issues. Conclusions of the discussion responded exactly to holistic approach principle of the sustainable design, defined as described here earlier in reference to architecture. Formulated by ergonomist, those conclusions sounded like choir with architects' postulates: "In essence, today's successful (sustainable) design implies synergy of the best technical, environmental, ergonomic, economic and social solutions for present and future all the more complex systems." "It is obvious that ergonomists need to promote a holistic approach and multidisciplinary teamwork for successful sustainable design, to make available relevant general guidelines and specific design recommendations to all stakeholders, and to monitor and actively participate in issues and policies for sustainable development" [8].

From the point of view of the practising architect, sustainability imposes new aspects and scopes of issues, which have to be taken into consideration during the design. Architectural design practice is a multilevel process, crucial for the whole investment because it sets a program for the whole life-cycle of a building. In terms of technical data, time and finances, design embraces all stages: initial programming phase and conceptual design, building project and working drawings, construction, use, and dismantling or demolition. In terms of human factors it is decisive for the health, safety, wellbeing and comfort of users. The design stage is also crucial for optimization of all aspects, because a virtual project is representation of a real object to be physically built and used for years in specific climate and site conditions by users, who can additionally influence performance of a building [9]. So in comparison to traditional set of initial parameters, targets and demands driving the investment, sustainability in design is a shift in priorities, time scale considered, it also raises ethical issues in reference to values [10]. To deal with these new aspects, i.e. to be sustainable, contemporary architects, the same as all participants of a building (investment) process – and especially investors – need to follow in understanding and choice of sustainable development principles while all technical, social and financial aspects are coordinated. Investment process is so complex and interdisciplinary nowadays as never before; thus without applying human factors to it and without optimising of all design issues it is difficult to assure intended effect or success. Ergonomic solutions in particular scope of design make possible further optimisation of all issues at higher levels, allowing the building to be sustainable in all aspects. At the level of constant verification and adjusting for further feasibility, ergonomics has its place also as a rule for the very process of design in terms of time, cost and human resources, which also has to be optimised to assure the construction to commence. As seen from the point of view of ergonomist: "A design undergoes many iterations before it is complete. Therefore, an ergonomist's role is to guide this process, and part of this is incorporating sustainable ideas and practices." [11]. Another reason why the ergonomics and optimisation are crucial is the economic cost connected with making decisions comprising many more issues extending beyond a building and its users: relation of a building to its surroundings and the natural environment. Nowadays, mutual relations between user and building and between users are complex and changing. Ergonomics is a condition for optimisation, which in turn assures financial feasibility and human oriented acceptability and comfort. This is also indispensable condition to achieve the postulate of synergy of design aspects. If any of them is a failure, there will be no proper or full effect of synergy. And this synergy, achieved by ergonomic optimization applied to all areas of a design, conditions the three pillars of sustainability: social, ecologic and economic. In practical approach, these three categories intermingle. Each action in one influences another or both of the other two. For example, it is additional cost to build more technically advanced object for better environmental performance; but the cost is calculated in connection to return on investment period. However, it may happen, that the users' inability to adapt to the building's performance and improper use of functional schemes or technical systems can squander the intended gains. Or despite intended performance and costs, the users may not find the building's internal comfort of spatial organization suitable, so the very basic purpose of a design is missed. So "the role of ergonomics might be applied to behavior change for conserving, preserving and restoring other natural resources such as water, air quality or biodiversity", this way extending the focus of Human Factors/Ergonomics profession: effectiveness, safety and ease of performance to take natural systems into account in its (ergonomics') new green alternative [12].

3 Examples of Synergy in Sustainable Ergonomic Architecture

Ergonomic sustainable design, as consistency of complex solutions coming from concept to detail, is presented on two examples of different kinds: recently built housing estate and 3D printing factory in Poland, in the area of city of Wrocław. In the first example, large scale development is based on combined housing and offices complex; in the form of a new city district, it derives advantages of location in close proximity to the city centre and recreational potential of adjacent river harbour. Intensive floor to area ratio is compensated by the added value of utilizing the river banks for beaches, pedestrian and cycle paths, this way bringing back to life the vast area of a post-industrial, downtown barren plot. In this development, housing and office quarters are spatially organised in a form of regular quarters, offering more intimate, internal spaces for recreation, at the same time allowing flow of pedestrians, bikes and air. Revitalization of abandoned area in the middle of the city, making use of the recreational potential of the river and existing city infrastructure (roads, public transport, shopping centre) raises this fairly simple spatial development to the level of achievement in terms of important aspect of sustainability – economic use of space in the big city. Connecting a few thousand new inhabitants and a set of new office quarters to existing services and infrastructure in the dense surrounding of the city fulfils the indication for efficiency and respect in using space and resources (no need for building new road connections or infrastructure services as in case of, for example, distant suburbs or new allocation plots).

Fig. 1. Aerial photo of housing-office development "Promenady Wrocławskie" by developer Vantage Development in Wrocław, Poland. 1c – office quarters, construction, 1e – office quarter (ZITA, EPSILON buildings), existing, 2 – housing quarters, existing and construction, 3 – retail unit, 4 – river Odra embankment with pedestrian and cycling paths. 5 – railroad. Photo: www. google.pl/maps/@51.1312952,17.0465399,582a,35y,244.08h,29.78t/data=!3m1!1e3

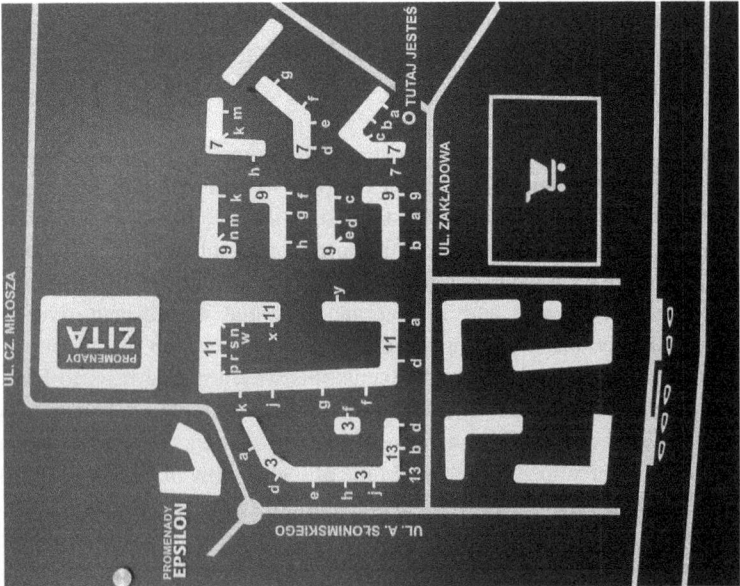

Fig. 2. Info post, photography by the author (image rotated left to match Fig. 1).

Development 'Promenady Wrocławskie' is an example of ergonomics in urban scale, giving the chance for spatially and economically efficient urban complex. This new housing and offices mix responds to the need of accommodation, employment and free time/entertainment opportunities with the advantage of proximity and reduction in commuting time (Figs. 1 and 2). The estate represents ergonomic approach also in the scale of buildings versus their immediate neighbourhood. Housing units are designed individually in terms of stylistics, and the quarters have courtyards with sport and play facilities (basketball playing field, benches, children's playgrounds), compensating large scale of housing blocks. Similar concerns office quarter with recreational court-yard and a restaurant (Fig. 3). Ergonomic optimization coming from urban scale is concluded at architectural scale, preventing big blocks of flats from being unfriendly and inhospitable. In prediction of residents' perception of satisfaction and safety based on research [13] and practical experience (author's own observations of other similar developments, which retain quality and do not show signs of improper use or van-dalism over time), this development seems promising to be a success. However, as a fairly recently finished construction it will show to be a success or a failure.

Fig. 3. Housing estate 'Promenady Wrocławskie': On the left: Office quarter aligned with housing quarters located at the same street; building line, scale and character kept as unifying parameter. On the right: interior of office quarter ZITA, human scale detailing of recreational courtyard. Photographs by the author.

The other example also shows intermingling of various aspects which due to ergo-nomic optimisation allow the built object to be recognised as sustainable. The aim of the design of a new 3D printing factory for the Polish branch of Belgian company Materialise was primarily to create highly functional production unit, that was also expected to be energy efficient. The architect proposed, as added value, high quality of representative office part and all internal spaces: not only production but also offices and rest: foyer, lounge, canteen, lunch rooms (Figs. 4 and 5). Ergonomics of production process was important from the applied human factor point of view. The machines and the system of their operating and control are highly advanced as the production relies on digital data input as a base for 3D printing. Spatial limitations coming from polyamide powder[3] circulation requirements for installations system had to be equally treated as the ergonomics of workspace and technical regulations for production building. Even though the 3D printing process requires very little direct manual operation, the workspace needed to be adjusted to human needs. Further finishing of ready elements (printed models) is another part of production with different spatial and ventilation demands. Thus production process seen as the entirety requires highly advanced installations – material supply, ventilation, electrical and IT - in the whole of the building, consisting of the offices and a production, connected by communication/lounge space.

Fig. 4. Entrance to the office building. 3D printing factory, Materialise, Bielany Wrocławskie. Design Horn Architects, photograph by the author.

Fig. 5. Lounge and communication space, before moving in. 3D printing factory, Materialise, Bielany Wrocławskie. Design Horn Architects, photograph by the author.

---

[3] The process of so-called 3D printing relies on selective laser sintering of polyamide powder. For more details go to http://www.materialise.com/en/manufacturing/3d-printing-technology.

The installations system project places the building among the most technologically advanced in Poland; it is designed to meet the specific requirements of production process, allows energy savings and pleasant internal microclimate for employees. Not merely the installations play environmental role. Sustainability is achieved in environmental aspect optimised with social and economic. Apart from safety in terms of not affecting the neighbourhood area (noise, exhaust, garbage), it is energy efficient due to recovery of waste heat coming from the production process, but also through the design of the building itself and its orientation on the site. In terms of human factor and technical regulations windows and roof skylights allow daylight access to all parts of the building, where people work or walk (Fig. 6). Offices are facing east to protect from overheating, while the sun shading blinds are calculated to allow enough of daylight, what altogether gives savings on energy consumption (artificial lighting, ventilation and air conditioning).

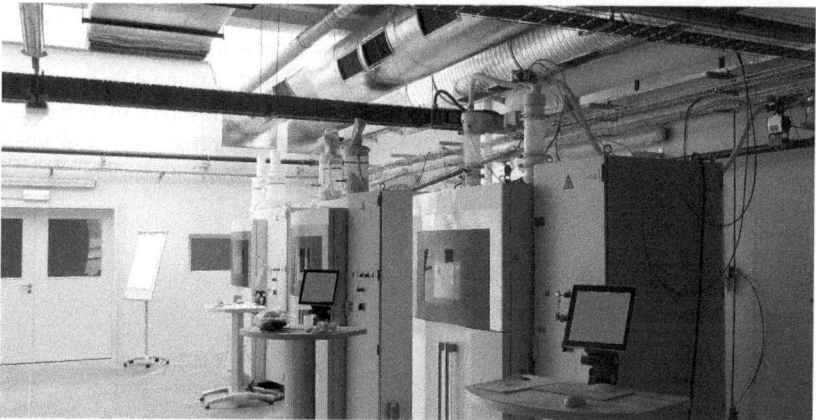

Fig. 6. Laboratory character of production rooms. 3D printing factory, Materialise, Bielany Wrocławskie. Design Horn Architects, photograph by the author.

Fig. 7. Canteen in the office building – common and recreational space open equally to all staff. 3D printing factory, Materialise, Bielany Wrocławskie. Design Horn Architects, photograph by the author.

The objects' orientation on site also acts as natural screening of the offices from the noise coming from the motorway running alongside the site's two borders. These environmental aspects were given priority to the possibility of advertising of the company to people driving on the motorway as the office part of the factory is aesthetically representative (Figs. 7 and 8). However, independent of the direction the object is approached or seen, the integral design of all parts of the factory allows it to be visual sign in the area dominated by monotonous and repetitive logistics and production sheds of uniform look. Visitors to the factory have the possibility to watch processing through windows in the wall between main corridor and the 3D printing rooms or glimpse from outside. The process of optimizing of program, technology and functional layout concept took nine months, and the designer introduced an added value to the project changing the necessary space division between production and office parts (buildings) into a common lounge space for all staff at no additional financial expense. Also interior design follows the concept of laboratory feel of the new type of contemporary production based on emerging technologies and displacement of a human from a machine stand to a comfortable office. This way the Materialise factory in Bielany Wroclawskie (adjacent to the city of Wrocław) is the implementation of the Investor's demand for a highly functional, human oriented, sustainable and fine production building and goes hand in hand with Investor's philosophy of materializing great ideas in high tech way. The building is at the same time recognised in the local commune as a benchmark in terms of top technologies housed in an attractive container of environmental performance close to zero-energy demand.[4]

Fig. 8. 3D printing factory, Materialise, Bielany Wrocławskie. Design Horn Architects, photograph by the author.

These two projects of high complexity and wide spectrum of technical and social issues show the role of ergonomics to achieve synergy effect in optimising all design solutions, which in turn conditions sustainability, independent of a type of a building.

---

[4] The opening of the building was an event at regional scale, with invited officials from Polish administration authorities at local and regional levels, press, investor's and contractor's CEOs and top level management. For more information go to http://www.ugk.pl/fabryka-druku-3d-w-bielanach-wroclawskich.

4 Summary

Optimisation in architectural design process means to achieve the best possible and financially feasible solution of all problems and tasks in a given situation. To understand that, one needs to keep in mind that each architectural object is unique – even in repetitive objects the set of site conditions may very significant to have influence on final result. The architect is the key person to translate an investors' commission into a feasible building project, within the budget, time and financial constraints. As a coordinator of a team of branch engineers – designers and consultants, architect not only comes with proposal of a building's functionality and aesthetics, but is decisive in reference to structure and materials to be used and should be able to foresee aspects of construction in terms of accessibility of chosen technology and local constructors' skills and competences. The future users' participation in design and designer's referring to similar built objects in terms of investigation of their performance and use compliance with design is however still new in architectural practice in Poland at the moment. These additional procedures are connected with new methods of research and obtaining data for design [14]. Thus the job of an architect is highly multi-sided and relies on interdisciplinary cooperation during planning, project and construction phases. However, the aspects that have to be considered are not only material. In architectural design important are also mental comfort of users, individual taste of client, traditional or preferred ways of behavior, cultural background, and many other related to particular client/user/location; also each designer is a human with personal set of social and professional skills, level of knowledge and experience, taste (style) and philosophy (values). In the light of this short practical resume of how architecture works in terms of changing the idea (a project) into reality (a built object) - architectural design is not merely a process of preparation of information for a physical object to be made. It has the aspects of artistic creation combined with precise methodology, referring to professional now-how, vast interdisciplinary reference plane based on cooperation with client and other engineers in a team of designers and contractors. Cooperation in investment process is based on information exchange, managing it, incorporating data, professional knowledge and experience and making decisions during team work of finding solutions. But also it relies on social skills, sharing values, priorities and mercantile targets. The sustainable architectural design itself is subject to many scientific theories, studies and publications, extending purpose of this article. The author hopes that this article helps in understanding discussion of dependences of ergonomics and sustainability in architectural design. Optimization of all design aspects requires an implementation of ergonomics, and allows achieving the highest possible level of sustainability, which is feasible in particular case in terms of cost, available materials and technical solutions, social acceptability, adaptability and consciousness. So ergonomics is inseparable from sustainable design on the level of the very basic definition of each and the convergence of both: the synergy in striving for goals in 'areas of critical importance for humanity and the planet', stated in "Transforming our world: the 2030 Agenda for Sustainable Development":

"People". We are determined to end poverty and hunger, in all their forms and dimensions, and to ensure that all human beings can fulfil their potential in dignity and equality and in a healthy environment.

"Planet". We are determined to protect the planet from degradation, including through sustainable consumption and production, sustainably managing its natural resources and taking urgent action on climate change, so that it can support the needs of the present and future generations.

"Prosperity". We are determined to ensure that all human beings can enjoy prosperous and fulfilling lives and that economic, social and technological progress occurs in harmony with nature. [15]. The aim of this article was to show ergonomics with its special interests in human wellbeing and health and a valuable set of human measurement tools as a vital condition to fulfill targets of sustainable development and a common denominator in environmental design for sustainable architecture.

References

1. Brief for GSDR 2015 The Concept of Sustainable Development: Definition and Defining Principles. https://sustainabledevelopment.un.org/content/documents/5839GSDR%202015_SD_concept_definiton_rev.pdf
2. Baranowski, A.: Sustainable design in architecture. (original title: Projektowanie zrównoważone w architekturze), pp. 93–107 Wydawnictwo Politechniki Gdańskiej, Gdańsk (1998)
3. Duerk, D.P.: Architectural programming. Information Management for Design. Van Nostrand Reinhold, New York (1993)
4. Niezabitowska, E.D.: Methods and research techniques in architecture. (original title: Metody I techniki badawcze w architekturze) Wydawnictwo Politechniki Śląskiej, Gliwice (2014)
5. Van der Voordt, T.J.M., van Wegen, H.B.R.: Architecture in Use. An Introduction to the Programming, Design and Evaluation of Buildings. Architectural Press, Oxford (2005)
6. The Constitution of the Republic of Poland of 2nd April 1997. http://www.sejm.gov.pl/prawo/konst/angielski/kon1.htm
7. Radjiyev, A., Qiu, H., Xiong, Sh, Nam, K.: Ergonomics and sustainable development in the past two decades (1992–2011): Research trends and how ergonomics can contribute to sustainable development. Appl. Ergon. **2015**(46), 67–75 (2015)
8. Pavlovic-Veselinovic, S.: Ergonomics as a missing part of sustainability. Work **49**(3), 395–399 (2014)
9. Baborska-Narożny, M.: POE and BPE evaluations - a postulated standard in British design practice in the transformation period to zero-emission architecture. Chapter in: the Lower Silesian energy-saving house. Conference materials. Collective work edited by Jacek Kasperski (original title: Oceny POE I BPE – postulowany standard w brytyjskiej praktyce projektowej w okresie transformacji do architektury zero-emisyjnej. Rozdział w: Dolnośląski dom energooszczędny Materiały konferencyjne Praca zbiorowa pod redakcją Jacka Kasperskiego), pp. 22–29 Oficyna Wydawnicza Politechniki Wrocławskiej, Wrocław (2011)
10. Idem, R.: Ethical aspects of the sustainable architectural design. Architectus, No. 2(30), pp. 43–46 (2011)
11. Martin, K., Legg, S., Brown, C.: Designing for sustainability: ergonomics - carpe diem. Ergonomics **56**(3), 365–388 (2013)

12. Thatcher, A.: Green ergonomics: definition and scope. Ergonomics **56**(3), 389–398 (2013)
13. Weidemann, S., Anderson, J.R., Butterfield, D.I., O'Donnell, P.M.: Residents' perceptions of satisfaction and safety. A basis for change in multifamily housing. Environ. Behav. **14**(6), 695–724 (1982)
14. Fross, K.: Qualitative research in architectural design on chosen examples. (Original title: Badania jakościowe w projektowaniu architektonicznym na wybranych przykładach). Wydawnictwo Politechniki Śląskiej, Gliwice (2012)
15. United Nations Resolution adopted by the General Assembly on 29 July 2016 70/299. Follow-up and review of the 2030 Agenda for Sustainable Development at the global level. http://www.un.org/ga/search/view_doc.asp?symbol=A/RES/70/299&Lang=E

New Challenges for the Industrial Architecture. Ergonomics on the Edge of a New Era of IT Technology and Deep Learning

Pawel Horn[✉]

Department of Housing, Industrial Architecture and Interiors,
Faculty of Architecture, Wroclaw University of Science and Technology,
B. Prusa Street 53/55, 50-317 Wroclaw, Poland
horn@hornarchitekci.pl

Abstract. Presently we observe a shift of human activity from the traditional methods of manufacturing products for increasingly specialized and evolving robotic and IT systems. For obvious economic and technological reasons this change is first strongly visible in industrial production. Along with technological advances is a dramatic shift of man's place in the production process from the position at the machine to the back of this process as designer, supervisor and controller of information systems that manage production process. This seemingly obvious change results in completely new challenges for both industrial architecture but also the wider built environment as it dramatically reduces the number of jobs with completely new requirements for workplace and its architecture. The author of this article discusses the above issue on the example of the design of technologically advanced 3D printing plant from the point of view of the designer.

Keywords: Industrial architecture · 3D printing · Design of a factory

1 Introduction

From the point of view of an architect, each commission is a challenge. The professional designer of a building or an interior always has to meet investor's requirements in terms of functional needs, economic foundations and immaterial aspects of personal/company preferences or philosophy regarding the activities, which the designed object is to house. In light of this, each functional concept, which is the base for a formal building project is an individual solution responding to the particular set of challenges. In case of projects of objects of one type, for example industrial, some factors are constant like technical and fire safety regulations for this particular type of a building, and some are specific depending on the very production process. Optimization of constant and variable aspects needs ergonomics, especially in industrial architecture, where requirements for human-objects interaction, in this case between factory workers and production machines, impose human measure on technology. Of course, architecture embraces such a wide range of aspects that optimization must take

J. Charytonowicz and C. Falcão (Eds.): AHFE 2018, AISC 788, pp. 400–407, 2019.
https://doi.org/10.1007/978-3-319-94199-8_38

into account many other than only technical and social aspects, like national regulations for fire protection and safety of working conditions and the needs of people in terms of safety and comfort during work time. Very important are cost, time schedule, construction feasibility. The author's observation derived from a practicing architect's experience is that often as important as the previously described factors is immaterial sphere of investment: for example added value for architecture as a representation of enterprise's aspirations or philosophy, or the policy of integrating employees into a team by creating a user friendly and motivating working environment. Another interesting phenomenon which can be observed nowadays is that often a philosophy of a given firm remains while the progress of technology implies new requirements in a short time, or new technologies emerge giving rise to brand new industrial companies. In this article, the author wants to discuss such a situation on the example of a factory designed by his architectural office for the process of additive manufacturing – so called 3D printing (Figs. 1, 2, 3 and 4). In the face of contemporary rapid changes of production technologies, which are not always followed by immediate change in building regulations, designers of industrial objects need ergonomics, as human measure remains constant and can help solving new problems, arising from new demands or the lack of precedent/reference objects.

Fig. 1. 3D printing factory for Materialise, design Horn Architects. On the left – office building with canteen for all staff, on the right – production building with scanning unit and technical roof behind aluminum mesh screen. Both buildings are joined by a connecting space of communication/lounge with green wall and the sculptural, 3D printed tree. The article was written in the final period of construction of the factory (2016–2017). Photo: finish of construction, July 2017. Photograph by the author.

In the design of this industrial unit there was no ready functional scheme[1] – the overall sequence of production stages had to be spatially organized in physical

---

[1] Architect's practice is based not only on professional skills but also experience. However, qualitative research of examples of buildings of the same type as the one currently being designed [1] is still not very popular among professionals. In this case of 3D printing factory, the author and the team of engineers visited main factory in Belgium, to become acquainted with the production and company's

container of a building at the particular site and with respect to Polish climate, building law and building regulations. The discussion of the subject is intended to show how the ergonomic optimization in design process of this factory was influenced by new challenges of this specific, new and developing technology of production. At the same time, the author intention is to show how ergonomics extends material limits of a functional scheme. It happens in aesthetic sphere, for example, when a building is designed to be visual sign of the company[2] (Fig. 1) and to meet company's philosophy for organization of work and integrating of staff (Fig. 2). Presentation of this case study is intended to show, that it often can be done within the same economic budget owing to architects' skills, foresight as well as an invention in approach to investors' brief for the building.

Fig. 2. 3D printing factory, design Horn Architects. Canteen with kitchenette in the office building and common spaces serve the whole of staff including production, managing and IT employees. Windows have colorful "spacer" panes, fully visible depending on the daylight or internal lighting, canteen is connected with outdoor terrace. There are also conference and meeting facilities and lunch rooms. Photo by the author.

2   New Challenges Coming from New Technologies

The example of 3D printing factory for Polish brand of Belgian company Materialise [3] in Bielany Wroclawskie (adjacent to the capital of the region – city of Wroclaw) is intended to show the architectural response to new challenges. The company needed space for multi stage production joined with offices for directors and managers, accountancy and "brain" of production – offices for IT specialists. It is where the

---

(Footnote 1 continued)

philosophy and policy. The knowledge had to be adopted to Polish conditions in individual, innovatory solution, as no other buildings for production of this particular type are yet built in Poland.

[2] In this article the author focuses on ergonomic optimization of multiple aspects of design of discussed building as model example of new challenges, which architects face in typical design of an utilitarian unit of production plant. For thorough in-depth study of interdependencies of aesthetic and technical aspects of industrial architecture on many examples of built objects refer to e-book "Aesthetics and technics in industrial architecture" [2].

production starts – after receiving a commission digital models are prepared in a form either of a new computer aided design of a product or adjusting an existing one. Part of the production is based on scanning objects to allow design of new parts to fit in. Digital models are sent to printers where laser sintering of polyamide powder takes place for few hours layer by layer using laser rays. After this stage is finished, the 3D printed items are extracted from the residue of the powder, cleaned and given fine finishes. This multi stage process needs adequate rooms with various conditions, which additionally have to be adjusted not only for the proper functioning of printing machines but in respects to health and safety of workers. This is achieved by special systems of cooling and ventilation. Economic optimization was introduced by such a design of the installations' system to allow maximum heat recovery for heating purpose and meeting high demands for air change and air conditioning (climate in Poland is typical for central Europe – with hot summers and cold winters, with periods of intensive sunshine and wet weather with snow in winter). The production imposes demands for working conditions connected with this special kind of production – operating and servicing of both 3D printing machines and their accompanying devices like coolers, powder supply, etc. and the "machinery" – enginery of the building itself. Consequent stages of finishing (mechanical, chemical and manual) and shipping of ready products require appropriate working conditions, and similar to all former stages, finishing, control, packing and shipping must have access to daylight required by Polish regulations. Apart from special requirements concerning ventilation or heating, such rooms are typical in terms of required facilities for the workplaces, where the laborer is physically present and needs cloakroom, bathroom and canteen. The building is exposed to natural conditions and weather, which influence the internal environment, considered in the design. For example, offices are protected from direct sunshine by sun blinds, the technical roof is prepared to snow fall. New challenges come from parameters of new production conditions like high temperatures due to the heat coming from laser sintering, or presence of the powder in the air, which have to be controlled and eliminated. Especially thermal performance of the building is important due to the process of introducing sustainable development and energy efficiency regulations in Poland as a member of European Union.[3] To comply with technical regulations, parameters of the building structure and thermal insulation together with the concept for heating and cooling have to comply with allowed values for energy demand for the building. The design of installations' systems not only meets this requirement through recovery of the heat from machines. The system is prepared for future adjustment – all devices have possibility to be joined in BMS (sophisticated building management system). Another parameter of a production imposing new challenges is its constant and rapid development – new machines are introduced, so the space for them has to be flexible in terms of spatial organization and adjustability of installations. As the building technologies are still fixed – once the building is built it cannot be easily changed – it is the design of interior and installations, which can allow for adopting to new requirements of production process. It is a new

---

[3] Poland as a member of European Union has to tune national policy of development with the common indications. However, this process takes time, due to the fact that richer countries which formed EU before Poland's access began introducing sustainability in nineties of 20th century [4].

challenge indeed, as the calculations for the operating of the building installations are based on defined values of production machines' parameters and operating conditions and technical devices in building's installation system. The design of new object for Materialise matches current needs while allowing future adjustments. This was very difficult to achieve, working on a building design and optimizing it in terms of present technical demands while keeping reasonable openness of the system for future amendments, based on predicted level of changes.

Despite the fact that expected flexibility and adjustability could seem to be primary difficulty in the design, ergonomics introducing human factor into the design as foundation turned out to be core of the design. The needs of a human, either connected with the body and mentality do not change so rapidly as the technology. Ergonomics of the design and technical regulations for building with spaces for production take into account not only functioning of the machine but also servicing and operating it by human, with the focus on health and safety. Rapid change of production requirements, which is not immediately followed in building regulations at national level, is one of the biggest problems for designers. This is nowadays a problem for an architect and the team of engineers working on a project, when some old regulations do not match current needs of a modern factory. For example, old technologies were more dangerous and safety regulations were appropriate, and now they are still very restrictive though technologies are safer and less harmful, but the designer must comply with these restrictions what sometimes dramatically limits the design. Another example - spaces of cloakrooms for different groups of workers according to the level of contamination in production are precisely defined and there is requirement for direct daylight in workplace while in fact most stages of production tend to involve less workers, and the time

Fig. 3. Photo of the 3D printing factory under construction – skylights in production room, photo by the author.

needed for particular activities is constantly reduced due to automatization and digitalization of production (Fig. 3).

3 Ergonomics Facing Dehumanization and Depopulation of Production Plants

The example of 3D printing factory for Materialise shows many-sided aspects of new challenges for industrial architecture today. Automatization and digitalization leads to limiting the number of necessary workers operating machines and increases the amount of office work. Also the office work of IT specialist can be done in any place on Earth as long as there is connection to internet so the information to production machines could be sent or the files kept on servers located somewhere else than the actual printing plant. This leads to depopulation of workplaces and dehumanization as well. The less people employed, the more difficult to keep focus on these aspects of working environment which refer to physical and mental needs of a human, like comfort, motivation, feeling of community, aesthetics, because to achieve this, material means have to be involved at the expense of greater cost of the investment (the object and its operating costs). Dehumanization makes it even more difficult to introduce ergonomics to a workplace when an employee works remotely. The factory operating personnel is reduced to necessary number of specialists, working in place as long as direct contact with the machine is needed and further treatment of raw product required. Finally, space is needed for storage of ready products and containers for shipping to destination addresses, and work conditions for people dealing only with this stage are as important as the ones for production or office work.

The ergonomics in the design of a factory is not limited to interior. The new industrial unit needs proper relationship with the surroundings in terms of protection from nuisance and dangers for people living and working around and generally for environment. This must be assured independent of the number of staff employed. Subsequently the optimization of combination of technical requirements and cost in reference to the budget influences the design of the building – the way that it is shaped, what materials are used, etc. From architectural point of view, fitting an industrial unit in urban or rural context is a challenge as well, keeping in mind all the technical, economic, aesthetic and functional aspects to be joined in one design, because they are subject to local conditions of the site, local master plan indications and general, national and local environmental restrictions, building regulations and law. However complicated and complex is the design process, it has to take into account that a building becomes part of a bigger organism – a city or a land around it. The building also overcomes physical borders of its location when particular company's assets are located in different places or countries. Through the network of physical, spatial and logistical connections, the new object is acting on bigger group of people than the specified number of employees of that particular factory. This global scale of reference comes not only from the aspect of becoming a part of a branch network or international enterprise. Today design takes into account influence of the building on environment, which is seen as a whole, contrary to approach in the past, when industry caused heavy pollution to local environment, what was not recognized as a global threat. In terms of built environment, industrial unit, which meets all the required internal and external standards, has no negative environmental impact on surroundings, but it can also contribute to an overall image of a place. Positive impact of the factory's design as a kind of spatial communique of a company's presence and activity is often a marketing decision. In this way, ergonomics of a workplace is no longer limited only to relation of worker to a machine but a helping tool for the design to cover issues of a bigger scale than the actual building. This shift comes primarily from the consciousness of global character of environmental impact that the buildings have. Nevertheless, the same important is the change of the role of a worker in production process: nowadays a human becomes a controller of electronic systems, which are responsible for physical production instead of manual operating and handicraft. This is the way in which ergonomics becomes related to more immaterial aspects of work, like comfort or motivation, because nowadays production relies on office work of information managing or creating virtual representations of items to be physically produced by contemporary machines. The more sterile and sophisticated is production, the more sublime becomes design of work spaces. Unfortunately, this technical sophistication is often related to high costs of the technology and providing the required conditions for it. Thus a new challenge for architects is to keep this human factor in architecture, in terms not only of correct measures but aesthetics and human oriented environment as well. In the discussed building, architect introduced this value in the design of common spaces. The aim was to help integrating production and office staff: canteen has an open

Fig. 4. Visualization of the 3D printing factory, design by Horn Architects. Special design of canteen with open terrace and colorful glass panes in glazing.

terrace (Fig. 4), representative entrance hall has opening to all office floors up to skylight, and meeting hall is a lounge space with the green internal wall and a sculptural tree, printed by the company as an example of their production profile.

4   Summary

At present day, we face technical progress on unknown scale. The IT systems are designed to replace a human in many aspects of life, mainly in production. Moreover, artificial intelligence is intended to learn and in parallel it reaches miniaturization.

Despite this amazing potential of so-called deep learning, it is rather impossible that machines equipped with artificial intelligence will reach level of self-consciousness and emotional personality introducing new level of interaction with human, and the author is convinced that needs of a human, both physical and immaterial will remain the basis for design and ergonomics of contemporary buildings and industrial units in particular. The aspects of context, environment and direct or indirect influence on people impose human measure on technology and its container – in this case an industrial unit, even though contemporary production becomes automated with hardly any operating personnel. The factory staff, either in production and office section remains main subject of the design in reference to architecture and the technology as well. Industrial production in terms of a process of making things and physical presence of a factory building is always subject to rules assuring wellbeing and health, with ergonomic optimizing of all design aspects under architects' vision based on professional knowledge, experience and aesthetic beliefs. And the author is convinced that these immaterial aspects remain constant despite technology and deep learning flourishing. No matter how revolutionary the new era of human civilization will be, the new challenges come from fitting technology to humans. So far our vital needs and physical characteristics remain unchanged to the level which can be a base for rules and regulations for technology, building and architecture. Without placing a human in a center and defining of constant repetitive interdependencies between human and artificial objects, ergonomics could not be possible, as well as design in general, and in reference to industrial architecture in particular.

References

1. Fross, K.: Qualitative research in architectural design on selected examples (Badania jakościowe w projektowaniu architektonicznym na wybranych przykładach.), pp. 140–145. Wydawnictwo Politechniki Śląskiej, Gliwice (2012)
2. Baborska-Narożny, M., Brzezicki, M.: Aesthetics and technics in industrial architecture. (Estetyka i technika w architekturze przemysłowej.), pp. 36–37. Oficyna Wydawnicza Politechniki Wroclawskiej, Wroclaw (2008)
3. Discover the Transformative Potential of 3D Printing Innovators You Can Count On. http://www.materialise.com
4. Gauzin-Mueller, D.: Sustainable Architecture and Urbanism Concepts, Technologies, Examples, pp. 12–31. Birkhaeuser – Publishers for Architecture, Basel (2002)

The Latitudinal Vomitories at the Stadium Stands - The New Concept versus Classical Solutions

Zdzislaw Pelczarski[✉]

Bialystok University of Technology, Bialystok, Poland
pelczarski.z@wp.pl

Abstract. The classic vomitory is a local opening in the structure of spectators' area, which provides a spatial connection with the external functional zones of the stadium. Major problems of these solutions are the complications resulting from the continuity interruption of the structure of the terraced surface of the stands. This fact necessitates the densification of the main carrier elements. The research conducted by the author, supported by his practice in the architectural design of stadiums, shows that classical vomitory can be replaced by other, more beneficial solutions. One of them is the idea presented in this paper, called *latitudinal vomitory*. The aim of the research were comparative analysis concerning the two systems, described above. The theoretical studies revealed that the new concept in comparison to the classical one has the numerous functional, structural and safety benefits, pointing to the need for its further research.

Keywords: Stadium · Architectural design · Spectator zone
Classical vomitory · Latitudinal vomitory

1 Introduction - Definition of the Research Problem

The author's interest in unconventional vomitory solutions at stadium stands is strictly related to his function as general architect of reconstruction of the *Silesian Stadium* in Chorzow. Due to the fact of the construction of a new eastern stands while maintaining the stadium's use activity, it was required to build a temporary, portable stand for the representatives of the mass media (Fig. 1). The structural concept of this object assumed the use of steel trusses and girders to shape the terraces of the stand. The trusses were based on sloping steel girders, the inclination of which resulted from the needs of the visibility profile. In this situation, there was the only one possibility to design the necessary vomitory. The space between two adjacent girders had to be fitted with stairs, running parallel to the rows of seats. In this situation, there was the only one possibility to design the necessary vomitory. The space between two adjacent girders had to be fitted with stairs, running parallel to the rows of seats. The application of this uncommon way solved a problem of a communication relation between the surface of the tier and the external zone. Such a concept has been named a *latitudinal vomitory* because of the nature of its links with parallel arrangement of the stand terraces. The idea has been known for a long time. This is evidenced by the over 200-year-old

© Springer International Publishing AG, part of Springer Nature 2019
J. Charytonowicz and C. Falcão (Eds.): AHFE 2018, AISC 788, pp. 408–418, 2019.
https://doi.org/10.1007/978-3-319-94199-8_39

example of its use from Warsaw. It is the historic, still in use, classicist, modelled on the archetype from ancient Herculaneum, the open-air theatre (Fig. 2). It is about, situated in *Royal Lazienki Park, Theatre on Water* alias *Theatre on the Isle,* designed by architect Jan Chrystian Kamsetzer, erected in 1790, which is recognized as one of the most interesting examples of 18th century theatre architecture in the world.

Fig. 1. Portable, temporary grandstand for media with *latitudinal vomitory*; *on the left* - assembly of steel structure at the *Silesian Stadium* in Chorzow, Poland; *on the right* - new location 300 km from the original one (the arrow indicates the vomitory position); (Photos: author)

As seen in Fig. 2 the stairs leading from the lobby, located beneath the auditorium, are placed in the slot resulting from reduction of the fragment of one row of seats. Because of such a concept, the run of stairs has an arched shape, which is identical with the shape of the chosen terrace of amphitheatre.

Fig. 2. *The Theater on Water* in the Royal Park *Lazienki* in Warsaw - a historical example of the use of a *latitudinal vomitorium* (1790); *on the left*: amphitheater with visible stairs of the *vomitorium*; *on the right* – lobby level with the entrance to the *vomitorium*; (Photos: author)

The purpose of the undertaken research was to analyse the characteristics of conventional vomitories [1–3] and the proposed new solution. The paper presents a comparison of the advantages and disadvantages both of these methods of communication between the spectators area located on the surface of terraces and external zones of the stadium. The final aim of these activities was to achieve the answer to the question whether the classical vomitories can be replaced by other alternative solutions and what are the arguments for this.

2   Explanations of Applied Terminology

It is noticeable that in the architectural literature in English, concerning the context of spectator area in the designing of stadiums identical components are frequently called by using completely different terminology. Taking into account the above, for the sake of discourse clarity, the author considered as necessary to present and explain all of such cases used in further part of work. This applies to the following terms:

- **Stadium** (plural stadiums or stadia) – a place or venue for (mostly) outdoor sports, concerts, or other events and consists of a field or stage either partly or completely surrounded by a tiered structure designed to allow spectators to stand or sit and view the event.
- **Spectator areas of a stadium** – referred to as *terraces* or *stands*, especially in the United Kingdom and as *tiers*, especially in the U.S.
- **Grandstand** – the main seating area, usually roofed, providing the best view for spectators at racetracks or sports stadiums.
- **Stand** – means the same as the grandstand but used for objects less representative.
- **Tier** (plural tiers) – means the same as the stand, used especially for multi-storey structures (a tier of seats, upper tier, two-tier, tier of benches); originally set out for standing places only, they are now usually equipped with seating.
- **Vomitorium** – each of a series of entrance or exit passages in an ancient Roman amphitheatre or theatre.
- **Vomitory** (plural **vomitories**) – the entrance or exit passages connecting spectator areas of a contemporary theatre, amphitheatre or a stadium with their external functional zones.

3   The Different Logic of Two Types of Vomitories

Diagrams (Fig. 3) presents two different ways of shaping the spatial connection of the stadium interior with its external zones. Arrangement of the commonly used classical vomitory consists in cutting out the hole in the stepped surface of the stand. Lateral lines of this opening intersects crosswise the slabs of terraces with rows of seats and perpendicularly to their edges. This causes certain structural and functional problems, details of which will be explained in further parts of the text.

The functional and structural logic of the *latitudinal vomitory* is intended to maintain the continuity of the supporting beams of individual steps of the tiers. This is achieved by creating a sufficiently wide gap in a compact arrangement of terraces and placing stairs in it. The runs of these stairs are parallel to the rows of seats and lead viewers from the space under the stand directly to its surface. The described gap divides the stand into two independent construction parts, i.e. lower tier and upper tier. These parts are also characterized by full functional autonomy. The third, completely structurally separated element, consist of stairs and platforms built into a slot.

The proposed solution of the vomitory problem characterizes itself by the full geometric coherence with the stand terraces by maintaining the parallel layout with the steps and the rows of seats. At the same time, it shows coherence with the elements of the load-bearing structure, what has a significant impact on costs.

Fig. 3. The diagrams of two alternative vomitory concepts; *on the left* - the classical vomitory; *on the right* – the *latitudinal vomitory*; (Source: author)

4   The Characteristic of the Classic Vomitories

As mentioned earlier, the solution of classic vomitories generates two serious problems. One of them is of a structural nature and results from the need to increase density of a supports in the region of the opening of the vomitory. The consequences of this fact are presented in the below photos (Fig. 4, *on the left*). One of the reasons for the use of two modules of deferent spans was the solution of the load transfer problem at vomitories area. The width of the vomitory depends on its efficiency, which must provide the required evacuation time from the given area of the audience. According to the reg- ulations, this time should be within 5–8 min, assuming that 40 persons pass through a

Fig. 4. The vomitories system designed for the new eastern stands of the *Silesian Stadium* in Chorzów; *on the left* - reinforced concrete supporting structure of the terraces which is doubled in order to create the vomitories; *on the right* – the view of the vomitory and the gangways served by it (Photo: author)

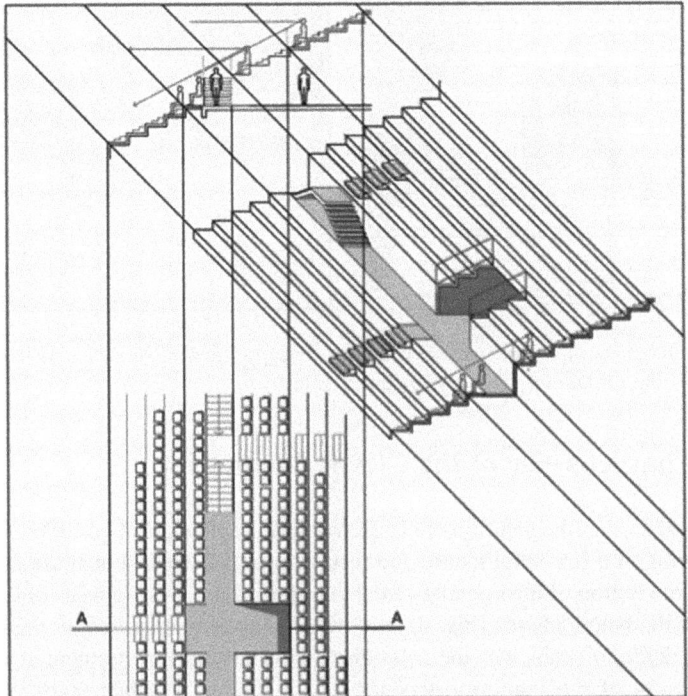

Fig. 5. The rules of functioning of the classic vomitory - an example of one of many possible solutions; a visible break in the continuity of several rows of terraces; (Source: author)

Fig. 6. The vomitories and gangways system on the *Stadium Slaski* eastern stands - aerial view. The upper tiers' stairs were moved to the middle of the rows between the vomitories; the level of the first row of this tier has been significantly raised, improving the field of view of spectators occupying these places; (Photo: author)

Fig. 7. The spectators communication and evacuation system applied to the tiers of *Wembley Stadium*. An example of classic vomitories with a gangways running tangent to their balustrades. Notice that the width of the vomitories has been decreasing on higher and higher tiers, while tire's capacity is growing; (Photo: M. Pelczarski)

lane with a width 0.6 m within one minute [4, 5]. The architectural result of the structural decisions can be seen in (Fig. 4, *on the right*). Compliance to the functional requirements caused a reduction in the field of places for spectators with a width of six seats in five rows. The implications of the proper arrangement of vomitory go back much further. First, elements of structural support, visible on both sides of the vomitory, have to descend downwards through many floors until to the foundations. This fact may result in serious limitations in the design of the functions on these levels. Secondly, the edges of the vomitory opening need protection by using balustrades, which presence can interfere with the field of view. Thirdly, the proper organization of the movement of viewers in the immediate vicinity of the vomitory is very difficult and creates significant problems. These problems are partially solved by the method shown in Figs. 5 and 6, which consists in shifting the stairs of the upper tier on the sector axis, far from the vomitory. Still however remains the task to distribute the movement at the outlet of the vomitory for efficient filling of the lower and upper tiers. Real design problems are caused by cases shown, as an example in Fig. 7, when the viewers move on paths that are tangent to vomitory balustrades. They are caused by geometrically complex places where the course of stairs changes from a perpendicular for a parallel to the terraces, and vice versa [6].

5   The Characteristic of the Latitudinal Vomitories

As has been said before, noticed by the author disadvantages of classical vomitory solutions and practical experience in the implementation of the temporary stand for journalists during a redevelopment of *Slaski Stadium* in Chorzow have become an impulse for seeking an alternative, more favorable solutions. The most interesting turned out to be, presented in this paper, the proprietary solution under the name *latitudinal vomitory*. The explanation of this concept is presented in Fig. 8, as the selected example representing many other variants.

The idea, which is the basis of this architectural proposal, consists in the juxtaposition of two autonomous, structurally independent stands in such a way that creates a free space between them. Obtained in this way space is used to build in it the stairs, leading from the utility level, located under the surface of the upper tier, directly to the level of the highest row of the lower tier. The run of these stairs is parallel to the edges of the stand terraces. The minimum width of these stairs is 1.2 m and 1.5 m, taking into account the handrails. The same dimension can have the gap between the lower and the upper stand. Figure 8 shows also the cross-section of the tier with the visibility profile. By following the line of the focus point view from theoretical point of eye of the first row of upper tier it can be stated that it runs above the figures of people moving along the latitudinal corridor. This is a great advantage of the designed profile, which provides excellent visibility of the field of observation, eliminating interference caused by the movement of viewers along communication routes. It should also be emphasized that these effects are obtained with the same inclination of both the upper and lower tier, what has a significant impact on a cost reduction.

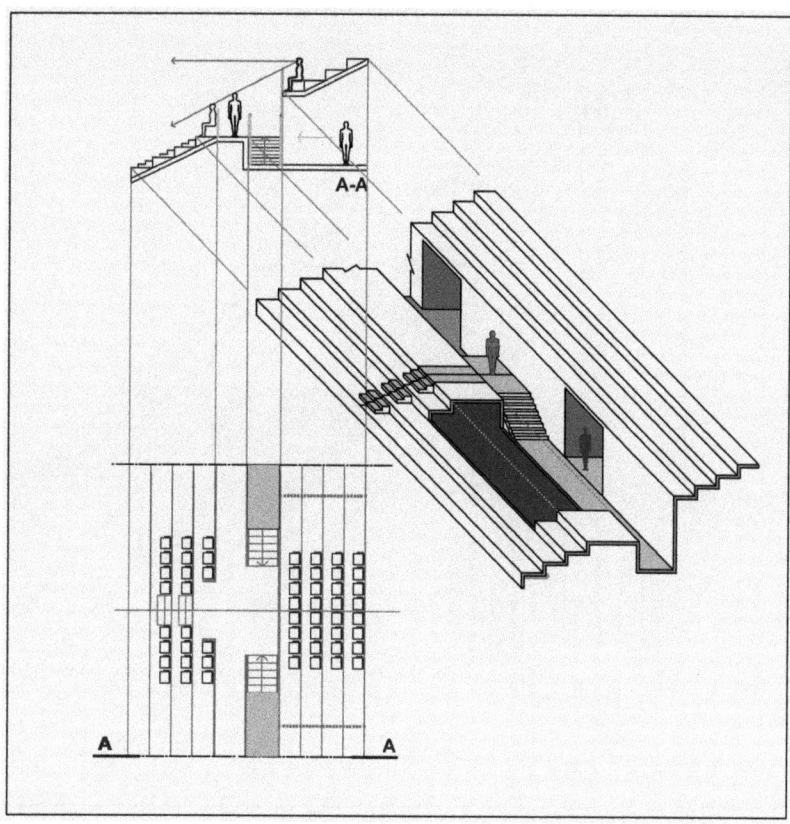

Fig. 8. The *latitudinal vomitory* - ideological, functional, spatial and structural assumptions of the alternative solution to the vomitory problem on the stadium stands. The simplest version when the vomitory serves only the lower tier; (Source: author)

Another advantage of the proposed solution is the fact that the above-mentioned functional benefits are obtained by simple methods and without consequences for the main supporting structure. The difference between the level of the external lobby and the level of the latitudinal corridor on the surface of stand is just six steps of stairs. The only determinant of the relationship between these levels is a height of the passage under the first row of the upper tier. The greatest advantage resulting from the use of *latitudinal vomitory* is the total structural independence of the vomitory system itself and the stand support system.

Because of the above, there are no disturbances in the continuity of the load-bearing beams of the tier terraces. This independence also results from the considerable freedom of layout of supports, and hence, greater flexibility in designing functions beneath the surface of the audience.

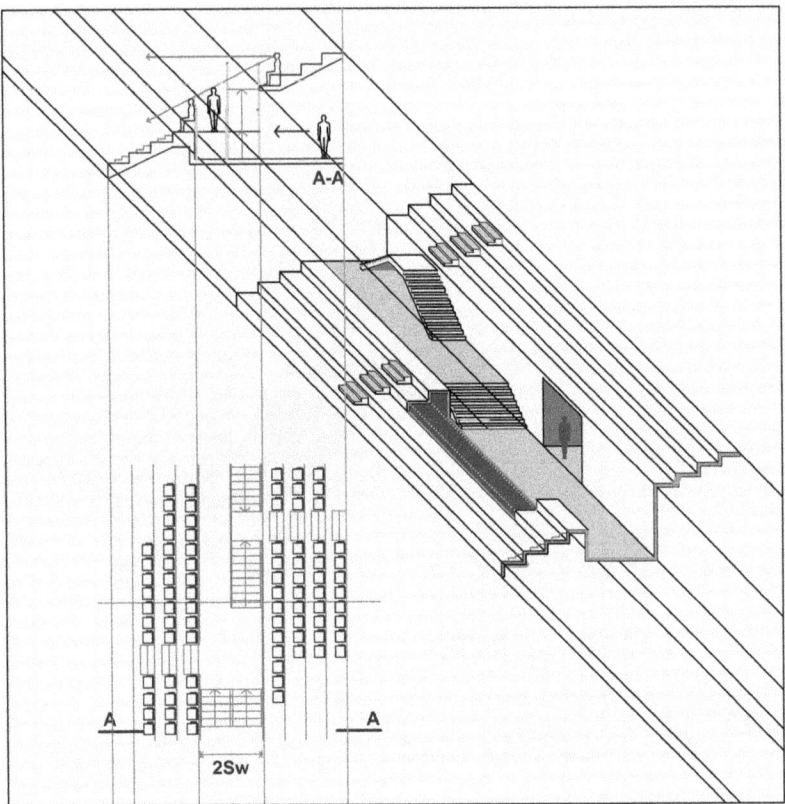

Fig. 9. Advanced version of *latitudinal vomitory*, serving both lower and upper tiers and providing the transverse gangway connecting all sectors of the spectator zone. It should be emphasized that the upper and lower tiers are based on the same profiles of visibility and their structures are independent each other and separated from the vomitory structure; (Source: author)

The described example belongs to the least complicated types of *latitudinal vomitories*. It allows filling the tier only from the top. The upper tier visible in Fig. 8 needs to be serviced by another vomitory, located at the level of its highest row of seats.

Figure 9 presents a diagram of the functioning of the vomitory, which can serve simultaneously both tiers, lower and upper. It also provides a latitudinal passage that links all the vomitories. Its presence serves to increase safety by providing additional escape routes through neighbouring exits. It also plays an important role as part of the communication system for the technical and security services. In comparison to the first solution, the gap between lower and upper tiers is wider. In comparison to the first solution, the gap between lower and upper tiers is wider because must accommodate two parallel runs of stairs. One of them serves the lower stand and the second the upper one. In addition, the passage opening to the lobby is enlarged accordingly.

Fig. 10. The concept design of the *Hetman Stadium* in Bialystok (2007) - example of the stands with *latitudinal vomitories*; (Authors: G. Dabrowska-Milewska, J. Kabac, Z. Pelczarski, J. Zarnowiecka.; Structure: M. Pelczarski; Visualization: A. Babula, B. Dudzinski).

6 Conclusions

The vomitories are an essential element of communication and evacuation systems serving mass audiences gathered in interiors of a large and extra-large capacity stadiums and similar venues. In the architectural design practice of this type of objects, traditional vomitory, known since ancient times, are commonly used. These solutions, however, generate serious structural and functional problems, resulting in reduction of usability and aesthetic, increased investment costs and reduced efficiency in terms of communication and evacuation of the spectator area.

The way of communicating the audience with the spaces below the surface of stands can be solved on quite different way than the classic one described above. It is, designed by the author and tested in practice, a slit vomitory that has been named *latitudinal vomitory*. The results of studies and analyses presented in the paper prove that the new type of vomitory is in no way inferior to classical, and some of their variations are in many respects more beneficial. The greatest advantages of the proposed new system include the complete functional and structural separation of the stands and the vomitories.

According to the author, the attractiveness of this concept lies also on the side of its high architectural values. This method, thanks to the complete elimination of repetitive, schematic perforation holes in stands, allows obtaining a new architectural expression of the stadium interior, compared to the traditional solutions (Fig. 10). The horizontal direction becomes here the dominant of composition, what is the result of spatial and functional concept of these vomitories. The architectural form of the sequence of the touching each other stands creates the frontage closing the space around the arena. Thanks to the *latitudinal vomitories* application, this form takes the character of a multi-layered structure. The reduction of the number of rows per one layer of this structure, leads to a significant increase in comfort and safety within such a type of sectors. The balustrades on the edges of *latitudinal vomitories* emphasize the horizontal divisions of the audience area, what positively affects the viewers' psyche. Firstly, it reduces the discomfort resulting from the subconscious acrophobia accompanying the location of the viewer on the steep and highly elevated surface of the stands. Secondly, the division into smaller, more autonomous enclaves, assigned to a countable group of people, fosters greater identification of spectators with a fragment of the interior space of the stadium, giving a sense of territorial affiliation. This contributes to reducing the sense of anonymity, resulting from the fact of melting into one mass with a crowd of thousands individuals.

References

1. Nixdorf, S.: Stadium ATLAS. Technical Recommendations for Grandstands in Modern Stadiums, Ernst & Sohn, Berlin (2008)
2. FIFA/UEFA: Technical Recommendation for the Construction and Modernisation of Football Stadia, Zurich (1994–2005)
3. John, G., Sheard, R.: Stadia. A Design and Development Guide. Architectural Press, Oxford (1997)
4. EN 13200-1: Spectator Facilities - Part 1: Layout Criteria for Spectator Viewing Area – Specification (2003)
5. The Green Guide: Guide to Safety at Sports Grounds. HMSO, London (1990, 1997)
6. Pelczarski, Z.: Grandstands of the Contemporary Stadiums. Determinants and Design Problems, (Widownie wspolczesnych stadionow. Determinanty i problemy projektowe), pp. 197–224, Oficyna Wydawnicza Politechniki Bialostockiej, Bialystok (2009)

Civil Projects (Participatory), Company Funds and Small Grants as Factors Integrating Local Communities

Jerzy Charytonowicz[1,2]([envelope]) and Alicja Maciejko[3]

[1] Angelus Silesius University of Applies Sciences in Walbrzych,
Zamkowa 4, 58-300 Walbrzych, Poland
jerzy.charytonowicz@pwr.edu.pl
[2] Faculty of Architecture, Wroclaw University of Science and Technology,
Wybrzeze Wyspianskiego 27, 50-370 Wroclaw, Poland
[3] University of Zielona Gora,
ul. Prof. Z. Szafrana 1, 65-516 Zielona Gora, Poland

Abstract. The growth of social involvement in the projects of development of public space has recently been noticeable in Poland. Social involvement takes place on a few levels. There may be social participation, in which citizens are invited to conversations related to projects and their demands may be taken into consideration as well as civil projects (participatory) in which citizens suggest projects, their ideas and solutions. Company funds and small grants constitute an entirely new model. The communities participate in preparing the project; decide about exploiting new places as well as they participate personally in their execution. All can be important factors integrating local communities. The aspects of implementation such projects in Poland are discussed in the paper.

Keywords: Urban planning · Participatory budgets · Company funds
Small grants

1 Introduction

The urban planning system in Poland is probably one of the most democratic forms of planning. According to the polish law, all persons, even those who do not live in a given territory, may participate in shaping the local development plan. However, generally, large, urban investments counted in millions of Polish zlotys are beyond the scope of discussion and decisions of citizens. Neither can they propose them nor have the possibility to enforce the implementation. Only a few percent of the budgets invested in the city are spent on realizing investments according to citizen's proposals. Sometimes social consultations are used. Citizens are invited to conversations related to projects and their demands may be taken into, but does not have to be finally entered into the project. In 2010 a new model of the real impact of residents on the public space appeared in Poland. This was participatory budgets called also civil projects. However, this is not a high share of urban investment. The size of the participatory budget for 2019 in Warsaw is PLN 64.7 millions. It is 0.4% city budget for 2018 and less than 3% of expenditure on urban investments. The idea of participatory budgets was born in the

© Springer International Publishing AG, part of Springer Nature 2019
J. Charytonowicz and C. Falcão (Eds.): AHFE 2018, AISC 788, pp. 419–429, 2019.
https://doi.org/10.1007/978-3-319-94199-8_40

last decade of the twentieth century thanks to Porto Alegre, a city in Brazil, which solved in this way - as supporters of civic budgets - problems with citizens participation in government, corruption and social discontent [1, 3, 4, 6].

The practice of the last years of their application in many Polish big cities also in and small towns has shown that it is not easy to reconcile interests of investors, residents expectations and directions of urban development. There is no direct legal basis for using the participatory budget in Poland. The procedure of creating and adopting the budget of local governments does not take into account the mechanisms of co decision by the inhabitants. There are no regulations that would oblige the authorities and the commune council to co-create the budget project with the participation of residents or consult with them important financial decisions, although there are also no legal regulations that prohibit it. It is included in the law as a social contract concluded between residents and city councilors, so it can be legally introduced in municipalities, districts and other self-government units. Councilors can determine the amount allocated from the general budget, which will be allocated in accordance with the will of residents expressed in an open voting. The decision-making procedure can be flexibly adapted to local conditions, so that it can select those proposals of residents who have the greatest public support and can be entered into the budget [2]. This form of public participation will be discussed in general by presenting data on the whole of Poland.

Corporate grants funded to improve the living space of residents are a commercial initiative, independent of the city authorities. For this reason, it is practically impossible to formulate a general formal basis for their use. Pursuant to the regulations, the founder defines the procedure and detailed criteria for the assessment of applications for the implementation of a public task as part of a local initiative. The detailed assessment criteria should take into account, above all, the contribution of social work to the implementation of the local initiative. The evaluation of the application remains at the discretion of the founder who, when assessing the application, should take into account the detailed criteria for the evaluation of the application and its purpose in terms of the needs of the local community. If the submitted application is approved, there is a contract for the implementation of the local initiative with the applicant. Residents or organizations that apply for the implementation of a local initiative may commit to their own contribution to the implementation of the project, which will be the social work of the initiators, a contribution in cash or in kind [2]. Company grants, implemented on a much smaller scale than participatory budgets, will be discussed on the example of the program *Good ideas change our world*, realized in Walbrzych by company Toyota Manufacturing Poland [9].

2 Participatory Budgets in Poland

The idea of introducing the first participatory budget in Poland appeared in 2010 thanks to the Sopot development initiative. The residents of the city were given the right to spend about PLN 3 million, which was slightly below 1% of the budget. In 2012, participatory budget introduced Bydgoszcz, Elblag, Plock, Poznan, Radom, Tarnow and Zielona Gora. A year later, participatory budget was implemented in at least 40 localities, the following communes 25,000 inhabitants, Gostyn, Gorlice, Mragowo, Pulawy, Sochaczew, Swiebodzice, and Zdunska Wola. In budgets for 2018, the total population could decide to allocate over PLN 261 million - in Warsaw 61.5 million PLN. The smallest amount was set in Opole, the pool allocated for the implementation of projects amounted to PLN 2.5 million there. In total, 1901 ideas will be implemented in the next year. The voter turnout is not the best, it rarely exceeds 10%. Kielce is favorite in this edition, where almost 20% of the city's inhabitants voted [5, 6].

In this kind of investments implemented in public space, each of the groups, city residents, authorities, private entrepreneurs or members of nonprofit associations has an impact on how cities look like. The number of urban residents is constantly growing, which is associated with countless problems that need to be resolved. People are very important, if not the most important force in shaping the cities of the future; however, the participation of people in this process is a very complex problem. The most popular social action is the participatory budget. However, such initiatives are still insufficient. Quite often, the inhabitants are not right and a professional initiative for shaping the city is needed. An example may be the ecological aspect. This is primarily the use of cars in cities. Many European cities have limited car traffic and can be modeled on them. In Poland, there are a few cases of moving cars out of city centers. Especially if the inhabitants have an influence on it. For example, they want to satisfy their own needs and postulate to increase the number of car parks at the expense of green areas. In this case, specialist help is needed from urban planners and architects who will design solutions that are good for all residents and for the city (Fig. 1).

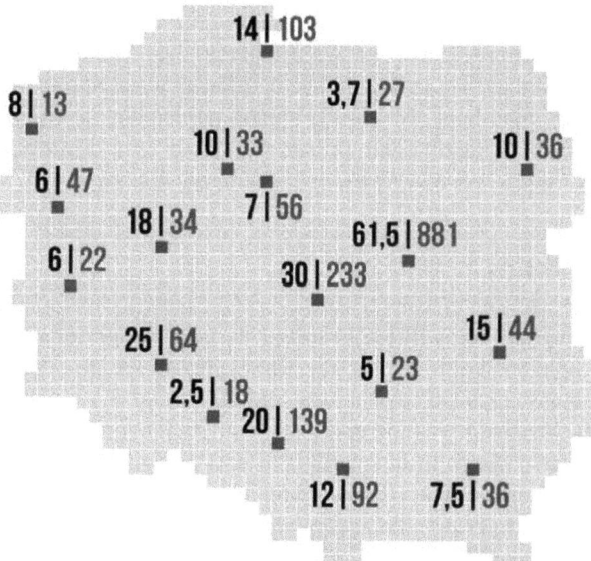

Fig. 1. Summary of participatory budgets for 2018 in all 18 provincial cities in Poland. Budget in million PLN/number of selected projects [6].

General rules of the participatory budget are:

- Every resident can take part - sometimes you do not have to be of age and you do not have to be registered.
- The process follows the rules that are known before the process begins.
- Residents decide on what specific public funds will be spent on - they present ideas, discuss them and choose those that they think are needed when voting.
- Projects are selected as a result of voting.

The city implements ideas selected by the residents. The procedure in various municipalities is most often similar. Residents report specific projects to be financed from a separate part of the budget. Then, the voting takes place and in the next budget year projects that received the most votes are implemented until the pool of money is exhausted. The typical course of participatory budget includes the following stages: (1) development of procedures and acceptance of the regulations; (2) reporting projects, (3) verification of submitted projects; (4) selection of projects for voting; (5) implementation of projects in the next financial year. Specific solutions, however, differ significantly.

The model organization of the participatory budget, as a rule, should consist also of three phases: (1) information, during which the municipal authorities present the idea of organizing a civic budget and educate residents in this regard, (2) implementation, during which the submitted projects are verified, voting, then their selection and implementation, (3) evaluation, during which projects are monitored, and the entire process is evaluated with the intention of improving it in subsequent editions. Before introducing a civic budget, it is worth discussing with residents about this matter, to see if there really is any sense in undertaking such an initiative. It may happen that there is no need to introduce a participatory budget in a given local community and at a specific time. Internet has more and more importance in informing and voting - as evidenced by the evaluation more than half of the civic budget participants learn about it from the internet (websites, social media) and on average, about 75% of residents vote on the Internet instead of in the version traditional, stationary and paper [5].

3   Company Funds

The idea of company funds will be presented on the example of the unique Toyota Fund program implemented in Walbrzych, which created the possibility of bottom-up impact on the shaping the local environment and the quality of residents life. Author of this paper, Alicja Maciejko was the coordinator of the Toyota Fund, thanks to which she was able to assess this form of small grants and their real impact on the quality of life of local communities. The Toyota Fund was established by company Toyota Manufacturing Poland to enable the implementation of interesting and original projects undertaken for the benefit of the local community and the natural environment in the Walbrzych region. For last six years (2011–2017 and it is continued), it was implemented in cooperation with the Foundation for European Education and the State Higher Vocational School in Walbrzych. Landscape architecture students drew the projects. In the preparation and implementation of projects in 2017, the organizations are assisted also by landscaping technical school students from School No. 5 in Walbrzych. The competition is open to non-profit organizations, such as foundations, associations, schools, universities, sports clubs, and local government communities. The co-financing amounts are not high but activate the local community to submit applications [7, 8] (Fig. 3).

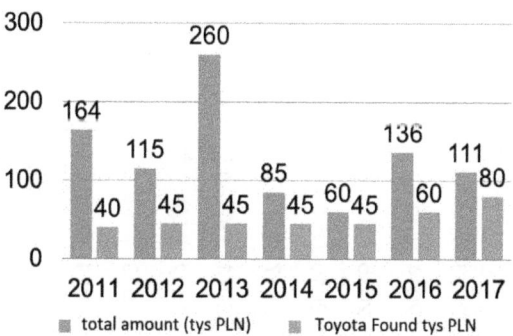

Fig. 2. Summary of general co-financing projects against the cost of all investments in a given year in (thousands PLN). Years 2011–2017. Total amount/Toyota found, [8].

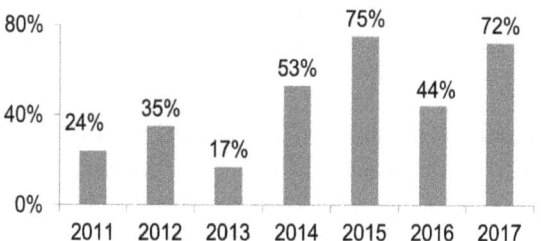

Fig. 3. Percentage share of Toyota Fund co-financing in the total budget for projects in given years, [8].

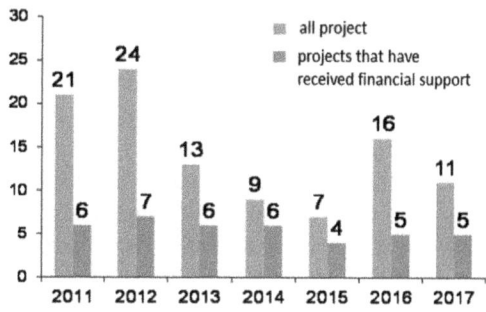

Fig. 4. Number of projects submitted and those that received co-financing, [8].

In the 7th edition (2017), the maximum amount of project support increased from PLN 17,000 to PLN 20,000. On the other hand, the total pool of funds for distribution increased from PLN 40,000 in 2011 to PLN 80,000 in 2017, (Fig. 2). Over 100 projects have been submitted in 2011–16, 39 of which have been subsidized, (Fig. 4). A record number of 24 projects was submitted in 2012. Seven of them received co-financing for a total of PLN 45,000. In the years 2011–2017 320 partners, non-profit organizations and companies took part in the implementation of projects, (Fig. 5). Also 250 Toyota factory employees were involved in implementing Toyota projects [7, 8].

Fig. 5. Number of partners per project. Non-profit organizations and company, [8].

Each organization can benefit from consultations regarding the principles of preparing the application and subsequent settlement, which are provided by the Foundation for European Education. In each edition, detailed rules were attached that defined the thematic scope, content and financing method [9]. The 7-year grant program has revealed many possibilities of such forms of social participation but also problems related to the implementation of projects without the participation of an architect. According to the authors, it shows also the architect's role in an organized society. Can architects be ignored in small-scale projects?

The Toyota Fund is in principle rewarding original and interesting ideas for small revitalization, with a focus on ecology, education and promotion of a healthy lifestyle, which assume a lasting change in the environment and are implemented in the Walbrzych. For example, in 2016, co-financing was granted to bicycle track construction projects, an ecological tourist trail, a playground, a weather station at a primary school, or a bicycle producing electricity. One of the most interesting was project on Bystrzyckie Lake in Sowie Mountains (Fig. 7). On the rock located in the area of the dam called Zagorzanski Bulder there were 7 climbing routes of varying difficulty. In addition, especially for children, three roads have been prepared with an upper assurance. Old roads on this object have also been inventoried. The rock is located on a slope, about a hundred meters from the dam. Climbing area with road descriptions is also described on information boards made as part of the project. These are not the only areas for practicing this sport around the lake. In previous years, the Walbrzych club traced and secured roads on two rocks near the suspension bridge: Sloping and Krucza Rock. From the beginning, the project partners were the Walim Commune and the Swidnica Forest District [7, 8] (Fig. 6).

Fig. 6. Design concept and photo for the climbing wall project on the Bystrzyckie Lake in the Sowie Mountains. The project received funding in 2015 and was implemented [7, 8].

Projects implemented under the Toyota Fund are an example of a different implementation process from participatory budgets. They are addressed to small local communities and implemented in small scale. It is independent of the participatory budget implemented in Walbrzych with a much larger budget. It is a private grant program founded by Toyota factory in cooperation with external partners addressed to non-profit organizations. The lack of possibility to supervise the projects carried out by the construction low administration is connected with several important problems. In practice, architects are formally eliminated from the design process of urban space. In the participatory budget, the citizen influences the concept of space, but ultimately the project is implemented by professionals and contractors selected by formal procedure, which, however, not carry out any implementation works. In the Toyota Fund the citizens implement the projects themselves. The designer's participation is not assumed, but only the help of landscape architecture students in drawing visualizations of projects. On the other hand, the question as to how much an architect, as a "donor of form" should design and propose aesthetic solutions, to what extent individual solutions should be and how much designed by amateur citizens and how using products standard. For example, in the field of small architecture, they were mostly solutions given by production companies. Cooperation on projects showed that the average citizen cannot cope with solving design problems, many projects required professional cooperation, but he was not rewarded. The organizations benefited from the help of family and friends. It was also problematic to use individual solutions that require designer's intervention (static calculations, technical problems, connections, details, etc.). Each place and every project in practice generated implementation problems, although in principle, they could not be projects requiring a permit for construction. Should a low budget justify the average aesthetic quality and the lack of a professional designer's participation? (Fig. 8).

Fig. 7. Design concept of project located in the Public Primary School No. 5 in Walbrzych. Laboratory "I am part of nature": wind turbine with backlit educational display case, wind direction and speed recorder, solar clock, portable measuring devices, i.e.: rain gauge, ground gauge and soil meter. Archives Angelus Silesius University of Applies Sciences in Walbrzych [7].

Fig. 8. Laboratory "I am part of nature". Pictures from the official opening. Archives Angelus Silesius University of Applies Sciences in Walbrzych [8].

The competition criteria support projects that: (1) are innovative, (2) solve local problems, (3) leave permanent changes in the environment, (4) increase the awareness of the local community in educational aspect, (5) assume financial, organizational or material involvement of many partners, (6) are intended for the widest possible range of beneficiaries, (7) are implemented in cooperation with the local community and employees of the Toyota factory (Fig. 9).

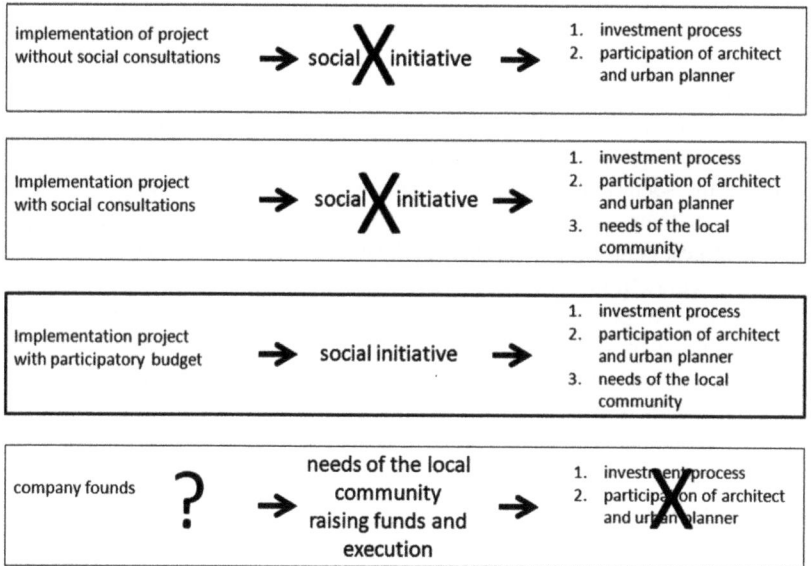

Fig. 9. Ways of transforming a place with the help of social involvement.

4 Summary

Possibility of occurring with initiatives for socially needed implementation undertakings by local entities, based on appropriate procedures and criteria, enables residents of particular municipalities to participate in co-deciding on the directions of spending public funds and sharing costs and responsibilities for the implementation of individual activities. As in the case of the legal bases discussed above, the good will of the local government authorities is also necessary in this case. From this will, it depends on whether applications will be submitted, implemented and whether they will contribute to the construction of permanent basis for civic participation in shaping the budget [1].

The process of shaping cities without people participation does not make sense. You can create beautiful architecture and urban plans, but without a man who has to use it, it will not do it well. Designing must have its origins in analyzing the needs of users. Taking care of locality is an extremely important aspect of urban design. In the future, housing councils will be very important. At a time when globalization is proceeding, we must start to return to the local. In the era of fast travels and the global world our local environment must offer us something really special and qualitative.

Participation does not solve all problems, because it is not always possible to reach a consensus. Finally, design decisions will not always take into account all interests. But, according to the research, citizens are able to accept conflicting decisions with their expectations, provided their arguments remain heard and seriously considered.

In the case of company grants, the low budget caused the use of lower quality materials but contributed to greater creativity, searching for solutions economic and aesthetic as well as ecological, made of natural materials and available locally. How can an uncommon, original idea and subject matter of the place affect residents and build space instead of, for example, high-quality solutions and materials? The completed projects speak for themselves. There was no project that could be rated as outstanding in modern design, however, projects have been created which have significantly changed the perception of the place and are liked by the residents. The most important factor for people in this initiative was freedom from building bureaucracy.

Involving the society in the aspects of urban development planning can contribute to the quality of their lives. But also makes it easier to identify the most important needs of the largest part of the population, allows to effectively respond to expectations, supports the integration of the local community, supports self-government community and raises the level of public trust in local authorities. Magistrates have to put in a lot of effort to encourage more people to participate in each of the stages of the budget procedure, whether through regulatory changes or reducing formalities. Rapid implementation of projects is also important - in many cities it has been going on for several years. Participatory budget means greater transparency of local government activities and involving citizens in the process of exercising power in local governments. It performs a very important decentralization function. People involved in the life of their local communities, operating mainly in housing estates, have a much better understanding of local problems and needs than municipal officials who look at many macro-sized issues. Therefore, by engaging in social participation, they are able to most effectively counteract these inconveniences and try to solve them. However, the personal involvement has a positive impact on integration of local communities. The observation related to utilizing the implemented places has exhibited that physical participation in the implementation process influences the emotional bonds with the projects long after they are finished as well as greater care about space. It may develop the sense of belonging to the place.

References

1. Ćwiklicki, M.: Partycypacja społeczna w Polsce: atlas dobrych praktyk. Fundacja Gospodarki i Administracji Publicznej (2013)
2. Podstawy prawne budżetów osiedlowych/obywatelskich. Centrum Promocji i Rozwoju Inicjatyw obywatelskich Opus (2016)
3. Kraszewski, D., Mojkowski, K.: Budżet obywatelski w Polsce. Fundacja im. Stefana Batorego (2014)

4. Szaranowicz-Kusz, M.: Budżet partycypacyjny w Polsce. Infos. Biuro Analiz Sejmowych (2016)
5. Partycypacja Obywatelska (2018). http://partycypacjaobywatelska.pl
6. Urbnews (2018). http://urbnews.pl/budzety-obywatelskie-2018-podsumowanie/
7. Archives Angelus Silesius University of Applies Sciences in Walbrzych
8. Archive of the company Toyota manufacturing Poland in Walbrzych
9. Foundation for European Education. http://www.fee.pl

Ergonomics in Functional and Spatial Shaping of Bedrooms

Przemyslaw Nowakowski[(✉)]

Faculty of Architecture, Wroclaw University of Science and Technology,
Wrocław, Poland
przemyslaw.nowakowski@pwr.edu.pl

Abstract. Households currently serve as people's main living environment since householders spend approximately two thirds of their time there. Fulfilling of basic human needs and, in particular, higher level needs requires providing of sufficient living space. Some of the housing needs are not considered or are omitted by architects and designers as a result of various reasons. Many of the inconveniences emerge as early as at the designing stage, while others appear during the exploitation of a dwelling. Optimal usage qualities of a dwelling place are attainable not only thanks to its considerable size, elevating of the technical standard of the finishing and equipment, but also by increasing its functional efficiency. This paper provides an overview of current practices and take an aim at addressing functional ergonomics and spatial needs.

Keywords: House design · Bedroom · Accessibility · Human factor in design

1 Introduction

Leisure area (together with individual bedrooms for adults and children) belongs to the private part of a dwelling. An uninhibited realization of individual needs requires providing of, among others: appropriate usable, social and symbolic space, comfortable psycho-physical conditions, appropriate furnishing and technical equipment and favourable conditions connected with guaranteeing of safety and privacy. The above tasks are realized by i.a.: meeting of spatial and movement requirements of users, as well as selection and appropriate arrangement of both fixed and mobile equipment.

2 Space Covered by the Elements of Equipment and Space Required for Their Usage

All appliances and devices located in a house cover certain area whose measurements (covered space and volume) result directly from forms, sizes and placement of particular pieces of equipment. The majority of furniture requires additional space connected with its usage and handling. In a bedroom those are in particular beds and units used for storage (mainly of clothes). The freedom of their usage results from their mutual placement and sizes as well as mobility of their users [5]. Optimal furniture arrangement aims at providing comfortable movement in a room and freedom of

© Springer International Publishing AG, part of Springer Nature 2019
J. Charytonowicz and C. Falcão (Eds.): AHFE 2018, AISC 788, pp. 430–437, 2019.
https://doi.org/10.1007/978-3-319-94199-8_41

performance of routine activities. However, it is necessary to assume spatial requirements of tall, well-built users (anthropometric features of men measuring in 95th centile). Useful data concerning body shapes can be found in anthropometric atlases published in numerous countries. Nevertheless, data presented there is not full, as it lacks the measurements of people assuming dynamic positions (in motion), especially positions leaning forward, sideways, squatting, kneeling, reaching on tiptoes, grabbing of items located on various heights with one or two hands [4, 5].

When designing for users with disabilities the extreme variant is assumed – conditions for movement of a tall man on a wheelchair – as it requires the most space in order to move freely. What is more, the choice of a wheelchair model also influences the way of manoeuvring and movement (turning radius, ease of turning).

3 Bed Accessibility

During the initial stage of designing, the assumed measurements for a single bed are 100 × 200 cm, while for a double bed 200 × 200 cm. The space in front of the bed and on its both sides should amount to 75 cm, which enables, among others, assuming a squatting or kneeling position while making a bed and is known as 'service space' which, together with the space covered by the bed, is necessary for the usage of this piece of furniture [1, 4].

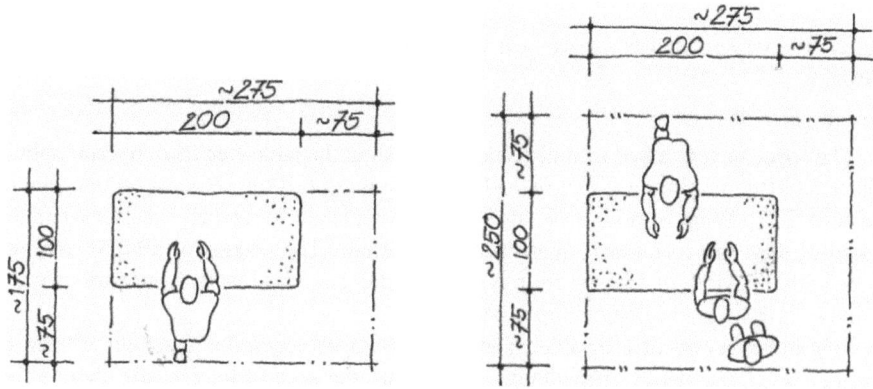

Fig. 1. The size of available space next to and around a single bed in various arrangements (author's drawing)

Single bed should be accessible from at least both sides. Such solution enables an easy access to the bed and making the bed (usually in leaning or kneeling position). Therefore, the floor space connected with the use of a single bed requires the measurements of approx. 175 × 275 cm (Fig. 1). For people who are chronically ill and require constant care it is necessary to grant the access to the bed from three sides (similarly to the hospital beds). Thanks to such a solution free access to the bed is provided from both longer sides of the bed. In such case, the required service area has the dimensions of approx. 250 × 275 cm (Fig. 1).

The most space of a bedroom it taken by a double bed, which may have dimensions of up to 200 × 200 cm, which covers 4 m^2 of the space of a bedroom. Such bed should be accessible from three sides (Fig. 2). Meeting of this criterion provides a possibility to lie down and get up without disturbing the other person. Required dimensions of a bed together with the access (service) area may amount to 350 × 275 cm (approx. 9.6 m^2 of space), which, in many cases, takes up the entire room [1, 4].

During the process of designing a double bedroom it is recommended to provide a possibility to place there one double bed or two separate single beds. Placing beds apart might be recommended in case of an illness or disability of one user of a bedroom. The rules of flexible shaping of the bedroom zone should enable placing of beds in various arrangements within one bedroom (Fig. 2).

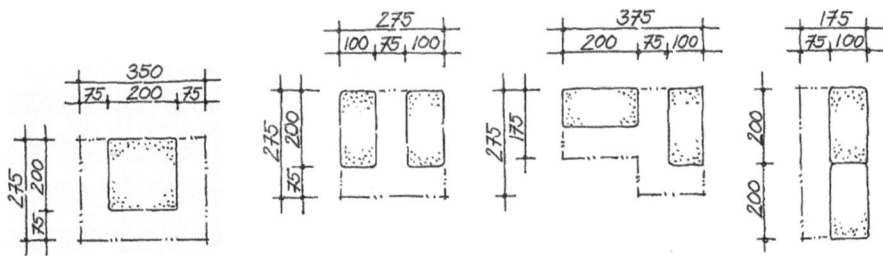

Fig. 2. The size of available space next to a double bed in various arrangements (author's drawing)

The space next to a bed of a disabled person should enable free maneuvering with a wheelchair (the minimum width of 150 cm, enabling a full turn at least on one side of the bed) [2, 3]. What is more, moving from the wheelchair to the bed is easier when the height of the bed is similar to that of the wheelchair. The need of availability of free space with the width of 150 cm from two perpendicular sides results from various ways of moving from a wheelchair into a bed (Fig. 3).

Oftentimes a bed of a disabled person needs to be accessible from three sides (a hospital bed). The width of available space along the second longer side should also amount to minimum 75 cm, which enables the access to a sick person from all sides (Fig. 4) [2, 3].

Some of space can be 'regained' thanks to use of i.a. sofa beds or wall folding beds. When folded, the furniture takes up little space. The space gained that way may be used for daily activity. However, the opened bed can take up almost the entire room and obstruct moving around it. Additionally, opening and closing of such a big piece of furniture is burdensome and usually requires great effort. That is why such beds are not fit for children, elderly people and people with disabilities.

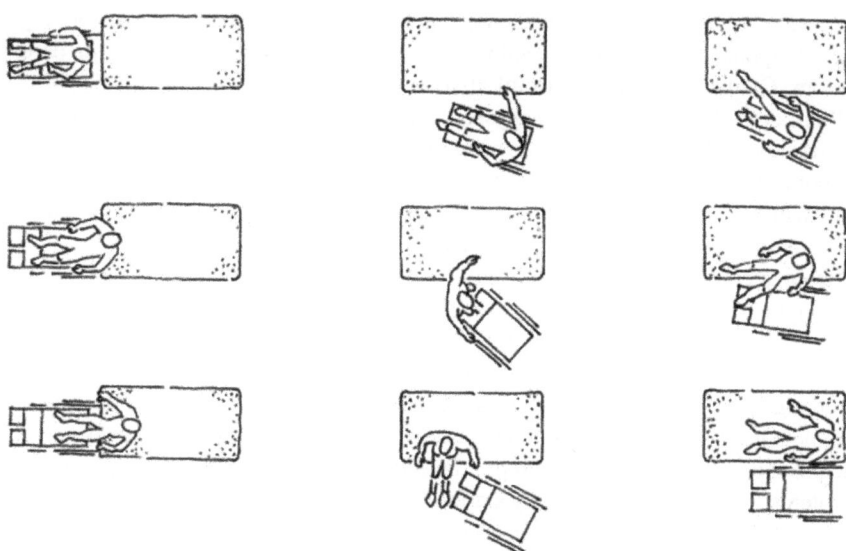

Fig. 3. Various ways of moving from a wheelchair into a bed (author's drawing)

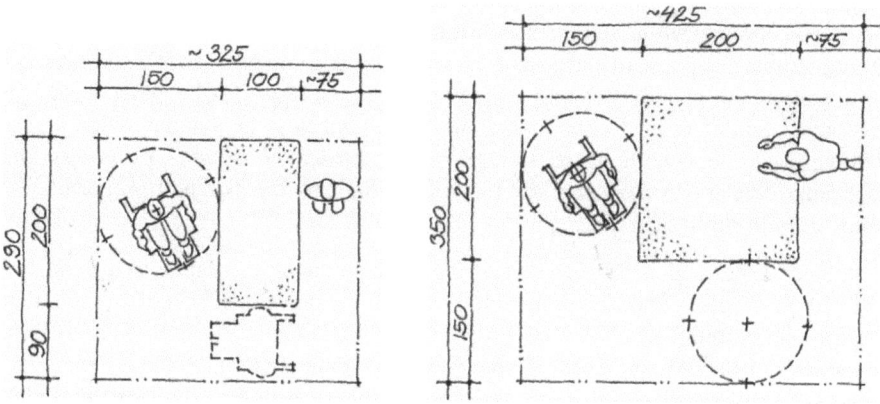

Fig. 4. The size of free space next to a single and double bed of a person on a wheelchair (author's drawing)

4 The Role of Bedrooms for Household Members and in the Structure of a Dwelling

Bedrooms form the private part of a dwelling, which is also often called the zone of night rest. Approximately 1/3 of a day is spent in a bedroom on relaxation and sleeping. The sense of privacy and even desired solitude plays an important role and realization of those needs is fulfilled by the possibility of isolation from other household members. That is why it is recommended for every member of the family (except spouses) to have a separate bedroom. Separate bedrooms are necessary for adolescents, not only for those of opposite sexes. Small children, on the other hand, may sleep in parents' bedroom. Common rooms are useful also for small children.

Bedrooms should be functionally separated from both other parts of a house and exterior noises. It is advised to place a wardrobe or a walk-in wardrobe and especially a bathroom in the vicinity of a bedroom. Thanks to such a solution moving between the rooms will not require walking through the so-called noise zone – living room and the area common for all householders and their guests.

Realization of the sleeping function is usually accompanied by such daily activities as: changing clothes (changing day to night clothing and the other way round), storing clothing and linen, taking out and putting away linen and making bed. What is more, bedroom in also a place of various additional activities: certain tasks (writing letters, crocheting, etc.), listening to the radio, reading books, doing makeup, etc. The equipment necessary for sleeping includes: a bed or a convertible sofa, duvets and other bedclothes and pillows, a night lamp, a bedside table or a shelf used for putting away such items as a plate, a book or an alarm clock, a piece of furniture used to put away clothes (a chair, a shelf), a container for sheets (a linen chest or a convertible bed). Among the additional equipment there are: a table and chair used for simple activities, a vanity, a mirror, a TV stand, wardrobes and walk-in wardrobes. As a result, bedrooms might have a really diverse equipment, which, together with limited space, often results in tight for space and limiting of free movement, as the additional equipment covers the necessary communication and service space.

Ergonomic designing of sleep space is a complex activity because it has to take into consideration the economical and construction limitations. On the other hand, the freedom of fulfilment of diverse and changing needs has to be assured. What is also important are the conditions of storage of a considerable number of possessions (especially clothes). Among the key factors deciding on the size and proportions of rooms of a dwelling there are:

- number, types and sizes of fixed and mobile equipment distributed in a room,
- freedom of access to the equipment related with the movement area of an individual (size and mobility),
- the way of distribution of equipment (furniture arrangement),
- providing of conditions for safe and comfortable usage of a room and individual pieces of furniture,
- providing of conditions for changing of the arrangement (changing of placement of equipment or its replacement),
- social, cultural and psychological requirements.

4.1   Double Bedroom

The possibility of flexible arrangement of beds together and separately, as well as additional equipment oftentimes requires more space than is usually made available (especially in multi-family buildings). In a small bedroom a double bed is placed in the corner of a room (Fig. 5) and it is accessible only form one side, which makes it difficult to lie down and get out of the bed, as the person lying next to the wall has to move over the other person. Furthermore, the bed is located directly next to the window, which is especially not recommended during winter time, when the cold air coming out of the window causes discomfort and oftentimes even results in getting common colds.

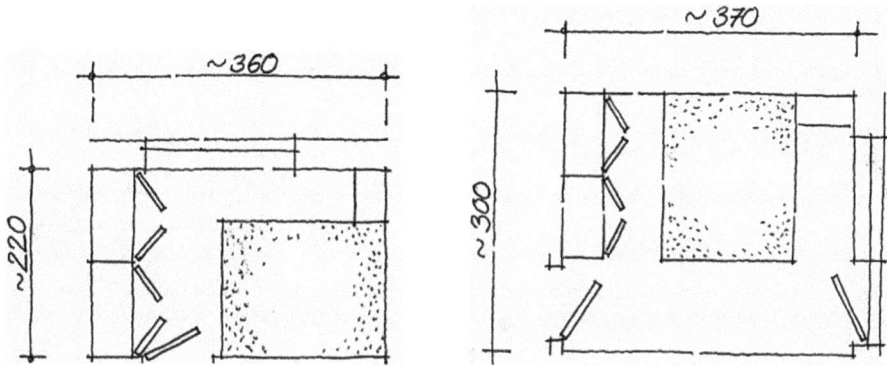

Fig. 5. Examples of faulty and correct arrangements of a double bedroom (author's drawing)

Another key factor is the size of the room. Appropriate placing of a bed together with adjacent service area requires also designing right proportions of a room as well as correct placement of windows and doors (Fig. 5), which also influences the location of additional equipment, e.g.: a wardrobe, a chair, a bedside table or a desk. Nevertheless, the size and proportions of a double bedroom are mainly determined by the possibility of placing of two beds together or separately.

4.2 Single Bedroom

Single bedroom is usually used by a child or an elderly person. It is usually the smallest room in a house. Nevertheless, children and elderly people (also people with disabilities) have additional needs connected with enlarged movement space. A child needs more space in its room for e.g.: playing, hobbies, storage of toys and entertaining friends. While an elderly person may need more space for rehabilitation exercises or for having guests.

Typical bedrooms are usually small and narrow; their width often amounts to merely 2.2 m. As a result, in a narrow room the bed has to be placed perpendicularly to the longer wall (parallel to the window) or along the side wall. A bookshelf is placed parallel; when entering a room its side is visible (Fig. 6). Free space is limited to a narrow passage along the furniture. Softening of such 'passageway' arrangement requires placing the bed perpendicularly. In such case the bed may be placed only under the window.

A room with correct proportions should cover space enabling a free placing of a bed parallel or perpendicularly to the longer wall (Fig. 6). In an appropriately wide room the parallel placing of a bed enables, among others, to free a considerable space

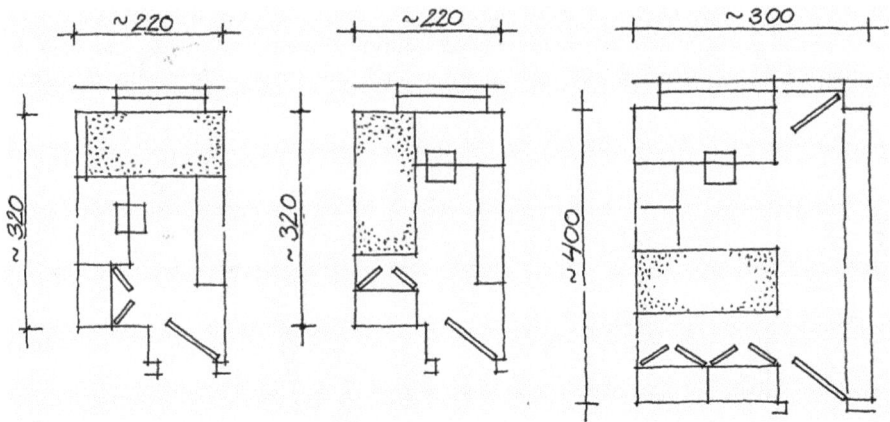

Fig. 6. Examples of faulty and correct arrangements of a single bedroom (author's drawing)

in the middle. Perpendicular placement may lead to a similar effect or enable a division of a room into smaller functional zones.

In a small room for a teenager it might be favourable to use a bunk bed. The space freed in that way might be used for e.g.: storage of clothing and toys or placing of a desk with a chair.

5 Summary

Functional and spatial designing of dwellings is a challenging task for architects, as it requires taking into consideration the economical and construction limitations, as well as specific requirements of the users. Such difficulties concern mainly the smallest rooms and functional zones of a dwelling (i.a. bedrooms). Space limitations may be alleviated by better proportions of rooms. Appropriate arrangement of furniture may lead to, among others: effective and economical management of space devoted for their placement and usage. Necessary free (service) space around the furniture might be reduced by its overlapping, for instance in case of the following pairs of pieces of equipment: bed – wardrobe, wardrobe – table, book case – desktop. Therefore, determining the areas and sizes of rooms during the designing stage should be a derivative of, among others: various arrangements of furniture together with consideration of necessary space for its handling. That is why it is necessary to remember about the following postulates:

- freedom of placement of a single bed and grating its accessibility from at least two sides (three sides in case of an ill person),
- freedom of placement of a double bed and its accessibility from three sides,
- possibility of various arrangements of one double bed or two separate single beds,
- accessibility of storage furniture also in squatting and kneeling positions,
- limiting of inaccessible space and space with a limited access (especially for users with disabilities).

Spatial and mobility requirements of people with various levels of fitness should be assumed as benchmarks used for shaping of living space and ways of distribution of elements which fill it. What is vitally important is taking into consideration of particular needs of people with lower mobility (disabled people, children, the elderly and people suffering from obesity). The possibility of rearrangement of furniture within single rooms or functional zones determines the level of flexibility of adjustment of dwellings to changing needs of their users.

Spatial limitations of contemporary dwellings result in a necessity of rational management of space required for arrangement of furniture. Since, in average houses there are a lot of pieces of furniture, which results from numerous needs of house-holders as well as owning numerous possessions.

Free arrangement of sleep zone together with placement of a door and a window influences the way of arrangement of other pieces of furniture (wardrobe, desk, etc.). It also influences the shape and size of free space in a room, which has not only a practical and usage (performing of routine activities) meaning, but also plays an important psychological (privacy, detachment from the environment) and social (spending time in solitude as well as with other people) roles.

References

1. De Chiara, J., Panero, J., Zelnik, M.: Time Saver Standards for Housing and Residential Development, pp. 384–391. McGraw-Hill Education, Columbus (1994)
2. Loeschke, G., Pourat, D.: Integrativ und Barrierefrei, pp. 283–285, 303. Verlag Das Beispiel, Darmstadt (1984)
3. Marx, L.: Barrierefreie Wohnungen, pp. 57–59. Beyerische Architektenkammer, Bayreuth (1992)
4. Panero, J., Zelnik, M.: Human Dimension & Interior Space. A Source Book of Design Reference Standards, pp. 150–152. The Whitney Library of Design, New York (1979)
5. Tilley, A.R.: The Measure of Man and Woman, pp. 53–59. The Whitney Library of Design, New York (2001)

From Industry 4.0 to Nature 4.0 – Sustainable Infrastructure Evolution by Design

Leszek Świątek[✉]

Faculty of Civil Engineering and Architecture,
West Pomeranian University of Technology Szczecin,
ul. Żołnierska 50, 71-210 Szczecin, Poland
swjoan@poczta.onet.pl

Abstract. The Grey Infrastructure is a result of engineering design in the Anthropocene. Globalized culture created systems destroying our relationship with Nature. Trade off sustainability is not enough for ecological crisis solutions, even if the invisible infrastructure of Industry 4.0 will replace Grey Infrastructure of old industry. Autistic Metaverse pushes humankind to digital words offering fake nature conversions - hypernature. IT landscapes are influential for human behaviors - different for demographic groups. Industry 4.0 becomes the Next Nature, attractive for the youngest generations - digital natives infected with the Nature-deficit Disorder. To evolve Sustainable Infrastructure by design "Human-Nature" systems reintegration and transformation from pathogenic to salutogenic are needed. Intelligent Green-Blue Infrastructure will enlarge bio-productive lands, to sustain biodiversity, to inforce regenerative abilities of coexisting ecosystems. The transgression from Industry 4.0 to Nature 4.0 is a long - term evolution focused on members of Society 5.0 vitality and regenerative ability.

Keywords: Hypernature · Metaverse · Transformative design
Networking · Regenerative Infrastructure · Coevolution

1 Introduction

The dominant human populations on the Earth are living in urbanized world. To exist and to develop economic, industrial or commercial activities we have had to create Black and Grey Infrastructure of *Homo Urbanus* as Rifkin [1] described present generation of the city dwellers in domain of metropolis artifice. According to Webster's New World Dictionary: "Infrastructure is the substructure or underlying foundation… on which the continuance and growth of a community or state depends" [2]. Uncontrolled growth enhanced the development of different types of infrastructure, visible and invisible one, hard and soft, integrated and intelligent. The infrastructure space and its transformations focus growing attention on urban planning and architectural design. To keep a high level of municipal consumption different forms of infrastructure evolves to maintain material, energy or mass information flows. Cities take up more or less only three percent of the world's land surface but explore the rest of natural world ecosystems in the light of the sixth great extinction event, massive deforestation process or degradation of water resources. With our human geological force we have

© Springer International Publishing AG, part of Springer Nature 2019
J. Charytonowicz and C. Falcão (Eds.): AHFE 2018, AISC 788, pp. 438–447, 2019.
https://doi.org/10.1007/978-3-319-94199-8_42

entered a new geologic era - the Anthropocene epoch which is partly a result of engineering activities and their infrastructure building as well as the false and changing representation of nature in contemporary urban planning and architecture. At present time the human race will need to decide between our material world and the natural world. "One world or the other will have to change" [3]. Therefore Sustainable Infrastructure design should evolve into Regenerative Infrastructure planning with emphasis on restorative abilities, high level of resilience, self-organization and adaptability. Methodology of design is in the transition process too, where anthropocentric vision of environment is being replaced with Biocentric approach to the planet Earth. Therefore William McDonough expressed as follows: "In the end the question is not, how do we use nature to serve our interests? It's how can we use humans to serve nature's interest?"

2 Models of Sustainability and Design Implementations

Anthropocentric, globalized culture created dichotomic systems: artificial – natural, mechanical – organic, empiric – noumenal. Such attitude destroyed our relationships with nature. Present, trade off sustainability model with its associated infrastructure (coexisting social, technical and natural one) is not enough to solve or to correct our global problems (Fig. 1).

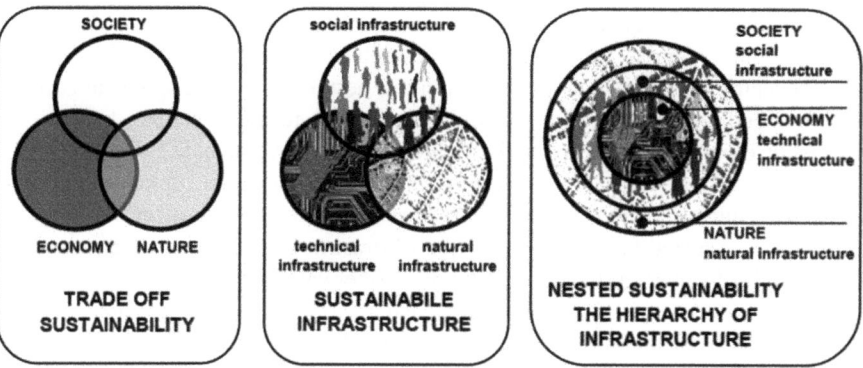

Fig. 1. Two models of sustainability and associated infrastructures. Changing development visions from trade off sustainability to nested sustainability.

Today's ecological crisis is increasing, our urbanized generation has to face with the end of nature syndrome connected with climate changes agenda or ecosystems services inflation. People need to keep and develop an integrative and inclusive approach to natural – artificial relationships to be compatible with natural systems. We do not understand consequences of massive material, energy and pollution flows which require regeneration activity. Hopefully still this days nature and the metropolis are fully intertwined. Agriculture, aquaculture, garden art or forest management are examples of human bioculture which reflects "man – environment" relations and a level of their vitality. Present urban civilization weakens our relation with nature – in globalization era we became the biosphere society. The ecological reform of bioculture, implementation of ecological renovation and ecosystems restitution are necessary today. As an important part of the global need for natural infrastructure restoration - design should base on dynamic development processes planning while considering social and natural capital as reviving symbiosis. Regenerative architecture, ecological engineering, urban ecology, bioregionalism are vibrant elements of regenerative development ideas, based on ecosystems dynamics may influence a new co-evolutionary bioculture dimension. Therefore Nested Sustainability model represents a transformative development approach with fuzzy logic and appropriate hierarchy of infrastructure, where law of nature is a basic rule in our common future rebuilding (Fig. 1). The hybrid system of overlaid and blurred social, technical (engineered) and natural infrastructure, where the last one plays a dominant role, is a new perspective looking beyond sustainability to thriveability, from coexistence to cohabitation and coevolution.

3   Nature and Its Infrastructure Meaning as a Basis of Contemporary Design

A civilization exists in a natural milieu. Human understanding of nature was culturally generated but strongly connected to Biophilia concept phenomenon. Is our notions of nature a real nature? "Nature" refers to physical features and processes of nonhuman origin that people ordinarily can perceive, including the "living nature" of flora and fauna, together with still and running water, qualities of air and weather, and the landscapes that comprise these and show the influence of geological processes [4, 5]. Anthropocentric perspective associates nature as aesthetic, as a metaphor, underlining the "eco-friendly" brand creation, naturally made with recalling natural, healthy feelings. On the other hand as quoted by Heraclitus: "Nature loves to hide". We can't ignore the changing definition of nature in the 21st century. The important component of nature is ecosystem infrastructure - defined as practices, services and industries that support a healthy and variegated biosphere [6]. Natural Infrastructure such as forests, meadows, agricultural lands, rivers, estuaries, swamp areas offer various benefits with high level of ecological services. Dynamic composition and relationships or interactions between Natural, Green and Gray Infrastructure are important constituents of ecosystems matrix. The Green Infrastructure (GI) is defined here as "an interconnected network of green space that conserves natural ecosystem values and functions and provides associated benefits to human populations" [7]. Green Infrastructure is a kind of biomimicry which could also include such technical aspects as light public transport or renewable energy systems, generally defined as sustainable or green technology. Broader variation of GI is the Green and Blue Infrastructure (G-BI) facilitating environmental hydro - biological cycles and enhancing resilience or stabilizing ecological dynamics in natural or semi natural systems such as bio retention areas, wetlands, floodplains or sustainable drainage systems, as our natural life support system. The capacity of drainage and rain water systems with the Green - Blue Infrastructure should be increased and areas for natural retention of rain water (biotops, swamps) as well as systems of water recycling be created. Rise of green surfaces in workplaces, streets, residential areas, generally in cities reduces an effect of city heat island (using permaculture and xeriscaping rules), improves local climate and hydrologic conditions and strengths biodiversity. Introducing more high green extend shadow places in the city, which work as a natural shield for rapid heat waves. Cities should promote actions of greening roofs, green ways or bioclimatic corridors location, urban forestry development and bio-productive land cultivation which can play a valuable role of climatic buffers, essential for Green and Blue Infrastructure, which brings an incremental social benefits. The urban structure of designed spaces should tend to a compact form, eliminate traffic preference and enforce passers-by and bikes tracks (Walkable City) [8]. An important element in increasing level of cities resilience and regenerative abilities is a matter of promotion and use of local products, living building materials and regional labour force as well as local food production (so called Urban Farming or City Aquaponics) being a form of social base mixed with Green and Blue Infrastructure. Holistic perception of the built environment and comparisons to complex, living systems should be recognized as a fundamental element in contemporary design. In the urbanized areas called the red zone, affected with strong social imperative we have to consider how to improve design to enhance the human – environmental experience. In the present urban fabric an effective integration of Green and Blue Infrastructure (rooted in the Natural Infrastructure) with the existing Black and Gray Infrastructure (incorporated into Technical Infrastructure) is needed with effective support of Social Infrastructure development. Urban areas facilitated with exploited and accumulated Technical Infrastructure are more affected with natural disasters, intense storms, typhoons, floods, heat waves and drought or high precipitation. The risk of large urban structures demolition or communal education and healthcare infrastructure damages is high, as well as inhabitants living conditions are in danger. Municipal living and working conditions are related to technical and moral span of properties and their Gray Infrastructure such as energy and communication grids, sewage piping or roads and highway networks, railways systems, when all of them emit pollution and generate strong emissions level. The social pathology or vitality, cultural disintegration or cohesion are shaping Social Infrastructure. Its components as sport or education facility, healthcare or neighborhood centers, parks and allotment gardens are considered as an important part of resilient, urban community. Thinking about a city as a living organism or the smart ecosystem – a Biotope City - we have to come from ego systems infrastructure to eco systems infrastructure. The wise transformation design methodology will create sophisticated network infrastructure, kind of the omnipotent hyper-infrastructure which might give an impression that manufactured landscapes and manufactured nature could be better than the real one.

4   Industry X.0 and Technological Emancipation from Forces of Nature

We are technological beings by nature [9]. Today people has to cope with technological change, with innovation and technology push that everything gets smart or virtual. "Throughout human history we practiced technology to emancipate us from forces of nature" [10]. We are submerged by technology and experience it as a natural part of our lives. Humankind is intertwined in a co-evolutionary relationship with technology. The effects and scale of human interventions and technological transformations create the next, hybrid nature, which is slightly artificialized. "We are living through a unique period of human history, an intense period of flux, change, and disruption that may never be repeated. Aseismic shift in living and thinking is taking place due to the rapid and pervasive introduction of new technologies to daily life, which has changed the way we communicate, work, shop, socialize, and do almost everything else" [11]. Beginnings of the Anthropocene Epoch started during the first industrial revolution, defined as Industry 1.0 associated with the introduction of steam and water power, fabrication and manufacturing mechanization. Mass migration processes and effect of urbanization with new patterns of land use and economic development were influential for Black Infrastructure development as novel engineering systems (Table 1). The next revolution was fast, disruptive and destructive. Named Industry 2.0 was characterized with mass production and consumption, use of electrical power, locomotion and mobility, introduction of specialization and effective assembly lines. The next step of humankind evolution in industrial development was digital, IT revolution. Automated, computer aided production, programmable logic controllers and microelectronics were vibrant features of so called Industry 3.0. The "human – machine" interaction evolved on high level of sophistication, where emergence of distributed networks and mobile affected the daily life. The fourth industrial revolution – Industry 4.0 (I4.0) is a kind of novel technological ecosystem dedicated for smart, green and urban production, specified with concept of the smart factory. The invisible infrastructure of Industry 4.0, kind of modern cyber-physical systems is characterized by ubiquitous, cloud computing, virtualization, big data distribution or additive manufacturing which will replace used, Grey Infrastructure of old industry in frames of new materiality revolution of high developed, Posthuman Information Society [12]. The adaptive, intelligent technical systems are based on 360° information, self-optimizing, data mining and intelligent networking. The Artificial Intelligence (AI), machine learning and deep learning are seeping into virtually everything. Big Data analysis provided with cloud technology and intelligent apps will create systems that learn, adapt and potentially act autonomously, according to the top 10 strategic tech trends for 2018 by Gartner [13]. AI technology landscape can be described with autonomous systems based on neural networks and neuromorphic computing. Industry 4.0 environment is affecting the compete lifecycle of products, including resources control, maintenance, change management or material and immaterial flows logistic with supply chain management systems. Technology integration with the environment and automatic adjustment to it are typical for smart adaptive systems. Such hyper-converged media network components are not neutral for our natural environment. The Information Technology (IT) sector with its spectacular Metaverse[1] development is responsible for high level of energy consumption or massive e-wastes disposal as well as misuse of human real time as resource depletion. "The Internet is omnipresent, always delivering rich, stimulating content—all day, all night, always on. Between the years 2000 and 2015, the number of

---

[1] According to Smart, E.J., Cascio, J., Paffendorf, J.: Metaverse Roadmap Overview, (2007) the Metaverse is defined as the convergence of (1) virtually enhanced physical reality and (2) physically persistent virtual space. It is a fusion of both, while allowing users to experience it as either.

people with access to the Internet increased almost sevenfold—from 6.5% to 43% of the global population" [11]. Smart grids, infrascrum, agile infrastructure will push human activities to digital words with augmented reality or the Internet of Things and e-services engagement.

Table 1. Evolutionary stages of economic development, its infrastructure, design and nature understanding.

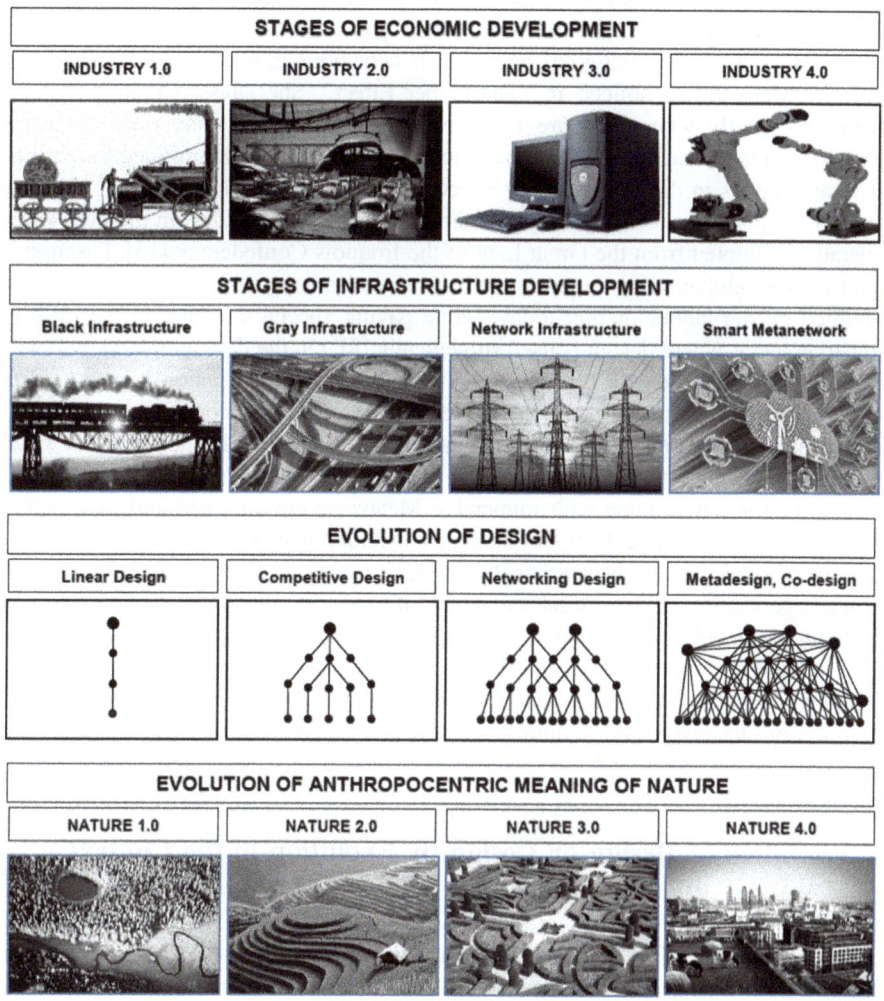

Progressive improvement in digital technology leads to effect of digital migration, enhanced by permanent connection between virtual and real world, where various items and things will communicate with each other. "The new freedoms allowed online are heady, thrilling, and enticing to billions of people, but also the new norms are rapidly being created due to an accelerated form of socialization, called cyber-socialization" [11]. This autistic digital environment will offer the hyperreality with addictive hyperpraxice in prosthetic, fake nature conversions, sometimes called the hypernature. Where "hyperreality is seen as a condition in which what is real and what is fiction are seamlessly blended together so that there is no clear distinction between where one ends and the other begins". "It allows the co-mingling of physical reality with virtual reality (VR) and human intelligence with artificial intelligence (AI)" [14]. The invisible IT infrastructure with digitally manufactured landscapes are influential for human daily behaviors, which are different for main demographic groups. "Our lives are changing, and human behaviour is evolving" as cyber psychologist Mary Aiken noticed in "The Cyber Effect". She stressed that: "...people behave differently when they are interacting with technology than they do in the face-to-face real world". Therefore "new norms created online can migrate to real life. So what happens in the virtual world affects the real world, and vice versa" [11]. "In our every deliberation we must consider the impact of our decisions on the next seven generations" quoted from the Great Law of the Iroquois Confederacy [15]. Discussion about present global development should be shared among at least four generations identified on the labour market today, while paying special attention to the creative industries sector where design is used strategically [8]. On the demographic level (with socio-psychological general perspective) a growing gap of alienation between every next entering generations is recognized. Newcomers in global economy, on competed design markets are presenting changing attitude to nature understanding and space organization. The demographic structure of the city and intergenerational relations are an important issues to cope with immersive Metaverse environment and cyber effect escalation as a corruptive factor of human cohabitation in real nature systems. The intergenerational gap is observed in life styles or ecological preferences, where the youngest generation called digital natives is infected with the Nature-deficit Disorder on the highest level [16]. Human embodied biophilia and empathy for the wellbeing of future generations should be drivers of change in contemporary design. It should be an imperative for a new approach to solve environmental global problems with systematic evolution of our design and planning systems from anthropocentric to biocentric perspective of development and the spatial coexistence.

5 Evolution of Design or Design by Evolution in the Context of Sustainable Infrastructure

Through the history of industrial development theory of design and space arrangement attitudes were changed, reflecting the Zeitgeist, the socio-economic conditions and "man-nature" relationship that transformed our built environment. From centralized, linear way of thinking focused on task oriented design specified with engineering disciplines in the 19th century to more complex, competitive decision-making processes spread between broad scope of actors participated in design and advantaged value-chain building in the 20th century. Multiplied problem-oriented design, with incoherent social and psychological aspects in correlation with ecological, circular way of thinking, based on theory of chaos and complex systems development pushed design to the new fields of perception. Explosion of computer aided design and distributed network of informatics and telecommunication advanced technologies created tremendous upside potential of digital tools application used in the creative industry. Different scales and time perspectives of design are mixed and interacted to shape cooperative, networking organization. Regional and urban planning penetrate architectural styles or fashionable trends in contemporary design. For example, following Ratti and Claudel [17] "Open Source Architecture (OSArc) is an emerging paradigm describing new procedures for the design, construction and operation of buildings, infrastructure and spaces. Drawing from references as diverse as open-source culture, avant-garde architectural theory, science fiction, language theory, and others, it describes an inclusive approach to spatial design, a collaborative use of design software and the transparent operation throughout the course of a building and city's life cycle. A contemporary form of open-source vernacular is the Open Architecture Network launched by Architecture for Humanity, which replaces traditional copyright restrictions with Creative Commons licensing and allows open access to blueprints". So we can experience the universal network building and its addiction and dependence. More and more aspects and speculative data are considered in a complex design process. Wise and effective implementation of collected data in further processing is a foundation of Metadesign concept "which was adopted in the 1980s regarding the use of information technologies in relation to art, cultural theories and design practices (from interactive art to biotechnological design). Metadesign presents an emergent design culture, calling for an expansion of the creative process in the new design space engendered by information technologies" [18]. Therefore we have to reintegrate human and natural systems, to evolve Sustainable Infrastructure by design, to improve quality of life and healing environment provision in the era of biogenetic capitalism and technophilic aesthetic. Based on a theory of ecosystems dynamics or ideas of autopoiesis our habitats have to be transformed from pathogenic to salutogenic.

6 The Next Generations of Nature

Human knowledge about Nature is an important urban resource, disproportion of knowledge transfer streams as well as information flows within the red zone associated with social imperative should be a matter of special scenarios. We have to create synergy for minimizing destructive anthropopression and increasing productivity of our Natural Network Infrastructure (hyper-infrastructure) by better flows of resources among actors and initiated positive actions in living city ecosystems. Through history of civilization anthropocentric meaning of Nature or its living ecosystems evolved (Table 1). In period of Hunters and Gatherers culture people adopted to the real Nature - Nature 1.0, being a part of natural ecosystems. The agrarian revolution was the beginning of Nature transformation process. When agriculture and animal domestication emerged – an area of cultivated lands increased, people modified their surrounding for cultivated second Nature – Nature 2.0. The art of cultivation and horticulture was a matter of primary conscious design decisions, gardening knowledge and experiences reinforced with novel aesthetic research, which created kind of Nature sensitive to beauty – Nature 3.0. Being a part of Nature in the past, industrial development and its design patterns initiated a state of war against non-human Nature and transformed our attitude to biosphere, creating artificial, built environment, alienated from vital ecosystems. After destroying nature in the industrial era, a new paradigm of relationship between human and ecosystems is needed in a global scale in the Anthropocene Age. The pace of modern live, the sense of urgency are accelerated by on-line communication readiness, rapid cyber culture lead our urbanized civilization to the fourth nature – Nature 4.0 - the uncertain future of next generations associated with posthuman, synthetic biology development, biohacking and genetic modifications or returning to the real wild and free nature. In this context a new Natural Network Infrastructure space should be reconsidered in designing process. Devastated natural capital should be restored with creation or restitution of the resilient natural infrastructure, with providing biodiversity potency in dense inhabited areas or poor, abiotic workplace environments. The invisible IT infrastructure should be transformed and interconnected into green, regenerative one. Development of intelligent Green and Blue Infrastructure integrated with smart remote sensing and software systems of environmental knowledge is needed to enlarge area of bio-productive land, to connect ecological corridors to mitigate climate changes, to sustain biodiversity richness and to inforce regenerative abilities of coexisting ecosystems. The process of transgression from Industry 4.0 to Nature 4.0 is a long - term perspective operation, evolution by design focused on problem how to regenerate our common environment for vital members of Society 5.0 cohabitation and how to avoid bios hybridization and mutant renegades for survival.

7   Conclusions

Because of the rapid development of globalization, present mankind status should be recognized as a biosphere community taking responsibility for nature in the scale of whole living planet and its natural ecosystems infrastructure health. Aspects of climate changes, resilience and ecological dynamics, CO_2 sequestration potency or biodiversity validation are still secondary in designing process measures [19]. "The City as a Factory" where the living quality is decreasing in uncertain times and unsafe environment should be replaced with an idea "The City as an Organism". Even enthusiastic "The City as a Smart Factory of sophisticated Industry 4.0" is not a natural solution. When urban areas are redefining as a novel living ecosystem, broad studies of the city metabolism, energy and materials flow analyses are required, as a mix of Grey – Green – Blue Infrastructure – part of Regenerative Infrastructure network. Industry 4.0 with supplementing invisible infrastructure becomes the Next Nature, attractive for the youngest generations of society, recognized as digital natives. Mitigation of depopulation, deindustrialization, shrinking cities processes, or urban degeneration are connected with climate changes and resources depletion as well as digital migrations, social alienation or the Metaverse escapism. Mentioned phenomena revealed with growing biophobia types or oikophobic attitudes [20] become a challenge for designers. We have to remember about importance of natural and social capitals building where appropriate designing system involvement is essential. To create and maintain "Living cities" or "Living habitats" like ecosystems we have to recognize different levels of resilience, to know ability to rebuild and to regenerate existing structures and networks. Disturbance sensitive local societies should follow dynamics of ecosystems, their ability of adaptation and regeneration and build resilient infrastructure while designing and creating programs of spatial regeneration. The 21st century bioculture will emanate, fertilize or mutate in urban areas where may comply with broad sense to the city aquaculture, aquaponics or urban farming, green belts, ecological corridors and rain gardens. Green Revival and Biocentric Worldview are nears the cutting edge of Metadesign coevolution. This is the potential to be used by local communities, self-governments, scientific communities and business which can favor synergy effect on the way to regenerative design and positive development.

References

1. Rifkin, J.: The Third Industrial Revolution: How Lateral Power is Transforming Energy, the Economy, and the World, p. 79. Palgrave Macmillan, New York (2011)
2. Webster's New World Dictionary, 5 edn. Houghton Mifflin Harcourt, Boston (2016)
3. McKibben, B.: The End of Nature. Random House, New York (1989)
4. Hartig, T., Mitchell, R., de Vries, S., Frumkin, H.: Nature and health. Ann. Rev. Public Health 2 (2014). http://publhealth.annualreviews.org
5. Coutts, C.: Green Infrastructure and Public Health, p. 4. Routledge, London (2016)
6. Plowden-Wardlaw, L.: Ecosystem infrastructure capital building ecosystem infrastructure for the 21st century. ALA (2013)
7. Benedict, M.A., McMahon, E.T.: Green Infrastructure: Smart Conservation for the 21st Century, p. 5, Sprawl Watch Clearinghouse, Washington (2002)
8. Świątek, L.: Regenerative ergonomic design – biocentric evolution. In: Charytonowicz, J. (ed.) Proceedings of the AHFE 2017 International Conference Advances in Human Factors, Sustainable Urban Planning and Infrastructure, pp. 96–105. Springer International Publishing AG, Cham (2017)
9. Gehlen, A.: Man, His Nature and Place in the World. Columbia University Press, New York (1988)
10. van Mensvoort, K.: Pyramid of technology: how technology becomes nature in seven steps. Eindhoven University Lecture Series, vol. 3, Eindhoven University, Eindhoven (2014)
11. Aiken, M.: The Cyber Effect: A Pioneering Cyberpsychologist Explains How Human Behaviour Changes Online, p. 5. John Murray Publishers, London (2016)
12. Braidotti, R.: Po człowieku. Wydawnictwo Naukowe PWN SA, Warszawa (2014)
13. Gartner: The Top 10 Strategic Tech Trends for 2018. https://www.gartner.com
14. Tiffin, J., Terashima, N.: HyperReality: Paradigm for the Third Millenium, pp. 4–5. Routledge, London (2001)
15. Seventh Generation. http://www.seventhgeneration.com
16. Louv, R.: Ostatnie dziecko lasu. Mamania, Grupa Wydawnicza Relacja, Warszawa (2016)
17. Ratti, C., Claudel, M.: Open Source Architecture, p. 120. Thames & Hudson Ltd., London (2015)
18. Giaccardi, L.: Metadesign as an emergent design culture. Leonardo **38**(4), 342 (2005)
19. McDonough, W., Braungart, M.: The Upcycle: Beyond Sustainability – Designing for Abundance. North Point Press, New York (2013)
20. Scruton, R.: Zielona filozofia. Jak poważnie myśleć o naszej planecie, Zysk i S-ka Wydawnictwo, Poznań (2017)

Post-occupancy Evaluation Research Method in Architecture - Conscious Creation of Safe Living Space

Joanna Zabawa-Krzypkowska[✉]

Faculty of Architecture, Silesian University of Technology,
M. Strzody Street 10, 44-100 Gliwice, Poland
joanna.zabawa-krzypkowska@pols.pl

Abstract. The article deals with the search for safety in urban housing architecture. Based on the author's own research a selected aspect was presented,
namely, it has been shown how important the accessibility issue is in the spatial
structure. On the basis of observational studies the current nature of changes in
the existing housing complexes has been illustrated.

Keywords: Safe living environment · Human needs · Behavior
Psychophysical abilities POE research method · Quality of living

1 Introduction

Safety in the Housing Environment. Safety of man living in the family and society
translated into the spatial aspect concerns the existence in the public and private,
domestic space. The space in question must assure the sense of security the lack of
which is related to such emotions as: fear, insecurity, anxiety, confusion, depression,
malaise. It is connected with the lack of privacy and the feeling of "being at home".
Such emotions have adverse impact on the quality of human life. Bańka writes
"Architecture should enhance the positive foundations and positive processes occurring
in human brain. It should not let the destructive feelings spread, therefore, anxiety on
the one hand and aggression on the other hand" [4].

Housing environment provides man with shelter and enables a number of activities
connected with both private and social sphere. It should protect them in the physical
and psychological sense alike, provide the need of shelter, privacy, intimacy and the
need of an individual to feel accomplished in the society, among other things by
facilitating interpersonal contacts, including the neighbourly ones [14].

2 Qualitative Research

Scientific research carried out up to the 1950s within the fields related to man,
including psychology, sociology and anthropology helped to understand the relationships between man and his environment. Knowledge derived from these fields enriched
the awareness of designers and made it possible to design the environment of life

© Springer International Publishing AG, part of Springer Nature 2019
J. Charytonowicz and C. Falcão (Eds.): AHFE 2018, AISC 788, pp. 448–456, 2019.
https://doi.org/10.1007/978-3-319-94199-8_43

according to the expectations of the recipients. It helped to move the physical environment closer to the expectations and motivations of the user. Environmental psychology and sociology developed in the 1960s were particularly important on that matter and were related to the criticism of American cities. Growing social problems required that attention should be given to the malfunctioning environment. Answers were sought to the questions bothering the scientists with respect to the causes of the problems connected with spatial solutions. It was then recognized how essential is the impact of architectural solutions on the feelings of the users, on their health and wellbeing. What was also noticed were the opportunities created by the environment built for the users by stimulating positive or negative behaviors. The attention was paid to the manner people behave in specific spatial cases.

Psychology as the field complementary to architecture is the means of conveying methods, techniques and information, which determine human needs [20].

Qualitative research, also known as environmental and behavioral, or as research of the developed environment are connected with environmental and social psychology. They include the analysis of relations between environment and human behavior. Knowledge of human needs, physical and mental comfort of use became helpful in architectural design [16].

In 1980s a universal evaluation method was developed known as POE (Post-Occupancy Evaluation), that is an overall evaluation of the quality of an architectural object in technical, functional and behavioral terms, and then in organizational and economic ones. The POE method enables to diagnose a given building, a group of buildings depending on the set and defined goals. It is the evaluation of the structure of the environment (of a building, group of buildings or a bigger urban complex) during its use. It enables to verify the compliance of the quality of a building with the user expectations and it includes both expert and participatory assessment.

Creators of the POE method Preiser et al. [17], envisaged three levels of evaluation, namely:

(1) Indicative evaluation – identification of maladjustments and defects permitting more detailed research.
(2) Investigative evaluation – in which the maladjustments identified during indicative evaluation will be diagnosed and detailed conclusions will be drawn necessary to design or modernize a building.
(3) Diagnostic evaluation, during which several buildings performing the same function are compared in order to seek wider generalizations concerning the architectural problems [21].

Therefore, the POE method helps to describe the existing practical problems in specific cases making it possible to determine the directives on improving the existing state. It is helpful to modify the spatial arrangement, improve the décor, functional changes as well as changes connected with solving behavioral problems, which appear in the behavior of people in architectural space. Method can be applied as case studies for comparative purposes, which may be helpful in modernizing and designing new projects [18, 19].

3   Qualitative Research in the Research on Safety and the Sense of Security

Qualitative research in respect of research on safety and sense of security includes:

- technical quality – analyzed as satisfactory state of upkeep and maintenance of buildings and the whole infrastructure of the surroundings. State of development and upkeep of land around the buildings in the context of ensuring repairs and removing the reported damage. Upkeep of a building in view of its age, durability of materials, materials used (insulation, removing asbestos cardboards, replacement of doors and windows, application of materials resistant to damage, etc., works connected with maintenance management). Technical quality embraces technical and technological facilities, condition of the surface, provision of the elements of small architecture, transport efficiency, number of car parks, transport system, pedestrian walkways, level of security, monitoring and condition of street lighting, pedestrian ways, urban interiors, building entrance zones;
- functional quality – is characterized by proper location, transport connections, structure of the nearest neighborhood, structure and composition of analyzed layout, including urban interiors, entrance zones, arrangement of forms of social activity, as well as the distribution of the occasion, problem generators. Functional solutions are analyzed paying special attention to adaptation of the size, composition and solutions of the common space, which enables the activity of users of the given space (simultaneously eliminating unfavorable behavior, that is, occurrence of antisocial behavior). The quality in question includes functioning of buildings and space resulting from grouping the function related to the hierarchy of space influencing its accessibility. Such elements are essential as access to collective services and transport, definition of the way of implementing the social needs of the users, that is: composition of greenery, playgrounds, recreation grounds for children and adults, availability for old and disabled people (elimination of architectural barriers in respect of providing community integrating places, because a better-consolidated community identifies with the dwelling place, cares for it and is more inclined to provide help). Research include safety – resulting from functioning in a wider structure – impact of neighborhood, potentially dangerous places, paths and driveways – safety of traffic which has effect on the frequency of making use of grounds, appropriation of places by cars, frequency at which people appear in the day/night space, possibility to make use of the land thanks to attractiveness of development, elimination of the poorly lit places,
- behavioral quality including also the aesthetic one – concerns human behaviors and individual emotions experienced under the influence of the surrounding space. It is examined as fulfilment of comfort, sense of security. In this case, the composition is essential, which creates the space hierarchy assuring privacy, territoriality. An important matter is the comfort of use, fulfilment of social needs, occurrence of integrating places for people residing in a given part of space, satisfaction, pleasure of staying in a given space. Under consideration are applied design solutions, equipment, small architecture facilities, quality of lighting and the number of light points, color, materials, and greenery, also the sense of climatic comfort (quality of

air, pollution, sound level, and transport nuisance). Appearance of the development concept is essential: applied form, color scheme, texture of buildings, and the state of upkeep and maintenance of buildings and the surrounding space in the context of impact on aesthetic impression. The scale of development concept has also a great significance in the feeling of the space. Sense of security, actually, its lack related to the existence of potentially dangerous places. Fear appears wherever a pedestrian feels uncertain. Any dark, poorly lit and remote paths and driveways can evoke anxiety. Neglected and desolate places are also hostile. Behavioral quality is connected with good orientation in space, ease of finding the way, elimination of the feeling of loss and fear.

The research method in this paper is based on the literature analysis, review of the current literature and own research - "in-situ" (especially multi-family housing development in Gliwice, in Poland).

4   Safety – Accessibility Issue

In today's world, we can observe many forms of seeking safety in the housing environment. One of them are gated communities, which appeared in Poland after the political transformation of 1989.

The author's observations reveal various forms of protection, or rather reduction of accessibility to the neighborhood and private space within the existing urban building development. It manifests itself in various sorts of enclosures, fencing as the most frequent form. On the one hand, closed structures appear. On the other hand, we can observe that some fragments are separated from a larger whole, i.e. from the space of a housing development. Such actions illustrate how important the issue of space accessibility is for the sense of security. It is because every man needs his own territory. Such place is determined by both the border and the whole system of distances and barriers, physical and symbolic ones.

The following kinds of space can be distinguished in the spatial environment:

- public, which are generally accessible, an individual can feel anonymous because he/she is not subject to close observation being one of many people,
- semipublic, these are spaces accessible to everyone, where the users' behavior, however, is under some public observation and control. These can be places of public contact, a bench in a park, a square in front of a building, pedestrian way, etc.,
- semiprivate are the spaces of limited accessibility assigned to a group of residents or employees, it can be a garden in front of a house, a staircase or a courtyard, it is a place of contacts where behaviors are subject to observation,
- private, which are inaccessible or where the access is controlled, it can be, for instance, the space of a house, flat or room [6, 7, 20].

Spatial structure should correspond to the nature of the function, both for public and private spaces. Easy accessible need to be open-access and public spaces. Accessibility should then be gradually limited in terms of the access towards private spaces. Such a division defines the scope and method of use, it introduces a certain order. Applying integration solutions, we stimulate interpersonal contacts, suggesting segregation solutions we reduce them. In the place of residence, social activity of residents is positive, but the presence of strangers is inadvisable because it often contributes to the decrease of a sense of security. It has been proved that a stress factor in the place of residence is an excessive number of people staying near a house, front door or under windows. Unwanted social contacts should be limited within the semiprivate and private spaces. Semiprivate spaces need to be free from too intensive traffic.

Accessibility to a certain space depends on the number of barriers and borders which are to be crossed and on the distance, thus their complexity can perform an isolating function [1–3, 5, 19].

Accessibility structure is made of the transportation system: roads, driveways, pedestrian ways, traffic routs, as well as the spatial structure and the composition of layout, which can be of less or more open nature. Symbolic barriers are also very important, namely, the color, type of flooring, and distinguishing elements, e.g. architectural detail, items of furnishing, size of the entrance to the building, its form [7–9].

Historical developments can be distinguished by hierarchical structure, which was the result of the separation of the private area from the public one; it was the result of the structure of ownership division. In traditional forms of architecture the ownership of buildings and land introduced the division into zones. The front elevation and a courtyard determined the accessibility of such places. Social order was translated into spatial order [6]. Nationalization of land ownership shattered this order. It turned out that common spaces usually remained no-man's land. That in turn resulted in adverse phenomena (dirt, rubbish, excrement), e.g. at the gateways to town houses, which remained open. At present, the urban structures are more and more open and undefined [20].

Examples. In order to find the irregularities in functioning of specific architectural development concepts (cases) it is necessary to carry out the qualitative research. It is essential to identify the existing problems. As a consequence, it enables to indicate the directions for modernization, improvement of the current state.

The research in question consist of several stages: it is the process of collecting data, information analysis, finding relationships occurring in the housing environment and negative behavior of users in the living space. Such research should be carried out each time for specific cases. As regards the safety research, the issue is multi-faceted and requires close analysis in specified social, spatial and cultural conditions.

Presented examples apply to typologically different groups. Four basic types can be distinguished in the housing development of Polish cities: social estates, residential complexes, large residential complexes (blocks of flats), old residential buildings mainly town houses, an estate of detached houses. Multi-family residential units generally prevail in the Polish urban development.

Accessibility issue is one of the elements creating a safe place of residence. It is connected with the spatial composition and structure of planning schemes, residential complexes and suitable spatial hierarchy. It concerns the symbolic and physical barriers, which ascribe certain buildings and surrounding land to specific group of users. Gradation of space determines possibilities of behavior suitable for a given structure [10, 11].

The need for determining one's own place by man is connected with the need for territoriality, which was mentioned by Oskar Newman, the creator of the defensible space theory [15]. Territoriality helps to recognize the outsiders, whose presence is often connected with the occurrence of negative phenomena such as loud conversations, making mess, etc [12, 13].

In historic town houses, it is clearly visible that the front doors and entry gates were locked. These elements gain a new aesthetic quality. They constitute an element, which facilitates orientation and identification with the place. Such a form is a non-invasive activity and it does not have a negative impact on the users of the space. It is a return to the old order (Figs. 1).

Fig. 1. Examples of limiting access to the private and semiprivate space.

There are also forms of closing a building or buildings and adjacent premises, which are often accidental actions and which introduce chaos in the space. The least advantageous is to separate smaller fragments from a certain whole, which hinders the daily lives of other users. Such a phenomenon can be observed above all in housing estates where the spaces between blocks of flats are open and accessible to outsiders. Smaller or bigger parts separated from a larger space appear dividing it in an irrational way. They perform the functions of car parks (Figs. 2).

In town house courtyards, in city blocks you can also see ever larger division of space into smaller yards adhering to specific buildings. It is connected with the change of ownership structure and management method. Each town house in a quarter usually has a different manager and a multitude of owners bring about difficulty in decision-making as regards management. It is followed by the incoherent fragmentation of space, which in addition is almost entirely taken over by cars. Social element becomes eliminated: opportunity to organize neighborhood meetings, recreation and rest in the place of residence (Figs. 3).

Fig. 2. Fencing backyard/front yard areas

Fig. 3. Fencing common space in high rise building estate.

5   Conclusion

Improvement of the quality of housing environment can be achieved through better functioning of development infrastructure and spatial solutions: location, availability of basic services, transport. The way of development and furnishing the places which perform integral functions, such as: courtyards, squares which enable to perform various needs/activities, including interpersonal contacts. You need to bear in mind the permanent improvement of practical qualities, care of the natural environment, because contact with nature is necessary for human physical, mental and social health. No less important element is attention to the protection of cultural heritage, which has a significant impact on a sense of place identity. Good environment is the one, which is not only functional but also comprehensible among other things by its historical continuity. It is such an environment, which reduces reprehensible behaviors to a minimum [6, 8].

Qualitative research makes it possible to understand the dependencies and problems associated with safety and security in specific spatial circumstances. Such research is helpful to diagnose the existing situation and to develop guidelines for modernization and design activities. Research on the adaptation of the living environment to the needs and especially the psychosomatic capabilities of people is necessary for improving the quality of life. The goal is to indicate designing methods, which will be able to fulfill a number of ergonomic requirements and ensure adequate physical and social conditions.

References

1. Alexander, C., et al.: A Pattern Language: Towns, Buildings. Construction. Oxford University Press, Oxford (1977). Copyright by Christopher Alexander
2. Altman, I., Stocols, D. (eds.): Handbook of Environmental Psychology. Wiley, New York (1987)
3. Altman, I.: The Environmental and Social Behavior: Privacy, Personal Space Territoriality and Crowding. Brooks/Cole, Monterey (1975)
4. Bańka, A.: Społeczna psychologia środowiskowa, p. 36. Wydawnictwo Naukowe Scholar, Warszawa (2000)
5. Bell, P.A., Greene, ThC, Fisher, J.D., Baum, A.: Environmental Psychology. Gdańskie Wydawnictwo Psychologiczne, Gdańsk (2004)
6. Czyński, M.: Architektura w przestrzeni ludzkich zachowań. Wybrane zagadnienia bezpieczeństwa w środowisku zbudowanym, pp. 110–130. Wydawnictwo Politechniki Szczecińskiej, Szczecin (2008)
7. Colquhoun, I.: Design Out Crime: Creating Safe and Sustainable Communities. Elsevier, Architectural Press, Oxford (2004)
8. Czarnecki, B.: Przestrzenne aspekty przestępczości. Metoda identyfikacji czynników zagrożeń w przestrzeni miejskiej. Oficyna Wydawnicza, Politechniki Białostockiej, Białystok (2011)
9. Czarnecki, B., Siemiński, W.: Kształtowanie bezpiecznej przestrzeni publicznej. Difin, Warszawa (2004)
10. Ellin, N.: Architecture of Fear. Princeton Architectural Press, New York (1997)
11. Flusty, S.: Building paranoia. In: Ellin, N. (ed.) Architecture of Fear. Prinston Architectural Press, New York (1995)
12. Gehl, J.: Life Between Buildings: Using Public Space. Arkitektens Forlag, Island Press (2001)
13. Hall, E.T.: Hidden dimention. Anchor: Reissue edition (1990)
14. Maslow, A.H.: Motivation and Personality. Harper & Row, New York (1954)
15. Newman, O.: Defensible Space: People and Design in Violent City. Collier Book, New York (1973)
16. Niezabitowska, E.: Metody i techniki badawcze w architekturze. Wyd. Pol. Śląskiej Gliwice (2014)
17. Preiser, W., Rabinowitz, H., White, E.: Post-Occupancy Evoluation. Van Nostrand Reihold, New York (1998)
18. Tymkiewicz, J.: Quality analyses of facades based on post occupancy evaluation. Research experience with students of architecture participation. In: ICERI Proceedings of the 9th International Conference of Education, Research and Innovation, ICERI 2016, pp. 8831–8838 (2016)

19. Tymkiewicz, J.: The advanced construction of facades. The relations between the quality of facades and the quality of buildings. In: 2nd International Conference on Advanced Construction. Kaunas University of Technology, Kaunas, Lithuania, 11–12 November 2010. Book Group Author(s): Kaunas University of Technology Press, Sponsor(s): Kaunas University of Technology, Lithuanian Academy of Sciences, State Fund Sci & Study Lithuania
20. Advanced Construction 2010 Book Series: Advanced Construction, pp. 274–281 (2010)
21. Zabawa-Krzypkowska, J.: Architectural research on semi-private space in multi-family housing development, Architecture, Civil Engineering, Environment, vol. 6, no. 1/2013, pp. 5–13. The Silesian University of Technology, Gliwice (2013)

An Investigation of Government Support Influence on Low-Income Housing Construction Quality in South Africa

Chikezie Eke$^{(\boxtimes)}$, Grace Akidi, Clinton Aigbavboa,
and Wellington Thwala

Sustainable Human Settlement and Construction Research Center,
Faculty of Engineering and the Built Environment,
University of Johannesburg, Johannesburg, South Africa
chykezie2002@gmail.com, graceakidi27@gmail.com,
{caigbavboa,didibhukut}@uj.ac.za

Abstract. Any form of support or help from the government for the betterment of the lives of people is known as government support. However, it literally means a financial help given by the government and most times, it does not always come in monetary form. The aim of the study is to investigate the extent housing construction quality of low-income housing is influenced by government support; to determine the most occurrence in the service government provide for low-income housing construction; and to determine how significant government support is to low-income housing construction. Quantitative research approach was adopted and the questionnaire survey was used in collecting the information. 220 questionnaires were distributed randomly to respondents who are working for/with Department of Human Settlement. Findings revealed that the relationship between the factor (government support) and the dependent variable (housing construction quality) was found to be statistically significant at ($p < 0.05$ or 5%).

Keywords: Government support · Quality · Low-income housing
Influence · Housing construction

1 Introduction

Literally, government support means a financial help given by the government. Most times, it does not always come in monetary form; hence any form of support or help from the government for the betterment of the lives of people. Government support or Grant is a public subsidy offered to a recipient or person for business or personal purposes. The subsidy is not expected to be paid back (except when terms of condition was violated), and may be used for research, business development, education or other endeavours that are anticipated to support a common cause. Government supports are vital to helping low-income families meet fundamental needs such as housing, nutrition, child care, and health care. For the most part, individuals who receive government supports find that the gap between their earnings and needs are generally met as a result of the government assistance they receive [1].

© Springer International Publishing AG, part of Springer Nature 2019
J. Charytonowicz and C. Falcão (Eds.): AHFE 2018, AISC 788, pp. 457–463, 2019.
https://doi.org/10.1007/978-3-319-94199-8_44

According to Lerner and Haber [4], the role of government support can influence the economic and non-economic opportunities that are essential to the creation of the conditions for developing low-income housing. A majority of individuals, however, do not receive all the government supports they are eligible for because of various barriers to accessing and using these resources. Government supports are limited in availability because of stringent eligibility criteria, insufficient funding, limited awareness, and lengthy and demanding application processes. For some government support programs, each additional income that the benefactor earns put the low-income house hood in danger of losing a substantial amount of valued financial assistance. When the loss of government support occurs it often decreases the household's total financial resources so drastically that it resembles the experience of falling off a cliff; this is, as a result, called the "cliff effect." However, not all government supports are fully cut off the moment the individual's income increases [1].

The South African Government provide supports to assist the inhabitants and also investors set up and run a business in South Africa. The form of assist varies depending on the individual, location and the nature of the business. For the people in business, the Government support is to encourage them to be more productive and have a sustainable enterprise [1–3, 5, 6]. The South African government have different support programmes and some of these programmes include:

- Employment
- Rural development
- Social and community development
- Justice, crime prevention and security
- Education
- Population registration
- Transport
- Energy
- Youth
- Women
- Housing
- Health
- Environment
- International relations
- Economic issues
- Science and Technology
- Agriculture
- Communications
- Public participation
- Local government

The above listed are some of the areas which South African Government supports and many more but in the context of the study, the research will be based on housing. Housing is one of the most important aspect of human settlements. The South African Government support programme on housing aims to eradicate informal settlements in South Africa. The aim of the study is:

- to investigate the extent housing construction quality of low-income housing is influenced by government support;
- to determine the most occurrence in the service government provide for low-income housing construction; and
- to determine how significant government support is to low-income housing construction.

2 Methodology

The methodology adopted in this study was a quantitative research approach. The study was done in the Gauteng Province of South Africa. The survey was carried out with the help of a structured questionnaire in collecting the information. 220 questionnaires were distributed randomly to respondents who are working for/with Department of Human Settlement and 100% response rate was achieved. The data analysis was conducted using descriptive analysis, multivariate correlational data analysis including exploratory factor analysis (EFA) and regression analysis using variance (ANOVA) to test whether the predictor is significantly better at predicting the outcome than using the Mean as a presumption.

2.1 Mean Item Score (MIS)

A five point Likert scale was used to rank the extent of the occurrence of government support factors. From the reviewed literature, the adopted scale was as follows;

1. = To no extent
2. = To a minor extent
3. = To a moderate extent
4. = To a large extent
5. = To a full extent

The five-point scale was transformed to mean item score (MIS) for each of the government support factors. The indices were then used to determine the rank of each item. The ranking made it possible to analyze the data collected from the questionnaires survey.

$$\text{MIS} = \frac{1n1 + 2n2 + 3n3 + 4n4 + 5n5}{\sum N}$$

Where;
n1 = Number of respondents for no extent;
n2 = Number of respondents for minor extent;
n3 = Number of respondents for moderate extent;
n4 = Number of respondents for large extent;
n5 = Number of respondents for full extent;
N = Total number of respondents.

After mathematical computations, the criteria are then ranked in descending order of their mean item score (from the highest to the lowest).

3 Findings and Discussion

Table 1 indicates the housing construction quality factors based on occurrence in terms of percentage responses on a scale of 1 (strongly disagree) to 5 (strongly agree), and a MS ranging between 1.00 and 5.00. All the MSs are above the midpoint score of 3.00, which indicates that the respondents agreed with the housing construction quality factors. Findings from the table reveals that majority of the factors have very high response above the midpoint score of 3.00, which shows that housing construction quality were strongly agreed for the six variables (MS > 4.00 ≤ 5.00) which ranges from reduces community chaos (MS = 4.17; SD = 0.629; R = 1) to ensures community culture (MS = 4.01; SD = 0.758; R = 5). Creates competitiveness amongst stakeholders (MS = 3.54; SD = 0.705; R = 6) was the only factor that has a neutral agreement which is above the midpoint score (MS > 3.00 ≤ 5.00).

Table 1. Housing construction quality based on occurrence

Variables	Strongly disagree… strongly agree (%)					MS	SD	Rank
	1	2	3	4	5			
Reduces community chaos				12.7	57.7	4.17	0.629	1
Creates healthy and safe living	0.5	0.9	20.6	39.4	38.5	4.15	0.807	2
Occupant satisfaction		0.9	8.2	72.7	18.2	4.08	0.543	3
Building sustainability		0.9	14.5	61.8	22.7	4.06	0.638	4
Attract investors and development	0.5		17.3	57.7	24.5	4.06	0.677	4
Ensures community culture	0.5	0.5	24.1	47.7	27.3	4.01	0.758	5
Creates competitiveness amongst stakeholders		5.9	40.9	46.8	6.4	3.54	0.705	6

Table 2 indicates the factors relating to service government provide which could influence quality of housing construction. Respondent were asked to rate the factors based on occurrence in the projects they are working/worked on. The factors in terms of percentage responses on a scale of 1 (to no extent) to 5 (to a full extent), and a MS ranging between 1.00 and 5.00. Findings from the table reveals that ranging from water supply (MS = 4.28; SD = 0.689; R = 1) and electricity supply (MS = 4.17; SD = 0.629; R = 2) had a very high response above the midpoint score of 3.00, which shows that government provided services occurrence were very high for the two variables (MS > 4.00 ≤ 5.00). Also, findings from the table reveals that six factors were above the midpoint score of 3.00, which shows that government provided services

Table 2. Factors government provide based on occurrence

Variables	Very poor... excellent (%)					MS	SD	Rank
	1	2	3	4	5			
Water supply		1.8	8.2	50.5	39.5	4.28	0.689	1
Electricity supply			12.7	57.7	29.5	4.17	0.629	2
Drainage system		0.5	42.7	51.8	5.0	3.61	0.590	3
Overall services provided by the government	0.5	0.9	45.5	50.9	2.3	3.54	0.584	4
Garbage and waste collection	0.5	9.1	37.7	50.5	2.3	3.45	0.710	5
Occupant safety (police services)		6.4	54.5	37.3	1.8	3.35	0.626	6
Sanitation in the vicinity		17.3	59.1	21.8	1.8	3.08	0.678	7
Fire protection services	0.5	13.6	69.1	15.0	1.8	3.04	0.614	8
Telephone service	20.5	70.0	7.7	0.9	0.9	1.92	0.629	9

occur moderately for the six variables (MS > 3.00 ≤ 5.00) ranging from drainage system (MS = 3.61; SD = 0.590; R = 3) to fire protection services (MS = 3.04; SD = 0.614; R = 8). Telephone service is the only variable with the lowest ranking less than the midpoint score of 3.00, which means that telephone service in low-income housing construction is minor or no existence with the variable MS < 3.00 ≤ 5.00.

Table 3. Frequency variables - normalized multivariate scores

Factors	Mean	Median	Mode	Std. D.	Skewness	Kurtosis	Minimum	Maximum
Housing const. quality	14.9676	15.0110	12.3	1.98444	−0.168	0.195	8.89	21.11
Government support	12.0031	12.0660	12.28	1.49411	0.672	4.481	7.01	19.14

This section in the questionnaire didn't give a good reliability when factor analysis was run and also cannot create constructs based on the reliabilities. Thus, the summation of factor Scores from principal components analysis was used as shown in Table 3. However, Table 4 shows the only predicting variable in the study which is government support (GS). Findings reveal that housing construction quality (HCQ) correlates moderately with the predicting variable (GS).

Table 4. Correlations of housing construction quality with government support

		Index of housing construction quality
Government support	Pearson correlation	0.254
	Sig. (2-tailed)	0.000
	N	217

3.1   Regression Analysis

Regression analysis is used when two or more variables are thought to be systematically connected by a linear relationship (Table 5).

Table 5. Model summary of HCQ

Model	R	R^2	Adjusted R^2	Std. error of the estimate	Durbin-Watson
1	0.668[a]	0.447	0.426	1.43238	2.195

a. Predictors: (Constant), GS
b. Dependent Variable: HCQ

Table 6. ANOVA of dependent variable – housing construction quality

Model		Sum of squares	df	Mean square	F	Sig.
1	Regression	312.861	7	44.694	21.784	0.000[b]
	Residual	387.775	189	2.052		
	Total	700.636	196			

a. Predictors: (Constant), GS
b. Dependent Variable: HCQ

The construction quality output containing the analysis of the variance (ANOVA) which tests whether the predictor is significantly better at predicting the outcome than using the Mean as a presumption. From the Table 6, the value of F is 21.784, which is highly significant ($p < 0.001$), interpreting that the predictor is significantly improves the ability to predict the outcome variable.

The beta (β) value has an associated standard error indicating to what extent the value would vary across different samples, and the standard error are used to determine whether or not, the b value differ significantly from zero (using the t-statistics). A t-test is a statistical hypothesis test. Therefore, if the t-test associated with a b value is significant (if the value on column labelled *sig.* is less than 0.05), then that predictor is making a significant contribution to the model. The smaller the value of *sig.* (and the larger the value of t) the greater the contribution of that predictor. However, for Table 7, GS [t (189) = 2.804, $p < 0.006$], are significant predictor of HCQ.

Table 7. Coefficients of dependent variable – housing construction quality

Model	Predictor	Unstandardized Coefficients		Standardized coefficients			Collinearity Statistics	
		b	Std. error	β	t	Sig.	Tolerance	VIF
	Constant	3.866	1.408		2.746	0.007		
1	GS	0.219	0.078	0.156	2.804	0.006	0.946	1.057

a. Dependent Variable: HCQ

4   Conclusion

Findings from Table 2 reveals that water supply was ranked the highest which shows the most occurrence in the services government provide in low-income housing construction. The results presented in Table 7 yielded support with the standardized parameter estimate for the general hypothesis. The relationship between the factor (Government support) and the dependent variable (Housing construction quality) was found to be statistically significant at $p < 0.05$ and $t = 2.804$. Although there was a significant correlation (0.254) or influence suggesting a moderate degree of linear association between the factor and the endogenous variable (Table 4). Inspection of the R^2 value (0.447 or 44.7%) and the adjusted R^2 value (0.426 or 42.6%) shows that (0.447 − 0.426 = 0.021 or 2.1%), approximately 2.1% less variance in the endogenous variable (Housing construction quality) shrinkage when the model were derived from the population rather than a sample. However, F-ratio (21.784) is highly significant ($p < 0.001$) in the improvement due to fitting the regression model which is much greater than the inaccuracy within the model meaning the model significantly improves the ability to predict the endogenous variable (Table 6). Hence, the score results suggested that the influence of government support on the housing construction quality (endogenous variable) was statistically significant.

References

1. Albelda, R., Shea, J.: Bridging the Gaps Between Earnings and Basic Needs in Massachusetts: Executive Summary and Final Report. Center for Social Policy Publication, Boston (2007). http://scholarworks.umb.edu/cgi/viewcontent
2. Co, J.M.: Perception of policy obstacles: a South African study. Small Bus. Monit. 2(1), 101–105 (2004)
3. King, K., Mcgrath, S.: Enterprise in Africa: Between Poverty and Growth. Intermediate Publications Ltd., London (1999)
4. Lerner, M., Haber, S.: Performance factors of small tourism ventures: the interface of tourism, entrepreneurship and the environment. J. Bus. Ventur. 16(1), 77–100 (2000)
5. Morrison, A., Teixeira, R.: Small business performance: a tourism sector focus. J. Small Bus. Enterp. Dev. 11(2), 1–12 (2004)
6. Rogerson, M.C.: Small enterprise development in South Africa. Afr. Insight 33(2), 109–115 (2002)

Human Factors in the Correlation with Aesthetics and Pro-ecological Technology in Modern Architecture

Anna Gumińska[(✉)]

Chair of Strategy of Design and New Technologies in Architecture,
Faculty of Architecture, Silesian University of Technology,
Street Akademicka 7, 44-100 Gliwice, Poland
anna.guminska@polsl.pl

Abstract. Contemporary problems of housing architecture focus mainly on saving the environment and sustainable spatial planning. An equally important aspect of the environment is its impact on people. One of the determinants of space is its aesthetics. The use of modern, energy-saving building materials often changes this aesthetics. The aim of the work is to analyse selected elements of architecture and space affecting this aesthetics. New objects and spaces located in Denmark, Germany, Sweden and Norway were selected for the research. The main elements of the research are: the size of buildings, construction of buildings and organization of space. The study was based on subject literature and photographic documentation. Examples of specific technological elements of renewable energy used to build facilities are discussed. Analysis of the examples showed the correlation between aesthetics and pro-ecological technologies and building materials. The technologies used have a large impact on people in the architectural space, affect the comfort of use and the quality of this space.

Keywords: Habitat · The beauty of useful · Modern housing complex
Sustainable living environment · Sustainable · Pro-ecological · Aesthetic

1 Introduction

The contemporary problems of housing architecture are mainly focused on issues of sustainable development of the environment. Legal regulations, the state of the human environment inspires the analysis of objects in terms of energy efficiency and their impact on the environment. Yet an equally important aspect of the environment is its aesthetics.

How do the changes affect the aesthetics of buildings in the use of building materials and technology for technical reasons and the adaptation of architecture to the demands imposed by climate change?

Research has attempted to analyze these aesthetic changes of selected contemporary facilities and housing estates adapted to current needs in Denmark, Germany, Sweden and Norway.

J. Charytonowicz and C. Falcão (Eds.): AHFE 2018, AISC 788, pp. 464–475, 2019.
https://doi.org/10.1007/978-3-319-94199-8_45

The main elements examined were: the size of buildings, housing estates, structure and structure of structures, development and layout of individual social spaces, adaptation to RES trends, climate change and how these factors influence the quality of aesthetics of objects and social spaces.

The study was based on an analysis of the literature of the subject and study visits in the analyzed objects and photographic documentation made in situ.

The analyzed examples point to the specific aesthetic elements used in housing due to the adaptation of these objects to the RES requirements. In this environment, there are many natural elements, as well as technical elements related to energy efficiency, energy production in both objects and space management.

There are correlations between aesthetics and pro-ecological technologies and construction materials. The materials used, the urban layout and the structure of the objects show naturalistic and ecological trends. Customization is different and depends on the size of the housing unit and its structure.

2   Examples of Qualitative Research Methods and Object Assessment

The study used scientific research on: qualitative research methods and object assessment, proprietary methods of qualitative research and applications in object assessment, pre-design research and quality analysis of post occupancy evaluation research.

Examples of qualitative research methods and object assessment are discussed in the literature of the following authors: Fross and Sempruch [7, 8].

Examples of authoritative methods of qualitative research and applications in the evaluation of objects are discussed in the literature of the following authors: Fross et al. [4, 5, 9, 21].

Examples of pre-design studies are discussed in the literature of the following authors: Tymkiewicz et al. [20, 22].

Examples of quality analysis of post occupancy evaluation research are discussed in the literature of the following author: Tymkiewicz [19].

Research and tools also apply to the students' architectural process by introducing a new subject into the curriculum about new technologies and methods in architecture design of the following authors: Masły and Sitek [6, 15].

Examples of Computer simulation-based building performance analyses (BPA): preliminary study of in the literature of the following author Masły [16, 17].

3   Technologies and Ecological Materials

When analyzing the links between sustainable development of the environment and the building process of buildings, it is important to note the tendency of changes in the use of building materials and technologies and to pay attention to the whole life cycle of the building. It is evident in modern housing architecture the consciousness of the pro-ecological development and maintenance of the environment built as an ecosystem, which is the main idea of sustainable development.

The analysis considers the criteria for the use of materials and technologies that contribute to the reduction of harmful effects on the environment or to harmful effects to a very small extent, utilize natural processes occurring in nature, and the structure and structure of an object is based mostly on a nature-inspired structure [14, 24].

4 Beauty and Usefulness

There is a wide literature of the subject dealing with the analysis of the principles of beauty, aesthetics in various areas of research. Researchers dealing with this issue should be mentioned in various aspects of research: theoretical, philosophical, architectural or urban planning: Tatarkiewicz, Gołaszewska, Schneider-Skalska, Böhm, Kosiński, Gzell [10, 12].

The study looked at beauty in terms of utility [2].

Research on this issue was conducted by, among others, Masaaki Kurosu and Kaori Kashimura, as well as Noam Tractinsky. They have proven by their research that there are correlations between perceived usability and aesthetics [8]. Tatarkiewicz also argued that the Sophists of architecture were included in useful arts as investigated objects of art [18].

Aesthetics is a subjective aesthetic sense, presented as reception, evaluation. Beauty, however, we associate with the form of works presented by order, harmony, legibility of the cognitive message, and understood meaning and value of the work [23].

Holistic thinking should be used in the analysis of objects because of the multifaceted nature of the problem.

5 Technologies and Materials for Ecology, Structure of the Object and Beauty and Utility - Examples

10-floor Mountain Dwellings building, Copenhagen, Denmark, PLOT + BIG Architects, 2008.

The 10-storey residential building is located in the suburban area of Copenhagen with 80 apartments with garden ("garden houses") on the roof with stunning views and parking directly under each apartment.

The roof gardens consist of a terrace with a garden that changes its character depending on the season.

Parking is cascaded under the building and remains in symbiosis with a residential building (480 parking spaces with sloping lift). In the north and west, the façades are made of perforated aluminum sheets that allow air and light to flow in the parking lot and create a massive reproduction of Mount Everest [25].

The object presents a different way of shaping a 10-storey building. With the use of terraced apartments and terraces, apart from the positive aesthetic effect of the buildings, the apartment was also a source of relief, and the residents also had access to the garden and the hidden garage. The use of such a tall building facade can contribute to lowering the temperature in the surrounding space (a component of sustainable development). The pictures below show the green roofs on the terraced, multi-family residential building, as well as the terrace views and the balcony in the side elevation (Fig. 1).

Fig. 1. The multi-family building in Copenhagen, Denmark, PLOT + BIG Architects, Mountain Dwellings, Residential Mountain 2008. Terrace gardens and the interior of the facility with garages, (photo A. Gumińska, 2016)

The 8 HOUSE building project, Bjarke Ingles Group, Bjarke Ingles, Thomas Chrostoffersen, 2010 (nominated for the Mies van der Rohe Award 2011), is located in Southem Orested on the Copenhagen Canal (Denmark, Copenhagen) overlooking the open air. It is a large apartment block, offering flats to people of all ages, with different needs (building area: 60,000 m^2, including 476 apartments).

The arch-shaped building creates two distinct spaces - intimate inner courtyards (with greenery and gravel-strewn), separated by central space, with communal facilities with a 9 m wide passage that connects two spaces: The street from the east side. 8 House creates a three-dimensional environment, different from typical high-rise buildings. Such a design of the building provides better insolation of the premises. The roofs of the building are covered with low green, which also contributes to the improvement of water management and the warming of the building. Various functions in the building have been spread horizontally, for example residential apartments located on the top of the building, and commercial premises, downstairs. Thanks to this solution the apartment has access to a lot of daylight, fresh air and beautiful view, and office-commercial access to the street (Fig. 2).

Fig. 2. Multifamily building in Copenhagen in Ørestad, Denmark, 8 HOUSE, Bjarke Ingels Group (BIG) archives project from 2010. (photo A. Gumińska, 2016)

Turning Torso is a residential and office building in Malmö, Sweden, designed by Santiago Calatrava, 1999–2005.

It is the tallest building in Sweden. The structure of the building was inspired by a spiral motif sculpted by a designer in marble torso with seven marble cubes twisted around the metal shaft. The entire building twists up every 10°, and the entire building twist is 90°. In addition to the skeleton function, the concrete core supports vertical circulation, providing support for mechanical parts, electrical and hydraulic systems, and building ventilation. The 190-m-high structure is built on 9 five-sided irregular lumps with 5 floors each around a concrete core of 10.6 m in diameter, with 54,400 m^2 of floor space. The two lowest modules of the skyscraper (12 lower floors) are pow. Office area of 4200 m^2, and from 3 to 9 segment is 147 flats of the area. 13.5 thousand. M2, each with a different layout, on the 53 and 54 floor there are conference rooms [26]. Due to the height of the facility, a reinforcing supporting structure was constructed in the form of a structure that is constructed of spinal structure, with similar mechanical properties and resistance to sudden weather changes.

The pictures below show the building, structural details - metal shaft, supports (Fig. 3).

Fig. 3. Office building-residential building HSB Turning Torso, Malmö, Sweden, Arch. Santiago Calatrava. (Photo A. Gumińska, 2016)

6 The New Pro-ecological Development in the Context of the Historical Environment

The Milan Residential Complex is located in the CityLife district, near the Fiera Milano Rho congress district and in the Milanese historic district of Milan, Italy. Project: Zaha Hadid, Patrik Schumacher, 2013 (2007–2018 CityLife building).

It consists of seven buildings of 5 to 13 storeys, with one-storey apartments and two-story penthouses.

The buildings are arranged in a meandering loop with a large courtyard in the middle and green areas throughout the estate. Each building is gently curved, with rounded corners, with transparent balconies wrapped around the building. The curvilinear movement of balconies and terraces results in smooth lines of buildings. The large windows allow visual contact with Milanese's historic district and the public park. Bright white facades were made using fibrous concrete slabs. Wooden panels on some walls and balconies represent, among other things, ecological architectural tendencies.

The mansion gardens consist of paved paths and grassy areas with small recesses that create resting and recreation areas, are safe and tranquil, and provide an intriguing view of the city and the park. The residence was awarded Italian Class A certificate [28].

It consists of seven buildings from 5 to 13 storeys, with apartments one storey two-storey penthouses.

Buildings have a height determined by the historical development of a nearby district. Each building has a slightly curved façade, and all corners have been rounded to alleviate sharp edges. Such a structure was designed to refer to the context of a nearby park. The buildings include environmental and comfort requirements (most of the apartments are south, east with terraces facing the city or public park) [28]. The development of recreational space is also aimed at isolating the inhabitants from wheeled transport, enabling active leisure and fun in the center of a large city. The photos below show the buildings, playgrounds and green spaces of the Residential Complex in Milan, Italy (Fig. 4).

Fig. 4. Multi-Family Buildings: City Life Residences in City Life, Milan, Italy, Proj. Zaha Hadid Architects (2004–2012), (photo A. Gumińska, 2015)

Daniel Liebeskind Architects (2013) has a completely unique geometry that accentuates the tops of buildings, fostering the integration of large-scale structures into the rich and varied surrounding urban fabric, and the next estate next to the above City Live housing estate in Milan, Milanese district, Italy.

Buildings utilize state-of-the-art design (i.e. state of the art refers to the highest level of overall development, equipment, technology, or scientific backgrounds achieved at a given time) and are certified according to the Italian A + Class - CENED.

Balanced functions include various aspects of sustainable development, e.g.: thermally adjustable ceiling radiators, energy efficient, programmable heating and cooling systems, modern insulation and durable tiles, fine textured, gray façade tiles Casalgrande Padana, corrugated sunblinds made of polymer composite Bamboo Soleils Brise and photovoltaic cells on the roof.

CityLive has a completely unique geometry that accentuates the tops of buildings, fostering the integration of large-scale structures into rich and diverse surrounding urban tissue.

The classic Italian courtyard layout of the courtyard: the courtyard and the naturalistic materials of the historic neighborhood of Milanese.

The complex of 5 residential buildings is located around the perimeter of an open courtyard connecting the north with the park.

The inner courtyard is without car transport (underground parking), with greenery as a social space [28] (Fig. 5).

Fig. 5. Multifamily Buildings: City Life Residences in Milan, Milanese District, Italy, Prospect Daniel Liebeskind Architects (2013), (photo by A. Gumińska, 2015)

Fig. 6. Multifamily building in Copenhagen in Ørestad, Denmark, 8 HOUSE, interior courtyard, green, gravel, Use of natural building materials: stone, wood and gravel; Bjarke Ingels Group (BIG) archives project from 2010. (photo A. Gumińska, 2016) and Detail, wooden bench and stone wall in front of the Museum of Denmark. (photo A. Gumińska, 2016)

7   Conclusion

The analyzed examples point to the specific aesthetic elements used in housing due to the adaptation of these objects to the RES requirements. In this environment, there are many natural elements, as well as technical elements related to energy efficiency, energy production in objects and space management (use of greenery on roofs, use of natural materials, nature-related structures of buildings).

There are correlations between aesthetics and pro-ecological technologies and construction materials. The materials used, the urban layout and the structure of the objects show naturalistic and ecological trends. Customization is different and depends on the size of the housing unit and its structure. However, the introduction of natural materials and human scale into large residential units in effect creates a friendly space with which users can identify and thus accept this space (Fig. 6).

In today's European construction there are pro-ecological tendencies, mainly showing the following directions: the use of increased amounts of greenery in the management of land and buildings (roofs, walls); shaping the structure of buildings using natural phenomena supporting energy conservation (ventilation, lighting, water storage, energy), shaping the structure and principle of operation for natural systems (shape and function of the tree), the principle of the spine operation, the use of natural or partly natural resources building materials (recyclable materials, conglomerates with natural materials), naturalistic shaping of the building environment.

Areas of active greenery (considering sudden and severe weather phenomena - heavy rains, droughts), using information technology to control energy consumption of the building, technologies using renewable energy sources (photovoltaic cells, geothermal energy).

The analyzed examples point to the specific aesthetic elements used in housing due to the adaptation of these objects to the RES requirements. In this environment, there are many natural elements, as well as technical elements related to energy efficiency, energy production in objects and space management (use of greenery on roofs, use of natural materials, nature-related structures of buildings).

There are correlations between aesthetics and pro-ecological technologies and construction materials. The materials used, the urban layout and the structure of the objects show naturalistic and ecological trends. Customization is different and depends on the size of the housing unit and its structure. However, the introduction of natural materials and human scale into large residential units in effect creates a friendly space with which users can identify and thus accept this space.

In today's European construction there are pro-ecological tendencies, mainly showing the following directions: the use of increased amounts of greenery in the management of land and buildings (roofs, walls); shaping the structure of buildings using natural phenomena supporting energy conservation (ventilation, lighting, water storage, energy), shaping. The structure and principle of operation for natural systems (shape and function of the tree), the principle of the spine operation, the use of natural or partly natural resources building materials (recyclable materials, conglomerates with natural materials), naturalistic shaping of the building environment.

Areas of active greenery (considering sudden and severe weather phenomena - heavy rains, droughts), using information technology to control energy consumption of the building, technologies using renewable energy sources (photovoltaic cells, geothermal energy).

Also, such buildings have a smaller, harmful effect on the environment.

It can therefore be said that the new environmental trends in construction do not adversely affect the aesthetics of urban space.

Contemporary construction strives for the least harmful impact on the environment while introducing the latest innovative technologies and building materials.

Architectural and urban space largely refers to and relates to nature. Such activities provide opportunities for greater user-friendliness, increased space functionality and, above all, less environmental destruction.

In research on contemporary facilities such as housing estates, residential buildings, office facilities, their size, division, structure, structure, management and placement of individual social spaces, adaptation to the trends of renewable energy sources and the aesthetic quality of objects and social spaces were analyzed.

Modern architecture strives to realize the principles of sustainable development while at the same time, through the high quality of innovative technology, ensures the living conditions for its users.

Much attention has been paid to many of the issues related to the quality of objects and space. Introduced in the reference architecture to traditional, natural construction, the principles guiding natural phenomena.

These references have a positive impact on users, while at the same time meeting the principles of sustainable development and contributing to the improvement of the quality of the environment.

References

1. Alexander, C., Ishikawa, S., Silverstein, M., Jacobson, M., Fiksdahl-King, I., Angel, S.: JĘZYK WZORCÓW Miasta - budynki – konstrukcja, Gdańskie Wydawnictwo Psychologiczne/GWP, Gdańsk (2008). ISBN: 978-83-60083-70-3
2. Andrzejewski, J.: Czy piękne oznacza użyteczne? http://www.webusability.pl/2008/04/23/czy-piekne-oznacza-uzyteczne/. Accessed 12 Jan 2017
3. Borsa, M.: Estetyka, funkcja i tożsamość przestrzeni miejskich. http://forumprzestrzenie miejskie.pl/images/Borsa%20M.,%20Estetyka%20funkcja%20i%20tozsamosc%20przestrzeni%20miejskich.pdf. Accessed 18 Oct 2016
4. Fross, K., Ujma-Wąsowicz, K., Wala, E., Winnicka-Jasłowska, D., Gumińska, A., Sitek, M., Sempruch, A.: Architecture of absurd. In: Antona, M., Stephanidis, C. (eds.) Universal Access in Human-Computer Interaction. Methods, Techniques, and Best Practices. Proceedings of the 10th International Conference, UAHCI 2016 Held as Part of HCI International 2016, Toronto, ON, Canada, 17–22 July 2016, Pt. 1, pp. 251–261. Springer, Cham (2016). bibliogr. 14 poz
5. Fross, K., Winnicka-Jasłowska, D., Gumińska, A., Masły, D., Sitek, M.: Use of qualitative research in architectural design and evaluation of the built environment. In: AHFE – HFSI 2015, Session: Ergonomical Evaluation in Architecture, Las Vegas (2015)
6. Fross, K., Ujma-Wąsowicz, K., Gumińska, A.: Teaching of architectural design – first steps. Driving course design methodology. In: International Conference WLCTA, Conference of Learning, Teaching and Educational Leadership, Paryż, 29–31 October 2016. Elsevier (2015)
7. Fross, K., Sempruch, A.: The qualitative research for the architectural design and evaluation of completed buildings – part 1 – basic principles and methodology. Archit. Civ. Eng. Env. ACEE 8(3), 13–19 (2015). (Silesian University of Technology)
8. Fross, K., Sempruch, A.: The qualitative research for the architectural design and evaluation of completed buildings – part 2 – examples of accomplished research. Archit. Civ. Eng. Env. ACEE 8(3), 21–28 (2015)
9. Fross, K., Winnicka-Jasłowksa, D., Sempruch, A.: "Student zone" as a new dimension of learning space. Case study in Polish conditions. In: The 8th International Conference on Applied Human Factors and Ergonomics, AHFE 2017, Los Angeles, USA, 17–21 July 2017. Elsevier (2017)

10. Gołaszewska, M.: Zarys estetyki: problematyka, metody, teorie. Wydawnictwo Naukowe PWN, Warszawa, Wydawnictwo (1986). ISBN: 830104795X

11. Gumińska, A. (Anna Brzezicka), Tożsamość miejsca przestrzeni miast różnej wielkości w aspekcie technicznym. In: ULAR 7. Odnowa krajobrazu miejskiego. Monografia = Urban landscape renewal. T. 1, Przyszłość miast średniej wielkości. Red. Nina Juzwa, Anna Szulimowska-Ociepka, pp. 451–460. Wydział Architektury Politechniki Śląskiej, Gliwice (2013). bibliogr. 17 poz. ISBN: 978-83-936574-1-4

12. Kosiński, W.: Preliminaria badań nad problematyką: Piękno miasta. http://www.pif.zut.edu.pl//images/pdf/pif-10_pdf/001%20KOSINSKI%20Wojciech%20OK.pdf. Accessed 18 Oct 2016

13. Lipiec, M.: Estetyka i ergonomia, czyli piękne jest użyteczne. http://magazyn.k2.pl/03/sub.php?art=4&page=1. Accessed 18 Oct 2016

14. Łaskarzewska, M.: Piękno i użyteczność "Polskiej Zielonej Ściany". http://cejsh.icm.edu.pl/cejsh/element/bwmeta1.element/. Accessed 18 Oct 2016

15. Masły, D., Sitek, M.: New ideas and tools in the educational process of students of architecture. The introduction of a new subject to the curriculum - new technologies and methods in architecture design. In: Gomez Chova, L., Lopez Martinez, A., Candel Torres, I. (eds.) Conference Proceedings of the 9th International Conference of Education, Research and Innovation, ICERI 2016, Seville, Spain, 14–16 November 2016, [Dokument elektroniczny]. IATED Academy, Valencia, pamięć USB (PenDrive), pp. 7374–7379 (2016). bibliogr. 11 poz. ISBN: 978-84-617-5895-1. Seria: (ICERI Proceedings; 2340-1095) Web of Science

16. Masły, D.: Computer simulation-based building performance analyses (BPA): preliminary study of an office building in Poland. In: Proceedings of the 10th International Conference of Education, Research and Innovation, ICERI 2017, Seville, Spain, 16–18 November 2017, pp. 3526–3535. IATED (2017). bibliogr. 12 poz. ISBN: 978-84-697-6957-7. Seria: (ICERI Proceedings; 2340-1095)

17. Masły, D.: Daylight in high-performance intelligent sustainable offices: simulation studies. In: Słyk, J., Bezerra, L. (eds.) Education for research, research for creativity, pp. 201–206. Wydział Architektury Politechniki Warszawskiej, Warszawa (2016). bibliogr. 6 poz. ISBN: 978-83-941642-2-5. Seria: (Architecture for the Society of Knowledge, vol. 1, 2450-8918)

18. Tatarkiewicz W.: Historia estetyki, t.1, ISBN-13 978-83-01-18085-0, t.2: ISBN-13 978-83-01-18087-4, t.3: ISBN-13 978-83-01-18089-8, Wydawnictwo Naukowe PWN (2009)

19. Tymkiewicz J.: Quality analyses of facades based on post occupancy evaluation. Research experience with students of architecture participation. In: Chova, L.G., Martinez, A.L., Torres, I.C. (eds.) Conference on 9th Annual International Conference of Education, Research and Innovation (ICERI), ICERI 2016, Seville, Spain, 14–16 November 2016, pp. 8831–8838 (2016)

20. Tymkiewicz, J., Winnicka-Jasłowska, D., Jastrzębska, M.: Pre-design studies on the example of modernization project of geotechnical laboratories. ACEE Archit. Civ. Eng. Environ. 10(2), 43–52 (2017). bibliogr. 14 poz

21. Ujma-Wąsowicz, K., Fross, K.: Beauty - aesthetics - senses. How to scientifically investigate the attraction and magic of the built environment. In: Charytonowicz, J. (ed.) Advances in Human Factors, Sustainable Urban Planning and Infrastructure. Proceedings of the AHFE 2017 International Conference on Human Factors, Sustainable Urban Planning and Infrastructure, Los Angeles, California, USA, 17–21 July 2017. Springer International Publishing, Cham (2018)

22. Winnicka-Jasłowska, D., Jastrzębska, M., Tymkiewicz, J.: Ergonomics of laboratory rooms - case studies based on the geotechnical laboratories at the Silesian University of Technology. ACEE Archit. Civ. Eng. Environ. **10**(2), 35–41 (2017). bibliogr. 17 poz

23. Witruwiusz: O architekturze ksiąg dziesięć. Prószyński i S-ka, Wydawnictwo (2004). ISBN 83-7180-972-7

24. Zielonko-Jung, K.: Kształtowanie przestrzenne architektury ekologicznej w strukturze miasta, Zeszyt "Architektura" nr 9. OWPW, Wydawnictwo (2013). ISSN: 1896-1630

25. http://www.world-architects.com/pl/projects/27105_Mountain_Dwellings/

26. http://www.constructalia.com/polski/galeria_projektow/szwecja/turning_torso/

27. http://www.korthtielens.nl/architecture/ijburg-46c/

28. http://www.dezeen.com/2015/06/03/zaha-hadid-daniel-libeskind-city-life-residences/

How Vertical Farming Influences Urban Landscape Architecture and Sustainable Urban Developments

Mo Zhou[1(✉)], Wojciech Bonenberg[1], Xia Wei[1], and Shoufang Liu[2]

[1] Faculty of Architecture, Poznan University of Technology,
ul. Nieszawska 13c, 61-021 Poznań, Poland
zhouxiaomo6141@hotmail.com
[2] Liaoning Urban and Rural Construction Planning Design Institute,
Wenhua Road, Heping District, Shenyang City 110003,
Liaoning Province, China

Abstract. In order to feed the growing population in the big cities, there is an increasing demand for sustainable food supply. Vertical farming as the effective way for growing in the city and it is supposed to provide a physical base to conduct research into sustainable urban food production. Nowadays vertical farming has been working as the skyscraper gardens or skyline farms in many cities. Acting as the community garden hub the urban vertical farming is providing the place where residence around can easily go to see, plant their lovely vegetables, take care of the gardens, working in the harvesting process of the vegetables and fruits. The vertical farms are usually located in the city and around the community, residents around have opportunities to access to the plants easily. With good design of the vertical farming as well as the proper management, vertical farming is playing an important role in the urban landscape architecture. In many cases, vertical farming can shape the urban landscape. Meanwhile, vertical farming is also considered as the symbol of wellbeing of the society and urban sustainable development by many people.

Keywords: Vertical farming-urban landscape · Sustainable development
Urban environment · Skyscraper garden

1 Introduction

By 2050, around 80% of world population is expected to live in urban areas, and the growing population will lead to an increasing demand for food. The efficient use of vertical farming may perhaps play a significant role in preparing for such a challenge.

In last 30 years the Vertical Farming industry has already developed a lot in many big cities around the whole world, which has been demonstrating a significant growth in the urban areas. Due to the big demand for safe and fresh food as by citizens as well as the benefits to urban landscape to the city, vertical Farming industry has been promoted to boost in the global expansions.

© Springer International Publishing AG, part of Springer Nature 2019
J. Charytonowicz and C. Falcão (Eds.): AHFE 2018, AISC 788, pp. 476–485, 2019.
https://doi.org/10.1007/978-3-319-94199-8_46

From an environmental perspective, there are also research have shown us that growing food indoors can create an opportunity for returning farmland to its original ecological function. In many cases, abandoned agricultural land results in the regrowth of hardwood forest [1].

The diffusion of urban farming reflects an increasing awareness of how food and farming can shape our cities. [2] 'vertical farms' represent the most far-reaching vision of farming in and on urban buildings. Another approach is building-integrated agriculture, i.e., greenhouse systems on buildings, to exploit synergies between buildings and agriculture [3].

The research was also carried by the scholars in Singapore. Findings in the research paper suggest that Singapore's public housing estates are suitable for rooftop farming. Implemented nationwide, such a scheme could result in a 700% increase in domestic vegetable production, satisfying domestic demand by 35.5% [4].

2   Projects Analysis from Urban Vertical Farms

The paper is based on the research in the vertical farming model in some cities and the simulation project of the vertical farming in the future. Through the analysis from the aspects as the available access to the nature, benefits to the environment, beautifying the landscape architecture in the urban developments, biodiversity in the urban area and food security for the cities.

Some research also clarify that the market potential with vertical farming in many countries in the world. Vertical Farming which undoubtedly is more sustainable than the form of agriculture, might have a market in these countries with big population. The purchasing power of the consumers as well as the population which mirrors the quantity of demand is the most important factor for assessing market potential, beside the needs for food sovereignty and incompatibility of agro-climatic factors for food production [5]. Many desert regions like big cities Dubai in Arabic countries as well as Australia, and the US which have substantial stretches of land that fall under the category of desert have big potential for developing the vertical farming.

Table 1. Asia markets with high potential in the big cities with big population.

The name of the city	Population	Market potential
Tokyo	38,001,000,000	388
Singapore	5,888,926	58
Shanghai	23,741,000,000	237
Hongkong	7,191,503	71
Taibei	2,666,000,000	266

Source: CIA (2017).

This method and statistic are basen on the research paper that if it is assumed that only 100 g of VF products is consumed per head per day, the design presented in this paper, can feed around 100,000 people round the year. The market for vertical farm were calculated based on this figure under the supply potential of a VF (100,000 people). Of course this is probably not the exact number of farms that would be built, but it gives us a rough idea of what is the potential for future plan in those big cities [5].

It lies ahead for this technology. As in the Table 1 described in Asian potential markets, we just make a statics with the big population cities because in many regions in Asia the food supply in the city is quite huge with the expansion of the urban developments. Thus, there are big potentials in those big cities with high population like Tokyo, Shanghai, Hongkong, Taipei and Singapore (the market potential number is relatively not high but with the high urbanization and no farm land, so in fact the potential is very big).

2.1 The Design of Vertical Farming is Well in Cooperating with the Modern Architecture in the Urban Areas

In Fact we recognize that the concept of vertical farming is not something new anymore, but we still are trying to do best to make the project more practical in the cities. As in the Fig. 1 showed us through the project Clepsydra Urban Farming we could see that it tries to incorporate the features of modern greenhouse and self-sufficiency into the modern city.

Fig. 1. The Clepsydra Urban Farming tries to incorporate the features of modern greenhouse and self-sufficiency into the modern city.

To ensure that the food production of the vertical farms, the greenhouse style is still very popularized by many projects. However, the innovate design with connection of the traditional greenhouse technology and new high-technology again lead the vertical farming coming to the modern cities.

2.2 The Skyscraper Orchards or Forests in the City Sky Again Combining the Urban Developments

Many orchards or forests called skyscraper orchards or forests are sprouting in the big cities as the vertical farming developments e.g. The sky forest is going to be built in Nanjing city in China as the project made for the vertical forest in Fig. 2 as below. It is reported that this Vertical Forest there are going to have 1100 trees from 23 local species, covering big plantable area. And it is estimated that it could provide a 25 tonnes of CO_2 absorption during a year and even produce about 60 kg of oxygen per day if it is working as the estimation. With the skyscraper forest or orchard in the skyscrapers construction in the city, we have to say it is changing the landscape of the world's green cities. It must especially apply to nature and human being as well.

Fig. 2. Photo by Stefano Boeri for Nanjing Vertical Forest

2.3   The Vertical Farm Developing also Shape the Urban Landscape Architecture

The vertical farming is coming to our city is according to the big demand and good access to the nature. There are needs to satisfying the residential as well as the local architecture. This vertical tower as in the Fig. 3 below is designed in the city Poznan where in this district around with tower offices and residential buildings. The tower group is forming the circle in this district in the city. With the vertical farming proposal, it is becoming more characteristics in this area with the tower buildings.

Fig. 3. Photo project designed by Filip Zielinski for sustainable vertical farms in the city

As in the Fig. 3 shown that the vertical farming promote the biodiversity in the urban area whiling combing the shape of the local architecture. Somehow with the sustainable green tower design. In my opinion the vertical farm developing also shape the urban landscape architecture.

2.4   The Vertical Faming is the Model Working More Efficiency and Saving Space for Landscape and Buildings in the Urban Area

As we know many places mentioned the vertical farming as the big food factory. Well during the process of sufficient food production it can save a lot of space for public buildings in the city. As shown from Fig. 4 that with the vertical shelf design in vertical farms that it can save a lot of space and good for management.

Fig. 4. Photo project designed by Filip Zielinski for sustainable vertical farms in the city

Many research already reported that the vegetable production is around 10 times more than the conventional agriculture. So in this way, if we calculate that with the same food production, vertical farm only occupy 1/10 of the land. Specially in many crowded cities, so vertical farming is no doubt saving a lot of space for other public service like urban landscape and residential buildings.

2.5   The Sustainable Vertical Farm with Modern Design Can Promote the Urban Architecture

The vertical farming industry in a large scale can work toward achieving its sustainability goals and potential growing in the big cities. The trend of vertical farming developments is corresponding to the urban architecture as the design shown in Fig. 5 above. With modern design project of vertical farming in the cities, it can also largely provide the platform for promoting the urban architecture in many instances.

Fig. 5. Photo project designed by *Alicja Michowska* for sustainable vertical farms in the city

Furthermore, vertical farming is the evolution the way how we grow in the traditional systems. Vertical farming systems can provide us a lot to increase productivity and value through increasing biodiversity, and by doing so furthering the benefit to local communities in our living cities. As the Fig. 6 below showed us the good design with sustainable energy application for vertical farms, which can realize maximally self-energy supply in some projects.

Fig. 6. Photo project designed by Alicja Michowska for sustainable vertical farms in the city with solar panels for self-sufficient energy supply

3   Conclusion

In general, the vertical farm developments in the urban areas are benefits to the environmental, economic and social aspects. On one side, it has offered the huge opportunities for demanding the food supply in the big cities. The other side by doing that it also tries to keep space to the local gardening according to the climate as well as the local urban landscape.

From the research and analysis with a few case studies with vertical farms project, we can get that the good design of vertical farming is well in cooperating with the modern architecture in the urban areas; Vertical farms or skyscraper gardens constitutes an alternative way for urban environment that give us possibilities to live closer to the nature in urban area.

Furthermore, the sustainable vertical farm with modern design can promote the urban architecture in many instances and it also can shape the urban landscape architecture.

Therefore, the sustainable vertical faring is boosting the urban environment of the landscape in aspect of providing more green space for citizens and improving the biodiversity in the cities as well.

Beside the vertical farming is giving more food production by the limited space, it can save more space for developing landscape or more other public service like recreation for citizens.

4 Discuss

There are the reports and paper researches on the advantages and disadvantages to the urban landscape and urban environment by vertical farming.

There are also reports from *Michal Surguladze* wrote that skyscraper farms, tree planting drones and vertical forests, they are changing the landscape of the world's green spaces. Then Exciting times for innovative green spaces for a remaindering people some of the most impressive urban green space landscapes that have set the benchmark to date [6].

Andre Viljoen wrote that urban agriculture provides a design proposal for a new kind of sustainable urban landscape: Urban Agriculture. By growing food within an urban rather than exclusively rural environment in the book Continuous Productive Urban Landscapes: designing urban agriculture for sustainable cities [7].

The paper concludes that, while urban agriculture is receiving a great deal of attention, the theory underpinning the design of productive landscapes and the rationale for developing policy to support its practice will require sophisticated cross-disciplinary work to articulate the full potential of concepts, such as CPUL, to make essential infrastructure within future sustainable cities [8].

In the paper with green roof garden on the vertical farms were described as the very good approaches for greenery the cities in urban area were mentioned [9, 10].

Acknowledgments. Thanks a lot to my colleagues, professors and students contribution for the research and design projects for the paper.

References

1. Smith, T.F., Waterman, M.S.: Identification of common molecular subsequences. J. Mol. Biol. **147**, 195–197 (1981)
2. Tak, K., et al.: The South Korean forest dilemma. Int. For. Rev. **9**, 548–557 (2007)
3. Thomaier, S., Specht, K., Henckel, D., Dierich, A., Siebert, R., Freisinger, U., Sawicka, M.: Farming in and on urban buildings: present practice and specific novelties of Zero-Acreage Farming (ZFarming). Renew. Agric. Food Syst. **30**(1), 43–54 (2015). (Innovations and Trends in Sustainable Urban Agriculture)
4. Caplow, T.: Building integrated agriculture: philosophy and practice. In: Heinrich Böll Foundation (ed.) Urban Futures 2030: Urban Development and Urban Lifestyles of the Future, pp. 54–58. Heinrich-Böll-Stiftung, Berlin (2009)
5. Astee, L.Y., Kishnani, N.T.: Building integrated agriculture: utilising rooftops for sustainable food crop cultivation in Singapore. J. Green Build. **5**(2), 105–113 (2010)
6. Banerjee, C., Adenaeuer, L.: Up, up and away! the economics of vertical farming. J. Agric. Stud. **2**(1) (2014). (ISSN 2166-0379 2014)
7. Appendix: Springer-Author Discount. https://michaelsurguladze.wordpress.com/2015/07/09/tree-]
8. Viljoen, A.: CPULs Continuous Productive Urban Landscapes: Designing Urban Agriculture for Sustainable Cities, Architectural Press. Oxford (2005)

9. Bohn, K., Viljoen, A.: The edible city: envisioning the Continuous Productive Urban Landscape (CPUL). FIELD **4**(1), 149–161. (ISSN 1755-0068)
10. Zhou, M., Bonenberg, W.: Application of the green roof system in small and medium urban cities. In: Advances in Human Factors and Sustainable Infrastructure: Proceedings of the AHFE 2016 International Conference on Human Factors and Sustainable Infrastructure, pp. 125–136, 27–31 July. Walt Disney World®, Florida (2016). (ISSN 2194-5357)

Construction Industry

Causes of Delay in Various Construction Projects: A Literature Review

Mbuyamba Mbala[⊠], Clinton Aigbavboa, and John Aliu

Sustainable Human Settlement and Construction Research Centre,
University of Johannesburg, Johannesburg, South Africa
jpmbala007@gmail.com, caigbavboa@uj.ac.za,
ajseries77@gmail.com

Abstract. The construction industry plays a vital role in the establishment of physical infrastructures to improve the quality of life today. However, one of the most common problems associated with the industry is the issue of delay in construction projects. Overtime, this subject has caused a multitude of negative effects on the project itself as well as the construction professionals. This paper aims to identify the various causes of construction projects delay and make recommendations to mitigate this concern. A review of relevant literatures was conducted from conference articles and journal from databases including Springer, Emerald, Taylor and Francis online, ASCE, Scopus amongst others. The distillation of literature through thematic analysis revealed that poor site management; shortage of skilled labour; unrealistic project scheduling; labour absenteeism; design changes/rework due to the construction errors and accidents due to poor site safety are some of the major causes of delay in the timely delivery of construction projects. This paper therefore recommends that in order to achieve timely delivery of projects, there is an increased need for teamwork between the client, contractors and consultants. The results of this study will be useful to construction professionals (clients, contractors and consultants), and academicians.

Keywords: Construction activities · Construction industry · Buildings
Construction management · Delay time · Timely delivery

1 Introduction

Delays in construction come at a costly price and achieving timely delivery of projects is beneficial to the project and construction team. Along with cost and quality, project schedule is often regarded as a significant aspect of the construction management life cycle, as well as one of the drivers for project success [7]. Despite its importance, most construction projects (both in developed and developing nations) face project delays, which makes it a severe problem [6, 13]. Therefore, the objective of this paper is to identify the various causes of project delays to significantly reduce corresponding expenses.

© Springer International Publishing AG, part of Springer Nature 2019
J. Charytonowicz and C. Falcão (Eds.): AHFE 2018, AISC 788, pp. 489–495, 2019.
https://doi.org/10.1007/978-3-319-94199-8_47

Various researchers have provided several definitions of construction delay for decades. Assaf and Al-Hejji [3] defines construction delay as time overrun beyond the completion date specified in the binding contract agreed by both parties for the project delivery. Zack [25] defined it as an event or act which prolongs stipulated time to complete a project of the contract itself as additional days of work. Delays in construction can also be defined as the execution of a project at a later date other than the projected date.

Construction delays have been discussed by various researchers in numerous manners. Some discussed the various causes of delay in various regions of the world and different project forms while other studies elucidated on the various ways to mitigate it. Sambasivan and Soon [22] highlighted the most significant causes of project delays in the Malaysian construction industry. They include: poor site management, improper planning, labour supply, inadequate finance of the client and payments for completed work, problems related to subcontractors, inadequate, experience of the contractor, material shortage, availability and failure of construction equipment, lack of communication between parties and errors during the construction stage. Likewise, Al-Kharashi and Skitmore [1] conducted a questionnaire survey to identify leading causes of construction project delays in Saudi Arabia. The most two significant causes of project delay were found to be lack of finance to achieve work completion and delay in progress payments by the client. In Pakistan, Haseeb et al. [12] conducted a similar study and the following were reported to be the most significant: financial and payment problems, improper planning, natural disaster, poor site management, shortage of materials and equipment and insufficient experience of contractors.

Doloi et al. [6] conducted a similar study in India through the use of factor analysis. The most influential factors were: inefficient site management, lack of commitment, improper planning, poor site coordination, lack of clarity in project scope; substandard contract and lack of communication. In Denmark, Lindhard and Wandahl [15] investigated causes of delays through the Last Planner System theory. They include: change in work plans, dynamics of workforce, external conditions, material and construction design. More so, Faridi and El- Sayegh [8] investigated more factors causing delay in construction projects based in the United Arab Emirates (UAE). They include delay in the approval of construction drawings, poor-pre-planning and a poor and slow decision making process. Similarly, Sweis et al. (2007), investigated the factors causing delay in construction projects based in Jordan and found that: equipment availability and failure; lack of communication between parties; errors during the construction stage; financial difficulties faced by the contractor and alterations to design by client.

These above researches identify various causes of project delays that hinders the timely delivery of construction projects. From various literatures, Table 1 was developed.

Table 1. Causes of project delays

S/No	Delay causes	1.	2.	3.	4.	5.	6.	7.	8.	9.	10.	Total	Rank
1	Poor site management/project complexities	X	X	X	X	X	X	X	X	X	X	10	1
2	Shortage of skilled labour	X	X	X	X	X	X	X	X	X	X	10	1
3	Unrealistic project scheduling	X	X	X	X	X	X	X	X	X	X	10	1
4	Labour absenteeism	X	X	X	X	X	X	X	–	X	X	9	2
5	Design changes/rework due to the construction errors	X	X	X	–	X	X	X	X	X	X	9	2
6	Accidents due to poor site safety	X	X	X	X	–	X	X	X	X	X	9	2
7	Subcontractor delays	X	X	X	X	X	X	X	X	X	–	9	2
8	Shortage of materials on site	X	X	X	X	X	X	X	X	–	X	9	2
9	Late delivery of construction materials	X	X	–	–	–	–	–	X	X	X	5	3
10	Effects of bad weather on construction activities	–	–	X	–	X	X	–	X	–	X	5	4
11	Price fluctuations	X	–	X	–	–	–	–	–	X	X	4	4
12	Late payment by the owner for the completed work	X	–	X	–	X	X	–	–	–	–	4	4
13	Lack of labour supervision	X	–	X	–	X	X	–	–	–	–	4	4
14	Project size	X	–	–	–	–	–	–	–	X	X	3	5
15	Price fluctuations	X	–	X	–	–	–	–	X	–	–	3	5
16	Poor communication and coordination	–	–	–	–	X	X	X	–	–	–	3	5
17	Frequent breakdowns of construction plant and equipment	–	X	–	–	–	X	–	X	–	–	3	5
18	Personal conflicts among labours	X	–	X	–	–	–	X	–	–	–	3	5
19	Legal disputes	–	–	–	–	–	–	–	–	–	–	0	6
20	Project complexities/construction method	–	–	–	–	–	–	–	–	–	–	0	6
	Total												

Authors: (1) - Assaf and Al-Hejji [3]; (2) – Sambasivan and Soon [22]; (3) - Haseeb et al. [12]; (4) – Orangi et al. [20]; (5) - Kazaz et al. [14]; (6) - Ogunlana et al. [20]; (7) - Sunjka and Jacob [24]; (8) - Santoso and Soeng [23]; (9) - Lindhard and Wandahl [15]; (10) - Doloi et al. [6].

2   Research Methodology

This paper was conducted with reference to extant literature published in conference papers, government reports and journals articles in order to highlight the various causes of delays in construction projects. Specifically, majority of the relevant literature were extracted from Scopus. Scopus is the largest abstract and citation database of peer reviewed literature: it is also the widest interdisciplinary database for science, technology, engineering and medicine, and it includes over 16,000+ peer-reviewed journals from over 4000 publishers and searches over 430 million scientific web resources and 23 million patents. Compared to other databases, it allows for more focused multiple search options such as SpringerLink, Ebsco and Web of Science. This study used Scopus to collate relevant literature based on the search criteria that included the following combinations of keywords: 'Construction AND activities', 'Construction AND industry', 'Buildings'; 'Construction AND management'; 'Delay AND time'; 'Timely AND delivery'. The searches were restricted to keyword combinations in the title, abstract and/or keywords of the retrieved articles. From the 87 shortlisted articles, the works of: Assaf and Al-Hejji [3]; Sambasivan and Soon [22]; Haseeb et al. [12]; Orangi et al. [20]; Kazaz, et al. [14]; Ogunlana et al. [19]; Sunjka and Jacob [24]; Santoso and Soeng [23]; Lindhard and Wandahl [15]; and Doloi et al. [6] were considered.

3   Findings and Discussions

As seen from the rankings on Table 1, The most ten influential delay factors are: poor site management; shortage of skilled labour; unrealistic project scheduling; labour absenteeism; design changes/rework due to the construction errors; accidents due to poor site safety; subcontractor delays; shortage of materials on site; late delivery of construction materials and effects of bad weather on construction activities.

With a perfect score, poor site management/project complexities are ranked the most significant cause of project delays according to Table 1. This results corroborate the findings of Miterev and Nedelcu [18], which asserts that project complexities can be responsible for project delays. The complexity of a project can be defined in terms of design specifications and project size. It is well known that construction projects with a certain level of complexity usually have more detailed and complex plans and schedule. This increase the need for construction professionals to be adequately skilled, and anything less can ultimately lead to project delay. To mitigate the effect of project delay due to this factor, careful planning should be carried out to cover every significant aspect of the project scope. This can be achieved through the use of program schedules and project reports to track the success of the constriction project. Also ranked with a perfect score is the shortage of skilled labour. Though, the construction industry has been influenced by technological driven ideas, the industry continues to be labor-intensive which remains a vital cog in the success of a project. Improving the quality of labor or workforce, will definitely lead to construction project success. Unrealistic project scheduling is also ranked highly as one of the significant factors responsible for construction project delays. This factor stems from the acceleration in the project schedule as a result of delays in activities and may cause frequent disruptions in site management due to the delays in tools, equipment and material supply. This result agrees with the findings of [10, 21]. This can also be remedied through the use of realistic program schedules which can improve output and deliver projects timely. Labour absenteeism was also ranked as a significant factor. This result is tenable as the absence of workforce on the construction site will of course lead to delays and low performance, which will affect the expected completion time. Design changes and rework due to the construction errors were ranked fifth as factors that can hinder the progress of construction projects. These factors occur after the award of a contract and ultimately leads to delays in project schedules [16]. These factors can also originate from various reasons including: extra requirements and modifications by the client; poor communication between parties, as well as conflicts and disputes among professionals on-site [4].

Further assessment on Table 1 shows that the next ranked factors were accidents due to poor site safety. The construction industry is one of the most hazardous as a result of site conditions, poor safety management, amongst others [9]. Accidents can reduce the work speed and efficiency which can lead to project schedule delay as corroborated in the study of [5]. Subcontractor delays is ranked seventh and this agrees with the findings of Haseeb et al. [12]. Subcontractors are mostly viewed as the extension of the main contractor which make their roles significant in the success of construction projects. Therefore, any delays from the subcontractors will definitely affect the project completion. It is always recommended that subcontractors are chosen based on their job experience and qualification to prevent project delays. Shortage of materials on site is ranked eighth and this corroborates the findings of Fugar and Agyakwah-Baah [11]. The success of a construction project is dependent on the availability of various construction materials such as cement, sand, reinforcements amongst others. As such, when there is a shortage of materials, the project completion time is adversely affected. Another cause of delay is the late arrival of construction materials which can hamper the timely delivery of construction projects. Finally, the effect of bad weather on construction activities cannot be over stated. In the cases of rains and floods, transportation of materials become difficult, hence, projects cannot be executed at their full performance. It is recommended that the development stage of a project should be well scheduled so that the key stages of projects are not affected by bad weather.

4   Lessons Learnt and Conclusion

Through a literature review, this study identified the various causes of construction project delays in our present day. Results of the literature review showed the ten most significant factors namely: poor site management; shortage of skilled labour; unrealistic project scheduling; labour absenteeism; design changes/rework due to the construction errors; accidents due to poor site safety; subcontractor delays; shortage of materials on site; late delivery of construction materials and effects of bad weather on construction activities. In light of these factors, the following ways to reduce project delays are recommended: providing a detailed and realistic work schedule on site so that con-struction activities and stages can be coordinated; timely delivery of materials on construction site; improvement of labor and workforce through continuous training; selection of appropriate professionals to control certain activities; usage of technology to monitor performance of workforce; proper planning of activities so as to prevail even during adverse weather conditions; defining plans and goals to limit design changes; employing subcontractors with adequate experience and improved qualification; improved site conditions to reduce the number of construction accidents and adequate funding from clients.

Furthermore, delays in construction projects can be minimized through joint efforts of players in the construction industry. Contractors, sub-contractors, supervisors, financial institutions, manufacturers and the government should cooperate to provide the necessary infrastructure required for efficient project management. This can be achieved through the formulation of a participatory programme to develop the industry through a national agency dedicated to the industry. As a result of this paper, further research can be conducted in different countries identifying causes for construction project delays. Comparing previous international literature on this same issue can attempt to identify the reasons for differences in causes based on geographic and socio-economic factors.

References

1. Al-Kharashi, A., Skitmore, M.: Causes of delays in Saudi Arabian public sector construction projects. Constr. Manag. Econ. **27**, 3–23 (2009). https://doi.org/10.1080/01446190802541457
2. Assaf, S.A., Al-Khalil, M., Al-Hazmi, M.: Causes of delay in large building construction projects. J. Manag. Eng. **11**(2), 45–50 (1995)
3. Assaf, S.A., Al-Hejji, S.: Causes of delay in large construction projects. Int. J. Proj. Manag. **24**, 349–357 (2006). https://doi.org/10.1016/j.ijproman.2005.11.010
4. Austin, S.A., Baldwin, A.N., Steele, J.L.: Improving building design through integrated planning and control. Eng. Constr. Architectur. Manag. **9**, 249–258 (2002). https://doi.org/10.1108/eb021220
5. Bronh, S., Cawdu, C., Choeung, T.: OSH status report – Cambodia. Asia Monitor Resource Centre (2012). http://www.amrc.org.hk/sites/default/files/Cambodia_1.pdf. Accessed 12 Aug 2016
6. Doloi, H., Sawhney, A., Iyer, K.C., Rentala, S.: Analysing factors affecting delays in Indian construction projects. Int. J. Proj. Manag. **30**, 479–489 (2012). https://doi.org/10.1016/j.ijproman.2011.10.004
7. Durdyev, S., Omarov, M., Ismail, S.: Causes of delay in residential construction projects in Cambodia. Cogent Eng. **4**(1), 1291117 (2017)
8. Faridi, A., El-Sayegh, S.: Significant factors causing delay in the UAE construction industry. Constr. Manag. Econ. **24**, 1167–1176 (2006)
9. Farooqui, R., Farrukh, A., Rafeequi, A.: Safety performance in construction industry of Pakistan. In: Proceedings of the 1st International Conference on Construction in the Developing Countries (ICCIDC-I), Karachi, 4–5 August 2008
10. Frimpong, Y., Oluwoye, J., Crawford, L.: Causes of delay and cost overrun in construction of groundwater projects in a developing country. Ghana as a case study. Int. J. Proj. Manag. **21**, 321–326 (2003). https://doi.org/10.1016/S0263-7863(02)00055-8
11. Fugar, F.D.K., Agyakwah-Baah, A.B.: Delays in building construction projects in Ghana. Australas. J. Constr. Econ. Build. **10**, 103–116 (2010). https://doi.org/10.5130/AJCEB.v10i1-2.1592
12. Haseeb, M., Xinhai-Lu, Bibi, A., Maloof-ud-Dyian, Rabbani, W.: Problems of projects and effects of delays in the construction industry of Pakistan. Austr. J. Bus. Manag. Res. **1**, 41–50 (2011)
13. Kaliba, C., Muya, M., Mumba, K.: Cost escalation and schedule delays in road construction projects in Zambia. Int. J. Proj. Manag. **27**, 522–531 (2009). https://doi.org/10.1016/j.ijproman.2008.07.003
14. Kazaz, A., Ulubeyli, S., Tuncbilekli, N.A.: Causes of delays in construction projects in Turkey. J. Civ. Eng. Manag. **18**, 426–435 (2012). https://doi.org/10.3846/13923730.2012.698913
15. Lindhard, S., Wandahl, S.: Scheduling of large, complex, and constrained construction projects – an exploration of LPS application. Int. J. Proj. Organ. Manag. **6**, 237–253 (2014). https://doi.org/10.1504/IJPOM.2014.065258

16. Mahamid, I., Bruland, A., Dmaidi, N.: Causes of delay in road construction projects. J. Manag. Eng. **28**, 300–310 (2012). https://doi.org/10.1061/(ASCE)ME.1943-5479.0000096

17. Mansfield, N.R., Ugwu, O.O., Doran, T.: Causes of delay and cost overruns in Nigerian construction projects. Int. J. Proj. Manag. **12**(4), 254–260 (1994)

18. Miterev, M., Nedelcu, R.: The nature of the relationship between project complexity and project delay: case study of ERP system implementation projects. Master thesis, Umeå School of Business and Economics (2011). http://www.projektakademien.se/pa/wp-content/uploads/2014/04/1.-Master-Thesis_2012_Miterev-and-Nedelcu.pdf. Accessed 21 Aug 2016

19. Ogunlana, S.O., Promkuntong, K., Jearkjirm, V.: Construction delays in a fast-growing economy: comparing Thailand with other economies. Int. J. Proj. Manag. **14**, 37–45 (1996). https://doi.org/10.1016/0263-7863(95)00052-6

20. Orangi, A., Palaneeswaran, E., Wilson, J.: Exploring delays in Victoria-based Australian pipeline projects. Procedia Eng. **14**, 874–881 (2011)

21. Ren, Z., Atout, M., Jones, J.: Root causes of construction project delays in Dubai. In: Dainty, A. (ed.), Proceedings 24th Annual ARCOM Conference, pp. 749–757. Association of Researchers in Construction Management, Cardiff, 1–3 September 2008

22. Sambasivan, M., Soon, Y.W.: Causes and effects of delays in Malaysian construction industry. Int. J. Proj. Manag. **25**, 517–526 (2007). https://doi.org/10.1016/j.ijproman.2006.11.007

23. Santoso, D., Soeng, S.: Analysing delays of road construction projects in Cambodia: Causes and effects. J. Manag. Eng. (2016). https://doi.org/10.1061/(ASCE)ME.1943-5479.0000467

24. Sunjka, B.P., Jacob, U.: Significant causes and effects of project delays in the Niger delta region. In: SAIIE25 Proceedings: Stellenbosch, South Africa © 2013 SAIIE, Cape Town (2013)

25. Zack, J.G.: Schedule delay analysis; is there agreement? In: Proceedings of PMI-CPM College of Performance Spring Conference. Project Management Institute—College of Performance Management, New Orleans, 7–9 May 2003

Construction Professionals Perception of Solid Waste Management in the South African Construction Industry

Ayodeji Oke[1], Clinton Aigbavboa[2], Douglas Aghimien[2(✉)],
and Nkululeko Currie[1]

[1] Department of Construction Management and Quantity Surveying,
University of Johannesburg, Johannesburg, South Africa
aoke@uj.ac.za, currienkululeko22@gmail.com
[2] Sustainable Human Settlement and Construction Research Centre,
Faculty of Engineering and the Built Environment, University of Johannesburg,
Johannesburg, South Africa
caigbavboa@uj.ac.za, aghimiendouglas@yahoo.com

Abstract. Proper management of the excess waste generated by the construction industry can lead to effective usage of same, and reduction in the pressure on the limited available earth resources consumed by the industry. This study presents the result of an assessment of the solid waste management (SWM) systems in the South African construction industry from the view point of construction professionals using a questionnaire survey. Data gathered were analyzed using appropriate descriptive statistics. Result reveals that construction professionals view SWM in the industry as the; re-use of waste, recycling of waste instead of dumping them at landfills, reduction of waste at source to lessens the cost of transportation and disposal, and the use of prefabricated components as a form of waste reduction. The major challenges of SWM in the industry include; inadequate formulation and application of policies, viewing implementation to be costly, and lack of mandatory waste management guidelines.

Keywords: Built environment · Construction industry · Solid waste
South Africa · Waste management

1 Introduction

South Africa is a growing country in the context of growth in population, which has a simultaneous contribution to the increase in resource consumption and subsequent waste generation. This growth in population also influences the need for infrastructure development such as buildings, roads, residential housing, etc. Judging by the amount of advertising given to green building in the media it is quite obvious that energy efficiency and water saving are the ones that have received greater recognition, forgetting that waste reduction is also a necessary and an imperative constituent [1]. Zurbrugg [2] observed that the real threat to the general conditions inhibited by humans is due to the human conduct when it comes to gathering, handling, keeping and putting

© Springer International Publishing AG, part of Springer Nature 2019
J. Charytonowicz and C. Falcão (Eds.): AHFE 2018, AISC 788, pp. 496–505, 2019.
https://doi.org/10.1007/978-3-319-94199-8_48

away of the surplus (waste). Thus, it is therefore necessary to have an effective way of managing these wastes in other not to pose any danger to the environment.

According to Van Beukering et al. [3] a lot of writing has been done on Solid Waste Management (SWM) in the past which back tracks even before the 1970's. Previous studies show that the focus on management of people in general authorities was the "quick waste removal and destruction" [4]. Schwarz-Herion et al. [5] however stated that amid the 1970's the focus moved towards matters of surplus usage, concentrating in specialized and financial concerns encompassing the portion and use of accessible assets. Cointreau et al. [6] and Diwekar [7] further distinguished that studies on recycling back in the day was most part mechanical and the focus was on the financial aspects to minimize manufacturing cost, not at all like the present prominence on reusing method of lessening discarded materials that are all over nature, and safeguarding resource consumption.

The construction industry is termed to be the major natural resources consumer and waste generator; therefore, every construction firm should have a SWM system in place. This is to ensure that generated waste are properly managed so as not to pose any danger to the environment, and also reduce the consumption of natural resources through the use of recycled waste materials. Materials regarded as construction waste; be it leftover during construction work or remains from demolition, can still be useful depending on the efforts made in finding and implementing ways to use it instead of throwing it away. Construction industry professionals and building proprietors can be taught and be instructed on waste management issues, such as, valuable reuse, viable techniques of identification and sorting of waste, economically practical methods for advancing earth, and socially proper methods for diminishing total waste. It is based on this knowledge, that this study assessed the SWM system being adopted in the South African construction industry, and its challenges, with a view to proffer possible solutions towards the increase of effective waste management for sustainable built environment in the country.

2   Literature Review

2.1   The Concept of Solid Waste Management

Zurbrugg [2] observed that solid waste management is much more than a technological issue - it also involves institutional, social, legal and financial aspects and involves coordinating and managing a large workforce. However, Sapuay [8] noted that the use of waste management methods such as recycling, reuse and resource recovery are not properly implemented, or are not given much attention like the maintenance of sanitary surroundings on construction sites. Wang et al. [9] suggests that reduction, reuse and recycling, are the most stressed systems in the waste management ranking.

Reduction. According to Yuan et al. [10] the prevention of construction waste generation can be managed in the activities performed in the stages of project design and building construction. Greater attention has been given to the investigation of the issue of reducing construction waste accumulation and the reason being that waste reduction is regarded as the most important amongst all waste management systems. Wang et al.

[9] explains that reduction, as one of the major systems for limiting the generation of construction waste, offers two noteworthy advantages: (a) avoiding the creation of building waste; and (b) lessening the price of greater charges for waste recycling, transference and removal. Osmani *et al.* [11] noted that the actual construction waste is produced throughout the phase of construction but there are waste reduction methods that can be applied before the start of construction, which can be more effective than the ones utilized during the construction stage. These stages are the design and materials procurement stages. It was further suggested that it is crucial to reduce waste from the beginning in relation to the design of building elements. However, Ajayi *et al.* [12] stated that there has been a lack of intent to look at waste reduction from the material procurement and logistics angle, since more attention in literature is focused on the design and construction strategies for reducing waste generation. It was further suggested that the improper management of materials procurement process is a major factor for construction waste. In terms of actual construction, the use of environmentally friendly construction methods has grown in terms of popularity in the attempt to manage waste accumulation. This can be achieved by utilizing panel system and prefabricated components for construction [13]. In addition, Tam *et al.* [14] argues that becoming accustomed to the use of prefabricated components can reduce waste accumulation by 100%, which can save up to 84.7% of wastage.

Re-use. Wang *et al.* [9] noted that reuse implies utilizing a similar material in development more than once - including utilizing the material again for a similar capacity and new life reuse where material is utilized as a crude material for another capacity. Sapuay [8] makes an example with cement mix left over in the concrete mixer which will be dumped on the ground trying to avoid it hardening inside the concrete mixer and this ends up having a negative effect on the environment like killing plants, affecting animals that live in the region or polluting running water. It was suggested that this cement mix can be utilized as lean concrete paving elsewhere. In addition, reuse are seen as the secondary options to reduction and elimination as this is confirmed on the waste management hierarchy. Reuse strategy can be used in two ways: Building Reuse and Material reuse. Building Reuse incorporates reusing materials from existing structures and keeping up specific rates of building structural and non-structural components, for example, inside walls, doors, floor covering and roofs. Material Reuse however is a standout amongst the best methodologies for limiting environmental effects, which should be possible by rescuing, renovating and reusing materials inside a similar building or in another building. A number of the outside and inside materials can be recuperated from existing structures and reused in new ones. Such materials will incorporate steel, dividers, floor covers, concrete, bars and posts, doorframes, cabinetry and furniture, block, and embellishing things. Reuse of materials and items will decrease the interest for virgin materials and diminish wastes.

Recycling. Napier [15] defines recycling as bringing a material into some procedure for re-make into another item, which might be the same or comparable item or a unique sort of item. Wang *et al.* [9] believes that by recycling construction waste it can produce a new item. According to Medina *et al.* cited in Ajayi *et al.* [12] while a few reviews assert that the nature of concrete diminishes with rise in recycled concrete aggregate, others contend that the nature of concrete stays the same when using recycled aggregate. Furthermore, recycled material market is reluctant in expansion as implied by existing practices. Tam [14] stated that as a key technique of construction waste management, recycling could numerous advantages. Some of them are diminishing the interest for new resources; eliminating transport and manufacturing energy cost; effective usage of waste, which might be disposed of at landfill sites; keeping safe areas of land, for subsequent urban construction; and enhancing the general condition of nature. Pietzsch *et al.* [16] tries to prove the lack of recycling by providing the following statistics, which show that only 15% of the 84% of solid wastes accumulated all over the world is recycled, and the rest of it disposed of at landfills.

2.2   Challenges in Implementing Solid Waste Management Systems

Pietzsch *et al.* [16] stated that the most challenging variables in the implementation of waste elimination are macro-environment related. These variables include the political, cultural, economic and technological issues.

Mandatory waste management guidelines, strong policy maker's commitment and support to environmental sustainability issues, the review of socio-political constraints and the creation of a consistent database for performance comparison, were all identified as significant in political challenges [16]. Konteh [17] states that within the political aspects the biggest difficulty is to formulate the "right balance between policy, governance, institutional mechanisms and resource provision and allocation". According to Marshall and Farahbakhsh [18] the failure of SWMS are triggered by weak policies which are a result of inadequate formulation and implementation of realistic policies. Furthermore, regional and state programmes creators and coordinators neglect SWMS, through extra attention given to problems of social and political importance, which results in little budget for waste problems.

About the cultural challenges, the consumption levels and conduct of humans when managing environmental issues are viewed as the hardest factors to change [16]. Factors that are moulded by the local cultural and social settings are the behaviour patterns and the subsequent attitudes of the public, which establishes the structure, and working of SWMS. The generous differing qualities of social and ethnic gatherings that frequently exists inside quickly growing urban areas, even inside individual private groups, significantly affects regions' abilities to execute SWMS. In addition, the technological challenges refer to the need for investments in the development of researches that improve waste management technologies, as well as their dissemination [18]. The spread of Zero Waste (ZW) is stalled by the inabilities of the available technological limitations. Utilizing waste management changes, particularly with respect to material effectiveness increment is very much reliant on technology [16].

Referring to economic challenges, [16] stresses the importance of properly planning tax collection and discounts considering the expenses to screen and to guide the effects of wastes after their generation. They further suggest that high charges of waste disposal at landfills would just support the preoccupation of waste to more remote landfills. As part of the economic difficulties, citizens will only be interested in recycling of waste only if it is financially worthwhile. According to Tam [14] high investment cost is regarded as the most notable difficulty, when it comes to providing funds for training courses. It was further stated that low financial incentive, and increase in overhead cost, are the challenges that will affect the investment cost by accumulating in the short-term, the reason being that working areas and equipment would have to be available on-site. In contrast, some public sectors have been unsuccessful in managing solid waste because of financial aspects. The provision of adequate waste management amenities is mainly affected by the improper use of economic resources [19].

3 Research Methodology

This study set out to assess SWM system in the South African construction industry. The study adopted a quantitative approach in which a questionnaire survey was conducted among construction professionals in Gauteng province, South Africa. These professionals include Architects, Engineers, Quantity Surveyors, Construction Managers, Project Managers and others. Fifty-Seven (57) construction professionals participated in the study. The instrument for data collect was a questionnaire, which is one of the most widely used social research techniques [20]. Tan [21] further stated that the questionnaire survey is a methodical technique of gathering data based on a sample. It is easy to use and have the ability to cover a wider range of participants. The questionnaire used was designed in section. The first section gathered information on the respondent's background and information gotten from this section provided quality check to the answers gotten from the other section. The second section was designed to harness information regarding the SWM system being adopted within the industry, and the challenges facing the effective implementation of SWM. For the second section a 5-point Likert scale was employed in measuring the identified variables. The circulation of the questionnaire was conducted through self-distribution in construction site and companies within the study area. Out of the 57 questionnaire that were distributed using convenient sampling approach, 46 were retrieved and deemed fit for analyses. This represents a response rate of 81% response rate, which is deemed adequate for the study.

The data received from the questionnaires were analyzed using Statistical Package for Social Science Version 24 (SPSS V24). Percentage was used in analyzing the data on the background information of the respondents, while Mean Item Score (MIS) was used in ranking the identified SWM system being adopted and the challenges of SWM. The premise of decision for the ranking is that the factor with the highest MIS is ranked first and others in such subsequent descending order. Since a Likert of 5-point scale was employed for the collection of data, the formula for MIS is written as:

$$\text{MIS} = \frac{5n^5 + 4n^4 + 3n^3 + 2n^2 + 1n^1}{n^5 + n^4 + n^3 + n^2 + n^1}$$

Where n is the frequency of each of the rankings.

4 Findings and Discussions

4.1 Background of Respondents

Analysis of the background information of the respondents reveals that in terms of the respondents' highest educational qualification, 11.6% have a Certificate Diploma, 32.6% have a National Diploma, 32.6% have a Degree, and 23.3% have a Master's degree. In relation to the profession, 16.3% were Site Agents, 9.3% were Architects, 11.6% were Project Managers, 9.3% were Construction Managers, 23.3% were Engineers and 30.2% of the participants were Quantity Surveyors. In terms of years of working experience, result reveals that 61% of the respondents were categorized in the range of Up to 5 years of experience, 16% were in the 6 to 10 years range, 19% were in the 11 to 20 years category and only 5% of the participants had experience more than 20 years. On the average, the respondents have 8 years of working experience. This indicates that the respondents of the study are well equipped in terms of happenings within the South African construction industry to give significant answers to the questions of the research.

4.2 Solid Waste Management System in the South African Built Environment

Having identified the significant waste management system from literature to include, reduction, re-use, and recycle, statements regarding these approaches were made and presented to the respondents to rate based on their level of agreement. A Likert scale of 5 to 1 was adopted, with 5 being very high, 4 being high, 3 being average, 2 being low, and 1 being very low. Result in Table 1 shows the ranking of these various SWM statements. A look at the table shows that all the 14 assessed dimensions of SWM system have a mean value of above average of 3.0. This means that to a significant extent the construction professionals understand and agree with these statements as regards SWM in the South African construction industry. However, the construction professionals tend to relate more with statements such as: Re-use of waste is a form of waste management system (mean = 4.21, S.D = 0.860); Recycling provides an alternate to waste being disposed at landfills (mean = 4.09, S.D = 0.840); Waste reduction at source lessens the cost of transportation and disposal (mean = 4.07, S.D = 0.910); and the use of prefabricated components is a form of waste reduction (mean = 4.05, S. D = 0.898).

This result shows that construction professionals within the study area recognises the three major approaches to SWM with more knowledge of re-use of waste product being the most used system of managing waste within the country's built environment. Wang et al. [9] has earlier noted that reuse implies utilizing a similar material in development more than once - including utilizing the material again for a similar capacity and new life reuse where material is utilized as a crude material for another capacity. While this result reveals the re-use of materials as a common SWM system, the result contradicts the submission of [8] who stated that re-use of solid waste is seen as a secondary option to source reduction. Regarding the results relating to recycling being an alternative to solid waste being disposed of at landfills, there was an

Table 1. SWMS adopted in the South African construction industry

Various SWMS	Mean	S.D	Rank
Re-use of waste is a form of waste management system	4.21	0.860	1
Recycling provides an alternate to waste being disposed at landfills	4.09	0.840	2
Waste reduction at source lessens the cost of transportation and disposal	4.07	0.910	3
The use of prefabricated components is a form of waste reduction	4.05	0.898	4
Re-use of materials decreases the consumption of raw materials	3.98	0.913	5
Waste reduction at source is the most effective system	3.86	1.000	6
Recycling enhances the general conditions of nature	3.84	0.974	7
Waste reduction prevents the generation of construction waste	3.77	0.947	8
Recycling eliminate transport and production energy cost	3.74	0.850	9
Materials procurement can assist in construction waste reduction	3.72	0.797	10
Re-use of waste materials prolongs their life span	3.65	1.067	11
Recycling preserves and enables certain areas for subsequent development	3.65	0.897	12
Recycled materials do not possess the same quality as raw materials	3.51	1.032	13
Recycling diminishes the interest for new resources	3.23	1.192	14

agreement with the study by Tam [14] which pointed out, using waste which might have been disposed of at landfill sites, as an advantage of recycling. Nevertheless, the results were not in agreement with the other advantages stated in these studies, as these were ranked in the lower half of the table. Wang *et al.* [9] also agrees that waste reduction lessens the high cost of transportation and disposal. The use of environmentally friendly construction methods has grown in terms of popularity in the attempt to manage waste accumulation. This can be achieved by utilizing panel system and prefabricated components for construction. Tam *et al.* [22] further affirmed that becoming accustomed to the use of prefabricated components can reduce waste accumulation by total of 100%, which can save up to 84.7% of wastage. Findings of this study agree with this submission, as respondents believe that the use of prefabricated components is a form of waste reduction being adopted in South Africa.

4.3 Challenges of Implementing Solid Waste Management

In determining the challenges facing the proper implementation of SWM within the South African built environment, some challenges were identified from the review of related literatures. Respondents were presented with these challenges and were asked to rank them based of on their level of significance. A Likert scale of 5 to 1 was adopted, with 5 being very high, 4 being high, 3 being average, 2 being low, and 1 being very low. Result in Table 2 shows the ranking of these challenges as perceived by the respondents. From the table it is evident that all the 13 assessed challenges have a mean value of above average of 3.0. This implies that they do have significant impact on the implementation of SWM in the county. Chief of these challenges are inadequate formulation and application of policies (mean = 3.93, S.D = 0.990), implementation is seen as costly (mean = 3.83, S.D = 1.022), lack of mandatory waste management guidelines (mean = 3.81, S.D = 1.075); changing the behaviour of citizens (mean = 3.81, S.D = 1.118), and lack of investments in development of improved waste management technologies (mean = 3.81, S.D = 1.074).

Table 2. Challenges of implementing solid waste management

Challenges	Mean	S.D	Rank
Inadequate formulation and application of policies	3.93	0.990	1
Implementation is seen as costly	3.83	1.022	2
Lack of mandatory waste management guidelines	3.81	1.075	3
Changing the behaviour of citizens	3.81	1.118	4
Lack of investments in development of improved waste management technologies	3.81	1.074	5
Lack of technical knowledge by professionals involved	3.78	1.036	6
Budget constraints	3.77	1.131	7
Changing the consumption patterns	3.74	1.002	8
Lack of proper training and education	3.74	1.049	9
The amount of time taken in implementing the strategies	3.63	1.113	10
Low financial incentive	3.58	1.006	11
Low disposal costs	3.49	1.077	12
Completing the project within time	3.33	1.210	13

This result reveals that issues relating to policies formulation and application are the greatest challenge of SWM implementation within the South African built environment. This finding is in agreement with the studies by Konteh [17] and Pietzsch et al. [16] which submitted that failure of implementation of solid waste management is due to poor execution and creation of genuine policy and the lack of compulsory waste management guidelines. Aside formulation of policies, the cost of implementing SWM system is seen as a major issue with the construction industry. This finding is in agreement with Tam [14] submission that increase in overhead cost is a challenge that will affect investment cost, which translates to high implementation cost, but the study differs with regard to low financial incentive being a challenge. The result is also in line with Pietzsch et al. [16] submission on the challenge of trying to change the conduct of citizens but was in total disagreement with changing consumption levels being a challenge.

5   Conclusion and Recommendations

This study set out to assess SWM system in the South African construction industry. Using a quantitative approach with questionnaire survey conducted among construction professional in Gauteng, South Africa, the study has been able to identify the SWM system used in the country's construction industry, and the challenges facing proper implementation of SWM in the industry. Based on the findings, the study concludes that using re-use of waste as a form of waste management system, recycling of waste as against dumping them at landfills, reduction of waste at source to lessen the cost of transportation and disposal, and the use of prefabricated components is a form of waste reduction, are the major SWM system being adopted. The major challenges of the effective implementation of SWM are inadequate formulation and application of policies, implementation is seen as costly, lack of mandatory waste management guidelines, changing the behaviour of citizens, and lack of investments in development of improved waste management technologies.

Based on the conclusion drawn, it is therefore recommended that the Government should put effective policies and guidelines that will encourage the reduction of waste from the onset of construction, and reuse of wasted materials in place. In addition, means of enforcing these policies should be created to ensure compliance. Further to this, construction stakeholders should be enlightened through seminars, workshops and trainings on the overall cost effectiveness of proper managing of waste, and its immense benefits to the environment and its resources. This will help in changing their attitude towards waste management.

References

1. van Wyk, L.: The Green Building Handbook: South Africa, vol. 5. Green Building Media, Cape Town (2012)
2. Zurbrugg, C.: Solid waste management in developing countries. SWM Introductory Text, p. 5 (2003). www.sanicon.net
3. van Beukering, P., Sehker, M., Gerlagh, R., Kumar, V.: Analysing urban solid waste in developing countries: a perspective on Bangalore, India. Collaborative Research in the Economics of Environment and Development (1999)
4. Melosi, M.V.: Garbage in the Cities: Refuse Reform and the Environment. University of Pittsburgh Press, Pittsburgh (2004)
5. Schwarz-Herion, O., Omran, A., Rapp, H.P.: A case study on successful municipal solid waste management in industrialized countries by the example of Karlsruhe city, Germany. J. Eng. Ann. Fac. Eng Hunedoara 6(3), 266–273 (2008)
6. Cointreau, S.J., Gunnerson, C.G., Huls, J.M., Seldman, N., Mitchell, P., Long, L.J., Bellassai, E.C., Mundial, B.: Recycling from Municipal Refuse: A State-of-the-Art Review and Annotated Bibliography, vol. 30. World Bank, Washington, D.C. (1984)
7. Diwekar, U.: Green process design, industrial ecology, and sustainability: a systems analysis perspective. Resour. Conserv. Recycl. 44(3), 215–235 (2005)
8. Sapuay, S.E.: Construction waste – potentials and constraints. Procedia Environ. Sci. 35, 714–722 (2016)
9. Wang, J., Yuan, H., Kang, X., Lu, W.: Critical success factors for on-site sorting of construction waste: a china study. Resour. Conserv. Recycl. 54(11), 931–936 (2010)
10. Yuan, H., Chini, A.R., Lu, Y., Shen, L.: A dynamic model for assessing the effects of management strategies on the reduction of construction and demolition waste. Waste Manag. 32(3), 521–531 (2012)
11. Osmani, M., Glass, J., Price, A.D.F.: Architects perspectives on construction waste reduction by design. Waste Manag. 28, 1147–1158 (2008)

12. Ajayi, S., Oyedele, L., Akinade, O., Bilal, M., Alaka, H., Owolabi, H.: Optimising material procurement for construction waste minimization: an exploration of success factors. Sustain. Mater. Technol. **11**, 38–46 (2017)
13. Begum, R.A., Siwar, C., Pereira, J.J., Jaafar, A.H.: Attitude and behavioural factors in waste management in the construction of Malaysia. Resour. Conserv. Recycl. **53**, 321–328 (2009)
14. Tam, V.W.: On the effectiveness in implementing a waste-management-plan method in construction. Waste Manag. **28**(6), 1072–1080 (2008)
15. Napier, T.: Construction Waste Management (2016). https://www.wbdg.org/resources/construction-waste-management
16. Pietzsch, N., Ribeiro, J., de Medeiros, J.: Benefits, challenges and critical factors of success for Zero Waste: a systematic literature review. Waste Manag. **67**, 324–353 (2017)
17. Konteh, F.H.: Urban sanitation and health in the developing world: reminiscing the nineteenth century industrial nations. Health Place **15**(1), 69–78 (2009)
18. Marshall, R., Farahbakhsh, K.: Systems approaches to integrated solid waste management in developing countries. Waste Manag. **33**(4), 988–1003 (2013)
19. Guerrero, L.A., Maas, G., Hogland, W.: Solid waste management challenges for cities in developing countries. Waste Manag. **33**, 220–232 (2013)
20. Blaxter, L., Huges, C., Tight, M.: How to Research, 2nd edn. Open University Press, London (2001)
21. Tan, W.C.K.: Practical Research Methods. Pearson Custom, Singapore (2011)
22. Tam, V.W.Y., Shen, L.Y., Tam, C.M.: Assessing the levels of material wastage affected by sub-contracting relationships and projects types with their correlations. Build. Environ. **42**(3), 1471–1477 (2007)

Benefits of Biomimicry Adoption and Implementation in the Construction Industry

Olusegun Aanuoluwapo Oguntona$^{(\boxtimes)}$ and Clinton Ohis Aigbavboa

Sustainable Human Settlement and Construction Research Centre,
Faculty of Engineering and the Built Environment, University of Johannesburg,
Johannesburg, South Africa
architectoguntona12@gmail.com

Abstract. Biomimicry, the novel field of discipline which studies and emulates nature's models to solve human challenges in a sustainable way is gradually becoming a global phenomenon. However, the paradigm is still in its infancy in the construction industry compared to other sectors. Despite its potential in providing outstanding innovative solutions, the adoption and implementation are impeded by several factors. This research sets out to address and establish the benefits of embracing biomimicry in the construction industry. A structured questionnaire survey was conducted with biomimicry practitioners and construction professionals as respondents. A quantitative approach to data analysis was employed using the mean scores of the factors identified. Creation of green market and services, protection of biodiversity, and conservation of natural resources are the top three benefits established. This systematic approach towards understanding the taxonomy of the benefits of biomimicry is imperative for aiding and reinforcing sustainable construction practices in the industry.

Keywords: Biomimicry · Construction industry · Innovative solutions
Nature · Sustainability

1 Introduction

Due to its ability to improve the economy and the human physical environment, the construction industry is attributed as one of the most essential sectors [1]. It has also been discovered that the industry remains a sizeable economic contributor, employment provider, and source of vital utilities globally. The use of construction investments as a tool by the government to stabilize the economy attest to the industry's key position in the national development strategies of many countries [2]. It can, therefore, be agreed that urbanization is closely linked to the industry due to its associated developments [3]. These include the provision of critical infrastructures like rail, water, bridges, roads, facility assets (office and residential buildings), and plants for production and transmission of energy amongst many others. Not only is the construction industry an integral part of the modernization process, its labor-intensive nature makes it particularly attractive as a means of creating employment, and stimulating economic growths of developing countries [4].

© Springer International Publishing AG, part of Springer Nature 2019
J. Charytonowicz and C. Falcão (Eds.): AHFE 2018, AISC 788, pp. 506–514, 2019.
https://doi.org/10.1007/978-3-319-94199-8_49

Statistics have shown that in the drive towards global economic development and rapid urbanization, the construction industry contributes heavily towards numerous environmental challenges. Such include pollution of the environment, excessive consumption of resources, waste generation and depletion of the global ecological integrity. Buildings and infrastructures which are the direct products of the industry have a long-term environmental footprint as they are known to continuously emit a large amount of pollution [5]. According to a study by the United States Environmental Protection Agency, indoor air levels of pollutants may be 2.5 times and occasionally more than 100 times higher than outdoor levels indicating [6]. Pollutant concentrations within the building space emanating from paints, finishes, backing materials and other components are responsible for making indoor pollutants level higher than those of the outside.

In the quest for solutions to the environmental challenges posed by the construction industry, engineers, architects, innovators, scientists and other sustainability proponents are now heading outside the circle to consult and learn from the natural world [7]. This paradigm is described as biomimicry, a novel and growing field of discipline which, which studies nature's models and then emulates their forms, systems, processes, and strategies to solve human challenges in a sustainable manner [8]. Hence, the objective of this study is to identify the benefits of adopting and implementing biomimicry in the construction industry. This study will also seek to identify the biomimicry principles for the sustainability of the construction industry and measures in bridging the knowledge gap.

2 Historical Background of Biomimicry

The emulation and application of systems in nature is no novel practice. Historically, the early man depended on the natural world for existence and survival, evident through the numerous records of native innovations. Few of these innovations cut across the fields of agriculture and food production; medical and pharmaceutical sciences; shelter architectures; manufacturing; and weapons and defense, including sensors, armors and alarm systems amongst others [9]. Examples of early nature-inspired innovations include Velcro (inspired by the re-attachable system of burrs from burdock plants), also known as hook and fasteners [10], London's Crystal Palace building in England (inspired by the huge leaves of the giant Amazonian waterlily) designed by Sir Joseph Paxton [11], and the Monoplane (also known as Avion III) designed built and first tested by Clement Ader [10].

Biomimicry as a term first appeared in the year 1982 as part of the words constituting the topic of Connie L. Merrill's doctoral thesis. It was titled 'Biomimicry of the Dioxygen Active Site in the Copper Proteins Hemocyanin and Cytochrome Oxidase' [12]. However, the term biomimicry became popularized and widely circulated in 1997 through a book titled 'Biomimicry: Innovation Inspired by Nature'. Janine M. Benyus, a biologist and co-founder of the Biomimicry Guild authored the book and is widely recognized as the founder of this novel field of study [13]. Biomimicry is therefore described as human's effort and quest towards the exploration of nature's masterpieces (natural selection, photosynthesis, self-assembly, self-sustaining ecosystems etc.) and the emulating these designs and manufacturing processes to solve their problems sustainably [14]. The idea is that nature has been found to have developed highly efficient systems and processes with the potential to propel and proffer solutions to the challenges facing humanity today [15]. Biomimicry proponents believed that the industry needs to study the highly successful Research and Development (R&D) lab that has been operational on earth for over 3.8 billion years in which 10 to 30 million species have learned to do everything humans want to do, without polluting the environment, or mortgaging the common future of generations to come [16].

3 Overview and Conceptual Delineation of Biomimicry

Nature has been found to be a robust source of knowledge, culminating in the discovery and progression of novel innovative solutions to present-day human challenges through the adoption and application of biomimicry [17]. These are challenges emanating from the unchecked global growth in industrialization and the resultant exploitation of natural resources [18]. As quoted by Angela Nahikian of Steelcase, "… nature is constantly innovating, endlessly experimenting and ever reinventing itself in the face of new challenges. From materials to products to business models, biomimicry offers a fresh lens for all the dreamers and doers remaking the man-made world" [19].

Throughout literature, multiple terms have been found to describe emulating and learning from nature. Biomimicry, biomimetics and other terms such as bionics, bio-inspired design, biomimesis, bioinspiration, bioanalogous design and biognosis are often used interchangeably to describe this novel paradigm [10, 20, 21]. However, despite the numerous terms, it has been established that there is no difference in their fundamental meanings [22].

Biomimicry originated from the combination of the Greek words *bios* (life) and *mīmēsis* (imitation), which literally means 'life imitation' or the 'imitation of life' [23–25]. Biomimicry (**bi•o•mim•ic•ry**) studies nature's models and then emulates their forms, processes, systems and strategies to solve human problems sustainably. It is defined as the examination of systems, processes, and elements of nature with the potential to solve human challenges [11]. In biomimicry, solutions are proffered to human challenges by emulating the mechanisms, principles and strategies unearthed within nature [26].

4 Principles of Biomimicry

Nature has managed to survive for 3.8 billion years with organisms as models that manufacture without heat, beat, and treat; ecosystems that are powered by sunlight; and create opportunities rather than waste [7]. Their resulting designs displayed are found to be functional, effective, efficient, sustainable, and aesthetically pleasing as well. However, in the book 'Biomimicry: Innovation Inspired by Nature', Benyus enumerates nine principles of nature, which are also the basic principles underpinning the concept of biomimicry [14]. They are the following:

Nature runs on sunlight;
Nature uses only the energy it needs;
Nature fits form to function;
Nature recycles everything;
Nature rewards cooperation;
Nature banks on diversity;
Nature demands local expertise;
Nature curbs excesses from within; and
Nature taps the power of limits.

5   Methodological Framework

This paper employed the combination of secondary data (review of literature) and primary data (survey questionnaire) to present informative evidence on the practitioner's perspectives for the adoption and implementation of biomimicry. A structured close-ended questionnaire survey which targeted biomimicry and construction professionals (i.e. architects, construction managers, construction project managers, quantity surveyors, structural engineers) was employed. The respondents adopted in the research are those registered with their various professional bodies in the SACI. The questionnaire survey was administered to one hundred and twenty respondents of which one hundred and four responses were received.

The first part of the questionnaire sought the background information of the respondents (i.e. age, educational qualification, professional qualification, years of experience). The second part sought the respondents' assessment of the biomimicry principles that can promote sustainability. The third part dealt with the barriers to biomimicry adoption and implementation and measures to bridge the knowledge gaps on biomimicry in the construction industry. Concerning the biomimicry benefits, the respondents were asked to indicate their level of agreement to the benefits on a five-point Likert scale (strongly disagree-1, disagree-2, neutral-3, agree-4, strongly agree-5). Statistical Package for Social Sciences Version 16 (SPSS V16) software was used to analyze the data obtained. Descriptive statistics, with the aid of mean and standard deviation, was employed to present the results of the analyzed data.

6   Results and Discussions

6.1   Background of Respondents

The distribution of the respondents according to their profession reveals that biomimicry professionals/specialists constituted 24%, quantity surveyors constituted 19.2%, architects constituted 18.3%, civil engineers constituted 15.4%, and construction project managers constituted 11.5%. The average years of experience of the respondents surveyed ranges between 10 and 20 years, implying that they do have significant experience in the construction industry. The result also revealed that all the respondents surveyed are duly registered and affiliated with their respective professional bodies. Majority of the respondents had master's degree representing 54.8%, 25% had bachelor's degree, 11.5% had diploma certificates while 8.7% had doctorate degree.

6.2 Application of Biomimicry Principles to Promote Sustainability

Table 1 reveals the level of agreement of the respondents to the application of biomimicry for promoting sustainability in the construction industry. All the 23 principles of biomimicry assessed have mean scores greater than 2.5 for the respondents [27]. This is an indication that the respondents concur that all the 23 biomimicry principles should be considered in all the construction stages if sustainable construction is to be achieved. The results further showed that the respondents consider 'harnessing freely available energy', 'using readily available materials', 'recycling all materials', 'using low energy process', and 'using multi-functional design' as the five most important biomimicry principles to be considered. The results, however, agree in its entirety with the studies of Goss [13], Polit [28], and Kennedy et al. [29] that listed biomimicry principles as important checklists through which sustainability can be evaluated and achieved.

Table 1. Biomimicry principles that can promote sustainable construction practices.

Biomimicry principles	Mean	Standard deviation	Rank
Harnessing freely available energy	4.89	0.309	1
Using readily available materials	4.83	0.380	2
Recycling all materials	4.62	0.658	3
Using low energy process	4.56	0.554	4
Using multi-functional design	4.49	0.521	5
Incorporating diversity	4.43	0.498	6
Replicating strategies that work	4.42	0.569	7
Fitting form to function	4.34	0.877	8
Cultivating cooperative/collaborative relationships	4.32	0.754	9
Leveraging cyclic processes	4.30	0.621	10
Using feedback loops	4.24	0.731	11
Self-organizing	4.18	0.619	12
Building from the bottom up	4.17	0.853	13
Doing chemistry in water	4.11	0.985	14
Maintaining integrity through self-renewal	4.08	0.809	15
Breaking down products into benign constituents	4.01	0.940	16
Embodying resilience through variation	3.92	1.121	17
Embodying resilience through decentralization	3.89	0.965	18
Building selectively with a small subset of elements	3.88	0.862	19
Combining modular and nested components	3.87	1.005	20
Integrating the unexpected	3.81	0.882	21
Reshuffling information	3.65	1.031	22
Embodying resilience through redundancy	3.31	1.215	23

6.3   Benefits of Biomimicry Adoption and Implementation

Mean scores and rankings of the benefits of adopting and implementing biomimicry are presented in Table 2. The results are based on the respondent's assessment of the listed benefits. The results showed that the respondents considered 'creation of markets for green products ad services', 'protection of biodiversity', 'conservation of natural resources', 'restoration of natural resources', and 'global warming reduction' as the first five most important benefits expected from the adoption and implementation of biomimicry in the construction industry. The results agree with the studies of Klein [30] and Zari [31]. It is believed that by embracing biomimicry, biodiversity and ecosystem services will be maintained, thereby mitigating greenhouse gas (GHG) emissions and enhancing adaptation to the impacts of climate change.

Table 2. Benefits of biomimicry adoption and implementation in the construction industry.

Biomimicry benefits	Mean	Standard deviation	Rank
Create markets for green products and services	4.56	.499	1
Protect biodiversity	4.48	.521	2
Conserve natural resources	4.43	.693	3
Restore natural resources	4.41	.495	4
Reduce global warming	4.40	.493	5
Improve air quality	4.38	.578	6
Reduce waste streams	4.38	.685	7
Expand markets for green products and services	4.28	.451	8
Optimize life-cycle economic performance	4.22	.682	9
Improve overall quality of life	4.19	1.255	10
Improve water quality	4.19	.925	11
Create new business opportunities	4.07	1.026	12
Enhance occupant comfort and health	3.98	.945	13
Minimize strain on local infrastructure	3.96	1.254	14
Create employment opportunities	3.78	1.307	15
Improve occupant productivity	3.77	1.081	16
Improve the image of the building	3.68	1.143	17
Reduce operating costs	3.61	1.101	18
Reduce maintenance costs	3.52	1.115	19
Heighten aesthetic qualities	3.52	1.140	20
Reduce the civil infrastructure costs	3.25	1.283	21
Minimize occupant absenteeism	3.13	1.204	22

6.4   Measures of Bridging the Knowledge Gap

Mean scores and rankings of the drivers of adopting and implementing biomimicry are presented in Table 3. The results are based on the respondent's assessment of the listed measures to bridge the knowledge gap. The results showed that the respondents considered 'providing biomimicry education and training', 'increasing client and stakeholders awareness', 'improving availability of biomimetic technology', 'improving availability of biomimetic materials', and improved affordability of biomimetic materials' as the first five most important drivers of biomimicry adoption and implementation in the construction industry. It is imperative for the government and other stakeholders in the construction industry to encourage and facilitate a multidisciplinary collaboration which will result in remarkable and sustainable solutions to human challenges. By promoting awareness, training, workshops and education, biomimicry will be well propagated across in the construction industry, thereby encouraging its adoption and implementation.

Table 3. Drivers of biomimicry adoption and implementation in the construction industry.

Biomimicry drivers	Mean	Standard deviation	Rank
Providing biomimicry education and training	4.69	.464	1
Increasing client and stakeholder's awareness	4.51	.607	2
Improving availability of biomimetic technology	4.45	.500	3
Improving availability of biomimetic materials	4.39	.645	4
Improved affordability of biomimetic materials	4.38	.685	5
Increasing client demand	4.35	.498	6
Providing economic incentives	4.30	.652	7
Improving multi-disciplinary collaboration	4.19	.712	8
Improving government support and intervention	4.18	.983	9
Improving availability of biomimetic framework/measurement standard	4.11	.787	10
Providing biomimicry innovation and certification	4.05	.989	11
Developing a policy monitoring system	3.96	1.004	12
Developing a legal and regulatory framework	3.91	1.158	13
Providing motivation and commitment (self and corporate)	3.90	.807	14

7   Conclusion and Recommendations

Biomimicry has the potential to offer sustainable solutions to identified human challenges, especially in the construction industry. They can also be beneficial in the invention of a novel and innovative materials, products and technologies with sustainable attributes. However, this study has shown that there is a low level of awareness and knowledge among the stakeholders in the construction industry on the concept of biomimicry. This has impeded the adoption and implementation of biomimicry to optimize sustainability in the industry. Biomimicry principles also play a key role in evaluating for sustainability as they are creative common tools and important checklists to be strictly adhered to when sustainability is in focus. Awareness, training and education of professionals and stakeholders in the SACI on biomimicry should be encouraged for its adoption and practice to be widely accepted. As the whole world is now feeling the effects of climate change, to which the construction industry is known to be the highest contributor, it is imperative to adopt and embrace biomimicry in its entirety for mitigation and adaptation purposes. This will be justified through the benefits of biomimicry which include protection and conservation of biodiversity and natural resources, the creation of employment opportunities and markets for green products and services, reduced waste streams and effects of global warming, restoration of natural resources, and improved air and water quality, amongst others.

Awareness, training and education of professionals and stakeholders in the SACI on biomimicry should be encouraged for its adoption and practice to be widely accepted. As the whole world is now feeling the effects of climate change, to which the construction industry is known to be the highest contributor, it is imperative to adopt and embrace biomimicry in its entirety for mitigation and adaptation purposes. This will be justified through the benefits of biomimicry which include protection and conservation of biodiversity and natural resources, the creation of employment opportunities and markets for green products and services, reduced waste streams and effects of global warming, restoration of natural resources, and improved air and water quality, amongst others.

Acknowledgments. The University of Johannesburg is acknowledged for providing the Global Excellence Stature Scholarship through the Postgraduate School. The authors also wish to acknowledge the participants who responded to the survey.

References

1. Moavenzadeh, F.: Global Construction and the Environment: Strategies and Opportunities. Wiley, Hoboken (1994)
2. Giang, D.T., Pheng, L.S.: Role of construction in economic development: review of key concepts in the past 40 years. Habitat Int. **35**(1), 118–125 (2011)
3. Shi, L., Ye, K., Lu, W., Hu, X.: Improving the competence of construction management consultants to underpin sustainable construction in China. Habitat Int. **41**, 236–242 (2014)
4. Ramsaran, R., Hosein, R.: Growth, employment and the construction industry in Trinidad and Tobago. Constr. Manag. Econ. **24**(5), 465–474 (2006)
5. Wang, N.: The role of the construction industry in China's sustainable urban development. Habitat Int. **44**, 442–450 (2014)
6. Pearce, A., Ahn, Y.H.: Sustainable Buildings and Infrastructure: Paths to the Future. Routledge, Abingdon (2012)
7. Benyus, J.M.: A biomimicry primer. The Biomimicry Institute and the Biomimicry Guild (2011)
8. Rao, R.: Biomimicry in architecture. Int. J. Adv. Res. Civil, Struct. Environ. Infrastruct. Eng. Dev. **1**(3), 101–107 (2014)
9. Murr, L.E.: Biomimetics and biologically inspired materials. In: Handbook of Materials Structures, Properties, Processing and Performance, pp. 521–552. Springer (2015)
10. Vincent, J.F., Bogatyreva, O.A., Bogatyrev, N.R., Bowyer, A., Pahl, A.K.: Biomimetics: its practice and theory. J. Roy. Soc. Interface/Roy. Soc. **3**(9), 471–482 (2006)

11. El Din, N.N., Abdou, A., El Gawad, I.A.: Biomimetic potentials for building envelope adaptation in Egypt. Procedia Environ. Sci. **34**, 375–386 (2016)
12. Merrill, C.L.: Biomimicry of the Dioxygen Active Site in the Copper Proteins Hemocyanin and Cytochrome Oxidase: Part I: Copper (I) Complexes Which React Reversibly with Dioxygen and Serve to Mimic the Active Site Function of Hemocyanin. Part II: Mu-Imidazolato Binuclear Metalloporphyrin Complexes of Iron and Copper as Models for the Active Site Structure in Cytochrome Oxidase, Doctoral dissertation, Rice University (1982)
13. Goss, J.: Biomimicry: Looking to Nature for Design Solutions. Corcoran College of Art and Design. ProQuest Dissertations Publishing, Ann Arbor (2009)
14. Benyus, J.M.: Biomimicry: Innovation Inspired by Nature. William Morrow & Company, New York (1997)
15. Hargroves, K., Smith, M.: Innovation inspired by nature: biomimicry. Ecos **2006**(129), 27–29 (2006)
16. Strategic Direction.: Nature's inspiration: solving sustainability challenges. Strateg. Dir. **24** (9), 33–35 (2008)
17. Nychka, J.A., Chen, P.: Nature as inspiration in materials science and engineering. JOM J. Miner. Metals Mater. Soc. **64**(4), 446–448 (2012)
18. Rinaldi, A.: Naturally better. Science and technology are looking to nature's successful designs for inspiration. EMBO Rep. **8**(11), 995–999 (2007)
19. Ask Nature: Biomimicry taxonomy. The Biomimicry Institute. https://asknature.org/resource/biomimicry-taxonomy/
20. Shu, L., Ueda, K., Chiu, I., Cheong, H.: Biologically inspired design. CIRP Ann. – Manuf. Technol. **60**(2), 673–693 (2011)
21. Gamage, A., Hyde, R.: A model based on biomimicry to enhance ecologically sustainable design. Archit. Sci. Rev. **55**(3), 224–235 (2012)
22. Aziz, M.S., El Sherif, A.Y.: Biomimicry as an approach for bio-inspired structure with the aid of computation. Alexandria Eng. J. **55**, 707–714 (2015)
23. Pronk, A., Blacha, M., Bots, A.: Nature's experiences for building technology. In: Proceedings of the 6th International Seminar of the International Association for Shell and Spatial Structures (IASS) Working Group (2008)
24. De Pauw, I., Kandachar, P., Karana, E., Peck, D., Wever, R.: Nature inspired design: strategies towards sustainability. In: Knowledge Collaboration & Learning for Sustainable Innovation: 14th European Roundtable on Sustainable Consumption and Production (ERSCP) Conference and the 6th Environmental Management for Sustainable Universities (EMSU) Conference, Delft, The Netherlands, 25–29 October, Delft University of Technology, The Hague University of Applied Sciences, TNO (2010)
25. Arnarson, P.O.: Biomimicry: New Technology. Reykjavík University. http://olafurandri.com/nyti/papers2011/Biomimicry%20-%20P%C3%A9tur%20%C3%96rn%20Arnarson.pdf
26. Badarnah, L., Kadri, U.: A methodology for the generation of biomimetic design concepts. Archit. Sci. Rev. **58**(2), 120–133 (2015)
27. Field, A.: Discovering Statistics Using IBM SPSS Statistics. Sage Publications, Thousand Oaks (2013)
28. Polit J., David C.: Regreening nature: turning negative externalities into opportunities, Master's thesis, Delft/Delft University of Technology (2014)
29. Kennedy, E., Fecheyr-Lippens, D., Hsiung, B., Niewiarowski, P.H., Kolodziej, M.: Biomimicry: a path to sustainable innovation. Des. Issues **31**(3), 66–73 (2015)
30. Klein, L.: A phenomenological interpretation of biomimicry and its potential value for sustainable design, Doctoral dissertation, Kansas State University (2009)
31. Zari, M.P.: Biomimetic design for climate change adaptation and mitigation. Archit. Sci. Rev. **53**(2), 172–183 (2010)

Measuring Labour Productivity in Labour Intensive Works on the Road Construction in Ghana

Emmanuel Bamfo-Agyei[✉], Clinton Aigbavboa,
and Thwala Welligton Didibhuku

Department of Construction Management and Quantity Surveying,
University of Johannesburg, Johannesburg, South Africa
kwaminabamfoagyei@gmail.com,
{caigbavboa,didibhukut}@uj.ac.za

Abstract. Reports from earlier studies indicate that low productivity is common in the construction industry in developing countries like Ghana and this could be attributed to poor labour productivity. The aim of this research is to develop a model for estimating the optimal Construction Labour Productivity (CLP) which can aid construction industry in determining and measuring labour productivity in labour intensive public works in Ghana. The objective are to identify the trend in measuring labour productivity; to identify the factors that affect construction labour Productivity on the site and to evaluate the effect of identified factors that affect construction labour productivity in Ghana. Content analysis was adopted to measure the productivity and the factors that affect productivity of labour during labour intensive works. The findings indicated that productivity; profitability; performance; efficiency and effectiveness contributed to the tipple model and productivity from literature was measured under work measurement which included time study; work sampling and activity sampling. The factors identified were grouped into five components, Management and control component; Work characteristics component; Equipment component; Materials component and workers component.

Keywords: Labour · Productivity · Construction sites · Ghana

1 Introduction

Construction labour productivity is regarded as an effective indicator of the efficiency of the industry activities. It is an important aspect of the construction industry that may be used as an index for measuring the efficiency of production. Consequently, it can also serve to measure the status of economic growth and related production from industrial and corporate perspectives.

The construction industry is labour-intensive and relies heavily on the skills of its workforce. [12] noted that the workforce is the industry's most valuable asset, which, at the very least, accounts for over a quarter of the total project cost. More often than not, owing to its volatile nature, this workforce can significantly influence the cost, schedule, and quality of a construction project [8].

© Springer International Publishing AG, part of Springer Nature 2019
J. Charytonowicz and C. Falcão (Eds.): AHFE 2018, AISC 788, pp. 515–523, 2019.
https://doi.org/10.1007/978-3-319-94199-8_50

Labour is considered to be the most uncertain factor among costly project components (materials, equipment, and labour). The other components, materials, and equipment, are predominantly determined by market prices, and are consequently beyond the influence of project management. The management of labour and its productivity is therefore crucial for determining the success of a construction project [9]. Labour productivity is considered to be one of the best indicators of production efficiency. [12] asserted that higher productivity usually yields superior profitability.

2 Definition of Productivity

Productivity is defined as the ratio of the quantity of output to the quantity of input. Productivity is an important aspect of the construction industry that may be used as an index for measuring the efficiency of production.

Construction labour productivity is influenced by various factors whose impact can be quantified in productivity models. These models play an important role in estimating cost, in scheduling, and in planning. A number of models have been developed using regression analysis to provide a qualitative evaluation of the impact of different factors on construction labour productivity [20].

3 Measurements of Productivity

In construction, productivity is measured at different levels of detail for different purposes [21]. Measurements of productivity at a company level or at a project level provide internal and external benchmarks for comparison with company or project norms [20]. For detailed estimation and project scheduling, productivity is measured at an activity level. Because construction is usually labour intensive, productivity at the activity level is frequently referred to as labour productivity; it measures the input in terms of hours of work and the output in terms of the material value of the work achieved [4].

As cited by [17] summarized three main measurements of construction labor productivity at activity, project and industry levels; the major differences between these measurements are the source of data, the level of aggregation, the definition of the production process and the completeness, as indicated in Fig. 1. The input and output proxies used to reflect construction labor productivity vary from detailed physical quantities to general economic indicators, as the aggregation of interest moves from the micro-perspective to the macroperspective.

A successful construction project is one that is completed on time, within budget, meets specified standards of quality, and strictly conforms to safety policies and precautions. All of these are feasible only if the premeditated levels of productivity can be achieved. All the same, productivity or lack thereof, is one of the construction industry's most prevalent problems. Due to the nature of construction projects, its importance to society and the existing economic resources, more emphasis should be given to improving productivity.

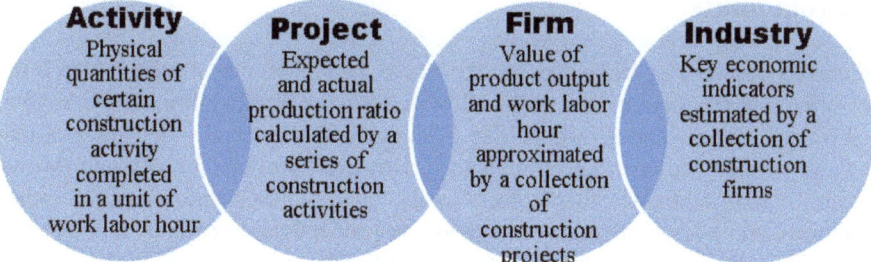

Fig. 1. Measurements of construction labour productivity.

4   Aim of the Study and Objectives

Generally, the study is an attempt to develop a model in estimating the optimal Construction labour Productivity (CLP) in Ghana. More specifically, the study attempted to examine the trend in measuring labour productivity and to identify the factors that affect construction labour productivity on the site.

5   Research Methodology

The study focus more on existing studies and content analysis was employed to find the trend in measuring labour productivity and to identify the factors that affect construction labour productivity on the site.

6   Terminology of Productivity

[19] argued that productivity is one of the basic variables governing economic production activities, perhaps the most important one. However, at the same time as productivity is seen as one of the most vital factors affecting a manufacturing company's competitiveness, researchers argue that productivity is often relegated to second rank, and neglected or ignored by those who influence production processes [22].

Productivity is therefore, on the one hand, closely connected to the use and availability of resources. This means in short that productivity is reduced if a company's resources are not properly used or if there is a lack of them. On the other hand, productivity is strongly linked to the creation of value. Thus, high productivity is achieved when activities and resources in the manufacturing transformation process add value to the produced products.

The literature show that the concept of productivity should be distinguished from four other similar terms: profitability, performance, efficiency and effectiveness, which will now be explained.

7 Profitability

Perhaps the reason why companies tend to ignore the importance of productivity is that they often link productivity and profitability as one issue. Profitability is the overriding goal for the success and growth of any business, and is generally defined as a ratio between revenue and cost (i.e. profit/assets). However, profitability as a performance measure mainly addresses shareholders as the interest group and many researchers therefore claim that using monetary ratios as productivity measures will result in several shortcomings, for instance, induce short-termism and discourage the customer perspective. Profitability can change for reasons that have little to do with productivity, such as inflation and other external conditions that may bear no relationship to the efficient use of resources.

The term profitability clearly has a productivity component, but it is strongly influenced by the prices a company pays for its input and receives for its output.

8 Performance

Many people who claim to be discussing productivity are actually looking at the more general issue of performance. While productivity is a fairly specific concept related to the ratio between output and input, performance is a term which includes almost any objective of competition and manufacturing excellence such as cost, flexibility, speed, dependability and quality.

9 Efficiency and Effectiveness

The two terms effectiveness and efficiency render the terminology even more complicated. There is no single accepted view about these terms; however, effectiveness is usually described as 'doing the right things', while efficiency means 'doing things right' [19]. Most researchers agree that efficiency is strongly linked to the utilisation of resources and mainly influence the input of the productivity ratio. This means that efficiency in manufacturing can be seen as the minimum resource level that is theoretically required to run the desired operations in a given system, compared to how much resources are actually used.

Jackson [11] argues that effectiveness, on the other hand, is a more diffuse term and in most cases very difficult to quantify. It is often linked to the creation of value for the customer and affects the output of the productivity ratio. In conclusion, a single focus on efficiency does not seem to be a fruitful way to increase productivity.

However, it is the combination of high values of both efficiency and effectiveness in the transformation process that leads to high productivity. Thus, is it possible for an effective system to be inefficient; it is also possible for an efficient system to be ineffective.

10 Productivity Measurement

The most basic case to consider is that of a single firm that produces one output (y) using a single input (x), the ratio of these two (y/x) yields a measure of the level of productivity. Both output and input are measured in real units (i.e. their level reflects the real quantity of either output or input).

Work Measurement. Work measurement is the application of techniques designed to establish the time for an average worker to carry out a specified manufacturing task at a defined level of performance. It is concerned with the length of time it takes to complete a work task assigned to a specific job.

Revealing existing causes of ineffective time through study, although important it is, is perhaps less important in the long term than the setting of sound time standards, since these will continue to apply as long as the work they refer to continues to be done. They will also show up any ineffective time or additional work which may occur once they have been established.

At the activity level, labour productivity is widely reflected by hourly outputs, where a labour hour and physical quantity of work completed are used as input unit and output, respectively [9]. [17] noted that external factors can barely control the actual physical quantity and working hours, the hourly output measurement of productivity can provide an accurate proxy for construction activity efficiency. At the project level, the aggregation is composed of a series of construction activities with different units of measurements. Therefore, the efficiency of the whole construction process is reflected by the ratio between expected productivity and actual productivity, which is determined by the work days and quantities, installed under consideration and practice, respectively [17].

To provide information on which the planning and scheduling of production can be based, including the plant and labour requirements for carrying out the programme of work and the utilization of available capacity. To provide information on which estimates for tenders, selling prices and delivery promises can be based. To set standards of machine utilization and labour performance this can be used for any of the above purposes and as a basis for incentive schemes. To provide information for labour-cost control and to enable standard costs to be fixed and maintained. Work measurement provides the basic information necessary for all the activities of organizing and controlling the work of an enterprise in which the time element plays a part. [3] assert that, "you can't manage what you can't measure and mentioned that measurement is crucial before a management activity is executed. However, it should be noted that an incorrect measurement of labour performance can lead to incorrect and warped decisions by the management team. Therefore, the importance of measuring the right thing at the right time to reflect the true conditions of a company cannot be over emphasized. The measurements should offer the management an opportunity to make effective and accurate decisions. The various techniques available to supply information need to be explored so as to identify the right measurement technique for prevalent conditions. These techniques include: time studies, work sampling, subjective evaluations and) personal recording of activities [2]. Use of technological advances makes it easier to efficiently acquire accurate results.

Time Study. Time study is a method used to determine the actual time required to complete a task and takes into account time allowances and delays [1]. Time studies enable the management to make effective decisions aimed at improving the efficiency of the entities operating within the system. It is important to realize that accurate time studies yield positive results and inaccurate time studies can create many problems [6] as mentioned earlier.

Work Sampling. On the other hand, work sampling was developed for the first time in 1935 [5]. Work sampling is the activity of taking randomly distributed observations of activities; these activities can include both humans and machines, with the objective of determining their utilization [5]. The fundamental principle of work sampling is that it is based on the laws of probability [6]. Work sampling only takes samples rather than continuous observation of the object being studied. This allows for the acquisition of reasonably accurate representations of the work under study, without the drawbacks of continuous monitoring. Work sampling is also observed as a low cost alternative method for determining of workforce utilization [5]. In work sampling, the accuracy of the results is linked to the number of samples taken in the study [19], as well as the time period during which the analysis is performed. It is, therefore, important that the analyst understands the operations of the company so as to identify an appropriate time frame for the study. The analyst needs to select a sample size that represents the true conditions of the system accurately, while also understanding the capabilities of the observer. Some standards have been developed to aid in making selections about the number of samples in a study. One of such guidelines is not to exceed eight observations per hour [19]. Traditionally work sampling does not only determine the effectiveness of workers whether they are working or not. It does this by recording the amount of time workers spend on certain activities [5]. Thorough planning needs to be done prior to the start of the work sampling study hence the following procedure covers the work sampling process from the objective identification step through to the analysis of the results [5].

Activity Sampling. Activity sampling is a technique through which information can be obtained not only quickly and economically but also to predetermine levels of accuracy. It is a method that measures the time labour spent in various categories of activities [23].

A sequence of project-based studies were carried out to investigate the factors influencing construction labour productivity, with the purpose of maintaining construction development [7, 11, 15]. Based on a collection of substantial cross-sectional data, detailed insights into construction labour productivity were provided at the project level. The productivity ratio can reflect performance efficiency, where differences between various construction activities can be diminished.

Activity sampling can be defined as a technique in which a large number of instantaneous observations are made over a period of time of workers, machines, or processes. Each observation records what is happening at that instant and the percentage of observations recorded for a particular activity or delay is a measure of the percentage of time during which that activity or delay occurs. Activity sampling study provides the necessary information to help determine how time is being employed by the workforce, identifies the problem area that causes the work delay, and sets up a base line measure for productivity improvement. The main advantage of using activity sampling is that it allows a larger number of machines or men to be studied at one time that can be managed using a continuous time study (Table 1).

Table 1. Selecting factors and components by previous studies.

Authors	Factors	Components
Kim and Kim [14]	Percentage of prolonged working hours (work delay) crew size (crew size and composition)	Management and control component Work characteristics component
Thomas and Sakarcan (1994)	Crowdedness (working space), crew size (crew size and composition), percentage of workers (crew size and composition), work size (work quantity), job type (work method) supervision (manager's capability, management system), information (information technology and integration), re work (rework), work continuity (work continuity), overtime work (work delay) equipment (factors in equipment component), tools (factors in equipment component), concrete pumping (factors in equipment component) materials (factors in materials component), factory conditions (material condition, materials transport environment, material procurement delay)	Work characteristics component Management and control component Equipment component Materials component
Sonmez and Rowings (1998)	Percentage of overtime work (work delay) crew size (crew size and composition), the quantity of the completed work (work quantity), job type (work method), percentage of labourers (crew size and composition) concrete pouring (factors in equipment component)	Management and control component Work characteristics component Equipment component
Hanna et al. [10]	Order changes (work order), work sequencing (work continuity), work shifting (work continuity), schedule compression (field work plan), overtime work (work delay) absenteeism and turnover (sense of responsibility), labour problems (teamwork, communication, work attitude), trade stacking (teamwork) material problems (material condition, material procurement delay, material applicability)	Management and control component Workers component Materials component
Lu et al. (2000)	Administration/project manager/superintendent (factors in management and control component), duration of them construction work (field work	Management and control component

(continued)

Table 1. (*continued*)

Authors	Factors	Components
	plan), order changes (work order), drawing and specifications quality (defects in design documents), safety (safety/accidents), inspection (permission/approval delay), location classification (field work plan), extra work (work delay) prefabrication/field work (prefabrication/standardisation/field work), average crew size and peak crew size (crew size and composition), total work quantity and installation quantities (work quantity), method of installation (work method), site working conditions (work environment), overall degree of work difficulty (work difficulty) equipment (factors in equipment component) materials (factors in materials component), type of materials (material applicability) crew ability (capability)	Work characteristics component Equipment component Materials component Workers component
Rojas and Aramvareekul (2003)	Management systems and strategies (factors in management and control component) manpower (capability) and external conditions (work environment)	Management and control component Workers component Work characteristics component
Jang et al. [12]	Rework, work continuity and information technology	Work technique

11 Conclusion

The long-run trend of the growth in construction productivity at an industry level was explored, suggesting that exogenous technological progress and the existence of capital are the most important factors influencing construction labor productivity changes (Mills 2016). This research has giving insight to other research works on labour output in the construction industry in other countries.

The long-run equilibrium and dynamics of construction labor productivity across the Australian states and territories have been revealed. The developed models have been further used to simulate regional construction productivity in order to discover the regional clusters for Ghanaian construction labor productivity.

The results derived from the study will help in developing a model that will focusing in forecasting the labour output in Ghana. At the site management level, the results derived will provide a daily progress report which will contain the required information necessary to obtain the baseline productivity to develop a benchmarking standard for each construction firm in Ghana, which may lead to an improvement in the national construction productivity.

References

1. Amaoh Mensah, K.: Building Estimating Manual for West African Construction Practices. Parcom Ghana limited, Kumasi (1995)
2. Bamfo-Agyei, E., Kotey, S.: Establish the labour output constants for rendering, painting and terrazzo works using Activity Sampling technique. In: Commonwealth of Polytechnic Conference, Accra, Ghana, 23 November 2009
3. Cornwell, J.R., Cornwell, R.W.: You Can't Manage What You Can't Measure – Assessment of Learning in a Connected World (2006). http://tcfir.org/opinionretrieved. Accessed 20 May 2015
4. Dozzi, S.P., AbouRizk, S.: Productivity in construction. Institute for Research in Construction, National Research Council, Ottawa (1993)
5. Fitzgerald, B.: Open source software implementation: anatomy of success and failure. Int. J. Open Source Softw. Process. **1**(1), 1–19 (2009). http://www.brian-fitzgerald.com/publications/journal-paperse. Accessed 10 February 2015
6. Freivalds, A.: Niebles Standards Methods and Work Design. McGraw Hill, New York (2009). http://tcfir.org/opinionretrieved. Accessed July 2015
7. Goodrum, P., Haas, C.: Partial factor productivity and equipment technology change at activity level in US construction industry. J. Constr. Eng. Manag. **128**(6), 463–472 (2002)
8. Han, S.H., Park, S.H., Jin, E.J., Kim, H., Seong, Y.K.: Critical issues and possible solutions for motivating foreign construction workers. J. Manag. Eng. **24**(4), 217–226 (2008)
9. Hanna, A.S., Taylor, C.S., Sullivan, K.T.: Impact of extended overtime on construction labor productivity. J. Constr. Eng. Manag. **131**(6), 734–739 (2005)
10. Hanna, A.S., Russell, J.S., Nordheim, E.V., Bruggink, M.J.: Impact of change orders on labor efficiency for electrical construction. J. Constr. Eng. Manag. **125**(4), 224–232 (1999)
11. Jackson, M.: An analysis of flexible and reconfigurable production systems. Dissertation No. 640, Linköping University, Sweden, pp. 85–104 (2000). (Chap. 6)
12. Jang, H., Kim, K., Kim, J., Kim, J.: Labour productivity model for reinforced concrete construction projects. Constr. Innov. **11**(1), 92–113. http://www.emeraldinsight.com/doi/abs/10.1108/14714171111104655. Accessed 5 May 2016
13. Jarkas, A., Bitar, C.: Factors affecting construction labor productivity in Kuwait. J. Constr. Eng. Manage. **138**(7), 811–820 (2012)
14. Kim, S., Kim, Y.: A study on the construction labor productivity model using neuro-fuzzy network. In: Conference of the Architectural Institute of Korea, Ansan, Korea, vol. 21, no. 1, pp. 493–496 (2001)

An Assessment of Lean Construction Practices in the Construction Industry

Olusegun Aanuoluwapo Oguntona[(✉)], Clinton Ohis Aigbavboa,
and Gloria Ndalamba Mulongo

Sustainable Human Settlement and Construction Research Centre,
Faculty of Engineering and the Built Environment, University of Johannesburg,
Johannesburg, South Africa
architectoguntona12@gmail.com

Abstract. The performance of the construction industry is hampered by waste ranging from material resources, cost, time, and manpower amongst others. Hence, the need to embrace the implementation of Lean Construction (LC) practices which encompasses culture, plans, tools, and concepts to maximise value while also minimising all forms of waste. This research paper aims to assess the use of LC practices to effectively improve productivity and performance of the construction industry. A well-structured questionnaire was administered, with construction professionals as respondents. A quantitative approach to data analysis was adopted using the mean item scores of the identified variables. Findings revealed waste reduction, effective administration of materials on site, improved lifecycle cost, good project coordination, improved safety on site, and greater productivity as the top six benefits of implementing LC practices. By adopting LC practices, productivity will be increased thereby leading to successful delivery of construction projects.

Keywords: Construction waste · Construction industry · Productivity
Lean management · Sustainability

1 Introduction

The construction industry contributes a lot to a country's economic stability by providing employment, physical facilities and infrastructures. In South Africa, the industry is responsible for the realization of civil and building [1]. The government's effort towards economic development and urbanization has created a great demand for physical developments which in turn placed a significant demand on natural resources [2]. These activities have therefore resulted in having a negative impact on the environment, human health and natural species [3]. It is widely known that the industry enormously consumes natural resources; consuming more than half of the mineral resources excavated from nature and thus contributes the largest portion of waste to landfill [4]. Waste generation has, therefore, become a serious drawback that the construction industry is suffering from [5], and one of the major challenges affecting its performance. Hence, the introduction of Lean Construction (LC) as a waste reduction, supply management, value for money and project quality management mechanisms in the construction industry.

© Springer International Publishing AG, part of Springer Nature 2019
J. Charytonowicz and C. Falcão (Eds.): AHFE 2018, AISC 788, pp. 524–534, 2019.
https://doi.org/10.1007/978-3-319-94199-8_51

LC has become one of the primary performative improvement recipes for the construction industry [6]. It emphasizes and focuses on improvement of relationships among project participants to ensure productive and quality project outcomes by initiating a positive attitude, mindset and behavior of workers [7]. Firstly, LC emphasizes on waste reduction from the technical and operational point of view and secondly on the elimination of harmful relationships in the process while promoting teamwork between the supply chain managers [8]. The essence of LC will, however, be effective in enhancing project productivity when supported from the upper management level and implemented by the workforce in the use of lean tools. Hence, the objective of this study is to assess the level of use of LC practices in the construction industry. In other to achieve this objective, the study will seek to identify the challenges encountered in the implementation, the benefits and lastly the measures of improving its use in the construction industry.

2 Waste in the Construction Industry

Globally, the issue of waste in the construction industry has been the subject of several research studies with growing concerns about its adverse effect on the human environment. These are wastes generated from all the activities involved in the construction processes. Waste is assumed to be the only physical waste by construction professionals; however, there have been records of noticeable wastes in the construction industry which are generated through what is known as "non-value adding activities" [5].

Construction and demolition waste are described as not only physical waste but also include waste produced from construction activities such as transportation, waiting times, excess materials, rework, delays and defects [9, 10]. Waste is also defined as any construction activities or process that incur costs which do not directly or indirectly add value to the construction project [11]. These construction activities can be value-adding activities (conversion of raw materials to the final product) and non-value adding activities (wastes, wasteful operations). Construction waste could be as a result of errors in design, modifications, redoing of work, defects and the use of excess materials [9]. It is therefore important to ensure waste generation is minimized in construction processes and activities.

3 Possible Types of Waste During Construction Processes

Construction waste comprises of waiting times; lack of safety; rework; unnecessary transportation trips; excess inventory; quality costs, set up, handling, inspections, expedition, prioritizing, improper choice of management method; and lack of constructability amongst others [11]. These wastes are generated at one stage or the other in a construction process.

3.1 Waste from Overproduction (Unnecessary Work)

Overproduction is related to the production of a quantity greater than required or than necessary [12]. Moreover, companies sometimes produce more than they have sold or might sell because they want to build inventories or because they want to keep their equipment and facilities running to achieve high-level resource utilization [13]. However, making products for which there is no demand is wasteful, therefore waste from overproduction is difficult to identify, unless what is produced is compared with what is sold and shipped otherwise nothing appears wrong [13]. This kind of waste is also in prefabricated construction components.

3.2   Waste from Rejects (Defects/Unsatisfactory Work)

Waste of defects occurs when the final or intermediate product does not fit the quality specifications. A defect is an error in a process that makes a service less valuable to a client, or that requires additional processing to correct the defect [13]. Defects can occur through a wide range of reasons such as poor design and specification, lack of planning and control, poor qualification of the teamwork, lack of integration between design and production among others [11]. These will add additional rework, inspection, design changes, process changes, and machine downtime [13]. Therefore, the new procedures to handle these wastes had to be implemented.

3.3   Waste in Transportation (Materials Movement/Conveyance)

In construction works, people and materials being processed or serviced must be moved from one location to another over several stages to carry out the project [13]. This transportation is concerned with the internal movement of materials on site where workplace layout or a lack of process flow creates many stops and start in a production cycle [14]. Every movement should have a purpose for items being moved incur a cost. Excessive handling, the use of inadequate equipment or bad conditions of pathways can also cause this kind of waste [11].

3.4   Waste in Processing (Over Processing)

Waste in over-processing is described as the extra processing of the job beyond the requirements of the client [14]. The waste of processing also includes the waste of intellect which refers to any under-utilization of manpower. This is also referred to as unused employee creativity by not making the most use of employee skills, creativity, and knowledge [15].

3.5   Waste from Inventory

It has been suggested that excess inventory can be found in construction products and raw materials (in deterioration, losses due to inadequate stock conditions on site, robbery, and vandalism) and monetary losses due to the capital that is tied up [14]. Companies always order more than required to fulfil an order and as a result, larger quantities are ordered to cover for defects in materials on-site, finished products and in event of equipment breakdowns or delivery delays. Also, inventory resulting from large production runs, also lead to time and cost overruns [13]. It might also be as a result of lack of resource planning or uncertainty on the estimation of quantities.

3.6   Waste from Waiting (Delays)

Unlike waste from overproduction, waste generated due to waiting is easy to identify. Waste from waiting is related to turn-around time or cycle time of any process in the project. It includes waiting for materials, information, equipment, and tools among others. This means that whenever goods are not moved or being processed, waste occurs due to waiting. Some companies minimize waste from waiting by operating a policy of keeping workers busy and machines running, irrespective of demand [13]. Waste of waiting is necessarily about jobs waiting to be processed. This can be drastically reduced by linking up the processes together to one which feeds directly into the next [11].

3.7   Waste from Movement (Motion)

Waste of motion is generated through the movement of resources including human being and equipment not necessary for the successful completion of the operation. This might also be caused by inadequate equipment, ineffective work methods, or poor arrangement of the working place [14] resulting in processing delays. It is therefore important to ensure waste generation is minimized in construction processes and activities. This can be achieved by introducing new management practices which aim to minimize waste generation and improve tools in minimizing the destruction of the environment by construction activities [16, 17]. Hence, the introduction of LC practices which will ensure a significant reduction in waste generated during construction process/activities.

4   Historical Overview of Lean Concept in Construction

The term lean is a word that originated from the manufacturing industry as a concept since the beginning of the 1900s [18, 19]. Lean production management principles were developed around the year 1950 by Taiichi Ohno, the promoter of the Toyota motor company in Japan, who concentrated his efforts on finding ways to convert waste (also known as 'muda') to value [20]. LC originated based on the concepts of lean manufacturing [21]. It is pioneered by Koskela with the sole aim of minimizing waste and maximizing value in the construction industry [22, 23].

LC helps to eliminate materials and time waste in order to produce value in possible ways [25]. It also provides a cohesive supply chain that reduces time frame for project completion [23]. LC is different from project management method as it clears project objectives [24], and the purpose is to maximize project performance by eliminating waste throughout the lifespan of the project from design stage to completion and delivery stage. This mechanism of waste reduction leads to improvement in the construction process. Objectively, LC aims to meet client requirements, reduce the value stream and pursue perfection in construction projects with the application spanning across construction project design through delivery and use. Since the introduction of LC, there have been records of benefits when reducing waste on site during execution of the project.

5 Principles of Lean Construction

Lean thinking (LT) is a concept that organizations engage to change their way of thinking. LT combines different philosophies such as waste elimination, continuous improvement, availability of resources, teamwork and supply chain management cooperatively [26]. These ideas have become a concept for ensuring success in projects. Also, LT identifies five principles to implement LC. These lean principles (LP) helps stakeholders to manage their companies and projects flexibly in other to meet their client's needs. To reduce waste, LP is simplified and categorized into five groups as listed below: such as:

- stipulating **value** from the customer perspective;
- recognizing the **value stream**;
- produce **flow**;
- allow demand of customer to pace and **pull** production; and
- achieve continuous improvement and **purse perfection** [15].

5.1 Value

Specifying a value from the client's perceptive is the first principle of LC [27]. When a company starts implementing lean, the redefinition value of its services from clients' perspective is the first step. This leads to the recognition of waste as everything that adds no value from the clients' perspective [15]. Time has also been recognized as having the utmost value to the client. Therefore, these steps will be identified along the value stream through a process called Value Stream Mapping (VSM).

5.2 Value Stream

Value Stream Mapping (VSM) is an assessment and planning tool that lean practitioners use to apply lean thinking. VSM helps to determine inefficiencies in an end to end process. It also monitors all the activities being done by looking through the time aspect. People, materials and equipment are managed using a chart which tracks down the flow of information along the process flow. Some of the benefits accrued to the construction industry using VSM are: ability to visualize the production flow, foreseeing waste in the system, preventing focus on large improvement opportunities with little impact, creating framework for designing complete system, demonstrating interaction between information and material flow and developing an implementation plan for future lean activities [15].

5.3 Flow

Flow requires a focus and a vision on construction activities. When a flow is produced along the stream of value, the process which occurs between the client order and the delivery is shorten therefore reduce lead time. However, waste can happen if services are generated without any order. Flow, however, deals with developing value without interruptions as well as eliminating lead time and waste when producing new products [15].

5.4 Pull

Pull is the signification of shortening response to client's orders so that overproduction do not occur. Pull and flow are main features of LT. Pull system generates services when the client demands it. This system is useful in cases where the company is amid challenges of sustaining the flow along value stream continuously.

5.5 Perfection

Working to perfection in a project does not only means it is defect free but also means delivering exactly what the client wants, at a reasonable price. In the construction industry, perfection is the most core element required, due to the project time management, delivering quality services to the contractor as described by the architect of that project [15]. It is regarded as one element to yield production in the construction industry.

6 Research Methodology

This research is placed within a deductive methodological reasoning approach, thus employing the combination of secondary data (review of literature) and primary data (structured survey questionnaire). This is used to present the respondent's perspectives on the assessment of LC practices in the South African construction industry (SACI). A structured questionnaire survey was adopted, targeting construction professionals (architects, quantity surveyors, civil engineers, and construction and project managers) duly affiliated with their respective professional bodies in South Africa.

Only close-ended questions were administered to the respondents. The questionnaire was divided into four sections. The first section sought information related to the background of the respondents (professional qualification and affiliations, years of experience, category of organization working with, and category of projects involved in). The second part of the questionnaire sought to identify the challenges of implementing LC practices in the SACI. The third part dealt with identifying the benefits of LC practices implementation in the SACI. The last part dealt with assessing the measures of improving the adoption and implementation of LC practices in the SACI.

Concerning the challenges, benefits, and measures of implementing LC practices, the respondents were asked to indicate their level of agreement on a five-point Likert scale (from 1 = 'Strongly Disagree' to 5 = 'Strongly Agree'). The questionnaire was administered through a face-to-face medium, which ensured that 45 questionnaires out of 60 were returned complete and analyzed, representing a 75% response rate. A quantitative approach to data analysis was employed and the Statistical Package for Social Sciences (SPSS) v16 was employed to analyze the data collected. Descriptive statistics (particularly mean and standard deviation) were used with the mean scores of the factors and their standard deviations compared in a tabular form. With a mean value of 2.5 or more, a factor is deemed to be significant to the study [28].

7   Results and Discussions

7.1   Background of Respondents

For the profession of the respondents, quantity surveyors constituted 33.3%, 22% were civil engineers, 20% were construction managers, 15.3% were project managers, and architects represent 8.9%. The result also showed that 48.9% of the respondents work with consultancy firms, 33.3% work with contracting firms, 15.6% work privately while 2.2% work with public/government institutions.

7.2   Challenges of Implementing Lean Construction Practices

The level of agreement of the respondents on the challenges of implementing LC practices in the construction industry is revealed in Table 1. All the factors listed have mean values of more than 2.5 indicating strongly that the respondents agree that they are barriers to LC practices implementation. The results further revealed that 'poor culture among project partners', 'lack of good policies', 'complexity of lean construction process', 'poor organizational knowledge', and 'lack of understanding of lean construction practices' are the top five barriers to LC practices implementation. The result resonates with the studies of Kim [29], Bicheno and Holweg [30], and Bashir et al. [31]. Their study revealed that poor knowledge of LC practices, the complexities involved and lack of policies to drive the use, hampers its implementation in the construction industry.

Table 1. Challenges of implementing lean construction practices in the construction industry.

Challenges	Mean	Standard deviation	Rank
Poor culture among project partners	4.62	5.921	1
Lack of good policies	4.00	0.798	2
Complexity of lean construction process	3.93	0.874	3
Poor organizational knowledge	3.91	0.848	4
Lack of understanding of lean construction	3.89	1.049	5
Takes time to adopt	3.87	1.057	6
Inherently knowledge-intensive	3.87	0.661	7
Lack of skill in lean construction process	3.82	0.747	8
Poor human attitude	3.78	0.795	9
Financial boundaries	3.73	1.031	10
Lack of resources	3.73	0.939	11
Poor time management	3.69	0.821	12
Ineffective management	3.67	1.022	13
Stress creation on contractors	3.62	0.860	14
Reduction of inventory	3.47	0.919	15

7.3   Benefits of Implementing Lean Construction Practices

Table 2 presents the mean values and rankings of the benefits of implementing LC practices in the construction industry. The results were based on the respondent's assessment of the listed factors. The results revealed that 'reduction of waste', 'effective administration of materials on site', 'improved life-cycle cost', 'greater client satisfaction', 'good project coordination' were the top five benefits of implementing LC practices. From Table 2, all the factors listed have mean values greater than 2.5 indicating that they are all highly beneficial to the construction industry. The findings are in concordance with the study of Horman et al. [32] which revealed that the most important benefit of implementing LC practices is the reduction of waste and achievement of value throughout the construction process.

Table 2. Benefits of implementing lean construction practices in the construction industry.

Benefits	Mean	Standard deviation	Rank
Reduction of waste	4.27	0.580	1
Effective administration of materials on site	4.22	0.599	2
Improved life-cycle cost	4.16	0.706	3
Greater client satisfaction	4.09	0.701	4
Good project coordination	4.07	0.837	5
Improved safety on site	4.02	0.812	6
Greater productivity	4.00	0.769	7
Better risk management	3.98	0.583	8
Effective communication between client and construction team	3.93	0.915	9
Reduced project schedule	3.91	0.821	10
High-quality construction	3.87	0.726	11
Efficient system with less cost	3.80	0.842	12

7.4   Measures of Implementing Lean Construction Practices

The mean values and rankings of the measures of implementing LC practices in the construction were presented in Table 3. The results revealed that 'appointment of lean expert/consultant', 'education and training of stakeholders', 'motivation and commitment of stakeholders', 'establishment of policy and regulatory system', and 'introduction of certification and measurement standards' were the top five measures of implementing LC practices. All the listed factors have mean values above 2.5 indicating their significance. These findings are in tandem with those obtained by Bashir et al. [31] and Olatunji [33], affirming the following (appointment of lean expert/consultant, education and training of stakeholders; motivation and commitment of stakeholders; and establishment of policy and regulatory system) as the most important ways to improve the adoption and use of LC practices in the construction industry.

Table 3. Measures of implementing lean construction practices in the construction industry.

Measures	Mean	Standard deviation	Rank
Appointment of lean expert/consultant	4.42	0.533	1
Education and training of stakeholders	4.40	0.618	2
Motivation and commitment of stakeholders	4.36	0.618	3
Establishment of policy and regulatory system	4.27	0.654	4
Introduction of certification and measurement standards	4.16	0.903	5
Government intervention and support	4.11	0.959	6
Introduction of economic incentives	4.09	0.701	7
Coordination and communication among stakeholders	4.04	0.673	8
Supply management	4.00	0.769	9
Stakeholders awareness	3.91	0.670	10
Change in organizational culture of stakeholders	3.82	0.720	11

8   Conclusion and Recommendations

Waste is a significant impediment to productivity in the construction industry and this occurs throughout the project lifecycle. Lean construction (LC) practices can, however, improve the performance of the sector by facilitating timely project delivery, provision of right resources of desired quantity and quality, and within the budgetary allocation for such project. The implementation of LC practices will also be a bold step towards achieving sustainability in the construction industry as it focuses on improving productivity while reducing waste in construction activities and processes.

However, there are many hindrances to the adoption and implementation of LC practices in the construction industry. As one of the sustainability practices, LC is also faced by the same barriers limiting the adoption and implementation of sustainable construction. Unwillingness to embrace change, lack of knowledge, understanding and education in LC practices, lack of policies and legislation to support LC, and the complexity in the implementation are few among the barriers. It is therefore important to embrace measures to overcome these barriers in other to enjoy the benefits which in turn has the potential of helping the construction industry realize its sustainability goals.

Training and education in the components of LC practices right from the higher education level to the continuous professional development level is imperative. Support from government and other stakeholders is recommended in fast-tracking the adoption and implementation of LC practices. Also, appointing a lean professional as part of the construction team is another way of promoting behavioral change and encouraging participants to adopt LC practices.

References

1. PWC: Highlighting Trends in the South African Construction Industry. http://www.pwc.co.za

2. Bohari, A.A.M., Martin, S., Bo, X., Melissa, T., Xiaoling, Z., Khairul, N.A.: The path towards greening the Malaysian construction industry. Renew. Sustain. Energy Rev. **52**, 1742–1748 (2015)
3. Nahmens, I., Laura, H.I.: Effects of lean construction on sustainability of modular homebuilding. J. Archit. Eng. **18**(2), 155–163 (2011)
4. Construction Industry Institute: Application of Lean Manufacturing Principles to Construction. https://www.construction-institute.org/scriptcontent/more/rr191_11_more.cfm
5. Hosseini, S.A., Nikakhtar, A., Wong, K., Zavichi, A.: Implementing Lean Construction Theory to Construction Processes' Waste Management (2012)
6. Alarcon, L.F., Loreto S.: Developing Incentive Strategies for Implementation of Lean Construction (2002)
7. Mwacharo, F.: Challenges of Lean Management: Investigating the Challenges and Developing a Recommendation for Implementing Lean Management Techniques (2013)
8. Green, S.D., Susan, C.M.: Lean construction: arenas of enactment, models of diffusion and the meaning of 'leanness'. Build. Res. Inf. **33**(6), 498–511 (2005)
9. Formoso, C.T., Lucio, S., De Cesare, C., Eduardo, L.I.: Material waste in building industry: main causes and prevention. J. Constr. Eng. Manag. **128**(4), 316–325 (2002)
10. Senaratne, S., Duleesha, W.: Lean construction as a strategic option: testing its suitability and acceptability in Sri Lanka. Lean Constr. J. 34–48 (2008)
11. Li, H., Guo, H., Li, Y., Skitmore, M.: From IKEA model to the lean construction concept: a solution to implementation. Int. J. Constr. Manage. **12**(4), 47–63 (2012)
12. Abdelrazig, Y.E.: Using Lean Techniques to Reduce Waste and Improve Performance in Municipal Project Delivery (2015)
13. Nicholas, J.: Lean Production for Competitive Advantage: A Comprehensive Guide to Lean Methodologies and Management Practices. CRC Press, Boca Raton (2011)
14. Aziz, R.F., Sherif, M.H.: Applying lean thinking in construction and performance improvement. Alexandria Eng. J. **52**(4), 679–695 (2013)
15. Salem, O., Solomon, J., Genaidy, A., Luegring, M.: Site implementation and assessment of lean construction techniques. Lean Constr. J. **2**(2), 1–21 (2005)
16. Tan, B.W., Hooi, H.L.: An analysis of dynamic linkages between domestic investment, exports and growth in Malaysia. Eur. J. Soc. Sci. **16**(1), 150–159 (2010)
17. Ogunbiyi, O., Jack, S.G., Adebayo, O.: An empirical study of the impact of lean construction techniques on sustainable construction in the UK. Constr. Innov. **14**(1), 88–107 (2014)
18. Common, G.E.J., David, G.: A Survey of the Take-Up of Lean Concepts Among UK Construction Companies (2000). Accessed 17–19 July 2000
19. Mossman, A.: Why isn't the UK construction industry going lean with gusto? Lean Constr. J. 24–36 (2009)
20. Sarhan, S., Andrew, F.: Barriers to implementing lean construction in the UK construction industry. The Built Hum. Environ. Rev. **6**, 1–17 (2013)
21. Rahman, H., Wang, C., Lim, I.: Waste processing framework for non-value-adding activities using lean construction. J. Front. Constr. Eng. **1**(1), 8–13 (2012)
22. Koskela, L.: Application of the New Production Philosophy to Construction, vol. 72. Stanford University, Stanford (1992)
23. Naim, M., James, B.: An innovative supply chain strategy for customized housing. Constr. Manag. Econ. **21**(6), 593–602 (2003)
24. Koskela, L., Greg, H., Glenn, B., Iris, T.: The foundations of lean construction. In: Design and Construction: Building in Value, pp. 211–226 (2002)
25. Green, S.D.: The Future of Lean Construction: A Brave New World (2000)
26. Antillon, E.I.: A Research Synthesis on the Interface Between Lean Construction and Safety Management. University of Colorado at Boulder (2010)

27. Womack, J.P., Daniel, T.J.: Lean thinking—banish waste and create wealth in your corporation. J. Oper. Res. Soc. **48**(11), 1148 (1997)
28. Field, A.: Discovering Statistics Using IBM SPSS Statistics. Sage, Thousand Oaks (2013)
29. Kim, D.: Exploratory Study of Lean Construction: Assessment of Lean Implementation (2002)
30. Bicheno, J., Holweg, M.: The Lean Toolbox: The Essential Guide to Lean Transformation, 4th edn (2009)
31. Bashir, A.M., Subashini S., David, G.P., Rod, G.: Barriers Towards the Sustainable Implementation of Lean Construction in the United Kingdom Construction Organizations (2010)
32. Horman, M.J., David, R.R., Anthony, R.L., Sinem, K., Michael, H.P., Christopher, S.M., Yupeng, L., Nevienne, H., Peter, K.D.: Delivering green buildings: process improvements for sustainable construction. J. Green Build. **1**(1), 123–140 (2006)
33. Olatunji, J.: Lean-in-Nigerian Construction: state, barriers, strategies and "Goto-gemba" approach. In: Proceedings 16th Annual Conference of the International Group for Lean Construction, Manchester, United Kingdom (2008)

Environmental Impacts of Construction Activities: A Case of Lusaka, Zambia

Chanda Musenga[✉] and Clinton Aigbavboa

Sustainable Human Settlement and Construction Research Centre,
Faculty of the Built Environment, University of Johannesburg,
Johannesburg, South Africa
chandamusenga.cm@gmail.com

Abstract. The construction industry plays a vital role in wealth creation through provision of infrastructure. However, the industry is known to be a major contributor to environmental degradation. This paper presents the results of the study, which assessed the environmental impacts of construction activities in Zambia. A quantitative approach was adopted for the study and a well-structured questionnaire survey was conducted on the construction professionals. Analysis of the data was done using the mean item scores. The study revealed that interface with the ecosystem, increase in various types of pollution and deforestation are the major impacts. Finally, there is need to adopt and implement sustainable construction practices as a way of reducing the impacts of construction activities on the environment. This can be achieved if all the stakeholders are involved, especially the government, which plays a cardinal role in the formulation of regulations and policies that support sustainable construction practices.

Keywords: Construction activities · Construction industry · Environment
Sustainability · Zambia

1 Introduction

Increasing economic development has led to increased attention on the environment and sustainability the world over. A lot of this attention is on the construction industry as it is a large consumer of resources and its activities have a significant impact on the environment [1]. On the other hand, the construction industry serves as a vehicle for economic development through its provision of facilities and infrastructure [2]. This is illustrated through studies that have found that the Zambian construction industry (ZCI) plays a vital role in the development of the economy. Activities in the sector in the last couple of years have been driven by public and private projects which include construction of hospitals, roads, schools, stadia and commercial and residential property. The industry has continued to grow over the last 12 years at a steady annual average rate of 17.5% [3]. This development comes at a price as it exerts pressure on natural resources and this has had a serious impact on all living organisms and the environment.

© Springer International Publishing AG, part of Springer Nature 2019
J. Charytonowicz and C. Falcão (Eds.): AHFE 2018, AISC 788, pp. 535–541, 2019.
https://doi.org/10.1007/978-3-319-94199-8_52

The degradation of the environment has captured the world's attention and has been one of the most discussed subjects locally, nationally and globally [4]. Du Plessis [5], discovered that environmental impact of the construction industry is highest in developing countries and Zambia falls in this category. This is because developing countries are still under construction and have a relatively low degree of industrialisation. The construction industry is a heavy contributor towards unsustainable development, and its impact on both the economy and the environment is tremendous. According to literature, the processes and products used in the traditional construction approach have a negative environmental and social impact [6]. This is because the activities in the construction process consume massive quantities of natural resources including numerous energy sources and water. Extraction of raw materials, manufacturing and transportation tend to lead to a reduction of resources and losses of biological diversity whilst acid rain and global warming are the result of high-energy consumption. It has been further identified that the impacts of the construction industry on the environment are significant because they occur throughout the project life cycle [4, 7].

With the current rate at which the earth's resources are being depleted, there is a growing urgency to restructure the construction industry globally. In response to this, sustainable construction (SC) was proposed as a way of making the construction processes, activities and practices more economically, socially and environmentally responsive [4]. This was motivated by the goal of securing the future generations ability to meet their needs through the application of sustainable development principles to meet the present needs. With the rising awareness of the need for sustainability, especially in the construction industry, it is important for Zambia to keep pace with this global movement by adopting SC practices. Hence, the goal of this study was to determine the impacts of construction activities on the environment in Lusaka, Zambia.

2   Literature Review

SC is defined as the formation and management of a healthy built environment through the sensible use of resources and ecological principles [8]. It is important to note that the terms high performance, green and sustainable construction are used interchangeably. Other terms that are synonymous with sustainable construction are green building and sustainable building [9]. Similarly, SC may be described as an all-inclusive process with the aim of re-establishing and maintaining harmony between the built and natural environments and the creation of settlements that assert human dignity and encourage economic equity [5]. This implies that SC takes a lifecycle perspective with emphasis on environmentally orientated design, operation and maintenance procedures. Another scholar identified SC to be an approach that addresses the sustainable needs of the built environment and it is furthermore a term that was conceptualised as a result of the sustainability paradigm shift that has gained momentum over the last few years [7, 8]. Sustainability is defined as the state or condition that would enable the continued existence of human beings and it is concerned with three key aspects namely; environmental responsibility, social awareness; and economic profitability [5].

It has become common knowledge that construction activities have an impact on the environment. Furthermore, it is now a requirement by law in many countries that an evaluation of these impacts be carried out before construction can commence [10]. This has been embraced by developed countries whilst in the developing ones the situation remains unchanged as more attention is placed on the construction and human development relationship whilst the environmental aspect is ignored. The construction industry is considered to be one of the largest exploiters of both renewable and non-renewable natural resources as it depends heavily on the environment for its source of raw materials like aggregates, sand and timber for the building process. Studies have shown that building construction consumes 40% of the world's raw stones, gravel and sand and 25% of the virgin wood per year. It also consumes 40% of the energy and 16% of water annually [3, 6]. The high energy consumption and use of finite fossil fuel resources have contributed significantly to carbon dioxide emissions. Research by Sterner [11] found that the construction industry was a primary source of environmental degradation through high energy consumption for heating and domestic electricity. The main source of this energy are non-renewable resources, such as coal and oil products.

The high extents of deforestation being experienced especially in developing countries can be traced back to the construction industry. A large amount of timber is harvested from indigenous forests for construction and for providing the energy required to produce building materials. It has been found that deforestation worldwide is occurring at a rapid rate, where about 0.8 hectares of the rainforest is lost every second. The effect of this is that about 1.8 billion tons of carbon is released into the atmosphere every year. In addition, removal of trees and their roots tends to result in landslides and avalanches [9, 12]. Other indirect impacts are soil erosion, siltation of watercourses, loss of biodiversity, global warming and desertification. Deforestation at a larger scale leads to alteration of the rainfall patterns due to a change in the earth's surface temperature and energy and the rate of absorption of surface water.

Soil Erosion is said to be a major cause of loss of valuable farmland. Statistics show that farmland around the world of 70–140,000 km^2 per year is lost through soil erosion [5]. The repercussions of this, especially in most African countries, are high as land is a vital component of most of the economic activities. This is because most areas have soils of poor quality and so the more fertile areas tend to be costly. Studies indicate that valuable farmland and forests are irreversibly trans-formed through construction activities into structures like buildings, roads and dams [4, 13]. Other activities like quarrying and mining for construction raw materials and dumping of waste generated during construction also aid to the loss of land.

Research has shown that the construction industry is responsible for the production of various types of wastes [14, 15]. The quantity and type of this waste depend on numerous factors such as the stage of construction, type of construction work and the construction practices on site. In addition, waste is produced from product generation and use of materials [13]. The statistics from some parts of the world show that there are large volumes of material being taken to the landfills. For instance, the USA contributes about 29% of waste with more than 50% coming from the UK and 20–30% in Australia. This waste more especially in developing countries is often dumped illegally into dams, river courses and any available hollow. Sterner [11], points out that most of this waste is unnecessary and has the potential to be recycled and reused.

The extraction of raw materials and construction activities are one of the main culprits of pollution to land, air and water. Extraction of these materials often leads to degradation of land and the ecosystem. The effects of processing of these raw materials and transporting the finished products to site results in air and dust pollution, high energy consumption and emission of toxic gases and effluents in water bodies which affect aquatic and marine life, as well as contributing to atmospheric pollution [5, 8]. The dust and other emissions that are released often contain toxic substances such as Nitrogen and Sulphur oxides. Pollutants are also released into the biosphere causing serious land and water contamination. In most cases, negligent operations on site result in toxic spillages which are then washed into underground aquatic systems and reservoirs.

3   Research Methodology

This study adopted a quantitative approach using a questionnaire survey conducted among construction professionals in Lusaka, Zambia. These construction professionals include; Architects, Construction managers, Construction Project Managers, Project Managers, and Quantity Surveyors. These professionals were selected from the private (contractors and consultants) and public sectors in Lusaka, the capital city of Zambia. The city was selected based on its central location and because it provides adminis-trative functions to the entire country. The instrument used for data collection was a structured questionnaire which was administered to the identified construction pro-fessionals using convenience sampling approach. The questionnaire was designed in two parts with the first part designed to collect demographic information like the level of education, profession and experience from the respondents. The second part sought to determine the major environmental impacts of construction activities. The respon-dents were asked to indicate their level of agreement on the environmental impacts of construction activities based on a five-point Likert scale (from 1 = 'strongly disagree' to 5 = 'strongly agree'). Out of the 75 questionnaires sent out, 44 were received back representing a 59% response rate, which was deemed adequate for the study. The data gathered were analysed using Statistical Package for Social Science Version 24 (SPSS V24). Data gathered on the background information of the respondents was analyzed using percentage, while mean item score (MIS) was used in ranking the identified impacts.

4   Results and Discussion

4.1   Background of Respondents

The findings revealed that the professionals involved in the study were represented as follows 14% architects, 30% quantity surveyors, 39% engineers, 9% construction managers, 7% construction project managers, 2% project managers. The study further indicated that 43% of the respondents were government employees whilst 59% of the respondents were employees of private organizations and furthermore, 11% of the respondents had a diploma, 57% had a degree, 30% had a master's degree, and 2% had a doctoral degree. The average years of experience of the respondents were 6 years and above. This result shows that the respondents have enough experience in terms of happenings with the Zambian construction industry, thus, making them capable to give reasonable answers to the questions of the research.

4.2   Environmental Impacts of Construction Activities

In the quest to determine the environmental impacts of construction activities in Zambia, literature was reviewed and the items that were identified were then presented to the respondents as indicated in Table 1. The respondents then ranked the impacts based on the five-point Likert scale. The results based on the mean item score and standard deviation indicate that the major environmental impacts are interference with the ecosystem, increase in various types of pollutions, deforestation and habitat destruction with mean values of 3.80, 3.64, 3.64 and 3.55 respectively. Furthermore, it can be seen that not all the variables were considered to be significant to the study as some were below the mean value of 3.0.

The results indicate that the respondents are in agreement with the findings of the studies conducted in Ghana, Uganda, South Africa and the UK [3, 4, 7, 13]. However, it can be seen from the table that some of the results are not in agreement with the study conducted by Kibert [8] in which it was found that construction activities would result in desertification due the loss of vegetation. In addition, the results further indicated that there was no agreement with the studies that found that there is a decrease in available land due to waste disposal [7, 14]. This implies that the respondents are partially informed about the environmental repercussion of construction activities from inception to demolition. It is therefore imperative that sensitisation on the environmental impacts of construction activities should be prioritised.

Table 1. Environment impacts of construction activities

Impacts	Mean item score	Standard deviation	Rank
Interference with the ecosystem	3.80	0.954	1
Increase in various types of pollution	3.64	1.080	2
Deforestation	3.64	1.163	2
Habitat destruction	3.55	1.190	4
Soil erosion	3.43	1.087	5
Depletion of non-renewable energy resources	3.39	1.039	6
Climate change	3.36	0.865	7
Decrease in the availability of arable land	3.18	1.147	8
Desertification	2.98	1.067	9
Loss of marine life	2.93	1.043	10
Decrease in the availability of land due to disposal of waste	2.82	1.018	11

5 Conclusion and Recommendation

The purpose of this study was to determine the environmental impacts of construction activities in Lusaka, Zambia. Through the questionnaire survey, it was revealed that the main impacts of construction activities on the environment in descending order were interference with the ecosystem, increase in various types of pollutions, deforestation and habitat destruction. The findings in relation to the literature revealed that the respondents are partially informed about the environmental repercussion of construction activities from inception to demolition. It is imperative that sensitisation on the environmental impacts of construction activities should be prioritised.

Therefore, it is essential that the various stakeholders be made aware of the sustainable practices that have been developed to ensure that the various projects that are being carried out in the country are sustainable. Thus, it is recommended that a course on sustainable construction should be introduced in the universities and in the school curriculum. The government should develop a framework for the adoption of SC in the country and it should show its commitment to the adoption of SC through the introduction of legislation and incentives. In addition, the government should encourage collaboration with the various professional bodies to ensure enforcement of regulations. Training and education of the various professionals should be encouraged through continuous professional development. Lastly, there is need for the country to collaborate with countries with established green building councils. The study provides direction for further study in the aspect of determining the level of awareness of sustainable construction practices in the Zambian constriction industry and the identification of the barriers and drivers to adopting these practices.

References

1. Ge, Z., Gao, Z.: Applications of nanotechnology and nanomaterials in construction. In: International Conference on Construction in Developing Countries, pp. 235–240 (2008)
2. Majdalani, Z., Ajam, M., Mezher, T.: Sustainability in the construction industry: a Lebanese case study. Constr. Innov. 6(1), 33–46 (2006)
3. Seventh National Development Plan 2017–2021. http://www.mndp.gov.zm/download/7NDP.pdf
4. Ametepey, S.O., Ansah, S.K.: Impacts of construction activities on the environment: the case of Ghana. J. Constr. Proj. Manag. Innov. 4(1), 934–948 (2014)
5. Du Plessis, C.: Agenda 21 for: sustainable construction in developing countries. CSIR Report BOU E 204 (2002)
6. Baloi, D.: Sustainable construction: challenges and opportunities. In: 19th Annual ARCOM Conference, University of Brighton, United Kingdom, pp. 289–297 (2003)
7. Son, H., Kim, C., Chong, W.K., Chou, J.S.: Implementing sustainable development in the construction industry: contractor's perspectives in the USA and Korea. Sustain. Dev. 19(5), 337–347 (2011)
8. Kibert, C.J.: Sustainable Construction: Green Building Design and Delivery. Wiley, Hoboken (2013)
9. Wang, L., Toppinen, A., Justin, H.: Use of wood in green building: a study of expert perspectives from the UK. Cleaner Prod. 65, 350–361 (2014)

10. Ayarkwa, J., Acheampong, A., Hackman, J.K., Agyekum, K.: Environmental impacts of construction activities in Ghana. Afr. Dev. Resour. Res. Inst. (ADRRI) J. **9**(2), 1–19 (2014)
11. Sterner, E.: 'Green procurement' of buildings: a study of Swedish clients' considerations. Constr. Manag. Econ. **20**(1), 21–30 (2002)
12. Uher, T.E.: Absolute indicators of sustainable construction. In: Proceedings of COBRA, pp. 243–253 (1999)
13. Pearce, A.R.: Sustainable Buildings and Infrastructure: Paths to the Future. Routledge, London (2012)
14. Muhwezi, L., Kiberu, F., Kyakula, M., Batambuze, A.O.: An assessment on the environment in Uganda: a case study of Iganga municipality. J. Constr. Eng. Proj. Manag. **2**(4), 20–24 (2012)
15. Mulenga, F., Kamalondo, H.: An investigation of waste management practices in the Zambian construction industry. J. Constr. Plan. Res. **5**, 1–13 (2017)

Author Index